Key Heterocycle Cores for Designing Multitargeting Molecules

Key Heterocycle Cores for Designing Multitargeting Molecules

Edited by

Om Silakari
Department of Pharmaceutical Sciences and Drug Research,
Punjabi University, Patiala, Punjab, India

Elsevier
Radarweg 29, PO Box 211, 1000 AE Amsterdam, Netherlands
The Boulevard, Langford Lane, Kidlington, Oxford OX5 1GB, United Kingdom
50 Hampshire Street, 5th Floor, Cambridge, MA 02139, United States

Library of Congress Cataloging-in-Publication Data
A catalog record for this book is available from the Library of Congress

British Library Cataloguing-in-Publication Data
A catalogue record for this book is available from the British Library

ISBN: 978-0-08-102083-8

For information on all Elsevier publications visit our website at
https://www.elsevier.com/books-and-journals

Publisher: Susan Dennis
Acquisition Editor: Emily McCloskey
Editorial Project Manager: Michelle Fisher
Production Project Manager: Nilesh Kumar Shah
Cover Designer: Christian Bilbow

Typeset by TNQ Books and Journals

Contents

3. Acridones: A Relatively Lesser Explored Heterocycle for Multifactorial Diseases

Rajesh Kumar, Sarita Sharma and Deonandan Prasad

4. Flavone: An Important Scaffold for Medicinal Chemistry

Manjinder Singh and Om Silakari

5. Thiazolidine-2,4-Dione: A Potential Weapon for Targeting Multiple Pathological Conditions

Navriti Chadha and Om Silakari

9. Triazoles: Multidimensional 5-Membered Nucleus for Designing Multitargeting Agents

Aanchal Kashyap and Om Silakari

10. Benzoxazolinone: A Scaffold With Diverse Pharmacological Significance

Himanshu Verma and Om Silakari

List of Contributors

Navriti Chadha, Punjabi University, Patiala, India

Shalki Choudhary, Punjabi University, Patiala, India

Aanchal Kashyap, Punjabi University, Patiala, India

Maninder Kaur, Punjabi University, Patiala, India

Rajesh Kumar, Shivalik College of Pharmacy, Nangal, India; I.K. Gujral Punjab Technical University, Jalandhar, India

Deonandan Prasad, Shivalik College of Pharmacy, Nangal, India; I.K. Gujral Punjab Technical University, Jalandhar, India

Sarita Sharma, I.K. Gujral Punjab Technical University, Jalandhar, India; Global College of Pharmacy, Anandpur Sahib, India

Om Silakari, Punjabi University, Patiala, India

Manjinder Singh, Chitkara university, Patiala, India

Pankaj K. Singh, Punjabi University, Patiala, India

Himanshu Verma, Punjabi University, Patiala, India

Preface

Research scholars of medicinal chemistry do not readily find books to guide their research ideas. Most of the books available in libraries either focus on organic chemistry of a class of molecules or provide pharmacological data regarding that class of molecules. Closer inspection of such books disappoints expectations of scholars. We wanted to write a book that reflects the ideas of researchers rather than a simple compilation of biological attributes of some heterocycles. Of all the current trends in medicinal chemistry, multitargeting molecules are an interesting case. The jury is still out on the clinical significance and efficacy of such molecules but they tend to lure several members of the medicinal research community. We were greatly intrigued by the medicinal potential offered by such molecules and believe that researchers would benefit more from a book that provides information regarding current status and overall work done on the multitargeting potential of several classes of heterocycles.

We also kept in mind that information regarding the general chemistry associated with such scaffolds is also imperative for the book to become a significant piece of literature. We aimed at an approach that would make sense and appeal to today's research scholars. Thus we incorporated a subsection in each chapter that specifically understated the current findings regarding the structure activity relationship of each heterocycle as multitargeting agent. We have provided graphical representations of pharmacological profiles attributed to specific substitutions on the scaffolds. To avoid complexity in discussions, we have deliberately omitted detailed discussion of obscure synthetic schemes of little value, or of variant reactions that simply repeat mechanistic logic utilized previously. We have also avoided vague pharmacological reports that do not justify the rationale of study, rather merely complete a study.

This book guides researchers working on medicinal, biochemical, and pharmacological aspects of multitargeting heterocycles. Heterocycles covered in this book include benzimidazole, oxindole, triazole, among others. Each chapter covers the multitargeting story of each heterocycle, except for Chapter 1, which discusses the basics about multitargeting agents, the need for multitargeting agents, strategies utilized for their design along with problems faced during their generation, different applications of these developed molecules, and a few recent reports of multitargeting agents from the literature. The following chapters discuss the multitargeting aspect of each heterocycle in detail.

The authors are indebted to the supportive, and at the same time critical, faculty members of the department of Pharmaceutical Sciences and Drug

Research. We would also like to acknowledge the support and guidance from Anneka Hess, Acquisitions Editor, Medicinal Chemistry and Environmental Science, Elsevier, and other members of the editorial team at Elsevier. Finally, the time spent on the preparation of this book was made available only with the forbearance of our families, friends, and research groups, and we thank all of them for their patience and understanding.

Chapter 1

Multitargeting Heterocycles: Improved and Rational Chemical Probes for Multifactorial Diseases

Pankaj K. Singh, Om Silakari
Punjabi University, Patiala, India

Chapter Outline

Key Heterocycle Cores for Designing Multitargeting Molecules. https://doi.org/10.1016/B978-0-08-102083-8.00001-7

1. INTRODUCTION

There are several heterocycle scaffolds reported in literature to possess significant potential as medicinal chemistry agents [1]. These heterocycles offer an optimal source for core scaffolds and fragments for designing libraries focused on a wide range of targets [2]. These heterocycles/scaffolds possess the ability to bind promiscuously with a number of pathological targets utilizing a variety of favorable structural and physicochemical attributes. Nevertheless, the phrase "privileged scaffold" represents a substructure or even a template or fingerprints, which upon integration in a molecule, provide an enhanced possibility of a druglike nature because of the presence of functionalities such as volume, electronegativity, polarizability, hydrophobicity, hydrogen-bonded potential, hybridization, and partial atomic charge, which are relevant and significant for ligand binding. Such privileged scaffolds often comprise an aromatic heterocyclic system capable of interacting with multiple hydrophobic residues present in predictable orientations of space [3]. These privileged scaffolds, depending on a variety of substituent, show varying affinities for different targets hinting toward an entopic mechanism of privilege. Compounds with such privileged heterocycles have an increased chance of being a bioactive entity, underlining their utility in pharmaceutical discovery process [4].

The term privileged structure was initially discussed by Evans in 1988, followed by Patchett in 2000, and reviewed by Welsch for its applicability in a library design and drug discovery in 2010 [5]. There are no specific collections of scaffolds or heterocycles as privileged scaffolds, but several reports have been published indicating different heterocycles such as flavones, indole, acridone, oxindole, thiazine, triazole, benzopiperazinone, quinazoline, benzimidazole, and so on as privileged structures.

2. COMPLEX DISEASES AND POLYPHARMACOLOGY

In recent decades, there have been enormous advancements and gain in the knowledge and understanding of the pathogenesis of various diseases, from demarcating genes to cellular pathways crucial for the development of diseases. This also led to disclosure of various unwanted and unexpected complex pathological scenarios. A disease condition usually involves various pathological processes that are interlinked through a complex network, thus multifactorial nature is observed in major chronic diseases [6]. Due to the diverse nature of causes, the general research viewpoint of targeting protein by one specific agent has been challenged [7]. The basic concept of hitting a single disease mechanism with the one drug–one target paradigm does not always affect complex systems efficiently, even when a complete change in behavior of the intended target is achieved. Thus, with time, a more holistic approach (i.e., polypharmacological approach) has emerged, which relies on simultaneous management of different etiological targets. The concept of polypharmacology constitutes both multiple drugs that act independently on different targets, and a single drug binding to multiple targets within a network [8].

3. COMBINATION THERAPIES

The use of a combination of drugs usually is referred to as combination therapy or polypharmacy. It has been utilized as the most basic and immediate approach to gain multiple modes of agitation to counter the pathogenic pathways. There has been a constant focus on this approach, as many times it has been proven effective in contesting several multifactorial diseases such as cancer and HIV infections. The basic concept of combination therapy involves administration of multiple drugs, intended to partially inhibit multiple targets rather than fully antagonize a single one. This approach is helpful in both obtaining drug synergies and preventing the development of resistance via unwanted compensatory mechanisms. One of the key advantages, by going for a weak perturbation of the biological system, is that lower doses of each drug is used, thus leading to better therapeutic selectivity [9].

In spite of these positives, however, combination therapy raises some serious concerns. The most significant problem is that therapeutic regimens become complicated and amplify compliance problems. One step to overcome this weakness involves the incorporation of different drugs into the same formulation, creating a multiple medication. This approach has also achieved few marketing successes. However, combining several active pharmaceutical ingredients into a single formulation is not an easy task. Problems such as the pharmacokinetic differences between the individual components are usually encountered [10]. Another significant issue with the coformulation that further hampers combination therapy is the drug–drug interaction issue. Two drugs, which upon individual administration are safe, cannot be considered safe when administered in combination. Such drug–drug interactions can occur at any level and usually involve multiple mechanisms, including competition for common metabolic pathways and chemical incompatibility. Error in the identification of drug–drug interactions can easily lead to over- or underdosing, resulting in severe clinical consequences. All these facts have directed researchers from the concept of polypharmacology toward the multitargeting approach.

4. MULTITARGETING

The advancements in probing tools and techniques have resulted in shifting the drug discovery process from completely being a human phenotype-based effort to a more advanced, current reductionist approach, based on molecular targets [11]. Thus, nowadays the development of drugs is focused on the identification and knowledge of potential targets at molecular levels, utilizing genomic and proteomic studies. With the technological advancement in genomics and high throughput screening, the drug discovery protocols have become more focused on the modulation of molecular targets involved in a disease condition. The basic protocol has shifted from animal models to simpler isolated proteins by the use of cellular models. Interestingly, this has decreased complexity but also reduced its relevance to the actual human condition [12].

Current research follows a basic archetype: searching for a target with clinical significance and then discovery of small molecules that are able to modulate the physiological functions of those protein targets considered to be fully responsible for a disease condition. To accomplish this, significant efforts are made to achieve selectivity for that given target, therefore many molecules nowadays are reported to possess outstanding in vitro selectivity. This one-molecule, one-target paradigm has led to the discovery of many successful drugs, and it will probably remain a milestone for years to come.

To its success, this approach of involving development of highly specific targeted agents has gained significant results [13]. However, in spite of all the efforts, the molecules that could be successfully developed as drugs are very few [14]. This is usually due to the ligand's failure to recognize the target or because the ligand does not reach the site of action. Another significant factor that could responsible for the failure of these drugs is the involvement of more than one molecular target in any given disease condition and thus, the interaction with the respective target does not have enough impact on the diseased system to restore it effectively. This led to the development of different targeting strategies for the disease involving more than one specific molecular target. Thus, network therapeutics have gained much attention in the last decade or so [15] as one of the potential solutions to diseases of complex etiology.

There are increased efforts toward modulating a collection of targets in the treatment of various disorders; that is, targeting more than one molecular target involved in the pathophysiology of the disease either via combination of drugs (i.e., polypharmacy [16]) or targeting all molecular targets via a single molecule simultaneously (i.e., polypharmacology). Polypharmacy is very commonly utilized in the treatment regimen of various diseases such as cancer and AIDS. Thus, the concept of developing multitargeting molecules, polypharmacology, is currently pursued by various research groups throughout the globe [17]. Another reason for multitargeting agents (MTAs) to gain recognition is the promiscuous nature of many approved, and/or in clinical trial, molecules. Initially, the promiscuous nature of molecules was considered a problem in the design and development of a new drug. However, nowadays this promiscuity provides a basic scaffold, which is then optimized to design and develop MTAs against a complex disease condition [18].

Failure of drugs that aim at only a single molecular target is not always due to low potency of the molecule; it might be due to another back-up signaling system available for disease pathophysiology. Cells often find ways to compensate for a protein whose activity is affected by a drug by taking advantage of the redundancy of the system (i.e., of the existence of parallel pathways [19]). So, even though the drug modifies the complex signaling systems in the required way, it could not produce the desired effects due to these back-up systems. Additionally, cellular networks are robust and usually avoid major changes in their responses regardless of significant alterations in their constituents. These considerations do not really depend on whether the molecule inhibits or activates its target [20].

The main improvement of the multitarget approach over a single-target approach is an increase in the overlapping between different signaling pathways, which improves the number of proteins that can be altered with a single druglike molecule. There are few hundred proteins known for their involvement in various disease conditions and also considered druggable [21], therefore a multitarget approach becomes significant since it provides an option to indirectly regulate proteins, which are involved in the same signaling pathway as an existing target protein. Conversely, these so-called multitarget agents have an overall low binding affinity, as it is not possible for a single molecule to possess a similar binding affinity toward each target protein. Thus, as the very same molecule is incorporated with features to bind with more proteins, the binding affinity drops very low, in the range of higher micromolar or even close to millimolar. The basic reason for the feasibility of any molecule to be able to bind with multiple targets is the possibility of multiple dynamic states of target proteins around their native state. So, during the process of binding, multiple targets having different conformers, different dynamic state, and a ligand can bind in one of the most favorable energetic states. However, whether binding with different proteins will occur simultaneously or eventually will depend on when and how those targeted proteins interact with each other. In both cases, considering targets and ligands more as dynamic, possessing different conformations and binding site shapes driven by different energetic states, explains how different target binding sites can fit the same single ligand [22].

5. MULTITARGET STRATEGY EXAMPLES

There are various examples of drugs that affect many targets simultaneously such as nonsteroidal antiinflammatory drugs (NSAIDs), antidepressants, antineurodegenerative agents, and multitarget kinase inhibitors [13]. Similar work is also being carried out to develop multitarget antibodies, which are utilized in cancer therapy to prevent/avoid resistance [23]. Commonly used phrases by the research community to report their molecules having multiple activities include balanced, binary, bivalent, dimeric, dual, mixed or triple with agonist, antagonist, blocker, conjugate, inhibitor, or ligand. The development of NSAIDs involved an interesting evolution from nonselective agents, like aspirin, which inhibits both COX-1 (cyclooxygenase) and COX-2, to selective COX-2 inhibitors, like celecoxib, and finally, designing multitargeting agents, such as dual inhibitors of COX-2 and 5-LOX (5-lipoxygenase) that is supposed to possess greater efficacy with reduced side effects of selective COX-2 inhibitors [24]. A similar approach is being utilized for antidepressants, coming from nonselective tricyclics such as amitriptyline, to selective serotonin transporter (SERT) inhibitors, and ultimately to dual SERT and norepinephrine transporter (NET) inhibitors, which were found to possess an improved onset of action with significantly enhanced efficacy [25]. Other examples following a similar trend include a number of designed molecules that have moved to later stages of clinical development; for example, omapatrilat [13], which is a dual angiotensin converting enzyme (ACE) and neutral endopeptidase (NEP) inhibitor, and neto-glitazone, which is a peroxisome proliferator-activated receptor (PPAR)-α and PPAR-γ agonist [26].

Another significant report of MTAs includes a review by Borkow and Lapidot, in 2005, in which they disclosed how the multistep nature of HIV-1 entry, which provides multisite targeting at the entrance door of HIV-1 to cells and prevents HIV-1 access to its host cells, has clear advantages over blocking the virus at later stages in the life cycle of the virus. There are previous reports in which several entry inhibitor combinations led to potent and synergistic inhibition of HIV-1 proliferation. They disclosed a new class of compounds, amino-glycoside-arginine conjugates (AACs), which may serve as lead compounds for the development of multitarget HIV-1 inhibitors [27].

6. STRATEGIES FOR DESIGNING MULTITARGETING AGENTS

The concept of multitargeting drugs has gained a lot of attention in the last decade with many molecules surfacing in the market, especially in the field of oncology [28] and neurological disorders [29]. Interestingly, many drugs in clinical use were found to have a multitarget profile, but their mechanisms of action have usually been discovered only retrospectively. Thus, the main challenge that remains with the concept of MTAs is the intentional and rational design of multitarget ligands with well-defined biological profiles. The crucial issue is maintaining the affinity balance toward different target proteins. Similarly, maintaining the right balance of target occupancy for achieving the desired in vivo efficacy profile is another key challenge. These aspects of the multitarget approach shift the research community toward designing dual rather than multitarget compounds, which could still show a better efficacy profile than single-target drugs, and are supposedly more feasible to design and synthesize than multitarget compounds in terms of affinity balancing and in vivo profiling [30].

Another significant aspect of designing MTAs is the selection of targets. The selection should be based on chemical and pharmacological considerations. The first and foremost point to understand is whether or not modulating the two selected targets could lead to an additive or synergistic effect. Next, the pharma-cophoric features essential for binding to the selected targets must be identified. Later, these selected key pharmacophoric features can be integrated in one dual or multitarget compound to obtain a so-called hybrid, fused, or chimeric compound. The selections to design hybrid, fused, or chimeric compounds should be driven by the nature of the targets, the availability of reference compounds, and the synthetic feasibility of the designed molecule [31].

From the preceding information, it clearly emerges that the basic aspect of designing an MTA may involve either of two strategies: (1) rational designing by a combination of pharmacophores, also known as the fragment-based approach and (2) involving the computer-assisted screening of known drugs libraries, which is more rigorous. These strategies can be further divided on the basis of methodologies employed for the desired outcomes (Fig. 1.1).

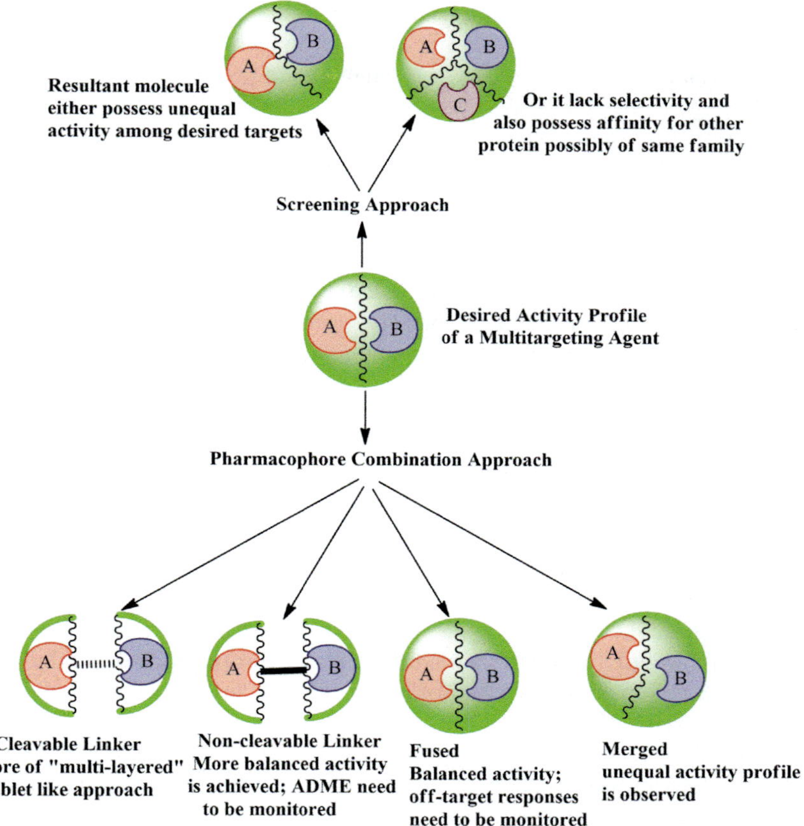

FIGURE 1.1 Comparative analysis of both approaches employed for the design of multitargeting agents.

6.1 Methodical Combination of Pharmacophore (Fragment-Based) Approach

Due to development of the technology to detect inhibitory activity sensitively in the micromolar range, a new range of possibility unwrapped for the designing of drugs by stepwise addition of different substructural units (i.e., functional groups to simpler low molecular weight chemical scaffold [32]). Since then, designing molecules using a fragment-based approach has become a well-established drug discovery tool for identification and optimization of small, highly efficient molecules as lead molecules [33]. Further, as the concept of multitargeting molecules has developed, the concept of fragment-based approaches is also being employed to design and disclose multitarget hits.

The existence of chemical scaffolds with a higher propensity for binding to different proteins (i.e., promiscuous ligands) has provided a newer opportunity to design MTAs. Such molecules could be judiciously modified to identify new and highly selective modulators. Interestingly, if target selectivity is considered the main goal, the propensity of some scaffolds to bind promiscuously becomes an inconvenient feature; however, the same feature becomes a boon if a multi-targeting molecule is to be developed.

Some of the initial work related to identification of MTAs was done by Hann and colleagues, using a simplified model that calculated the probability of interaction between proteins and ligands with diverse complexity [34]. They concluded that smaller molecules can act as a better reference point for their discovery since they have higher chances of hitting different biological targets due to lower complexity, especially signaling enzymatic proteins. Interesting work on multitarget ligands was done few years later by Hopkins and coworkers. They utilized Pfizer corporate screening data to extract information about the binding promiscuity of compounds against a statistically relevant number of diverse biological targets. Interestingly, they observed an inverse correlation between promiscuity and mean molecular weight (cut-off activity of 10 mM), claiming that smaller molecules have a greater tendency to interact with multiple biological targets as they have a fewer number of negative interacting features [35]. There are several reports where researchers found that molecules showing various degrees of promiscuity tend to fade upon progression of the inhibitors by the addition of chemical functionalities [36]. Little other research work disclosed the fact, such as analysis of Organon's SCOPE database, which disclosed a well-defined correlation between size and selectivity, and thus supported the hypothesis that the intrinsic simplicity of small compounds leads to their nonselective binding [37].

Small molecules having appealing potential against different targets disclose areas of overlap in the specificity landscapes, and lead to disclosure of various scaffolds capable of modulating the activity of two or more biomolecules. The basic advantage of using fragments instead of complete molecules is the reduction in the available chemical space; however, it remains challenging to determine the specific substructure and keeping its conformation in the new fused molecule. Another challenge involves prediction a priori whether two target proteins possess an area accessible to diverse ligands. To tackle this issue, in 2010, Miletti and Vulpetti reported a method based on identification of similar binding pockets and then evaluated a panel of protein kinases and, subsequently, on proteins from the Worldwide Protein Data Bank [38]. Interestingly, significant resemblances were reported at the subpocket level, also in between unrelated proteins. Medium-throughput techniques, such as ligand-observed nuclear magnetic resonance (NMR) or surface plasmon resonance (SPR), have a significant role in handling the risk of having false positives. Reports also disclose the use of X-ray methods in a relatively high-throughput manner as the primary method for hit detection [39]. The identification of hits is the starting point, followed by several steps that focus on balancing the activities toward the targets while keeping the molecule more druglike by stepwise incorporation of functional groups.

Currently NMR- or SPR-generated inhibition data along with X-ray crystallography and protein-observed NMR are being utilized as efficient tools in the design and optimization of multitarget molecules [40]. Information obtained from these experimental techniques regarding the protein sites could guide the modifications that are expected to be either beneficial or at least tolerated by both proteins. In further steps, the ligand efficiencies toward their targets must be monitored, to maintain a defined activity profile. Finally, in the advanced stages, the enthalpic and entropic binding contributions could be calculated to fine-tune the optimization process.

Work on MTAs by Morphy et al. disclosed the fact that majority of the multitarget molecules are designed using a designing-in approach—incorporating the different pharmacophoric features into a single target agent; while very few of them are designed using a designing-out strategy—from promiscuous ligand to maybe a dual-target inhibitor [13]. Thus, for a highly promiscuous substructure or fragment, one should focus more on a designing-out strategy rather than a designing-in approach to design an MTA. However, as expected promiscuous substructures possess a low molecular weight and thus provide a possible scope for the incorporation of other chemical entities that could fit within protein-binding pockets, one step of the designing-in approach basically focused on increasing the molecular weight can also be utilized.

Thus, extraction and coupling of pharmacophores from various selective ligands appears to be a more obvious and logical method for the generation of multitargeting molecules. These pharmacophores can be either coupled together, retaining almost complete structural integrity of all the pharmacophores, or fused together, thereby keeping only the essential features of all the contributing pharmacophores. Coupling, which can be performed by a cleavable or noncleavable linker forming conjugates and fused molecules, which are more commonly employed, involves overlapping of the pharmacophores by considering the structural features. The most interesting fact regarding this integration of pharmacophoric substructural units present in different selective ligands is that usually these consensus substructures are hydrophobic or basic ring systems and thus the structure–activity relationship (SAR) of these functionalities is really interesting.

6.1.1 Cleavable Conjugates

This approach is a rather simple method where drugs are coupled prior to their administration; that is, rather than administering two separate drugs, one single drug molecule consisting of both drugs is formed, similar to the multilayered tablet approach used in pharmaceutics. Cleavable conjugates usually consist of two individual selective drugs coupled together with a linker and most of these cleavable conjugates contain either an ester linkage or a disulphide linkage [41], which is later cleaved by plasma esterases or some other enzymes releasing both the drugs, which then act independently. However, in the case of these conjugates the pharmacokinetic–pharmacodynamic (PK–PD) relationship usually

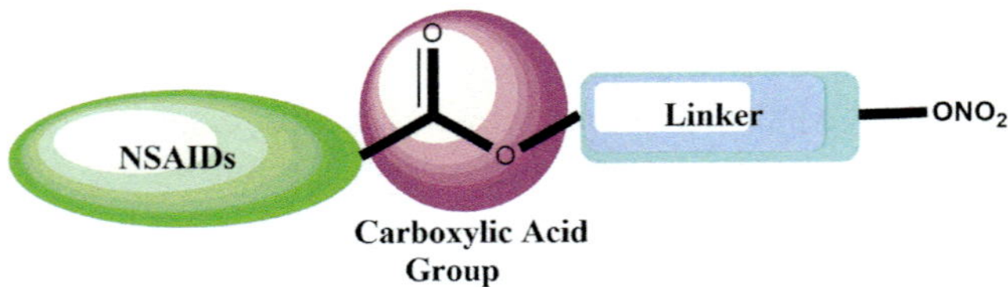

FIGURE 1.2 Graphical representation of cleavable conjugates of a nitric oxide (NO)-releasing functionality linked to NSAIDs.

gets very complicated after the cleavage of the linker, which requires extensive study. Various examples of cleavable conjugates include a nitric oxide (NO)-releasing functionality that is linked to aspirin (NCX4016) [42] and ibuprofen [43] as antiinflammatory agents (Fig. 1.2).

6.1.2 Conjugated Pharmacophores

Conjugated pharmacophoric drugs include molecules that include two individual drugs coupled with a noncleavable linker or a single agent in its dimeric form (i.e., a bivalent ligand). In 1999, Buijsman et al. reported a conjugate molecule consisting of a thrombin inhibitor linked to a pentasaccharide inhibitor. The conjugate was found to have a longer antithrombotic effect compared to an individual combination of the pentasaccharide and thrombin inhibitor [44]. Similarly, a bivalent ligand of an opioid was reported with improved potency and selectivity compared to its monomer [45]. Mechanism of action of this bivalent ligand revealed that initially univalent binding to one unit of the receptor homodimer occurs, which leads to an increased positive entropy that further enhances association of the second ligand unit to the second unit of the homodimer (Fig. 1.3).

6.1.3 Fused Pharmacophores

This approach involves overlapping of two selective pharmacophores, to develop a single molecule effective against both the targets. In 2002, Murugesan et al. developed an MTA for AT1 (angiotensin-1) and ETA (endothelin-A). Their design was based on the fact that both the AT1 and ETA inhibitors possess a biaryl system. Similarly, both the AT1 receptor and ETA receptor are reported to allow substitutions at the 2-position and 4′-position of the biaryl nucleus. Thus, they introduced a 2′-substituent that provided balanced activity at the AT1 and ETA receptors [46] (Fig. 1.4). Similarly, an MTA for H1 (histamine-1) and NK1 (neurokinin-1) and H1 and PAF (platelet activating factor) have also been reported [47,48]. Additionally, triple inhibitors of endothelin converting enzyme, ACE, and NEP were also reported in the literature [49]. Another example of MTAs acting on the members of same family include an orally active dual COX-2 and 5-LOX inhibitor [24].

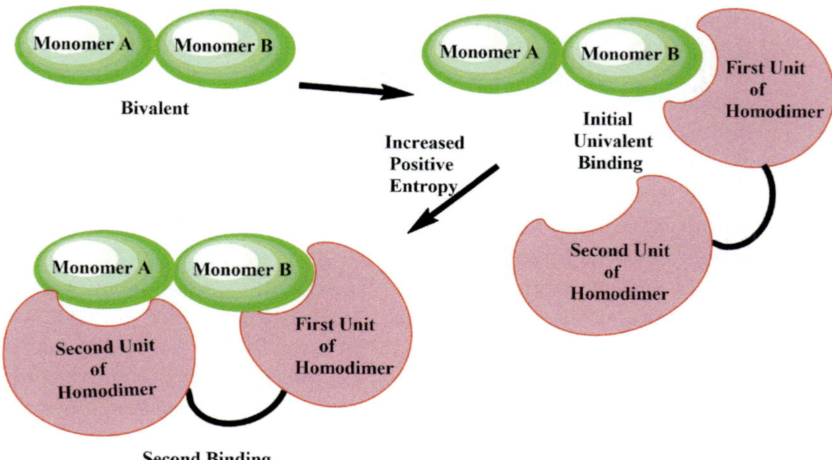

FIGURE 1.3   Graphical representation of dimeric ligands with homodimer proteins.

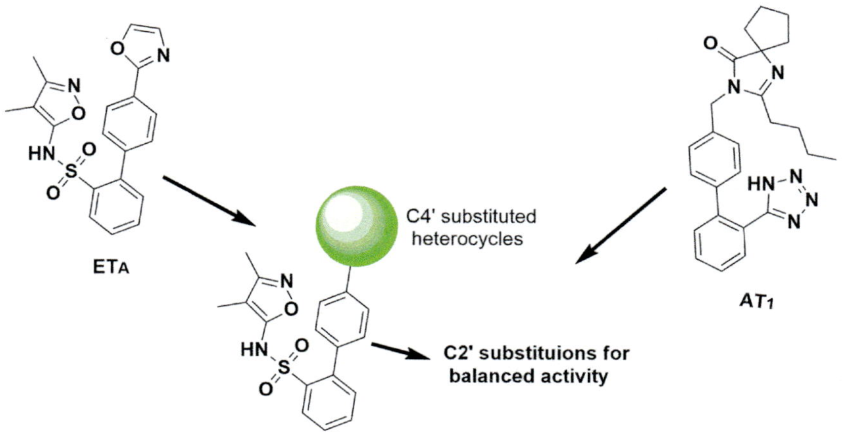

FIGURE 1.4   Design of dual inhibitors via 2′-substitution for balanced activity against angiotensin-1 and endothelin-A receptors.

A similar approach has also been applied to design and develop MTAs for apparently dissimilar receptor families. An initial work with this kind of approach was reported by Kogen et al.; they designed a dual inhibitor of AChE (acetylcholinesterase) and SERT for the treatment of Alzheimer's disease [50]. Another example is represented by a molecule designed by incorporating a PAF receptor antagonist into the selective TxS inhibitor by substituting the phenyl ring of the TxS inhibitor [51].

6.2 Computer-Assisted Retrieval of Molecules With Multiple Pharmacophores

Computational techniques have played a significant role in the standard protocol for drug discovery processes. Different tools have been constantly utilized to identify new hits or to improve the pharmacological profile of a drug candidate [52]. Utilization of computational tools and techniques in multitarget drug discovery projects are therefore a natural extension of these in silico strategies. Computational methods provide the most effective ways for mining an ever-increasing flow of data, and are useful in identifying target combinations that, if appropriately modulated, could provide a synergistic physiological response. There have been various advancements in the computational field; particularly, interactomics and pocketomics are emerging disciplines that are gaining increasing importance in polypharmacology. Interactomics studies involve analysis of networks of protein connections at the molecular level, further compiling and comparing the obtained interaction maps [53].

These studies require an interaction network, which can be considered a colored digraph—annotated nodes, which represent the biological pathways that connect receptors and enzymes [54]. Simple topographical representations are thus transformed into predictive models. Network analysis provides a simpler and easily understandable form of relevant signaling pathways. These networks strengthen the concept that the comparably weak but simultaneous inhibition or modulation of various nodes involved in the network can be a better strategy for triggering a desired physiological response than potent and focused inhibition of a single target [55]. This type of network analysis usually provides significant information regarding the crossroads in different pathways and branching points along a signaling pathway. This improvement in the in silico technique thus provides a key to overcome a network's strength and develop a multitarget molecule [56]. However, no conclusive results have yet been reported where a specific set of target combinations has been exploited using interactomics [37]. The basic reason for such failures is the fact that no clear structural information of a specific scaffold or chemical class could be gathered from relationships based solely on molecular biology and biochemistry [57].

Another significant drawback with network analysis is that only a small number of mapped nodes actually represent drug targets, thus implementations of such information lead to generation of false positive results. Another significant work in the network-based approach was reported by Kieser et al., known as the similarity ensemble approach (SEA). They developed a pharmacological network built by connecting nodes according to the similarity of their binders, independent of their experimentally tested cross-reactivity. SEA was able to provide an assessment on meaningful connections that reflect underlying similarities between pharmacological profiles [54].

However, even if using network analysis and a suitable target combination can be identified, the actual development of a multitarget candidate heavily relies on the possibility of developing a molecule that can actually interact with multiple proteins. This, in turn, requires a putative drug to actually fit in the binding regions of the target proteins. The development of tools and techniques to detect such features falls under pocketomics, which focuses on those regions of the target where interactions at the molecular level can take place [58]. Ligands have the tendency to bind to dissimilar pockets, on the basis of alternative conformations or driving interactions involving distinct substructures of the same molecule. Thus, the presence of significantly related binding pockets in different targets is not an indispensable requirement for promiscuous pharmacological activity; rather the opposite is usually true. The presence of very similar pockets probably indicates a certain level of cross-reactivity, establishing significance of pocket similarity predictive algorithms [38]. These similarity assessment methods rely on the same modular approach, which combines three-dimensional and evolutionary traits. Initially, they provide a simplified representation of the pockets, coding pharmacophoric features and structural determinants in compact data structures. Then, a pairwise matching procedure computes similarities between different descriptions. These predictive tools are particularly useful when complemented with a functional definition of drugability, namely the propensity of a cavity to accommodate not just every molecule but a molecule with druglike features [59]. Another approach proposed by Miranker and Karplus, multicopy simultaneous search method, can be particularly suited, at an upstream level of a multitarget drug design workflow, for the identification of common patterns between divergent targets involved in the same disease [60].

Metz and colleagues developed a statistically weighted map of the kinome by assembling information from sequence homology and ligand-binding affinity. This network provided an important insight into identifying target combinations: the strength of the connection between two nodes can be maintained, strengthened, or almost abolished by resorting to different chemotypes [61]. Therefore, networks complemented by information on structures and binding pocket similarities can not only help predict an optimal target combination but also can suggest novel chemical scaffolds likely to provide the sought effect. The recent account by Apsel and colleagues exemplifies the challenges facing those who seek to use advanced tools for predicting polypharmacology target combinations. The authors report the discovery of a series of aryl-substituted pyrazolopyrimidine inhibitors displaying activity against tyrosine kinases such as Src, VEGFR, and Hck, and lipid kinases, in particular the phosphatidylinositol-3-OHkinase (PI(3)K) family [62]. Tyrosine kinases and lipid kinases share only a limited sequence identity and loose structural similarity. In particular, there is a small gatekeeper residue that characterizes tyrosine kinases and that was exploited to achieve specificity for this subfamily.

Once a suitable combination of targets has been identified and validated, a rational drug design project can begin identifying multitarget hits. In a single-target endeavor, high throughput screening (HTS) would represent a straightforward strategy for identifying initial hits. Although HTS is powerful, it is costly in terms of resources, time, and personnel. This is true when just one target is involved. The costs increase sharply if multiple targets are to be considered simultaneously. Virtual ligand screening (VLS) represents a fast and efficient alternative to HTS for processing large libraries of compounds [63]. In single-target VLS every molecule in the library is tested against an ideal model of activity. This model can be based on pharmacophoric and physicochemical descriptors extracted from known ligands or on interactions at the target binding site. Ligand and structure-based models can be used independently or in combination. Each compound is assigned a predicted activity score and the library is ranked accordingly. Only the top-ranking fraction proceeds to further testing. The most straightforward way to apply VLS in the multitarget framework is to apply the screening protocol to each target independently [64]. Although VLS applied to multitargets can provide useful information for identifying hybrid, fused, or chimeric hit compounds, the most straightforward results coming from a multitarget VLS campaign could be the identification of hybrid molecules able to bind simultaneously to the selected targets. In fact, in the next step of a possible VLS-based workflow, the researchers must somehow combine and analyze the generated results to decide which molecules to prioritize for testing, based on the previously mentioned cornerstone of polypharmacology: a weaker activity, as long as it involves multiple targets, is preferable to an activity that is potent but limited to a single protein [65]. An experimental multitarget profile is likely to emerge from those molecules that, even if they do not reach the top-ranking fraction in any single run, score on average adequately well and never drop below a given threshold.

The screening approach is based on the concept of filtering the compounds against various targets to obtain an MTA. Although the screening approach appears to be a useful and simple method, it is a less utilized approach due to its low accuracy. Although ligands identified by screening approaches normally possess all the desired biological activities, it remains highly unlikely that they would have the desired activity profile. Molecules obtained after screening require balancing of features to maintain the desired multitargeting profile. Additionally, the molecules obtained after screening may also have the tendency to bind to other undesired proteins, and so, to reduce the risk of these side effects, few of the undesired features are required to be designed-out. The concept of balance will probably have a pivotal role in attempts to improve potency toward multiple targets without compromising the LE and the PK profile of the candidate. In this regard, we can envisage a new generation of computational tools that will automatically address the optimization of a multitarget drug in terms of scaffold morphing, introducing modifications that are beneficial not just for one target and tolerated by the others but enhance the compound profile with respect to all targets at once. Thus, this approach requires molecules to be further modified as MTAs. In 2002, Ryckmans et al. developed a multitargeting NK1 and SERT inhibitor by optimizing the lead molecule obtained from the screening approach [66]. In silico techniques, ligand-based and structure-based screening, are more commonly employed under this approach.

The work of Wei and coworkers represents a good case study in the practical application of multitarget VLS. The authors successfully identified novel anti-inflammatory candidates displaying activity against phospholipase A2 (PLA2) and human leukotriene A4 hydrolase (LTAH4-h). First, they devised a common pharmacophore that combined relevant features from both targets. Then, they carried out independent structure-based VLS runs, filtering out all conformations that did not match the common pharmacophore [67]. Notably, none of the compounds eventually reported to be active would have been identified by simply testing top-ranking molecules. Kernel methods and, in particular, support vector machines (SVMs) are versatile and efficient strategies already used in single target screening processes. They are emerging as particularly suitable tools for multitarget-oriented campaigns.

The main downside of combined approaches is that, because every screening paradigm is prone to errors, the combination of multiple screening runs is bound to increase the number of false positives significantly. Ma and colleagues demonstrated that, at least for structurally related targets such as kinases, it is possible to train multiple SVMs using only single-target inhibitors and to identify dual inhibitors by combining common hits [68]. They still used multiple runs, but they managed to discriminate between nonspecific pan inhibitors and inhibitors specific for a given target combination. Moreover, if the training is performed on known multitarget compounds, it is possible to bypass the need to process the same library multiple times. However, this straightforward approach strongly limits the applicability of SVMs when studying unprecedented target combinations. Ideally, it should be possible to devise a supervised learning approach to predict directly multitarget compounds that can be trained using sets of specific inhibitors only [69].

The application of in silico techniques after hit discovery will be another challenge for future CADD studies. So far, not much has been reported. Accounts of these kinds of studies are limited to inhibitors acting on closely related targets. Ligand docking and molecular dynamics were successfully applied in SAR studies on dual c-Src/Abl and dual EGFR/VEGFR2 kinase inhibitors. In 2001, Sternbach et al. utilized such information derived from in silico modeling to develop a library of lipophilic carboxylic acid-derivatives as dual agonists for the both PPAR-γ and PPAR-δ [70]. To identify such molecules, an algorithm, Cavbase, was developed, which predicts cross-reactivity between distinct target proteins by identifying the intersection in surface-exposed physicochemical properties [71]. Such in silico tools, which identify similar cavities, provide help in the identification of molecular targets that could be targeted simultaneously via an MTA.

In 2011, Koutsoukas et al. reviewed databases that are being or can be utilized for ligand-based target prediction. They also outlined the methods that exist for target prediction, both ligand-based, which rely on the knowledge of bioactivities of the ligands, and structure-based, which rely on the known protein structure. They also presented few success stories where attempts based partly or in whole on computational target predictions identified potential targets in the first instance [72].

Basically both the screening and methodical combinations of pharmacophore approaches are complementary in nature, and the most optimum strategy involves employing both approaches to improve the overall chance of success. The main advantage with the combination of pharmacophore approach is that it provides a basic structural point to design MTAs, which can be further enhanced by considering the detailed SAR information of the scaffold under consideration. Also, this strategy remains useful in a case where targets of a desired combination possess increasingly dissimilar active sites, reducing the probability of success with the screening approach. However, rationally incorporating the additional biological activity into a selective ligand, simultaneously maintaining the basic activity it was meant for and wider selectivity, remains a daunting task. Thus, there is a major advantage with the screening approach, as it provide a compound that already has, to an extent, the desired multiple activities. Since the pharmacophore combination approach is very useful in the design and discovery of dual ligands, discovering ligands that can bind to more than two targets usually requires the screening approach. Additionally, molecules obtained from the screening approach possess an improved ADME profile and pharmacokinetic properties compared to molecules obtained from the pharmacophore combination approach.

7. CHALLENGES IN MULTITARGETING STRATEGIES

Work focused on the design and synthesis of an MTA usually suffer from two major challenges, balancing the optimum level of desired activities along with limiting the undesired responses and simultaneously optimizing the pharmacokinetic parameters of the drug specifically for oral drugs.

7.1 Regulation of the Activity Profile

The first and foremost point is keeping in mind the optimal ratio of the bioactivities to gain efficacy and maintain safety. Usually the knowledge obtained from the clinical studies provides significant insight into an optimal profile for an MTA. However, in the case of designing novel agents, no prior clinical data is available and thus, researchers rely on preliminary laboratory in vitro activity data against each target to determine the activity profile for the agent, following the concept that similar kind of target modulation will be obtained during in vivo studies [73]. In a majority of cases the designed MTAs become more selective toward one target, making them dual inhibitors only at higher doses, as in the case of venlafaxine [25]. Considering such complications in balancing the activities against different targets, focus is on targeting the proteins and enzymes of the same families, which have conserved binding sites, thus limiting the complexities. Alternatively, there have been several examples where desired selectivity has been achieved. In 2003, Kogen et al. reported dual AChE/SERT blockers that showed high selectivity over several other members of the same family [50].

7.2 Physicochemical Aspects

An interesting point observed in reported MTAs is that usually these agents tend to be bulkier and more lipophilic in nature than marketed drugs or drugs in preclinical studies [74]. This leads to a major challenge as these features limit the ADME profile of the molecules and majorly hamper the oral bioavailability. However, even though there are drawbacks, the coupling of the pharmacophore approach has provided few oral drugs such as a dual D2/5-HT2 inhibitor that were synthesized by considering the framework of dopamine fused with a 5-HT receptor inhibitor [75]. For the pharmacophore fusion strategy, a vital point is to reduce the bulk and complexity of the component ligands by improving the overlap of the framework. In a screening-based strategy, less bulky molecules are obtained because initial molecules obtained from screening supposedly have multitarget affinity, so only the lead optimization step is performed on them, which does not lead to any significant increase in the size or bulk of the existing molecule in comparison to two pharmacophoric fused frameworks. However, a physiochemical parameters related issue does not hamper the main goal of drug discovery. Various examples of drugs are reported in literature where ADME parameters such as oral availability have been ignored to obtain an efficacious molecule [45].

8. APPLICATIONS OF MULTITARGETING AGENTS

There are three major possible applications of the multitargeting approach in the current scenario of drug discovery.

8.1 Management of Complex Disorders

Multifactorial disorders such as mood disorders, neurodegenerative diseases, chronic inflammation, or cancer usually involve a number of intrinsic and/or extrinsic components working simultaneously. In such cases, MTAs appear to be advantageous over conventional agents [76]. Several heterocycles, which have been developed as MTAs, underlining such advantages, are discussed in forthcoming chapters.

8.2 Drug Resistance

Another significant outcome of blocking multiple pathways includes overcoming drug-induced resistance, which alters single target proteins involved in activation pathways [77]. This aspect provides a significant application for MTAs as there is less probability of development of resistance linked to single-point mutations against multitarget than single-target agents [78].

8.3 Prospective Drug Repositioning

A different application, known as drug repositioning, has also been disclosed in literature. It is the concept of identifying a different medical use for a known therapeutic agent [79]; MTAs have become one of the contenders for drug repositioning. However, most reports of successful drug repositioning were serendipitous. Currently, pharmaceutical companies consider exploring repositioning possibilities for the drugs in the pipeline. The basic concept of prospective drug repositioning rely on the determination of the proper mechanism of action of the medicinal agent and further identify whether the same mechanism can be utilized to treat some other disease condition. One example of such a mechanism-based drug repositioning involves dronedarone [80]. As one future aspect, computational network-based predictions may provide significant data to unveil hidden similarities between various disease conditions and provide candidates for drug reposition.

9. HETEROCYCLES FOR MULTITARGETING AGENTS

A heterocyclic nucleus is the most commonly employed structural feature in a variety of clinically approved drugs for various disease targets. More than 80% of small molecule drugs currently in retail sale contain heterocyclic nuclei in their structures [81]. One of the main causes for incorporating hetero-aryl ring systems such as oxygen, sulfur, and especially nitrogen-containing ring systems in drug molecules is an attempt to mimic the physiological molecules involved in the pathways related to the disease condition. And as heterocycles are the core substructure in numerous physiological molecules and natural products such as nucleic acids, amino acids, carbohydrates, vitamins, and alkaloids, the drug discovery process often leads to heterocyclic structural motifs. Also, strategic incorporation of a heterocyclic nucleus into the drug modulates several properties of the molecule; for instance, potency and selectivity can be modulated through bioisosteric replacements; similarly, lipophilicity, polarity, and aqueous solubility can also be altered.

Heterocycles are frequently employed as bioisosteres for a wide range of functional groups in drug molecules [82]. While it is well-known that particular bioisosteric replacement would not yield the desired pharmacological effect for every SAR development, it is nonetheless expected that heterocycles with different ring size, variety of shapes, and electronic and physicochemical properties provide useful options for optimization of lead compounds. The pharmacological improvement of incorporating heterocycles for better potency and specificity can in many cases be explained by their

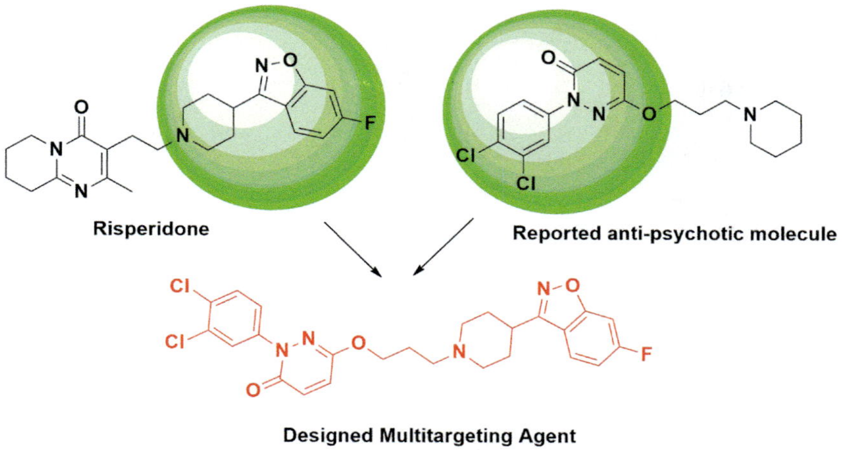

FIGURE 1.5 Molecular hybridization based design of multitargeting agents by Cao et al.

ability to participate in hydrogen and halogen bonding with the target protein [83]. Overall, heterocycles play a central role in the design of a therapeutic molecule. They are utilized to optimize potency and selectivity through bioisosterism and pharmacokinetic and toxicological properties by offering wide opportunities to adjust lipophilicity, polarity, and solubility of the target molecules. It can also be noted that sometimes the desirable result of the heterocycle incorporation into the molecule comes at the expense of negative changes of other parameters of the drug. Recognition of the most important properties for the particular target and their careful manipulation to achieve optimal balance between potency, selectivity, pharmacokinetic properties, and toxicity is the hardest part of the medicinal chemists' job [84].

Thus, similar to single targeting agents, MTAs are also based on heterocyclic ring systems as their core nucleus. More so than single targeted agents, the basic strategy involved in the design of these agents includes fusion of different pharmacophoric features, necessitating the need of different heterocyclic nuclei coupled together. Complying with the same need, a number of heterocyclic-based molecules are reported for multitargeting activities.

In 2016, Cao et al. reported pyridazinone substituted benzisoxazoles-based molecules as the multitargeting inhibitors for D2 (dopamine), 5-HT1A, and 5-HT2A (serotonin) receptors. They fused ligands of dopamine and serotonin receptors into a single molecule (Fig. 1.5), which was then evaluated via in vitro and in vivo analysis and were found to possess significant binding affinity toward D2, 5-HT1A, 5-HT2A, and 5-HT6 receptors, showing substantial potential for atypical antipsychotic condition [85].

In 2016, Singh and Silakari developed multitarget directed ligands to target the complex pathophysiology of Alzheimer's disease. They synthesized a series of 2-phenyl-1-benzopyran-4-one derivatives as inhibitors of acetylcholinesterases

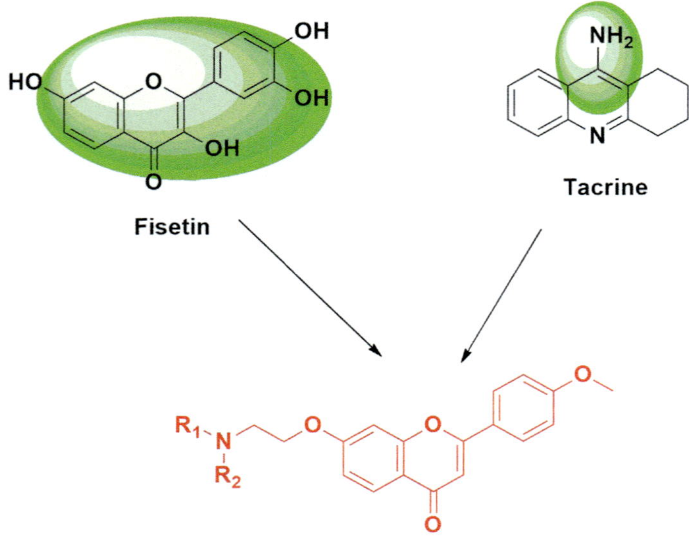

FIGURE 1.6 Design of multitargeting agents for VEGFR-2 and HDAC reported by Peng et al.

and advanced glycation end product formation. Additionally, on biological evaluation, their compounds exhibited radical scavenging properties (Fig. 1.6), thus making a promising claim of potential polyfunctional compounds to be more effective in the management of Alzheimer's disease [86].

In 2016, Kataoka et al. evaluated proline-type fullerene derivatives against hepatitis C virus (HCV) NS5B polymerase and HCV NS3/4A protease. All the compounds evaluated by the group showed the potential to inhibit both the enzymes, suggesting the dual/multitargeting nature of the fullerene derivatives against NS5B and NS3/4A, useful in the treatment of HCV infections [87]. In 2016, Jeon et al. disclosed chalcone derivatives as competitive inhibitors against μ-calpain and cathepsin B. Also, few of the evaluated compounds showed neuroprotective effects against oxidative stress-induced apoptosis in the SH-SY5Y cell line. Additionally, these compounds reduced p25 formation, tau phosphorylation, and insoluble Aβ peptide formation, establishing the multitargeting nature of these compounds [88]. In 2016, Peng et al. reported MTAs, inhibiting VEGFR-2 and histone deacetylase (HDAC) in cancer. They reported hybrids bearing *N*-phenylquinazolin-4-amine and hydroxamic acid moieties having potency against VEGFR-2 and HDAC. The most potent compound showed significant inhibitory potential against HDAC and strong inhibitory effect against VEGFR-2 (Fig. 1.7). It also possessed potent inhibitory activity against a human breast cancer cell line MCF-7 [89].

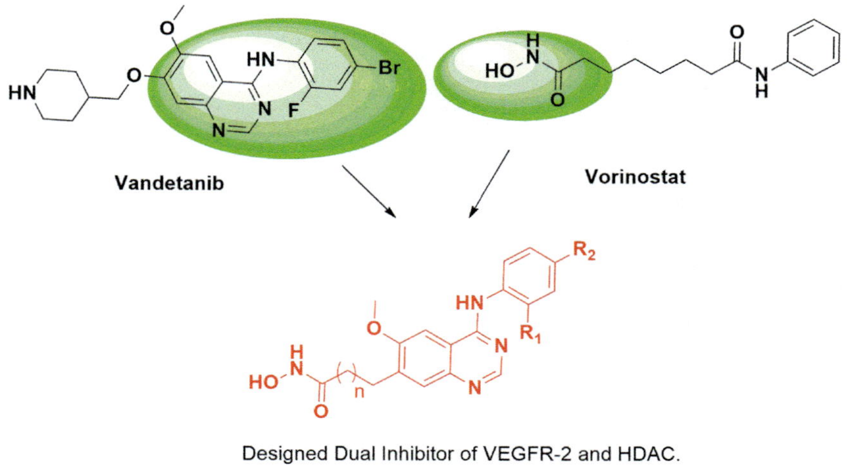

Designed Dual Inhibitor of VEGFR-2 and HDAC.

FIGURE 1.7 Podophyllotoxin derived multitargeting inhibitors of topoisomerase II and histone deacetylase.

In 2016, Osman et al. reported tetrahydroacridin-9-amines-based MTAs for the treatment of Alzheimer's disease. They synthesized a library of substituted tetrahydroacridin-9-amine derivatives and further evaluated them as inhibitors of cholinesterase and amyloid aggregation. Their findings clearly demonstrated the effectiveness of 3,4-dimethoxyphenyl substituent as a pharmacophore having dual cholinesterase inhibition and anti-Aβ-aggregation properties, which could be further exhausted to design and develop molecules with multitargeting potency for the treatment of Alzheimer's disease [90]. Similarly, in 2016, Aher et al. developed ethylenedisalicylic acid (EDSA) derivatives as a novel scaffold for inhibitors of protein tyrosine phosphatase (PTP) 1B and IkB kinase β (IKKβ). EDSA represents a modified form of methylenedisalicylic acid and consists of an ethylene moiety in place methylene. Further on performing the biological evaluations, many of the compounds showed inhibitory potencies in the low micromolar range against PTP1B and IKKβ, signifying them to be promising scaffolds for the development of MTAs in obesity [91]. In 2014, Jiang et al. reported a series of diaryl-1,2,4-triazole and *N*-hydroxyurea based hybrids as antiinflammatory agents. Biological evaluation of these molecules verified the claim of these agents being MTAs of COX-2 and 5-LOX. These MTAs were designed by fusing previously reported triazole-based selective COX inhibitors with reported LOX inhibitor Zileuton and were found to exhibit antiinflammatory activity comparable to the reference drug celecoxib [92] (Fig. 1.8).

In 2013, Zhang et al. designed and synthesized podophyllotoxin derivatives as multitargeting inhibitors of topoisomerase II and histone deacetylase. Further inhibitory activities against histone deacetylases, Topo II, and their cytotoxicities in HCT116 colon carcinoma cells were also evaluated (Fig. 1.9). Synthesized

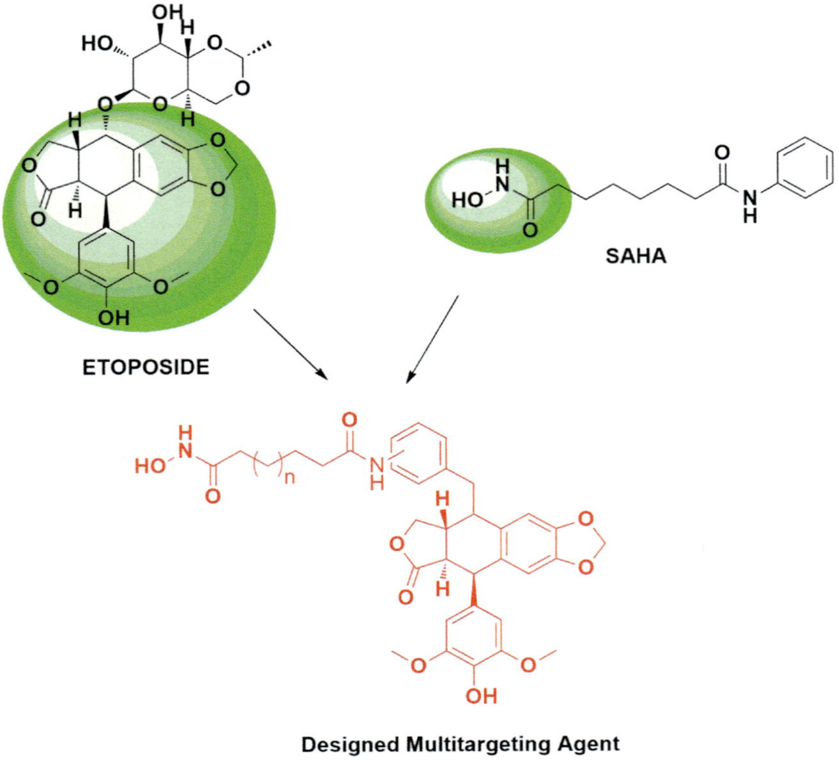

FIGURE 1.8 Diaryl-1,2,4-triazole and *N*-hydroxyurea based hybrids by Jiang et al., as antiinflammatory agents.

hybrid compounds showed potent HDAC inhibitory activity at a low nanomolar levels with antiproliferative activity at a low micromolar levels [93].

In 2012, Zhou et al. identified AL3810, a quinolone-based molecule, as a multitargeting inhibitor against various tyrosine kinases having significant roles in the process of angiogenesis. Angiogenesis is one of the most critical processes having a role in neoplastic transformation and progression, enhancing the metastasis process in most of the cancers. Their study disclosed that due to its multitargeting nature the overall pharmacological profiles of AL3810 were superior to sorafenib, an FDA-approved anticancer agent [94]. In 2010, Cai et al. discovered quinzoline-based CUDC-101 as an MTA by coupling HDAC inhibitory functionality into the pharmacophore of the epidermal growth factor receptor (EGFR) and human epidermal growth factor receptor 2 (HER2) inhibitors. There results suggested that a single molecule simultaneously inhibiting HDAC, EGFR, and HER2 offers more therapeutic benefits than single-acting agents in cancer [95].

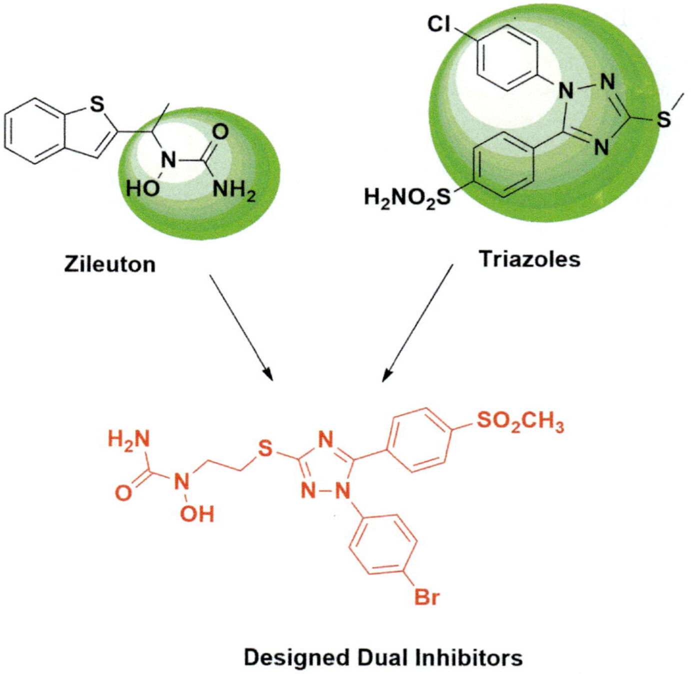

FIGURE 1.9 Podophyllotoxin derived multitargeting inhibitors of topoisomerase II and histone deacetylase.

10. CONCLUSION

MTAs appear to be promising to counter complex and multifactorial diseases and drug-resistant disorders. Thus, there is an increased interest in the design of MTAs as an alternative for developing drugs with superior efficacy and safety profiles in comparison to those selective for single target drugs. Exhaustive work is currently being carried out by various research groups using various rational and elegant approaches. One of the many approaches involves network analysis to identify targets, which play a key role in pathophysiology of the disorder and/or involved in mechanisms of resistance. The most significant role in designing these multitargeting inhibitors is played by the basic heterocyclic nuclei, which form the core of the molecule and are reported to possess significant activities against individual targets. Ignoring the oral availability, the basic bottleneck in designing MTAs remain the balance in the desired activities. Along with this, the major future perspective that remains is how to enhance the overall probability of success at an early stage of designing. Although the future for the discovery of MTAs appears lucrative, it still encompasses various difficult challenges.

REFERENCES

[1] D.A. Horton, G.T. Bourne, M.L. Smythe, The combinatorial synthesis of bicyclic privileged structures or privileged substructures, Chemical Reviews 103 (3) (2003) 893–930.

[2] S. Bongarzone, M.L. Bolognesi, The concept of privileged structures in rational drug design: focus on acridine and quinoline scaffolds in neurodegenerative and protozoan diseases, Expert Opinion on Drug Discovery 6 (3) (2011) 251–268.

[3] J. Poupaert, P. Carato, E. Colacino, 2 (3H)-benzoxazolone and bioisosters as "privileged scaffold" in the design of pharmacological probes, Current Medicinal Chemistry 12 (7) (2005) 877–885.

[4] J. Klekota, F.P. Roth, Chemical substructures that enrich for biological activity, Bioinformatics 24 (21) (2008) 2518–2525.

[5] M.E. Welsch, S.A. Snyder, B.R. Stockwell, Privileged scaffolds for library design and drug discovery, Current Opinion in Chemical Biology 14 (3) (2010) 347–361.

[6] A.-L. Barabási, N. Gulbahce, J. Loscalzo, Network medicine: a network-based approach to human disease, Nature Reviews Genetics 12 (1) (2011) 56–68.

[7] C. Ainsworth, Networking for new drugs, Nature Medicine 17 (10) (2011) 1166–1169.

[8] A.D. Boran, R. Iyengar, Systems approaches to polypharmacology and drug discovery, Current Opinion in Drug Discovery and Development 13 (3) (2010) 297.

[9] M. Rosini, Polypharmacology: the rise of multitarget drugs over combination therapies, Future Medicinal Chemistry 6 (5) (2014) 485–487.

[10] J.G. Monzon, J. Dancey, Combination Agents versus Multi-Targeted Agents–Pros and Cons, RSC Publishing, Cambridge, UK, 2012.

[11] A. Cavalli, M.L. Bolognesi, A. Minarini, M. Rosini, V. Tumiatti, M. Recanatini, C. Melchiorre, Multi-target-directed ligands to combat neurodegenerative diseases, Journal of Medicinal Chemistry 51 (3) (2008) 347–372.

[12] P.T. Lansbury, Back to the Future: The 'Old-Fashioned' Way to New Medications for Neurodegeneration, 2004.

[13] R. Morphy, C. Kay, Z. Rankovic, From magic bullets to designed multiple ligands, Drug Discovery Today 9 (15) (2004) 641–651.

[14] P. Szuromi, V. Vinson, E. Marshall, Rethinking drug discovery, Science 303 (5665) (2004) 1795.

[15] H. Zheng, M. Fridkin, M. Youdim, From single target to multitarget/network therapeutics in Alzheimer's therapy, Pharmaceuticals 7 (2) (2014) 113–135.

[16] M.M. Fulton, E. Riley Allen, Polypharmacy in the elderly: a literature review, Journal of the American Academy of Nurse Practitioners 17 (4) (2005) 123–132.

[17] G.R. Zimmermann, J. Lehar, C.T. Keith, Multi-target therapeutics: when the whole is greater than the sum of the parts, Drug Discovery Today 12 (1) (2007) 34–42.

[18] L.M. Espinoza-Fonseca, The benefits of the multi-target approach in drug design and discovery, Bioorganic and Medicinal Chemistry 14 (4) (2006) 896–897.

[19] S. Frantz, Drug discovery: playing dirty, Nature 437 (7061) (2005) 942–943.

[20] H.A. Kitano, robustness-based approach to systems-oriented drug design, Nature Reviews Drug Discovery 5 (3) (2007) 202–210.

[21] D. Brown, G. Superti-Furga, Rediscovering the sweet spot in drug discovery, Drug Discovery Today 8 (23) (2003) 1067–1077.

[22] B. Ma, M. Shatsky, H.J. Wolfson, R. Nussinov, Multiple diverse ligands binding at a single protein site: a matter of pre-existing populations, Protein Science 11 (2) (2002) 184–197.

[23] A. Todorovska, R.C. Roovers, O. Dolezal, A.A. Kortt, H.R. Hoogenboom, P.J. Hudson, Design and application of diabodies, triabodies and tetrabodies for cancer targeting, Journal of Immunological Methods 248 (1) (2001) 47–66.

[24] S. Barbey, L. Goossens, T. Taverne, J. Cornet, V. Choesmel, C. Rouaud, G. Gimeno, S. Yannic-Arnoult, C. Michaux, C. Charlier, Synthesis and activity of a new methoxytetrahydropyran derivative as dual cyclooxygenase-2/5-lipoxygenase inhibitor, Bioorganic and Medicinal Chemistry Letters 12 (5) (2002) 779–782.

[25] S.M. Stahl, R. Entsuah, R.L. Rudolph, Comparative efficacy between venlafaxine and SSRIs: a pooled analysis of patients with depression, Biological Psychiatry 52 (12) (2002) 1166–1174.

[26] O.P. Lazarenko, S.O. Rzonca, L.J. Suva, B. Lecka-Czernik, Netoglitazone is a PPAR-gamma ligand with selective effects on bone and fat, Bone 38 (1) (2006) 74–84.

[27] G. Borkow, A. Lapidot, Multi-targeting the entrance door to block HIV-1, Current Drug Targets-Infectious Disorders 5 (1) (2005) 3–15.

[28] R. Morphy, Selectively nonselective kinase inhibition: striking the right balance, Journal of Medicinal Chemistry 53 (4) (2009) 1413–1437.

[29] S. Giordano, A. Petrelli, From single-to multi-target drugs in cancer therapy: when a specificity becomes an advantage, Current Medicinal Chemistry 15 (5) (2008) 422–432.

[30] R. Morphy, Z. Rankovic, Designed multiple ligands. An emerging drug discovery paradigm, Target 1 (2005) 3.

[31] G. Bottegoni, A.D. Favia, M. Recanatini, A. Cavalli, The role of fragment-based and computational methods in polypharmacology, Drug Discovery Today 17 (1) (2012) 23–34.

[32] S.B. Shuker, P.J. Hajduk, R.P. Meadows, S.W. Fesik, Discovering high-affinity ligands for proteins: SAR by NMR, Science 274 (5292) (1996) 1531.

[33] S. Schultes, C. de Graaf, E.E. Haaksma, I.J. de Esch, R. Leurs, O. Krämer, Ligand efficiency as a guide in fragment hit selection and optimization, Drug Discovery Today: Technologies 7 (3) (2010) e157–e162.

[34] M.M. Hann, A.R. Leach, G. Harper, Molecular complexity and its impact on the probability of finding leads for drug discovery, Journal of Chemical Information and Computer Sciences 41 (3) (2001) 856–864.

[35] A.L. Hopkins, J.S. Mason, J.P. Overington, Can we rationally design promiscuous drugs? Current Opinion in Structural Biology 16 (1) (2006) 127–136.

[36] Y. Chen, B.K. Shoichet, Molecular docking and ligand specificity in fragment-based inhibitor discovery, Nature Chemical Biology 5 (5) (2009) 358–364.

[37] R. Morphy, Z. Rankovic, Fragments, network biology and designing multiple ligands, Drug Discovery Today 12 (3) (2007) 156–160.

[38] F. Milletti, A. Vulpetti, Predicting polypharmacology by binding site similarity: from kinases to the protein universe, Journal of Chemical Information and Modeling 50 (8) (2010) 1418–1431.

[39] M.J. Hartshorn, C.W. Murray, A. Cleasby, M. Frederickson, I.J. Tickle, H. Jhoti, Fragment-based lead discovery using X-ray crystallography, Journal of Medicinal Chemistry 48 (2) (2005) 403–413.

[40] C.W. Murray, T.L. Blundell, Structural biology in fragment-based drug design, Current Opinion in Structural Biology 20 (4) (2010) 497–507.

[41] B.P. Roques, "Mixed inhibitor-prodrug" as a new approach toward systemically active inhibitors of enkephalin-degrading enzymes, Journal of Medicinal Chemistry 35 (1992) 2473–2481.

[42] M. Bolla, N. Almirante, F. Benedini, Therapeutic potential of nitrate esters of commonly used drugs, Current Topics in Medicinal Chemistry 5 (7) (2005) 707–720.

[43] M.L. Lolli, C. Cena, C. Medana, L. Lazzarato, G. Morini, G. Coruzzi, S. Manarini, R. Fruttero, A. Gasco, A new class of ibuprofen derivatives with reduced gastrotoxicity, Journal of Medicinal Chemistry 44 (21) (2001) 3463–3468.

[44] R.C. Buijsman, J.E. Basten, T.G. van Dinther, G.A. van der Marel, C.A. van Boeckel, J.H. van Boom, Design and synthesis of a novel synthetic NAPAP-penta-saccharide conjugate displaying a dual antithrombotic action, Bioorganic and Medicinal Chemistry Letters 9 (14) (1999) 2013–2018.

[45] P.S. Portoghese, 2000 Alfred Burger award address in medicinal chemistry. From models to molecules: opioid receptor dimers, bivalent ligands, and selective opioid receptor probes, Journal of Medicinal Chemistry 44 (22) (2001) 3758.

[46] N. Murugesan, J.E. Tellew, Z. Gu, B.L. Kunst, L. Fadnis, L.A. Cornelius, R.A.F. Baska, Y. Yang, S.M. Beyer, H. Monshizadegan, Discovery of N-isoxazolyl biphenylsulfonamides as potent dual angiotensin II and endothelin A receptor antagonists, Journal of Medicinal Chemistry 45 (18) (2002) 3829–3835.

[47] G.D. Maynard, L.D. Bratton, J.M. Kane, T.P. Burkholder, B. Santiago, K.T. Stewart, E.M. Kudlacz, S.A. Shatzer, R.W. Knippenberg, A.M. Farrell, Synthesis and sar of 4-(1H-benzimidazole-2-carbonyl) piperidines with dual histamine H 1/tachykinin NK 1 receptor antagonist activity, Bioorganic and Medicinal Chemistry Letters 7 (22) (1997) 2819–2824.

[48] R.J. Vaz, G.D. Maynard, E.M. Kudlacz, L.D. Bratton, J.M. Kane, S.A. Shatzer, R.W. Knippenberg, Use of CoMFA in validating the conformation used in designing 4-(1H-benzimidazole-2-carbonyl) piperidines with H 1/NK 1 receptor antagonist activity, Bioorganic and Medicinal Chemistry Letters 7 (22) (1997) 2825–2830.

[49] N. Inguimbert, H. Poras, F. Teffo, F. Beslot, M. Selkti, A. Tomas, E. Scalbert, C. Bennejean, P. Renard, M.-C. Fournié-Zaluski, N-[2-(Indan-1-yl)-3-mercapto-propionyl] amino acids as highly potent inhibitors of the three vasopeptidases (NEP, ACE, ECE): in vitro and in vivo activities, Bioorganic and Medicinal Chemistry Letters 12 (15) (2002) 2001–2005.

[50] N. Toda, K. Tago, S. Marumoto, K. Takami, M. Ori, N. Yamada, K. Koyama, S. Naruto, K. Abe, R. Yamazaki, A conformational restriction approach to the development of dual inhibitors of acetylcholinesterase and serotonin transporter as potential agents for Alzheimer's disease, Bioorganic and Medicinal Chemistry 11 (20) (2003) 4389–4415.

[51] M. Fujita, T. Seki, H. Inada, K. Shimizu, A. Takahama, T. Sano, Approach to dual-acting platelet activating factor (PAF) receptor antagonist/thromboxane synthase inhibitor (TxSI) based on the link of PAF antagonists and TxSIs, Bioorganic and Medicinal Chemistry Letters 12 (3) (2002) 341–344.

[52] W.L. Jorgensen, The many roles of computation in drug discovery, Science 303 (5665) (2004) 1813–1818.

[53] J.T. Metz, P.J. Hajduk, Rational approaches to targeted polypharmacology: creating and navigating protein–ligand interaction networks, Current Opinion in Chemical Biology 14 (4) (2010) 498–504.

[54] M.J. Keiser, B.L. Roth, B.N. Armbruster, P. Ernsberger, J.J. Irwin, B.K. Shoichet, Relating protein pharmacology by ligand chemistry, Nature Biotechnology 25 (2) (2007) 197–206.

[55] T. Korcsmáros, M.S. Szalay, C. Böde, I.A. Kovács, P. Csermely, How to design multi-target drugs: target search options in cellular networks, Expert Opinion on Drug Discovery 2 (6) (2007) 799–808.

[56] A.L. Hopkins, Network pharmacology: the next paradigm in drug discovery, Nature Chemical Biology 4 (11) (2008) 682–690.

[57] J.D. Durrant, R.E. Amaro, L. Xie, M.D. Urbaniak, M.A. Ferguson, A. Haapalainen, Z. Chen, A.M. Di Guilmi, F. Wunder, P.E. Bourne, A multidimensional strategy to detect polypharmacological targets in the absence of structural and sequence homology, PLoS Computational Biology 6 (1) (2010) e1000648.

[58] R. Abagyan, I. Kufareva, The flexible pocketome engine for structural chemogenomics, Chemogenomics: Methods and Applications (2009) 249–279.

[59] M. Nayal, B. Honig, On the nature of cavities on protein surfaces: application to the identification of drug-binding sites, Proteins: Structure, Function, and Bioinformatics 63 (4) (2006) 892–906.

[60] A. Miranker, M. Karplus, Functionality maps of binding sites: a multiple copy simultaneous search method, Proteins: Structure, Function, and Bioinformatics 11 (1) (1991) 29–34.

[61] J.T. Metz, E.F. Johnson, N.B. Soni, P.J. Merta, L. Kifle, P.J. Hajduk, Navigating the kinome, Nature Chemical Biology 7 (4) (2011) 200–202.

[62] B. Apsel, J.A. Blair, B. Gonzalez, T.M. Nazif, M.E. Feldman, B. Aizenstein, R. Hoffman, R.L. Williams, K.M. Shokat, Z.A. Knight, Targeted polypharmacology: discovery of dual inhibitors of tyrosine and phosphoinositide kinases, Nature Chemical Biology 4 (11) (2008) 691–699.

[63] R. Abagyan, M. Totrov, High-throughput docking for lead generation, Current Opinion in Chemical Biology 5 (4) (2001) 375–382.

[64] E. Jenwitheesuk, J.A. Horst, K.L. Rivas, W.C. Van Voorhis, R. Samudrala, Novel paradigms for drug discovery: computational multitarget screening, Trends in Pharmacological Sciences 29 (2) (2008) 62–71.

[65] P. Csermely, V. Agoston, S. Pongor, The efficiency of multi-target drugs: the network approach might help drug design, Trends in Pharmacological Sciences 26 (4) (2005) 178–182.

[66] T. Ryckmans, L. Balançon, O. Berton, C. Genicot, Y. Lamberty, B. Lallemand, P. Pasau, N. Pirlot, L. Quéré, P. Talaga, First dual NK 1 antagonists–serotonin reuptake inhibitors: synthesis and SAR of a new class of potential antidepressants, Bioorganic and Medicinal Chemistry Letters 12 (2) (2002) 261–264.

[67] D. Wei, X. Jiang, L. Zhou, J. Chen, Z. Chen, C. He, K. Yang, Y. Liu, J. Pei, L. Lai, Discovery of multitarget inhibitors by combining molecular docking with common pharmacophore matching, Journal of Medicinal Chemistry 51 (24) (2008) 7882–7888.

[68] X.H. Ma, Z. Shi, C. Tan, Y. Jiang, M.L. Go, B.C. Low, Y.Z. Chen, In-Silico approaches to multi-target drug discovery, Pharmaceutical Research 27 (5) (2010) 739–749.

[69] L. Li, J. Li, M. Khanna, I. Jo, J.P. Baird, S.O. Meroueh, Docking small molecules to predicted off-targets of the cancer drug erlotinib leads to inhibitors of lung cancer cell proliferation with suitable in vitro pharmacokinetic properties, ACS Medicinal Chemistry Letters 1 (5) (2010) 229.

[70] K.G. Liu, M.H. Lambert, L.M. Leesnitzer, W. Oliver, R.J. Ott, K.D. Plunket, L.W. Stuart, P.J. Brown, T.M. Willson, D.D. Sternbach, Identification of a series of PPARγ/δ dual agonists via solid-phase parallel synthesis, Bioorganic and Medicinal Chemistry Letters 11 (22) (2001) 2959–2962.

[71] T. Krotzky, T. Fober, E. Hüllermeier, G. Klebe, Extended graph-based models for enhanced similarity search in cavbase, IEEE/ACM Transactions on Computational Biology and Bioinformatics (TCBB) 11 (5) (2014) 878–890.

[72] A. Koutsoukas, B. Simms, J. Kirchmair, P.J. Bond, A.V. Whitmore, S. Zimmer, M.P. Young, J.L. Jenkins, M. Glick, R.C. Glen, From in silico target prediction to multi-target drug design: current databases, methods and applications, Journal of Proteomics 74 (12) (2011) 2554–2574.

[73] R. Morphy, Z. Rankovic, Designing multiple ligands-medicinal chemistry strategies and challenges, Current Pharmaceutical Design 15 (6) (2009) 587–600.

[74] R. Morphy, Z. Rankovic, The physicochemical challenges of designing multiple ligands, Journal of Medicinal Chemistry 49 (16) (2006) 4961–4970.

[75] J. Lowe 3rd, T.F. Seeger, A.A. Nagel, H.R. Howard, P.A. Seymour, J.H. Heym, F.E. Ewing, M.E. Newman, A.W. Schmidt, J. Furman, 1-Naphthylpiperazine derivatives as potential atypical antipsychotic agents, Journal of Medicinal Chemistry 34 (6) (1991) 1860–1866.

[76] C. Lu, Y. Guo, J. Yan, Z. Luo, H.-B. Luo, M. Yan, L. Huang, X. Li, Design, synthesis, and evaluation of multitarget-directed resveratrol derivatives for the treatment of Alzheimer's disease, Journal of Medicinal Chemistry 56 (14) (2013) 5843–5859.

[77] L. Xie, L. Xie, S.L. Kinnings, P.E. Bourne, Novel computational approaches to polypharmacology as a means to define responses to individual drugs, Annual Review of Pharmacology and Toxicology 52 (2012) 361–379.

[78] D.G. Margineanu, Systems biology, complexity, and the impact on antiepileptic drug discovery, Epilepsy and Behavior 38 (2014) 131–142.

[79] N. Novac, Challenges and opportunities of drug repositioning, Trends in Pharmacological Sciences 34 (5) (2013) 267–272.

[80] G. Benaim, V. Hernandez-Rodriguez, S. Mujica-Gonzalez, L. Plaza-Rojas, M.L. Silva, N. Parra-Gimenez, Y. Garcia-Marchan, A. Paniz-Mondolfi, G. Uzcanga, In vitro anti-*Trypanosoma cruzi* activity of dronedarone, a novel amiodarone derivative with an improved safety profile, Antimicrobial Agents and Chemotherapy 56 (7) (2012) 3720–3725.

[81] T. Eicher, S. Hauptmann, A. Speicher, The Chemistry of Heterocycles: Structures, Reactions, Synthesis, and Applications, third ed., John Wiley & Sons, 2013.

[82] N.A. Meanwell, Synopsis of some recent tactical application of bioisosteres in drug design, Journal of Medicinal Chemistry 54 (8) (2011) 2529–2591.

[83] A. Gomtsyan, Heterocycles in drugs and drug discovery, Chemistry of Heterocyclic Compounds 48 (1) (2012) 7–10.

[84] G. Le, N. Vandegraaff, D.I. Rhodes, E.D. Jones, J.A. Coates, N. Thienthong, L.J. Winfield, L. Lu, X. Li, C. Yu, Design of a series of bicyclic HIV-1 integrase inhibitors. Part 2: azoles: effective metal chelators, Bioorganic and Medicinal Chemistry Letters 20 (19) (2010) 5909–5912.

[85] X. Cao, Y. Chen, Y. Zhang, Y. Qiu, M. Yu, X. Xu, X. Liu, B.-F. Liu, G. Zhang, Synthesis and biological evaluation of new 6-hydroxypyridazinone benzisoxazoles: potential multi-receptor-targeting atypical antipsychotics, European Journal of Medicinal Chemistry 124 (2016) 713–728.

[86] M. Singh, O. Silakari, Design, synthesis and biological evaluation of novel 2-phenyl-1-benzopyran-4-one derivatives as potential poly-functional anti-Alzheimer's agents, RSC Advances 6 (110) (2016) 108411–108422.

[87] H. Kataoka, T. Ohe, K. Takahashi, S. Nakamura, T. Mashino, Novel fullerene derivatives as dual inhibitors of Hepatitis C virus NS5B polymerase and NS3/4A protease, Bioorganic and Medicinal Chemistry Letters 26 (19) (2016) 4565–4567.

[88] K.-H. Jeon, E. Lee, K.-Y. Jun, J.-E. Eom, S.Y. Kwak, Y. Na, Y. Kwon, Neuroprotective effect of synthetic chalcone derivatives as competitive dual inhibitors against μ-calpain and cathepsin B through the downregulation of tau phosphorylation and insoluble Aβ peptide formation, European Journal of Medicinal Chemistry 121 (2016) 433–444.

[89] F.-W. Peng, J. Xuan, T.-T. Wu, J.-Y. Xue, Z.-W. Ren, D.-K. Liu, X.-Q. Wang, X.-H. Chen, J.-W. Zhang, Y.-G. Xu, Design, synthesis and biological evaluation of N-phenylquinazolin-4-amine hybrids as dual inhibitors of VEGFR-2 and HDAC, European Journal of Medicinal Chemistry 109 (2016) 1–12.

[90] W. Osman, T. Mohamed, V.M. Sit, M.S. Vasefi, M.A. Beazely, P.P. Rao, Structure–activity relationship studies of benzyl-, phenethyl-, and pyridyl-substituted tetrahydroacridin-9-amines as multitargeting agents to treat Alzheimer's disease, Chemical Biology and Drug Design 88 (5) (2016) 710–723.

[91] N.G. Aher, J.W. Park, B.H. Park, C.K. Kim, I.O. Han, H. Cho, Ethylenedisalicylic acid derivatives as dual inhibitors of PTP1B and IKKβ and their antiobesity and antidiabetic effects in mice, Bulletin of the Korean Chemical Society 37 (6) (2016) 855–863.

[92] B. Jiang, X. Huang, H. Yao, J. Jiang, X. Wu, S. Jiang, Q. Wang, T. Lu, J. Xu, Discovery of potential anti-inflammatory drugs: diaryl-1, 2, 4-triazoles bearing N-hydroxyurea moiety as dual inhibitors of cyclooxygenase-2 and 5-lipoxygenase, Organic and Biomolecular Chemistry 12 (13) (2014) 2114–2127.

[93] X. Zhang, B. Bao, X. Yu, L. Tong, Y. Luo, Q. Huang, M. Su, L. Sheng, J. Li, H. Zhu, The discovery and optimization of novel dual inhibitors of topoisomerase II and histone deacetylase, Bioorganic and Medicinal Chemistry 21 (22) (2013) 6981–6995.

[94] Y. Zhou, Y. Chen, L. Tong, H. Xie, W. Wen, J. Zhang, Y. Xi, Y. Shen, M. Geng, Y. Wang, AL3810, a multi-tyrosine kinase inhibitor, exhibits potent anti-angiogenic and anti-tumour activity via targeting VEGFR, FGFR and PDGFR, Journal of Cellular and Molecular Medicine 16 (10) (2012) 2321–2330.

[95] X. Cai, H.-X. Zhai, J. Wang, J. Forrester, H. Qu, L. Yin, C.-J. Lai, R. Bao, C. Qian, Discovery of 7-(4-(3-ethynylphenylamino)-7-methoxyquinazolin-6-yloxy)-N-hydroxy-heptanamide (CUDC-101) as a potent multi-acting HDAC, EGFR, and HER2 inhibitor for the treatment of cancer, Journal of Medicinal Chemistry 53 (5) (2010) 2000–2009.

Chapter 2

Benzimidazole: Journey From Single Targeting to Multitargeting Molecule

Pankaj K. Singh, Om Silakari
Punjabi University, Patiala, India

Chapter Outline

1. INTRODUCTION

Even though there have been continuous efforts to develop therapeutic agents for several multifactorial diseases, judicious use of heterocycles for the discovery of potent multitargeting molecules for multifactorial diseases has not been done yet [1]. A significant aspect governing the development of any such multitargeting agent (MTA) is the selection of suitable scaffold for designing multitarget drugs. The most logical and rational approach is to begin with privileged scaffolds, which provide an optimum core with their ability to promiscuously bind with multiple pathological targets depending on the nature of substituent present on them [2]. Amid the number of possible privileged scaffolds, the benzimidazole nucleus is one with an influential status in this unique family and

Key Heterocycle Cores for Designing Multitargeting Molecules. https://doi.org/10.1016/B978-0-08-102083-8.00002-9

offers a worthwhile initial core in the search of novel multitarget ligands against various multifactorial disease conditions. When appropriately substituted, it can modulate diverse receptors, pathways, and enzymes associated with the pathogenesis of complex inflammation.

Despite such remarkable propensity, including advantages offered on many fronts to the medicinal chemist, the literature covering the multitargeting capacity of the benzimidazole scaffold has been greatly limited. Although none of benzimidazole-based multitarget ligands have been transitioned into clinic, they might represent a valuable opportunity to identify biologically active multitarget compounds. Thus the main objective of this chapter is to offer some thoughts on the use of the privileged benzimidazole nucleus as a starting scaffold for a multitarget drug discovery program in the field of multifactorial inflammatory diseases and disorders.

Decades of research have led to the evolution of benzimidazole as a vital heterocyclic system and a spectrum of pharmacological properties are associated with this nucleus [3]. The benzimidazole nucleus has been therapeutically explored since the early 20th century. Initially researchers suggested that benzimidazole can elicit similar physiological responses as purines [4]. Later, upon studying degradation products of vitamin B_{12}, a substituted benzimidazole was obtained that showed vitamin B_{12}-like activity. Step by step exhaustive exploration of various substituents and derivatives of the benzimidazole nucleus led to the identification of a number of FDA-approved agents for a variety of disease conditions such as albendazole as antihelmintics, omeprazole as proton pump inhibitors, astemizole as antihistaminic, enviradine as antiviral, and telmisartan as antihypertensives.

There are several reports suggesting the significance of the benzimidazole nucleus as a single targeted agent for several disease conditions [5]. However, very few reports are available in literature that cover the multitargeting aspect of this molecule. This chapter describes the role played so far by the benzimidazole nucleus in development of multitargeting therapeutic agents and in management of multifactorial diseases in addition to the regular synthetic strategies and therapeutic uses of benzimidazole derivatives.

2. CHEMISTRY

Benzimidazole is a bicyclic heteroaromatic compound of fused benzene and imidazole. Benzimidazoles are weakly basic in nature, slightly less basic than imidazoles, therefore they are generally soluble in dilute acids. The pK value of benzimidazole was deduced a while ago and found to be $pK_{a1} = 5.30$ and $pK_{a2} = 12.3$. Benzimidazoles are also sufficiently NH-acidic to be generally soluble in aqueous alkali and form N-metallic compounds. The acidic properties of the benzimidazoles, like those of the imidazoles [6], seem to be due to stabilization of the ion by resonance. Benzimidazoles are distillable and can be distilled unchanged above 300°C.

Benzimidazoles with hydrogen at 1-position (i.e., imide nitrogen) are usually readily soluble in polar solvents and less soluble in organic solvents. Thus benzimidazole, which is readily soluble in hot water, is very poorly soluble in ether and insoluble in benzene and ligroin. However, upon incorporation of nonpolar substituents at varied sites on the benzimidazole nucleus, the solubility is enhanced in nonpolar solvents as observed in the case of 2-methylbenzimidazole, which is easily soluble in ether. Counterwise, upon introducing polar substituents on the benzimidazole nucleus its solubility in polar solvents increase; for example, 2-aminobenzimidazole is soluble in water. Overall, the solubility of different benzimidazole derivatives in alkaline solutions depends on the particular compound in question. Enough evidence is also available to indicate molecular association through N—H—N bonds in benzimidazoles with an unsubstituted NH grouping [7]. As expected, this bond strength is enhanced upon increasing resonance of the benzimidazole nucleus.

The dipole moment of benzimidazole has also been determined and was found to be 3.93 D (in dioxane) and 4.08 D [8].

2.1 Common Reactions on Benzimidazole

The benzimidazole ring possesses a high degree of stability and does not get affected either by concentrated sulfuric acid when heated under pressure to 270°C or by vigorous treatment with hot hydrochloric acid or with alkalis. Oxidation does cleave the benzene ring of benzimidazole only under vigorous conditions. Additionally, benzimidazole ring system is also considerably resistant to reduction except for few catalytic reduction methods [9]. Various positions of benzimidazole have been subjected to common chemical reactions to derive its derivatives for different medicinal purposes. Basic reactions given by benzimidazole involves reactions at the 1- and 3-position nitrogens. They readily form salts with acids such as monohydrochloride, monopicrate, mononitrate, monoacetate, and so on [10]. Benzimidazoles, upon alkylation with alkyl halides, yield 1-alkylbenzimidazoles and, under more vigorous conditions, 1,3-dialkylbenzirnidazolium halides [11]. Similarly, acylation reaction is also very commonly observed with benzimidazoles upon action of acid chlorides or anhydrides. The reaction is usually carried out in the absence of water. In the presence of water and especially in alkaline solution (Schotten- Baumann procedure), cleavage of the imidazole ring may occur [12]. Mannich reactions with benzimidazoles have been studied and well explained a long time ago by Bachman and Heisey [13]. Equimolecular amounts of benzimidazole, formaldehyde, and piperidine give a 97% yield of 1-(piperidinomethyl)benzimidazole.

The hydrogen at the 1-position plays a significant role in reactivity of benzimidazoles. It is sufficiently acidic in nature, which can be replaced by metals and thus yield N-metal benzimidazoles. Initial work was reported as early as the 1940s by Bamberger and Lorenzen. One such example includes 2,5(or 2,6)-dimethylbenzimidazole, which upon treatment with an ammonical silver nitrate solution yields the N-silver salt [12]. In addition, Hartmann and Panizzon, for the first time, reported that benzimidazoles can also be reduced to 4, 5,6, 7-tetrahydrobenzimidazoles with significant yield employing Adam's catalyst (platinum) in acetic acid. Interestingly, reduction was reported only for those benzimidazoles that possessed substitution on the 2-position [14].

Halogenation reaction is also well reported with benzimidazoles; upon treatment with a saturated solution of calcium hypochlorite at 35°C, 1-chloro-2,5(or 2,6)-dimethylbenzimidazole is obtained. However, such N-chloro compounds lose chlorine quite readily and are quite unstable even at relatively low temperatures. While on refluxing in benzene, rearrangement of the chlorine onto the benzene ring occurs, which can be utilized to obtain a completely chlorinated compound (1,4,5,6-tetrachloro-2,5-dimethylbenzimidazole) [12]. One of the more readily occurring reactions with benzimidazoles is the nitration reaction. Mostly, nitration occurs at 5- or 6-position; however, a nitro group may also enter at 4- or 7-position if the 5- or 6-positions are blocked [15].

2.2 Cleavage of the Imidazole Ring

The imidazole ring of benzimidazoles is highly susceptible to cleavage by various methods:

1. By treating with aroyl halides in the presence of water [16].
2. By reactions of pseudobases *N,N'*-dialkyl-o-phenylenediamines [15].
3. By treatment with acid anhydrides [17].

3. SYNTHESIS

Due to their immense pharmacological importance, much effort has been made to generate series of benzimidazoles and their derivatives. These synthetic strategies have been accordingly optimized to get products with varied substitutions, high yield, and purity. Several reports are available in literature presenting the synthesis of benzimidazoles and their derivatives. A few synthetic schemes reported after 2012 [18–26] are summarized in Fig. 2.1.

4. BENZIMIDAZOLE AS A PRIVILEGED SUBSTRUCTURE

Benzimidazole scaffolds possess several intriguing and versatile structural features that offer varyingly functionalized molecules with substitution at different positions. Due to being a member of the azaheterocycle class, molecules with a benzimidazole nucleus have an innate affinity for diverse and unrelated enzymes and protein receptors leading to multiple bioactivities. Another interesting characteristic of the benzimidazole framework that underlines its privileged nature is its structural similarities with naturally occurring nucleotides, which enable it to easily recognize the biopolymers of the human body such as proteins, enzymes, and receptors. Additionally, it has been a comprehensively

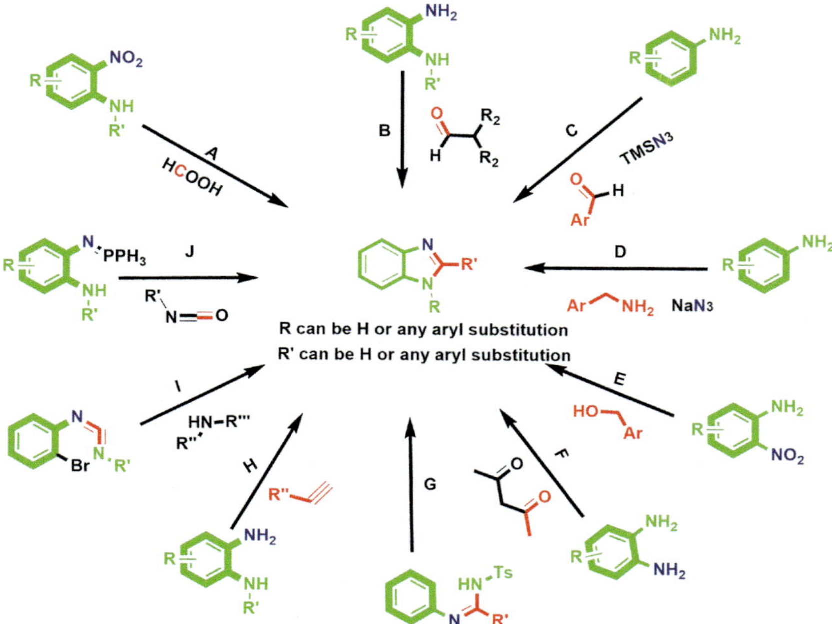

FIGURE 2.1 Recent synthetic strategies for benzimidazoles and their derivatives, (A) 10 eq. Fe, 10 eq. NH_4Cl; (B) $Zn(OAc)_2$, PMHS; (C) 0.1 eq. CuI, 1 eq. TBHP; (D) 0.1 eq. $Cu(OAc)_2$, 5 eq. AcOH, 2 eq. TBHP; (E) Na_2S (60%), 1 mol% $FeCl_3$; (F) 5 mol% TsOH, MeCN; (G) 0.2 eq. PhI, 1.5 eq. MCPBA, HFIP; (H) 1.1 eq. TsN_3, 0.1 eq. CuI, 1 eq. TEA, MeCN; (I) 0.2 eq. L-proline, 0.1 eq. CuI, 2 eq. Cs_2CO_3; (J) DCE.

explored pharmacophore in medicinal chemistry and has found applications in diverse therapeutic and clinical areas (Fig. 2.2); therefore it has a prominent place among various classes of organic compounds.

Due to its potential for altering different targets to elicit varied pharmacological properties, the benzimidazole nucleus is considered a "master key." Although all the positions of the benzimidazole nucleus can be substituted with a variety of chemical entities, most of the biologically active benzimidazole derivatives obtained up to now possess functional groups at 1-, 2-, and/or 5- (or 6-) positions and accordingly, the compounds can be mono-, di-, or trisubstituted derivatives of this nucleus. The major bioactivities establishing benzimidazole as a privileged scaffold include antihypertensive, antiinflammatory, antibacterial, antifungal, anthelmintic, antiviral, antioxidant, antiulcer, antitumor, and psychoactivity.

4.1 Antihypertensive Agents

The benzimidazole nucleus is present in many well-established antihypertensive drugs. Several reports have been published highlighting the utilization of

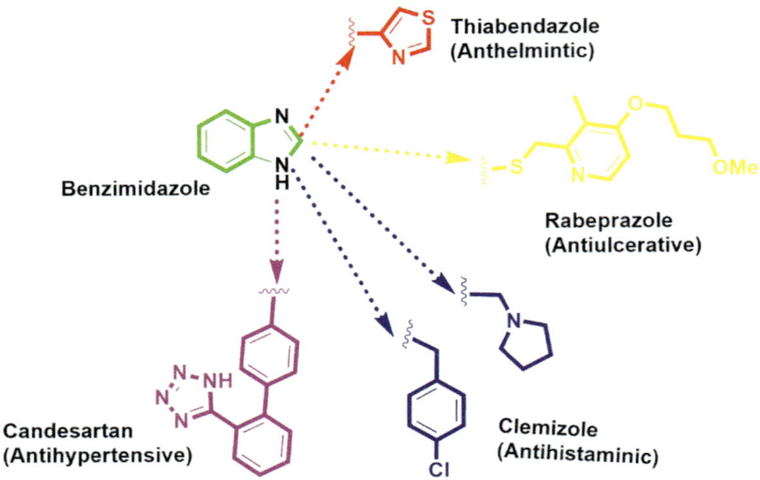

FIGURE 2.2 Few diverse therapeutic and clinical applications of benzimidazole nuclei.

the benzimidazole nucleus for the development of molecules that act as antihypertensives. A common mechanism followed by the majority of benzimidazole derivatives to show their antihypertensive potential involve blockade of renin–angiotensin system (RAS). RAS is a sophisticated physiological system that regulates vasoconstriction, thereby causing hypertension [27]. One of the initial reports, which revealed the role of benzimidazoles as antihypertensives, established that substitution of carboxylic acid moiety at 7-position leads to generation of molecules with potent angiotensin-1(AT1) receptor antagonist activity [28]. structure–activity relationship (SAR) studies of such benzimidazole derivatives led to the discovery of CV-11974, which acts on AT1 receptors in a noncompetitive manner, leading to reduced blood pressure in a dose-dependent manner [29]. Esterification of this 7-carboxylic acid substituted benzimidazole derivative led to the discovery of an orally active FDA-approved AT1 receptor blocker, candesartan cilexetil [30].

Overall exploration of the benzimidazole nucleus for its antihypertensive potential led to a well-documented SAR profile (Fig. 2.3), which disclosed that 4-position of the nucleus must remain unsubstituted for the favorable interaction of N-3 of the nucleus with H-bond residue in the AT1 receptor, while 1-position is reserved for biphenyl moiety. Substitution of a tetrazole ring system with a carboxyl group in the derivatives produces intractable and orally active antagonists [31]. A carboxylic acid group at 7-position provides potent compounds, upon esterification of which the oral bioavailability also improves [32]. The 5-position is also thoroughly explored and it is established that a group of optimum size and hydrophilicity increases the activity significantly. As for 2-position, substituted phenyl rings are reported to possess decent vasorelaxant activity.

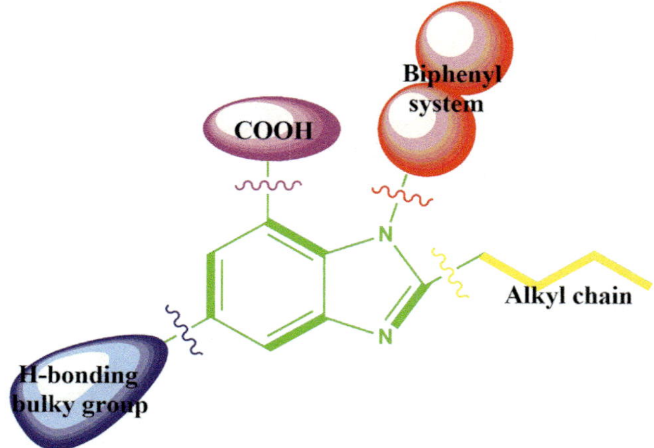

FIGURE 2.3 Structure–activity relationship profile of the benzimidazole nucleus for its antihypertensive potential.

4.2 Antiinflammatory Agents

A large number of heterocyclic nuclei are reported to inhibit or block the inflammatory process at a particular stage. Several antiinflammatory compounds derived from benzimidazole nuclei are reported in modern medical chemistry. Though a good number of research groups have reported various benzimidazole derivatives having good to excellent antiinflammatory activity, no such molecule has made its way to the clinics so far. Benzimidazole nucleus substituted at 1-position with varied heterocycles has produced potent antiinflammatory compounds. Pyrimidine substituted at 1-position on benzimidazoles were reported to elicit an antiinflammatory effect by blocking activity of Lck [33]. Another similar compound in which pyrimidine is replaced by thiophene has been reported to moderately inhibit IKK-3 kinase [34]. Further, substitution at 6-position in the benzimidazole has result in a potent inhibitor of JAK3, which is expressed in high levels in natural killer cells, platelets, thymocytes, mast cells, and inducible T and B cells. Similarly, 1,6-disubstituted benzimidazoles act as highly potent IRAK4 inhibitors along with good TNFα inhibition [34]. Based on the moderate antiinflammatory and analgesic activities of thiabendazole, the Pharmaceutical Research Centre at Kanebo Ltd. (Japan) developed a series of pyridinyl substituted benzimidazoles by isosteric replacement of thiazole ring [35].

A patent disclosed that *N*-acyl 2-aminobenzimidazole derivatives along with varied aroyl and heteroaroyl substituents at 2-position act as potent IRAK4 inhibitors. Another high throughput screening (HTS) at Abbott Corporate identified a series of 1,2-disubstituted benzimidazole derivatives through binding studies of CXCL10 to CHO cell membranes [36]. An interesting fact regarding benimidazole-based antiinflammatory agents was the significance of the

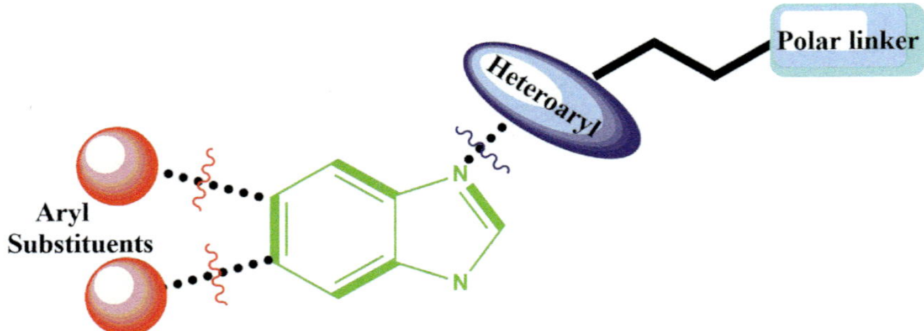

FIGURE 2.4 Structure–activity relationship profile of the benzimidazole nucleus for its antiinflammatory potential.

guanidine system. Replacements of amino with methylene at 2-position causes complete loss of activities supporting the importance of guanidine moiety for antiinflammatory activity.

Simultaneous substitutions at 2- and 5-positions of the benzimidazole nucleus have also provided several novel antiinflammatory molecules. Activity is significantly dependent on substituents at 6-position in the order of $Cl > OCH_3 > H > OCH_2CH_3 > NO_2$. The activity decreases with increasing distance of the carboxyl group from C-2 of benzimidazole. Substitution with benzyl group at 1-position further increases the activity. Many polysubstituted benzimidazole derivatives have also been reported to exhibit moderate to potent antiinflammatory activity. Substitution at piperidinyl nitrogen with 3(4-fluorophenylthio) propyl group increases the activity, significantly suggesting the importance of a bulky group at 1-position of benzimidazole, which is also supported by chemotactic activity of a series of similar compounds at Astra Zeneca [37]. The activity exhibited by compounds derived from varied substituents all around the benzimidazole nucleus has prompted many research groups to synthesize fused benzimidazole compounds.

The salient structural features of a typical benzimidazole-based antiinflammatory compound as suggested by critical analysis of chemical structures of various such compounds are given Fig. 2.4. Either of the 1- and 2-positions of the nucleus should bear a bulky, lipophilic aryl/heteroaryl moiety appropriately substituted with alkyl, electronic, or heterocyclic groups while the tolerable substituents at 5- or 6-position of the nucleus ranges from small electronic groups like halogens, nitro, amino, methyl, or lower alkoxy to mildly substituted or unsubstituted aryl or aralkyl groups.

4.3 Antimicrobial Agents

Antimicrobial agents constitute a diverse group of chemical entities acting against varied kinds of microbes including bacteria, protozoa, helminthes (worms), fungi, and viruses. Most of the research activities on development of antimicrobials from benzimidazole nuclei have occurred after the year 2000. Coupling of 2-alkylthiobenzimidazole with β-lactam ring has produced molecules with potent antibacterial and antifungal activities [38]. Literature suggested that 2-iminobenzimidazoles also have significant antibacterial potential acting through inhibition of trypanothione reductase [39]. 1-Substituted benzimidazole compounds have been found to exhibit poor antimicrobial properties. However, substitution at both 1- and 2-positions has produced potent antimicrobials. Semicarbazide, thiosemicarbazide, and carbamate substituent at 1-position along with a methyl group at 2-position have yielded potent bactericidal compounds [40].

Attachment of other heterocycles like chromane, β-lactam, thiadiazole, and oxadiazole to a benzimidazole nucleus resulted in hybrid compounds having potent antibacterial and/or antifungal properties [41]. Fusion of 5- or 6-membered heterocycle at N1—C bond of a benzimidazole nucleus has produced tricyclic derivatives of benzimidazole like triazino[1,2-a]benzimidazole bearing fluorophenyl group as moderately antibacterial, 1,2,4-triazolo[2,3-a]benzimidazole bearing short alkyl, and alkenyl groups as potent antimycobacterial and pyridobenzimidazole derivatives as antifungal. Varied 2,5- and 2,6-disubstituted benzimidazole derivatives have also been explored for antimicrobial activities.

Various 1,2,5-trisubstituted benzimidazoles have also been explored for antimicrobial and antifungal activities but without much success. Bis-benzimidazole derivatives have emerged as potent antimicrobial and antiviral agents. Any change in length of the linker methylene chain lowers the activity whereas substitution of a methoxy group in a benzimidazole nucleus increases the activity significantly [42]. A polysubstituted bis-benzimidazole compound has been synthesized and tested as a potent broad spectrum antibacterial equipotent to paromomycin. Complexation of benzimidazole compounds with varied metal ions like Fe^{3+}, Cu^{2+}, Zn^{2+}, Ag^+, and Co^{2+} is reported to increase antimicrobial activities of the compounds.

A critical analysis of these variedly substituted derivatives has revealed that either of the 1- and 2-positions of the benzimidazole nucleus should bear a bulky electronic and lipophilic group while the other should have a small alkyl substituent for optimum antimicrobial activity. Further, a small lipophilic group containing a heteroatom at 5- or 6-position incurs additional activity (Fig. 2.5).

Antiviral properties of various benzimidazole derivatives have been evaluated using different virus strains, such as human cytomegalovirus (HCMV), human herpes simplex virus, human immunodeficiency virus (HIV), and hepatitis B and C virus (HBV and HCV). Numerous nucleoside analogs of benzimidazole derivatives have been synthesized from 1950 to the 1990s as selective inhibitors of HCMV, among which 5,6-dichloro-1-(β-D-ribofuranosyl) benzimidazole (DRB) is the most explored nucleus. It inhibits viral RNA synthesis by blocking RNA polymerase II [43]. Incorporation of chloro and bromo group at 2-position of DRB provided molecules with a dramatically improved therapeutic index. A ribosyl moiety at 1-position proved to be very important for

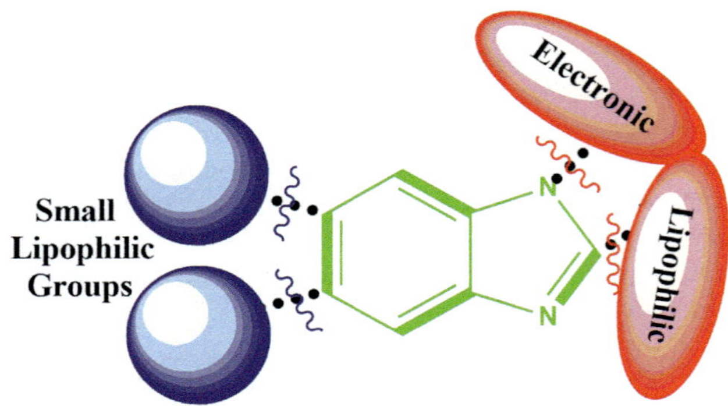

FIGURE 2.5 Structure–activity relationship profile of the benzimidazole nucleus for its antimicrobial potential.

the activity. The nonnucleoside derivatives of DRB prepared by replacing β-D-ribofuranosyl with a benzyl and phenethyl group were found inferior in activity against HCMV but active against HIV-1 [44].

For benzimidazole derivatives bearing amidino group at 5-position and varied heteronuclei such as pyridine, *N*-methyl-pyrrole, or imidazole, the compounds with the pyridine ring at 2-position show distinct and selective antiviral activity toward RNA replicating enteroviruses. In contrast, pyrrole substituted benzimidazole derivatives show prominent activity against other types of viruses, especially adenovirus [45]. A SAR study on 2-naphthyl benzimidazoles with varied substituents at 5-, 6-positions of benzimidazole ring and 4-position of a naphthyl ring suggests that electron releasing groups on benzimidazole enhances the activity. An amino group on a naphthalene ring yields a potent antiviral compound. Replacement of amino with nitro and acetyl groups significantly decreases the activity [46]. 1H-benzimidazole-4-carboxamide derivatives bearing furyl at 2-position and aryl moiety at carboxamide nitrogen possess good inhibitory activity [47].

Continual research efforts focused on benzimidazole derivatives led to 6-aza benzimidazolone derivative (BMS-433771) having good oral bioavailability and antiviral activity [48]. Attempts to further increase the activity culminated in 5-aminomethyl analog exhibiting potent antiviral activity toward wild-type RSV and excellent inhibitory activity toward a BMS- 433771 resistant viral strain [49]. Replacement of the benzimidazol-2-one moiety with benzoxazole, oxindole, quinoline-2-one, quinazolin- 2,4-dione, and benzothiazine revealed that intrinsic potency of 6,6-fused ring systems is generally less than that of 5,6-fused heterocycles [50].

Similarly modification at 5- and 6-positions of the benzimidazole core produced the most promising compounds against HBV. Hybrid molecules containing benzimidazole and coumarin with a methylene thio linker and the corresponding N-glucosides were found potent against anti-HCV compounds [50] (Fig. 2.6).

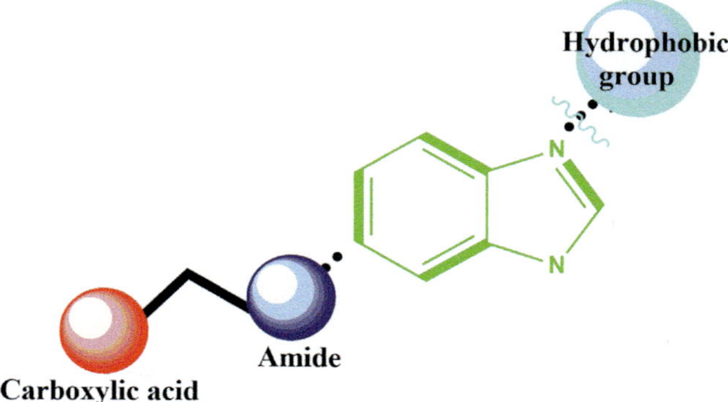

FIGURE 2.6 Structure–activity relationship profile of the benzimidazole nucleus for its antiviral potential.

4.4 Antioxidant Agents

There are several reports available that establish the role of benzimidazole nucleus in the development of various potent antioxidant molecules. Incorporation of thiadiazoles, triazoles, and their open chain counterparts (i.e., thiosemicarbazides at 1-position of benzimidazole) incurs antioxidant activity. Further placement of varied aryl and alkyl substituents on these heteronuclei at 1-position also yield potent antioxidants. Among these, semicarbazide derivatives produce stronger inhibitory effects on lipid peroxidation levels as well as DPPH model [51]. Fused thiazolo[3,2-a]benzimidazoles substituted at 3-position by an aminomethyl group inhibit the oxidation of adrenaline to adrenochrome by preventing the formation of a superoxide radical [52].

Similarly, hydroxyl groups in aroyl moiety on the imidazobenzimidazoles are reported to possess high antioxidant activity. However, 2,2,2-trichloro-1-hydroxyethyl group at 3-position weakens the antioxidant potential. A halogenophenyl group at 2-position incurs moderate antioxidant activity with fluorine producing the maximally active derivatives [53]. Further studies showed that Schiff's bases of benzimidazoles exhibit high lipid peroxidation inhibitory activity, which increases with lipophilicity. A 4-carboxamidobenzimidazole analog of Schiff's bases of benzimidazoles is identified to possess potent hydroxyl radical scavenging property through poly (ADP-Ribose) polymerase (PARP) inhibition [54] (Fig. 2.7).

4.5 Antitumor Activity

Cancer is one of the leading health hazards affecting a wide majority of world population. Various anticancer agents (also referred as antitumor, antiproliferative, and antineoplastics) reported for treatment of varied kinds of cancers act through different mechanisms. Benzimidazole being an isostere of a purine-based nitrogenous base and an important scaffold in various biologically

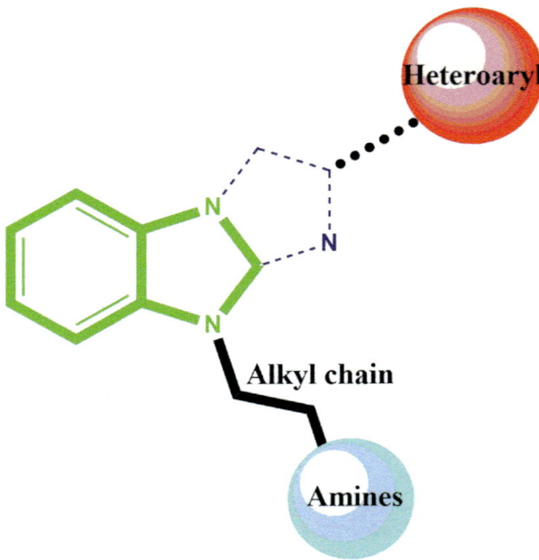

FIGURE 2.7 Structure–activity relationship profile of the benzimidazole nucleus for its antioxidant potential.

active molecules is widely explored for development of anticancer agents. Pyrrolo[1,2-a]benzimidazoles is one of the early classes of anticancer agents acting through cleavages of G and A bases and reductive alkylation of DNA [55]. The variedly substituted benzimidazole derivatives are known to be cytotoxic against lung and breast cancers. 2-(Substituted quinolinon-3-yl)benzimidazoles are reported as serine/threonine checkpoint kinase (CHK-1) inhibitors for treatment of cancer. Several 2-aminobenzimidazole derivatives are also known to be potent as Aurora kinase inhibitors [55].

Developments of 2-substituted benzimidazoles revealed that varied heterocycles at 2-position yield potent anticancer agents at various carcinoma cell lines. These include pyrimidine derivatives, pyrazoline derivatives, and thiazole derivatives. Further, 2-substituted benzimidazoles with chloro or carboxy group at 5-position having 4-amino-thioxothiazole, 4-oxothiazolidine, 4-fluorobenzylidene, and cycloalkylidene are reported as potent antitumor agents [56]. Evaluation of 2-methyl-5-nitro benzimidazoles substituted at 1-position with varied heterocycles revealed that thiadiazole ring linked through a methylene group at 1-position incurs the maximum antitumor activity. It is also revealed that nitro group at 5-position of benzimidzole is critical for the activity. Some 2,5,6-trihalogenobenzimidazoles are also established as androgen receptor antagonists finding their application in prostate cancer. Similarly, benzimidazole-5-carboxylic acid analogs have been evaluated as antileukemic agents. Planar fused benzimidazole analogs have the potential to get inserted into the space between the base pairs of DNA resulting in DNA cleavage. Based on

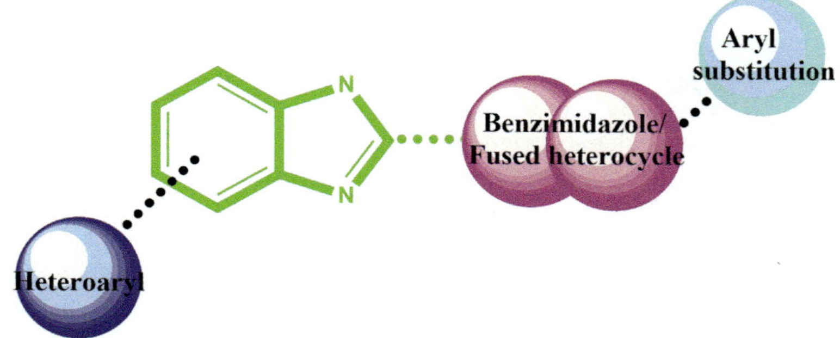

FIGURE 2.8 Structure–activity relationship profile of the benzimidazole nucleus for its antitumor potential.

this mechanism, a benzimidazo[1,2-a]quinoline derivatives have exerted potent activity on all cancer cell lines. Recently, more fused planar benzimidazole derivatives have been reported to exhibit potent cytotoxicity [54,57].

Bis-benzimidazoles is another class of compounds exploited for discovery of anticancer agents. Hoechst-33342 and Hoechst-33258 are the openers of this series exhibiting in vitro antitumor as well as DNA topoisomerase I inhibitory activities. Benzimidazolyl-1,2,4-triazino[4,5-a]benzimidazol-1-one is another bis(benzimidazole) analog having significant activity against multidrug-resistant P-glycoprotein expressing cell lines. Two benzimidazole nuclei linked through a thiophene ring are also well established as either moderate or strong antiproliferative agents [58].

Complexes of benzimidazoles (ligand) with transition metal ions possess antitumor activity. Cu^{2+} complex of benzimidazolylmethyl-1,3-diaminopropane has the ability to intercalate into the double helix of DNA. Assessment of Pt^{2+} complexes for antiproliferative properties showed potent activity against the human MCF-7 breast cancer cell line and HeLa cervix cancer cell lines [59]. Among the metal (copper, silver, iron, manganese) complexes of 2-methyl benzimidazol-5-carboxylic acid hydrazides, the silver complex is found to display cytotoxicity against two human cell lines. A Cu^{2+} complex of 2-pyridinyl-benzimidazole-5-carboxylic acid has been found to exert potent topoisomerase II inhibitory activity [60] (Fig. 2.8).

4.6 Psychoactive Agents

The H3 receptors in the central nervous system are associated with central disorders such as impaired cognitive functions. Molecules containing an imidazole ring connected through an alkyl spacer to a 2-aminobenzimidazole moiety are well reported H3 receptors antagonist. Similarly, 1,2-disubstituted-5-fluorobenzimidazole derivatives with aza-heterocycles are evaluated to have potent H3 antagonist activity [61].

2-Aminobenzimidazole scaffold has also been selected for development of H1-antihistaminic agents therapeutically used for insomnia. Involvement of specific subtypes of 5-HT receptors is well established in psychiatric disorders such as anxiety, depression, and memory loss. Azabicyclic benzimidazole derivatives are characterized as potent and selective 5-HT3R antagonists. Incorporation of additional pharmacophoric moiety in this molecule led to benzimidazole–arylpiperazine derivatives with mixed affinity for serotoninergic 5-HT1A and 5-HT3 receptors and selectivity over a1-adrenergic and dopamine D2 receptors [62]. Substituted arylpiperazines with 3-propoxy-benzimidazole or 3-propoxy-benzimidazole-2-thione groups are reported as atypical neuroleptics exhibiting good 5-HT2A/D2 pKi binding ratios. Andric et al. have found 4-halo-6-[2-(4-arylpiperazin-1-yl)ethyl]-1H-benzimidazoles to have affinity toward both D2-like and 5-HT1A receptors higher than their nonhalogenated analogs. Further, bromo derivatives showed a higher affinity than their chloro counterparts [63].

Another important category of the psychoactive drug is composed of the agents possessing GABAergic potential. The research group at Johnson and Johnson Pharmaceutical Research Institute has developed pyrido[1,2-a]benzimidazoles as potent GABA-A receptor agonists having potential anxiolytic activity. SAR studies have revealed that a polar group at N-1 of benzimidazole nucleus incurs maximum activity with an ethoxymethyl group found as the optimum substituent. N-substituted benzimidazoles are reported as potent in vivo sedatives due to their interaction with benzodiazepine and α1 receptors [64]. The NR2B subunit containing NMDA receptors are thoroughly studied targets in a wide range of CNS pathologies.

4.7 Lipid Modulating Activity

Of the various targets that modulate lipid levels, HMG-CoA reductase is the most widely explored target and many drugs inhibiting this enzyme are clinically available. Similarly, Liver X receptors (LXRs) are the gene transcription factors responsible for cellular lipid efflux. Literature suggests that benzimidazole derivatives possess significant affinity toward LXRs [65]. Farnesoid X receptor (FXR) is another kind of nuclear hormone receptor, which is expressed in liver, intestine, kidneys, and adrenal glands and is responsible for conversion of cholesterol to bile acids and hence decreases intestinal absorption of dietary lipids. Incorporation of varied polar functions in benzimidazole derivatives gives potent and partial agonist activity for FXR along with ADME properties [66].

Neuropeptide Y (NPY) causes orexigenic effects by binding with NPY receptors. The receptor subtype 5 (NPY Y5) is responsible for centrally mediated NPY-induced feeding response. Spiro[isobenzofuran-piperdine]-1-yl benzimidazoles are found as potent, brain-permeable, and orally available Y5 selective antagonists. Melanin-concentrating hormone (MCH) interacts with MCH receptor 1 (MCH R1) to regulate eating behavior in mammals. Thienopyrimidinone derivatives have been discovered as potent MCH R1 antagonists and further modifications have led to benzimidazole analogs having improved activity after oral administration in obese animals [67].

4.8 Anticoagulants

Thrombin causes proteolytic cleavage of fibrinogen, induces platelet activation, and triggers a wide range of effects secondary to thrombosis; for example, vascular smooth muscle cell and fibroblast proliferation, monocyte chemotaxis, and neutrophil adhesion. Inhibition of thrombin is an important mechanism for inhibition of coagulation. A benzimidazole nucleus acts as an appropriate template to place the varied substitutents required for interaction with thrombin. 1,2-Disubstituted benzimidazole derivatives possessing basic amine moieties have been reported as active site directed thrombin inhibitors [68]. Berlex Biosciences has reported tetrasubstituted benzimidazole with naphthylamidine group at 1-position as anticoagulant due to factor Xa (fXa) inhibition. The activity is found independent of the substituent at C-2, whereas substitution of a nitro group at 4-position on the benzimidazole template affords potent fXa inhibitors with excellent thrombin selectivity. Replacing the naphthylamidine with differently substituted biphenylamidines caused a disappointing change in in vitro profile. However, simplification of the naphthylamidine group to yield a propenylbenzene group dramatically improved the potency and selectivity over the unsubstituted naphthalene analogs. SAR studies have established that benzimidazole derivatives act as potent and selective factor Xa inhibitors possessing excellent anticoagulant activity with no fatal acute toxicity. Inhibitors of factor VIIa/Tissue Factor (fVIIa/TF) complex is another class of compounds for treatment of thromboembolic diseases. Further research into the compounds has led to development of selective dicarboxylic acid analogs with pharmacokinetic profile amenable to once daily subcutaneous dosing in humans [69].

4.9 Antidiabetic Agents

Diabetic mellitus is a metabolic disorder that is characterized by high blood pressure due to insulin resistance and relative insulin deficiency. Noninsulin-dependent diabetes mellitus (NIDDM) is the most prevalent type. The primary treatment goal of NIDDM is controlling the levels of blood glucose. The sodium-glucose cotransporters (SGLTs) in the proximal tubules are responsible for glucose reabsorption in the intestine (SLGT1) and kidney (SLGT2) and hence, provide a novel target for treatment of NIDDM through inhibition of renal glucose reabsorption. Phlorizin is a natural SLGT2 inhibitor and structural analogs of phlorizin having benzimidazole nucleus are reported to exhibit potent SGLT2 inhibitory activity. Glucagon receptor (GCGR) involved in hepatic glucose production has also emerged as a novel target for designing antidiabetics. GCGR antagonists inhibit glucagon-induced glucose production and decrease glucose levels, and varied derivatives of aminobenzimidazole are well documented as GCGR antagonists [70]. GK activity is also coupled to increase insulin secretion in beta cells and hence GK activators are expected to act as hypoglycemic agents. Systematic modification of novel 2-(pyridin-2-yl)-1H-benzimidazole identified from an HTS has led to discovery of potent and metabolically stable GK activators that demonstrate glucose lowering efficacy in a dose-dependent manner at 3 mg/kg [71].

4.10 Miscellaneous Activities

Bayer Yakuhin reports varied benzimidazole derivatives as luteinizing hormone-releasing hormone (LHRH) or gonadotropin-releasing hormone (GnRH) antagonists. Initially, 1-benzyl-2- ethylsulfanyl-1H-benzimidazole-5-sulfonamide was identified as a functional LHRH antagonist with potency in micromolar ranges [71]. The other related series of compounds were developed yielding the potency in nanomolar doses. Keeping the molecular core identical to that in the earlier series, a SAR revealed that presence of phenyl group at 2-position, t-butylurea at 5-position, and small alkyl groups at 1-position produce a potent LHRH antagonist [72]. Based on this substitution pattern, an alternative binding mode was also suggested. Pelletier et al. have reported 2-phenyl-4-piperazinyl-benzimidazoles as GnRH antagonists with nanomolar potency in in vitro binding and functional assays as well as excellent bioavailability. Among a series of compounds synthesized by appending small heterocycles to the 2-(4-tert-butylphenyl)-4-piperazinylbenzimidazole template, two imidazole analogs showed substantial in vitro potency at the target receptor as well as aqueous solubility [73]. Similarly, a series of novel 6-phenyl benzimidazolones is reported to exhibit a potent progesterone receptor antagonist activity in T47D cell alkaline phosphatase assay. A 2-(2,2,2)-trifluoroethyl-benzimidazole scaffold has also been explored as tissue-selective androgen receptor modulators that are agonists in muscles and antagonists in prostate to exhibit their therapeutic utility toward hypogonadism, cachexia, and the like [74].

Rho kinase, a serine/threonine kinase expressed in vascular tissues, plays an important role in essential signal transduction pathways and has potential utility in wide range of activities. In 2010, varied positions of benzimidazole nucleus were explored to optimize the structure for selectivity toward other protein kinases. Compounds developed with modification in the chromane ring showed Rho kinase inhibition in nanomolar doses whereas affinity toward other protein kinases is in micromolar concentrations [75].

5. BENZIMIDAZOLES AS MULTITARGETING AGENTS IN MULTIFACTORIAL DISEASES

Although literature suggests that benzimidazoles possess a variety of biological potency, there are only few example of successful utilization of this scaffold as an MTA. One of the many reasons could be failure in balancing the biological profile with alterations in the substitutions on the nucleus. Such a problem does occur with heterocycles having a vast pharmacological profile because it becomes quite hectic to control the selectivity of the molecule toward only selected multiple targets. However, some reports disclose that benzimidazoles indeed have been developed into MTAs, utilizing their pharmacological potential as provided in the preceding section.

Shi et al., in 2014, successfully inhibited both c-Met and VEGFR-2 (kinases targeted in cancer chemotherapy) by designing and developing quinazoline-benzimidazole coupled molecules. They could achieve micromolar range inhibition against both kinases. Designed molecules were having aryl substitution at 2-position of benzimidazole and quinazoline was coupled at the 5-position [76], complying with the pharmacophoric features explained in Section 4.5. Similar efforts in 2015 by Abdullah et al., led to the identification of benzimidazole derivatives as dual inhibitors of PARP and dihydroorotate dehydrogenase. Other than the conventional benzimidazole pharmacophore reported for anticancer activity, the molecules were having an additional substitution of carboxylic acid at 7-position [77].

In another report, Arshad et al. disclosed the multitargeting potential 5-bromo-2-aryl benzimidazoles against α-glucosidase and urease enzymes (key targets in diabetes as well as peptic ulcer) [78]. Developed molecules were effective in the micromolar range and expectedly, complied with the pharmacophoric features disclosed in earlier subsections. In another work, the multitargeting antimicrobial potential of benzimidazoles was disclosed. Jeyakkumar et al. fused tetrahydroprotoberberine, a natural product with benzimidazole scaffold, and obtained molecules with significant antimicrobial potential, effective against various strains along with resistant ones [79]. Recently, multifunctional benzimidazole-based molecules were reported for Alzheimer's disease. Designed molecules were found effective against butyrylcholinesterase and Aβ aggregation and consisted of piperidine-substituted aryl moiety at the 2-position of the benzimidazole nucleus [80]. An almost similar kind of work was reported by Unsal-Tan et al., where they designed and synthesized 2-aryl benzimidazole derivatives as MTAs for Alzheimer's disease. Molecules were reported to be effective against acetylcholinesterase/butyrylcholinesterase inhibition and amyloid beta (Aβ) antiaggregation, and structurally consisted of a 4-substituted aryl system at the 2-position of benzimidazole [81].

6. CONCLUSION AND PROSPECTIVES

Benzimidazoles are clearly established as significant chemical species with a wide range of applications and great potential; combining synthetic versatility and biological diversity, the possibilities for these unique compounds seem endless. Even though literature suggests benzimidazole to be a privileged scaffold in multitarget drug design, only limited reports are available regarding benzimidazole-based multitarget ligands against multifactorial pathological conditions, clearly indicating the lack of exploration of its vast potential. Numerous compounds derived from benzimidaozle nuclei are used in clinics for treatment of many diseases. However, despite the active, exhaustive, and target-based research on development of many compounds as antiinflammatory, immunomodulatory, lipid modulators, and so on, no molecule has made its way to the market and clinic. One reason could be lack of a comprehensive compilation of various research reports in each activity capable of giving an insight into the SAR of the compounds. Such information could be utilized to update knowledge on currently employed strategies to design and develop multitargeting agents, which could be further utilized by a medicinal chemist to design their own multitargeting agents. As highlighted in this chapter, benzimidazoles have the potential to provide uniquely featured biologically privileged derivatives that seem to be ideally suited for developing multitarget drugs in the field of multifactorial diseases/disorders.

REFERENCES

[1] R. Morphy, Z. Rankovic, Designed multiple ligands. An emerging drug discovery paradigm, Journal of Medicinal Chemistry 48 (21) (2005) 6523–6543.

[2] M.E. Welsch, S.A. Snyder, B.R. Stockwell, Privileged scaffolds for library design and drug discovery, Current Opinion in Chemical Biology 14 (3) (2010) 347–361.

[3] B. Narasimhan, D. Sharma, P. Kumar, Benzimidazole: a medicinally important heterocyclic moiety, Medicinal Chemistry Research 21 (3) (2012) 269–283.

[4] D. Woolley, Some biological effects produced by benzimidazole and their reversal by purines, Journal of Biological Chemistry 152 (2) (1944) 225–232.

[5] M. Boiani, M. González, Imidazole and benzimidazole derivatives as chemotherapeutic agents, Mini Reviews in Medicinal Chemistry 5 (4) (2005) 409–424.

[6] A.E. Remick, Electronic Interpretations of Organic Chemistry, 1949.

[7] L. Hunter, J.A. Marriott, 139. The associating effect of the hydrogen atom. Part VIII. The N–H–N bond. Benziminazoles, glyoxalines, amidines, and guanidines, Journal of the Chemical Society (Resumed) (1941) 777–786.

[8] Y.K. Syrkin, M. Dyatkina, The Chemical Bond and the Structure of Molecules, Goskhimizdat, Moscow-Leningrad, 1946, p. 219.

[9] T. Pavolini, Reactions with sodium nitroprusside, Bollettino Chimico Farmaceutico 69 (1930) 713–722.

[10] M. Straumanis, A. Cīrulis, Das Kupfer (II)-azid. Darstellungsmethoden, Bildung und Eigenschaften, Zeitschrift für Anorganische und Allgemeine Chemie 251 (4) (1943) 315–331.

[11] E. Bamberger, J. Lorenzen, Ueber Formazylmethylketon, European Journal of Inorganic Chemistry 25 (2) (1892) 3539–3547.

[12] E. Bamberger, B. Berle, Aufspaltung des Imidazolrings, European Journal of Organic Chemistry 273 (2–3) (1893) 342–363.

[13] G.B. Bachman, L.V. Heisey, The condensation of aldehydes and amines with nitrogenous five-atom ring systems, Journal of the American Chemical Society 68 (12) (1946) 2496–2499.

[14] I. Butula, 4, 5, 6, 7-Tetrahydrobenzimidazoles (Google Patents), 1976.

[15] J.B. Wright, The chemistry of the benzimidazoles, Chemical Reviews 48 (3) (1951) 397–541.

[16] O. Gerngross, Über die Benzoylierung von Imidazole-Derivaten, European Journal of Inorganic Chemistry 46 (2) (1913) 1908–1913.

[17] E. Wagner, W. Millett, Benzimidazole, Organic Syntheses (1943) 12.

[18] E. Łukasik, Z. Wróbel, Simple synthesis of 2-aminoaryliminophosphoranes from N-aryl-2-nitrosoanilines and their application in 2-aminobenzimidazole synthesis, Synlett 25 (02) (2014) 217–220.

[19] X. Lv, W. Bao, Copper-catalyzed cascade addition/cyclization: an efficient and versatile synthesis of N-substituted 2-heterobenzimidazoles, Journal of Organic Chemistry 74 (15) (2009) 5618–5621.

[20] J. She, Z. Jiang, Y. Wang, One-pot synthesis of functionalized benzimidazoles and 1H-pyrimidines via cascade reactions of o-aminoanilines or naphthalene-1, 8-diamine with alkynes and p-tolylsulfonyl azide, Synlett 2009 (12) (2009) 2023–2027.

[21] S.K. Alla, et al., Iodobenzene catalyzed C–H amination of N-substituted amidines using m-chloroperbenzoic acid, Organic Letters 15 (6) (2013) 1334–1337.

[22] M.S. Mayo, et al., Convenient synthesis of benzothiazoles and benzimidazoles through Brønsted acid catalyzed cyclization of 2-amino thiophenols/anilines with β-diketones, Organic Letters 16 (3) (2014) 764–767.

[23] T.B. Nguyen, L. Ermolenko, A. Al-Mourabit, Sodium sulfide: a sustainable solution for unbalanced redox condensation reaction between o-nitroanilines and alcohols catalyzed by an iron–sulfur system, Synthesis 47 (12) (2015) 1741–1748.

[24] D. Mahesh, P. Sadhu, T. Punniyamurthy, Copper (ii)-catalyzed oxidative cross-coupling of anilines, primary alkyl amines, and sodium azide using TBHP: a route to 2-substituted benzimidazoles, Journal of Organic Chemistry 81 (8) (2016) 3227–3234.

[25] D. Mahesh, P. Sadhu, T. Punniyamurthy, Copper (I)-catalyzed regioselective amination of N-aryl imines using TMSN3 and TBHP: a route to substituted benzimidazoles, Journal of Organic Chemistry 80 (3) (2015) 1644–1650.

[26] D.B. Nale, B.M. Bhanage, N-substituted formamides as C1-sources for the synthesis of benzimidazole and benzothiazole derivatives by using zinc catalysts, Synlett 26 (20) (2015) 2835–2842.

[27] P. Naik, et al., Angiotensin II receptor type 1 (AT 1) selective nonpeptidic antagonists—a perspective, Bioorganic and Medicinal Chemistry 18 (24) (2010) 8418–8456.

[28] K. Kubo, et al., Nonpeptide angiotensin II receptor antagonists. Synthesis and biological activity of benzimidazolecarboxylic acids, Journal of Medicinal Chemistry 36 (15) (1993) 2182–2195.

[29] P. Vanderheyden, et al., The in vitro binding properties of non-peptide AT1 receptor antagonists, Journal of Clinical and Basic Cardiology 5 (1) (2002) 75–82.

[30] T. Ogihara, et al., Persistent inhibition of the pressor and aldosterone responses to angiotensin-II by TCV-116 in normotensive subjects, Journal of Cardiovascular Pharmacology 26 (3) (1995) 490–494.

[31] N. Cho, et al., A new class of diacidic nonpeptide angiotensin II receptor antagonists, Bioorganic and Medicinal Chemistry Letters 4 (1) (1994) 35–40.

[32] Y. Bansal, O. Silakari, The therapeutic journey of benzimidazoles: a review, Bioorganic and Medicinal Chemistry 20 (21) (2012) 6208–6236.

[33] M. Sabat, et al., The development of 2-benzimidazole substituted pyrimidine based inhibitors of lymphocyte specific kinase (Lck), Bioorganic and Medicinal Chemistry Letters 16 (23) (2006) 5973–5977.

[34] P. Bamborough, et al., 5-(1H-benzimidazol-1-yl)-3-alkoxy-2-thiophenecarbonitriles as potent, selective, inhibitors of IKK-ε kinase, Bioorganic and Medicinal Chemistry Letters 16 (24) (2006) 6236–6240.

[35] K. Ito, et al., The studies of the mechanism of antiinflammatory action of 2-(5-ethylpyridin-2-yl) benzimidazole (KB-1043), Arzneimittel-forschung 32 (2) (1982) 117–122.

[36] M.E. Hayes, et al., Discovery of small molecule benzimidazole antagonists of the chemokine receptor CXCR3, Bioorganic and Medicinal Chemistry Letters 18 (5) (2008) 1573–1576.

[37] T. Eriksson, S. Ivanova, H. Lönn, Benzimidazol Derivatives Modulate Chemokine Receptors (Google Patents), 2008.

[38] K.G. Desai, K.R. Desai, Green route for the heterocyclization of 2-mercaptobenzimidazole into β-lactum segment derivatives containing–CONH–bridge with benzimidazole: screening in vitro antimicrobial activity with various microorganisms, Bioorganic and Medicinal Chemistry 14 (24) (2006) 8271–8279.

[39] G.A. Holloway, et al., Discovery of 2-iminobenzimidazoles as a new class of trypanothione reductase inhibitor by high-throughput screening, Bioorganic and Medicinal Chemistry Letters 17 (5) (2007) 1422–1427.

[40] Ö.Ö. Güven, et al., Synthesis and antimicrobial activity of some novel phenyl and benzimidazole substituted benzyl ethers, Bioorganic and Medicinal Chemistry Letters 17 (8) (2007) 2233–2236.

[41] K. Ansari, C. Lal, Synthesis and evaluation of some new benzimidazole derivatives as potential antimicrobial agents, European Journal of Medicinal Chemistry 44 (5) (2009) 2294–2299.

[42] H. Torres-Gómez, et al., Design, synthesis and in vitro antiprotozoal activity of benzimidazole-pentamidine hybrids, Bioorganic and Medicinal Chemistry Letters 18 (11) (2008) 3147–3151.

[43] L. Chodosh, et al., 5, 6-Dichloro-1-beta-D-ribofuranosylbenzimidazole inhibits transcription elongation by RNA polymerase II in vitro, Journal of Biological Chemistry 264 (4) (1989) 2250–2257.

[44] A.R. Porcari, et al., Design, synthesis, and antiviral evaluations of 1-(substituted benzyl)-2-substituted-5, 6-dichlorobenzimidazoles as nonnucleoside analogues of 2, 5, 6-trichloro-1-(β-D-ribofuranosyl) benzimidazole, Journal of Medicinal Chemistry 41 (8) (1998) 1252–1262.

[45] K. Starčević, et al., Synthesis, antiviral and antitumor activity of 2-substituted-5-amidino-benzimidazoles, Bioorganic and Medicinal Chemistry 15 (13) (2007) 4419–4426.

[46] K.P. Barot, et al., Novel research strategies of benzimidazole derivatives: a review, Mini Reviews in Medicinal Chemistry 13 (10) (2013) 1421–1447.

[47] Z.L. Zhang, et al., Design, synthesis and biological activity of some novel benzimidazole derivatives against Coxsackie virus B 3, Chinese Chemical Letters 20 (8) (2009) 921–923.

[48] K.-L. Yu, et al., Respiratory syncytial virus fusion inhibitors. Part 4: optimization for oral bioavailability, Bioorganic and Medicinal Chemistry Letters 17 (4) (2007) 895–901.

[49] X.A. Wang, et al., Respiratory syncytial virus fusion inhibitors. Part 5: optimization of benzimidazole substitution patterns towards derivatives with improved activity, Bioorganic and Medicinal Chemistry Letters 17 (16) (2007) 4592–4598.

[50] K.D. Combrink, et al., Respiratory syncytial virus fusion inhibitors. Part 6: an examination of the effect of structural variation of the benzimidazol-2-one heterocycle moiety, Bioorganic and Medicinal Chemistry Letters 17 (17) (2007) 4784–4790.

[51] C. Kuş, et al., Synthesis and antioxidant properties of novel N-methyl-1, 3, 4-thiadiazol-2-amine and 4-methyl-2H-1, 2, 4-triazole-3 (4H)-thione derivatives of benzimidazole class, Bioorganic and Medicinal Chemistry 16 (8) (2008) 4294–4303.

[52] V. Dianov, Synthesis and antioxidant properties of 3-methyl-substituted thiazolo [3, 2-a] benzimidazole, Pharmaceutical Chemistry Journal 41 (6) (2007) 308–309.

[53] V. Anisimova, et al., Synthesis and pharmacological activity of 9-R-2-halogenophenylimidazo [1, 2-a] benzimidazoles, Pharmaceutical Chemistry Journal 44 (7) (2010) 345–351.

[54] C.G. Neochoritis, et al., One-pot microwave assisted synthesis under green chemistry conditions, antioxidant screening, and cytotoxicity assessments of benzimidazole Schiff bases and pyrimido [1, 2-a] benzimidazol-3 (4H)-ones, European Journal of Medicinal Chemistry 46 (1) (2011) 297–306.

[55] R.C. Boruah, E.B. Skibo, A comparison of the cytotoxic and physical properties of aziridinyl quinone derivatives based on the pyrrolo [1, 2-a] benzimidazole and pyrrolo [1, 2-a] indole ring systems, Journal of Medicinal Chemistry 37 (11) (1994) 1625–1631.

[56] H.M. Refaat, Synthesis and anticancer activity of some novel 2-substituted benzimidazole derivatives, European Journal of Medicinal Chemistry 45 (7) (2010) 2949–2956.

[57] N.T. Gowda, et al., Synthesis and biological evaluation of novel 1-(4-methoxyphenethyl)-1H-benzimidazole-5-carboxylic acid derivatives and their precursors as antileukemic agents, Bioorganic and Medicinal Chemistry Letters 19 (16) (2009) 4594–4600.

[58] J. Stýskala, et al., Synthesis of 2-aryl-4-(benzimidazol-2-yl)-1, 2-dihydro [1, 2, 4] triazino-[4, 5-a] benzimidazol-1-one derivatives with preferential cytotoxicity against carcinoma cell lines, European Journal of Medicinal Chemistry 43 (3) (2008) 449–455.

[59] M. Gökçe, et al., Synthesis, in vitro cytotoxic and antiviral activity of cis-[Pt (R (–) and S (+)-2-α-hydroxybenzylbenzimidazole) 2 Cl 2] complexes, European Journal of Medicinal Chemistry 40 (2) (2005) 135–141.

[60] S.A. Galal, et al., Synthesis and antitumor activity of novel benzimidazole-5-carboxylic acid derivatives and their transition metal complexes as topoisomerease II inhibitors, European Journal of Medicinal Chemistry 45 (12) (2010) 5685–5691.

[61] R. Aslanian, et al., Benzimidazole-substituted (3-phenoxypropyl) amines as histamine H3 receptor ligands, Bioorganic and Medicinal Chemistry Letters 18 (18) (2008) 5032–5036.

[62] K. Lavrador-Erb, et al., The discovery and structure–activity relationships of 2-(piperidin-3-yl)-1H-benzimidazoles as selective, CNS penetrating H 1-antihistamines for insomnia, Bioorganic and Medicinal Chemistry Letters 20 (9) (2010) 2916–2919.

[63] D. Andrić, et al., Synthesis, binding properties and receptor docking of 4-halo-6-[2-(4-arylpiperazin-1-yl) ethyl]-1H-benzimidazoles, mixed ligands of D 2 and 5-HT 1A receptors, European Journal of Medicinal Chemistry 43 (8) (2008) 1696–1705.

[64] J.L. Falcó, et al., Synthesis, pharmacology and molecular modeling of N-substituted 2-phenyl-indoles and benzimidazoles as potent GABA A agonists, European Journal of Medicinal Chemistry 41 (8) (2006) 985–990.

[65] J.M. Travins, et al., 1-(3-Aryloxyaryl) benzimidazole sulfones are liver X receptor agonists, Bioorganic and Medicinal Chemistry Letters 20 (2) (2010) 526–530.

[66] H.G. Richter, et al., Optimization of a novel class of benzimidazole-based farnesoid X receptor (FXR) agonists to improve physicochemical and ADME properties, Bioorganic and Medicinal Chemistry Letters 21 (4) (2011) 1134–1140.

[67] A.J. Carpenter, et al., Novel benzimidazole-based MCH R1 antagonists, Bioorganic and Medicinal Chemistry Letters 16 (19) (2006) 4994–5000.

[68] K. Takeuchi, et al., 1, 2-disubstituted indole, azaindole and benzimidazole derivatives possessing amine moiety: a novel series of thrombin inhibitors, Bioorganic and Medicinal Chemistry Letters 10 (20) (2000) 2347–2351.

[69] H. Ueno, et al., Structure–activity relationships of potent and selective factor Xa inhibitors: benzimidazole derivatives with the side chain oriented to the prime site of factor Xa, Bioorganic and Medicinal Chemistry Letters 14 (16) (2004) 4281–4286.

[70] R.M. Kim, et al., Discovery of potent, orally active benzimidazole glucagon receptor antagonists, Bioorganic and Medicinal Chemistry Letters 18 (13) (2008) 3701–3705.

[71] M. Ishikawa, et al., Discovery of novel 2-(pyridine-2-yl)-1H-benzimidazole derivatives as potent glucokinase activators, Bioorganic and Medicinal Chemistry Letters 19 (15) (2009) 4450–4454.

[72] M. Tatsuta, et al., Benzimidazoles as non-peptide luteinizing hormone-releasing hormone (LHRH) antagonists. Part 3: discovery of 1-(1H-benzimidazol-5-yl)-3-tert-butylurea derivatives, Bioorganic and Medicinal Chemistry Letters 15 (9) (2005) 2265–2269.

[73] E.A. Terefenko, et al., SAR studies of 6-aryl-1, 3-dihydrobenzimidazol-2-ones as progesterone receptor antagonists, Bioorganic and Medicinal Chemistry Letters 15 (15) (2005) 3600–3603.

[74] R.A. Ng, et al., Synthesis of potent and tissue-selective androgen receptor modulators (SARMs): 2-(2, 2, 2)-trifluoroethyl-benzimidazole scaffold, Bioorganic and Medicinal Chemistry Letters 17 (6) (2007) 1784–1787.

[75] E.H. Sessions, et al., The development of benzimidazoles as selective rho kinase inhibitors, Bioorganic and Medicinal Chemistry Letters 20 (6) (2010) 1939–1943.

[76] L. Shi, et al., Discovery of quinazolin-4-amines bearing benzimidazole fragments as dual inhibitors of c-Met and VEGFR-2, Bioorganic and Medicinal Chemistry 22 (17) (2014) 4735–4744.

[77] I. Abdullah, et al., Benzimidazole derivatives as potential dual inhibitors for PARP-1 and DHODH, Bioorganic and Medicinal Chemistry 23 (15) (2015) 4669–4680.

[78] T. Arshad, et al., 5-Bromo-2-aryl benzimidazole derivatives as non-cytotoxic potential dual inhibitors of α-glucosidase and urease enzymes, Bioorganic Chemistry 72 (2017) 21–31.

[79] P. Jeyakkumar, et al., Novel benzimidazolyl tetrahydroprotoberberines: design, synthesis, antimicrobial evaluation and multi-targeting exploration, Bioorganic and Medicinal Chemistry Letters 27 (8) (2017) 1737–1743.

[80] K. Ozadali-Sari, et al., Novel multi-targeted agents for Alzheimer's disease: synthesis, biological evaluation, and molecular modeling of novel 2-[4-(4-substitutedpiperazin-1-yl) phenyl] benzimidazoles, Bioorganic Chemistry 72 (2017) 208–214.

[81] O. Unsal-Tan, et al., Novel 2-arylbenzimidazole derivatives as multi-targeting agents to treat Alzheimer's disease, Medicinal Chemistry Research (2017) 1–10.

Chapter 3

Acridones: A Relatively Lesser Explored Heterocycle for Multifactorial Diseases

Rajesh Kumar[1,2], Sarita Sharma[2,3], Deonandan Prasad[1,2]

[1]Shivalik College of Pharmacy, Nangal, India; [2]I.K. Gujral Punjab Technical University, Jalandhar, India; [3]Global College of Pharmacy, Anandpur Sahib, India

Chapter Outline

1. INTRODUCTION

Acridine and its analogues have created a great deal of technical and scientific interests since 1870, when Grabe and Caro discovered acridine in a high boiling fraction of coaltar. Ehrlich and Benda first proposed the use of aminoacridines as antimicrobial agents in 1912 and the first clinical use of these agents such as proflavine and acriflavine occurred in 1917. During World War I (1914–18) acridines were used as wound antiseptics. Mepacrine (quinacrine), an acridine-based antimalarial drug, was discovered in 1932, and was used during World War II (1939–45) due to scarcity of chloroquine. In 1944, penicillin superseded the acridine-based therapy as antiseptics [1]. In 1948, acronycine, a natural acridone alkaloid, was isolated from an Australian tree, *Acronychia baueri* [2,3]. However, its antitumor activity was evaluated by Eli-Lilly in 1966 against murine solid tumor models such as *S-180* and *AKR* sarcomas, *X-5563* myeloma, *S-115* carcinoma, and *S-91* melanoma. Hence the discovery of acronycine opened a new research area for the acridone derivatives as anticancer agents. During the 1970s two acridine-based anticancer drugs were discovered, 1-nitroacridine derivative nitracrine in Poland [4,5] and 9-anilinoacridine amsacrine (*m*-AMSA) in New Zealand [6,7]. Both of these drugs act by interaction with topo-II-DNA complex [8,9]. Asulacrine (CI-921), an *m*-AMSA derivative, displayed interesting antitumor properties and underwent phase-II clinical trial against non–small cell lung cancer (NSCLC) [10]. The continuous work on acridine led to the development of a new 9-aminoacridine-4-carboxamides derivative such as DACA in 1987, which act through interactions with topo I and II [11]. It underwent phase-II clinical trial against NSCLC and recurrent glioblastoma multiforme [12]. During the 1990s, triazoloacridone and imidazoacridone derivatives were synthesized at the Gdansk University [13–15]. Both of these derivatives act through inhibition of DNA topo II. The triazoloacridone derivative C-1305 showed potent antitumor activity toward a wide range of different experimental tumors in vitro and in vivo, including both murine and human colon carcinomas [16]. C-1311, the most prominent analog of imidazoacridone, underwent phase-II clinical trial in patients with advanced solid tumors [17]. The presence of the 8-OH group is found to be responsible for the antitumor activity of compounds of this type.

Further, its structural analogues such as thioacridone were first pursued for their antimicrobial properties against bacteria, parasites, and fungi. Some work in this area continues, particularly for the development of antimalarial, antiprotozoal, and anticancer agents [18]. However, more recent research has focused mainly on their use as anticancer drugs, due to their planar ring structure that confers the ability to intercalate within DNA and to inhibit topoisomerase or telomerase enzymes. The development of acridine and its structural analogue such as acridone employed for different pharmacological activities have been summarized (Fig. 3.1). Among them acridone has never been properly reviewed for multifactorial diseases in the literature available so far. So, there is a need to couple the latest information with the earlier information to understand the current status of acridone nucleus in medicinal chemistry research. In this chapter, SAR and various derivatives of acridone with different pharmacological activities have been described. The information furnished in this chapter may help researchers involved in synthetic medicinal chemistry to explore this nucleus for designing newer compounds with improved pharmacological profile.

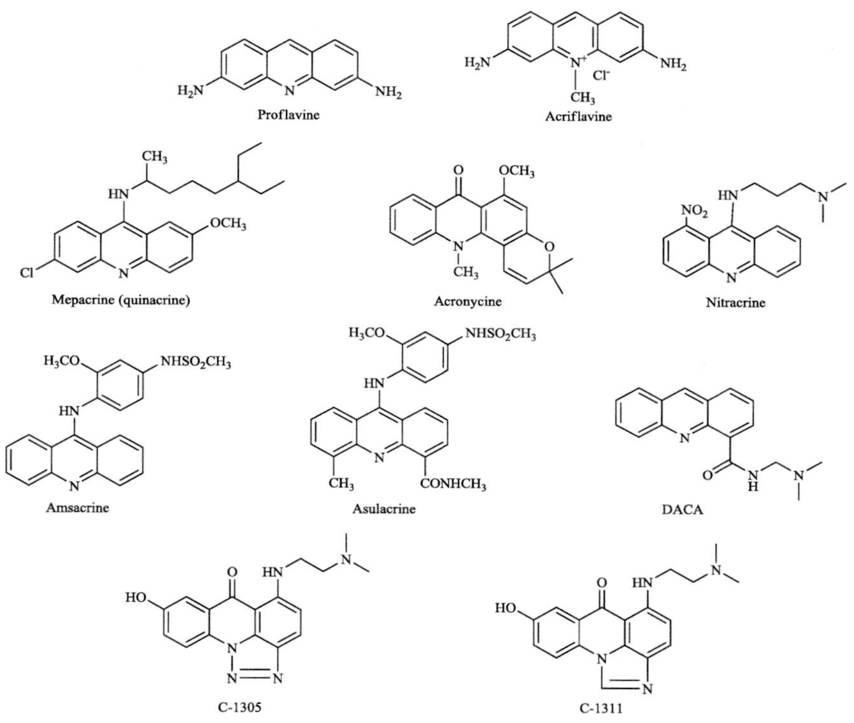

Proflavine

Acriflavine

Mepacrine (quinacrine)

Acronycine

Nitracrine

Amsacrine

Asulacrine

DACA

C-1305

C-1311

2. CHEMISTRY OF ACRIDONE

Acridone is a heterocyclic aromatic nucleus that contains a carbonyl and secondary amino group at its 9th and 10th position, respectively (Fig. 3.2), and moreover, it is an oxidized product of acridine nucleus. First of all, in 1880, 9-acridanone was synthesized, which was named acridone in 1892. The synonyms of acridone nucleus are 9(10H)-acridinone, acridine-9-one, 9-acridanone, 9,10-dihydro-9-oxo acridine, and 9-azanthracene-10-one.

Parent acridone is a pure yellow solid with a melting point of 354°C. It is insoluble in benzene, chloroform, ether, water, and ethanol except *N,N*-dimethylformamide (DMF) and dimethylsulfoxide (DMSO). In addition, it also dissolves in alcoholic potassium hydroxide to give a yellow brown solution of its potassium salt, which is completely decomposed by water. Acridone is highly fluorescent and stable against photodegradation, oxidation, and heat [19]. The intensities of fluorescence given by acridone in nonpolar solvents are extremely weak but become strong in polar solvents [20–22].

Due to acrid smell and irritating action of acridone on skin and mucous membrane, this substance was called "acridin" (acris = sharp, or pungent). Apart from the addition of a terminal "e" and the replacement of "c" by "k," in a few older papers, this name has not subsequently been changed. In the beginning,

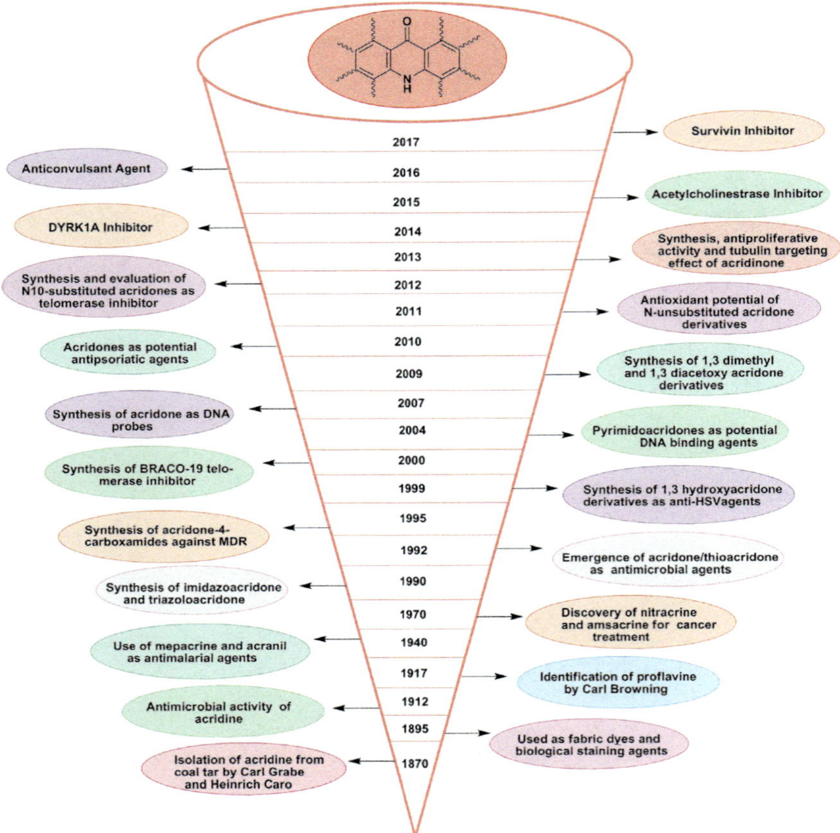

FIGURE 3.1 Development of acridines/acridones as a pharmacologically active compound.

FIGURE 3.2 Basic skeleton of acridone.

various ways of numbering the acridone nucleus have been adopted in different countries. In 1893, Graebe suggested a numbering system, **A**, based on the accepted numbering system used for anthracene, xanthene, and so on (Fig. 3.3). In early 1900, M. M. Richter used another system, **B**, which was changed by chemical abstracts in 1937, to **A**, and is still used at present [23].

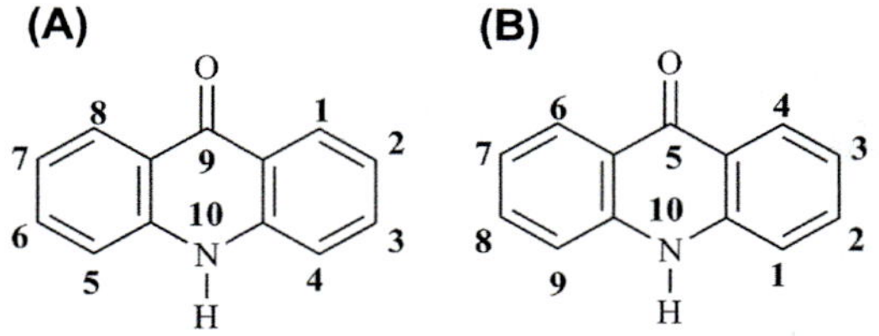

(A) **(B)**

FIGURE 3.3 Systems for numbering acridones. (A) Graebe numbering system. (B) Richter numbering system.

(**Ia**) X= H
(**IIb**) X= CH$_3$

(**Ia'**) Y= OH
(**IIb'**) Y= OCH$_3$

FIGURE 3.4 Tautomerism in acridones.

The molecular structure of the acridone molecule is planar with no atoms deviating by more than 0.02 Å from the molecular plane, defined by non-H ring atoms and the oxygen atoms. All torsion angles of the acridone molecule lie within ±1.5 degrees of 0 or 180 degrees, and it adopts a herringbone packaging arrangement very similar to that found in anthraquinone and quinacridone. There are two dominant forces that have been found to be responsible to control the crystal arrangement of acridone. The first consists of N—H—O hydrogen bonding between glide related molecules, with an N—O distance of 2.782 Å, such that each molecule is hydrogen bonded to two adjacent molecules. The second is π-π interactions between molecules stacked along the short crystal axis [24].

2.1 Tautomerism

Acridone can potentially exist in two tautomeric forms as shown in Fig. 3.4.

The acridone (**Ia**) and 9-hydroxyacridone (**Ia'**) exist in dynamic equilibrium with each other. According to the results of theoretical calculations, **Ia** is energetically and thermodynamically more stable than **Ia'** by 50 kJ/mol [25]. Thus acridin-9(10*H*)-one is the only form that exists in the solid phase. However, in solutions, since the two molecules differ in polarity, and by changing solvents, we can shift the equilibrium toward **Ia** in more polar media or toward **Ia'** in less polar ones. It may also be expected that the proton of the hydroxy group in **Ia'** can be removed in alkaline media, which would lead to the formation of a 9-oxoacridine anion. **IIb** and **IIb'** are the isomers and methyl-substituted derivatives of **Ia** and **Ia'**, respectively. In this case, too, the acridinone isomer

IIb is thermodynamically more stable than the acridine isomer **IIb'**. This is the reason when 9-acridinones are alkylated, it is 10-alkyl-9-acridinones that are predominantly formed [26].

3. SYNTHESIS

The range of physiological activities associated with acridone derivatives has stimulated a great deal of effort aimed at improving synthetic methods available for their construction. Most syntheses of acridone derivatives can be separated into two classes, illustrated retrosynthetically in Scheme 3.1. Disconnection of compound **1.1** at bond **a** adjacent to the keto function affords the *N*-anthranilic acid derivative (**1.2**), which is available through traditional cross-coupling methodology. Alternatively, disconnection of bond **b** adjacent to the ring nitrogen atom yields the amino benzophenone derivative **1.3** that is typically constructed using standard Friedel–Crafts chemistry.

3.1 Synthesis of Acridone via *N*-Arylanthranilic Acid Derivatives

Synthetic methods based on disconnection **A** (Scheme 3.1) have been most often employed for the construction of acridone derivatives. Traditionally, the prerequisite diarylamines have been generated by Ullmann condensation (Scheme 3.2), involving the copper-mediated coupling of an aryl halide (**1.4**) with an arylamine (**1.5**) [27–30]. On heating diphenylamine-2-carboxylic acid (**1.6**) in conc. sulfuric acid or polyphosphoric acid (PPA) at 85°C for 1 to 2h leads to cyclization that affords acridone (**1.7**) [27,31]. In Ullmann condensation, copper and copper oxides act as catalysts as well as halogen-acceptors from aryl halides. The reactivity of aryl halide grows in the order of PhCl<PhBr<PhI. Copper (Cu) commonly provides greater yields than Cu_2O and CuO and among the two copper oxides, Cu_2O has the better activity. The base (K_2CO_3) accepts the liberated HX (X=Cl, Br, I) and is necessary to promote the reaction.

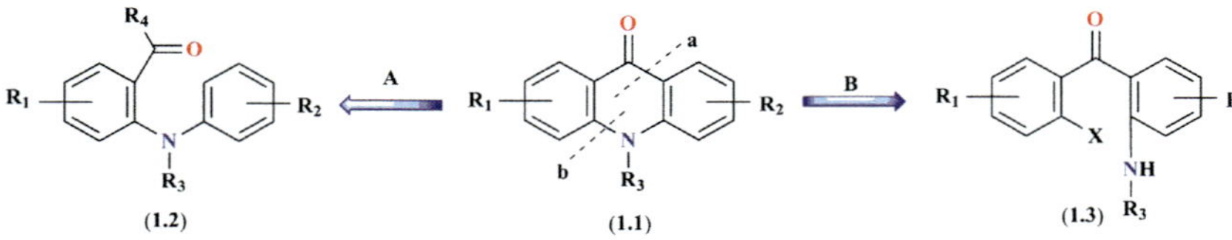

SCHEME 3.1 Retrosynthetically approach for the synthesis of acridones.

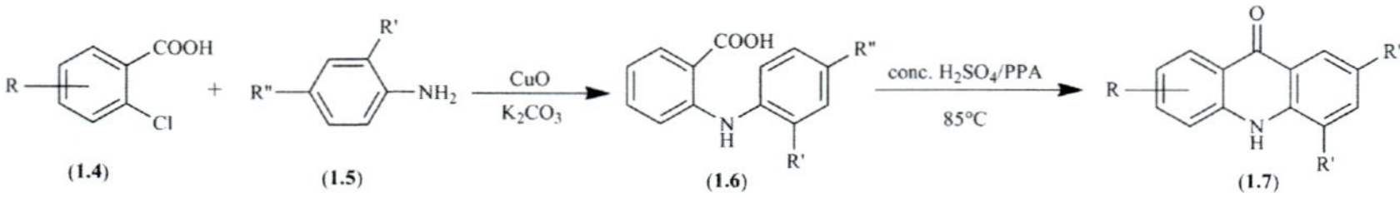

SCHEME 3.2 Ullmann reaction for the synthesis of acridone derivatives.

3.2 Synthesis of Acridone via Benzophenone Intermediates

The synthesis of acridone derivatives via benzophenone intermediates (Scheme 3.1, disconnection **b**) has been used much less frequently than the Ullmann condensation protocol. Primarily developed by Lewis, this method utilizes base- or thermally induced *ipso* substitution of alkoxy groups of benzophenone, which is most often prepared by standard Friedel–Crafts chemistry. The limited utilization of this method is likely a result of inherent regiochemical problems, both in benzophenone formation, dictated by electrophilic aromatic substitution rules, and in the substitution reaction in unsymmetrically 2*, 6*-disubstituted substrate [32].

Horne and Rodrigo in 1991 reported an interesting variant of this method. The required benzophenone intermediate (**1.9**) was prepared by a metal-halogen exchange initiated anionic Fries rearrangement of *N*-tosyl-*o*-iodobenzanilide (**1.8**) (Scheme 3.3). Hydrolysis and cyclization of benzophenone intermediate (**1.9**) by a classical nucleophilic aromatic substitution of fluorine occurred in a single step upon treatment with Et₄NOH affording acridone (**1.10**) [33].

3.3 Synthesis of Acridone From Nitrobenzaldehyde

In this reaction 2-nitrobenzaldehyde treated with benzene in presence of conc. H_2SO_4 gave 3-phenylanthranyls, which on further treatment with nitrous acid at room temperature undergo rearrangement to form acridone. This reaction is called the Lehmstedt–Tanasescu (L–T) reaction after the name of its researchers. Mechanism of reaction involves protonation of benzadehyde by H_2SO_4 followed by electrophilic addition to an electron-rich arene (benzene). Intramolecular attack of formed hydroxyl group to ortho nitro group gives anthranils (3-pheny-1,2-benzisoxazole). Anthranils on treatment with nitrous acid give *N*-nitroso acridone, which finally liberates the nitro group to form acridone (Scheme 3.4) [34].

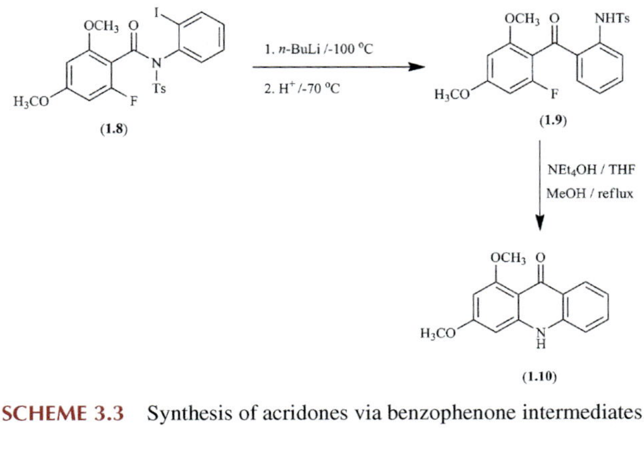

SCHEME 3.3 Synthesis of acridones via benzophenone intermediates.

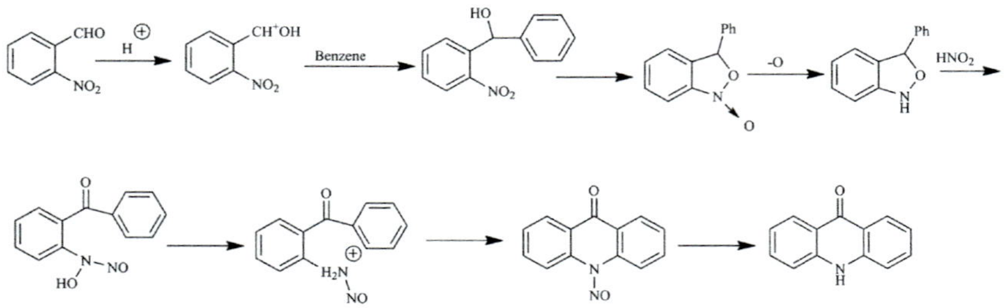

SCHEME 3.4 Lehmstedt–Tanasescu (L–T) reaction for the synthesis of acridone derivatives.

In the literature, several methods have been reported for synthesis to improve the yield, purity, and quality of acridone. But the most common method used for the synthesis of acridone is the Ullmann condensation method. Other synthetic protocols of acridone include benzophenone intermediate [35–39], L–T reaction [40], diazotization of anthranilic acid [41], hydroxylation of acridine [42], condensation of *N*-methylisatoic anhydride [43–45], and alternative methods [46–51].

4. DIVERSE BIOLOGICAL ACTIVITIES

Among a wide variety of heterocyclics that have been explored for developing pharmaceutically important molecules, acridone has played an important role in medicinal chemistry. Diversely substituted acridone and its derivatives embedded with variety of functional groups are important biological agents. A significant amount of research activity has been directed toward this class and they are used as anticancer, multidrug resistance (MDR) inhibitors, antiviral, antimalarial, antiparasitic, antimicrobial, antiinflammatory, antipsoriatic, fluorescent probes, acetylcholinestrase inhibitors, and miscellaneous (Fig. 3.5), which is discussed ahead.

4.1 Anticancer Activity

Cancer, the uncontrolled, rapid, and pathological proliferation of abnormal cells, is the leading cause of human death, after cardiovascular diseases. In 2018, 1,735,350 new cancer cases and 609,640 cancer deaths are projected to occur in the United States. Although there has been increasing sophistication of conventional therapeutic strategies such as surgery, radiotherapy, and chemotherapy, such approaches are capable of remediating only approximately half of cancer patients, while over 40% of patients are still likely to die from the disease [52,53].

Acronycine (**1**), a pyranoacridone alkaloid isolated for the first time in 1948 from the methanolic extract of *A. baueri* bark, exhibits antitumor properties against a large panel of solid tumor models, but its moderate potency and low water solubility severely hampered the subsequent clinical trials [54].

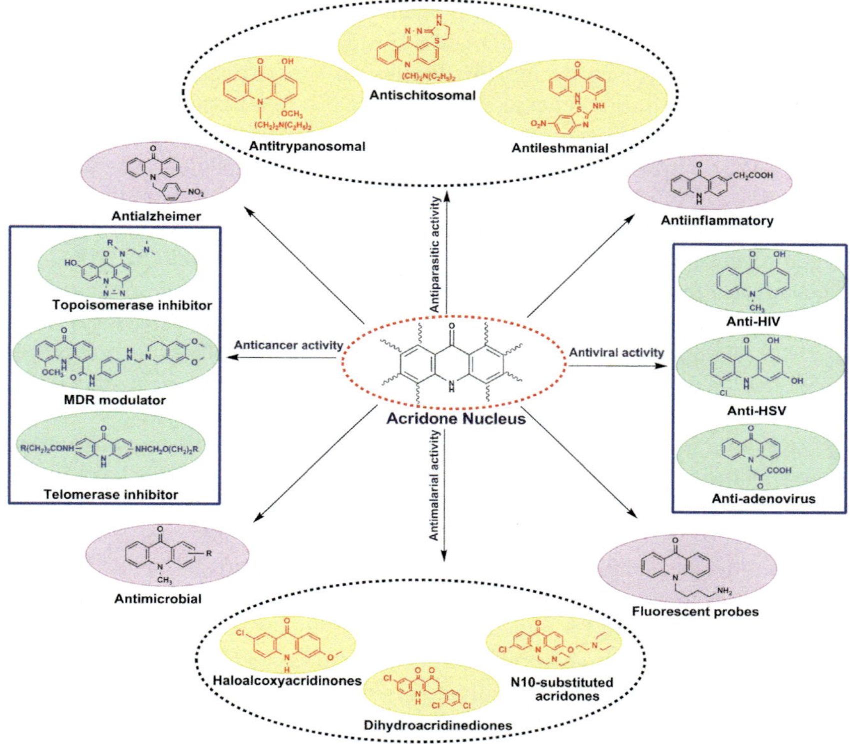

FIGURE 3.5 Applications of multifunctional acridones.

Substitution of the aromatic A ring by an electron-donating group at position 11 gave rise to active compounds. Indeed, when tested for human promyelocytic leukemic cell line HL-60 growth inhibition, both 11-methoxyacronycine (**2**) and 10,11-dimethoxyacronycine (**3**) displayed significant activity (IC$_{50}$ = 17.2 and 15.5 µM, respectively) compared with acronycine (**1**) (IC$_{50}$ = 26.2 µM) [55,56]. The dimethylchromene unit corresponding to the C and D rings was shown to play an essential role. Indeed, the substituent at C-6 on the C-ring had a dramatic influence on the biological activity. Noracronycine (**4**), bearing a phenolic OH group instead of a methoxy group at C-6, did not possess any significant antitumor activity [57]. Acronycine epoxide (**5**) was obtained in minute amounts from *Sarcomelicope argyrophylla* Guill. and *Sarcomelicope simplicifolia* (Endl.) Hartley ssp. *neo-scotica* (P.S. Green) Hartley. This compound was very unstable and its high reactivity led to speculation that it might be the active metabolite of acronycine in vivo, responsible for the alkylation of nucleophilic targets within the tumor cell [58]. When tested in vitro against L1210 cell proliferation benzo[*b*]acronycine (**6**) (IC$_{50}$ = 1.9 µM) were found to be c. 4- to 5-fold more potent than acronycine (**1**) (IC$_{50}$ = 10.4 µM). A comparative testing of acronycine (**1**) and compound (**6**) against several human cancer cell lines, including HCT-116, DU-145, MDA-MB231, and HL-60 confirmed the higher potency of benzo[*b*]acronycine (**6**). Preliminary experiments also suggested that this compound should possess significant DNA topoisomerase II inhibitory activity. However, this activity could not be directly correlated with the cytotoxicity [59]. Further it has been observed that *cis* diacetate (**7**) was the most active, since it inhibited growth by 96% at 6.25 mg/kg, with two out of seven mice being tumor-free on day 43. Due to its favorable pharmacological profile, which recently underwent phase I clinical trials under the code S23906-1, its mechanism of action implies elimination of the acetoxy group at 1-position to generate a carbocation that is responsible for the alkylation of NH_2 groups of DNA guanine units in the minor groove. The covalent binding to DNA of S23906-1 (**7**) was shown to induce a marked destabilization of the double helix, with the formation of single-stranded DNA, resulting in an increase of the cyclin E level in cells that were arrested in the S-phase and subsequently underwent apoptosis [60,61]. A series of benzo[*a*]acronycine and benzo[*b*]acronycine were synthesized in which compound **8** (benzo[*b*]acronycine) was more potent against L1210 ($IC_{50}=0.6\,\mu M$) and KB-3-1 ($IC_{50}=0.115\,\mu M$) cancer cell lines [62]. Acronycine and its analogues show their anticancer effects by alkylation of nucleophilic targets within the tumor cells. A number of efforts have been made to optimize acronycine for anticancer effects by modifying its structure. Such structural modifications and brief SAR is given in Fig. 3.6.

Another natural acridone alkaloid, glyfoline (**9**) and its congeners were synthesized by Su et al. These compounds were evaluated for their ability to inhibit human leukemic HL-60 cell growth in culture in which glyfoline displays potent activity ($IC_{50}=2.2\,\mu M$) (Table 3.1). The SAR studies showed that C-1 hydroxy acridones were more active than their corresponding C-1 methoxylated derivatives, indicating that the presence of the intramolecular hydrogen bond in 1-hydroxy-9-acridones played an important role for their cytotoxicity. 6-*O*-methylglyfoline was shown to be potent as glyfoline with an IC_{50} of $3.4\,\mu M$. Removal of 6-OH of glyfoline decreased the cell growth inhibitory activity. Replacement of the methyl function at the heterocyclic nitrogen with hydrogen or (diethylamino)ethyl side chain resulted in either total loss or dramatic reduction of activity, respectively. Glyfoline congeners with nitro function at the A-ring were inactive, while compounds with amino substituent were shown to be cytotoxic in vitro. These results demonstrated that the presence of hydroxyl group at C-1, C-6, and methyl group at N10 position are essential to preserve the cytotoxicity of glyfoline (Fig. 3.7) [63].

Acridone alkaloids isolated from the root barks of some rutaceae plants. The best ability was shown by atalaphyllinine (**10**) as a potent antiproliferative agent against a variety of several cancer cell lines. The presence of secondary amine, hydroxyl groups at C-1 and C-5, and a prenyl group at C-2 played an important role for antiproliferative activities of the tetracyclic acridones [64]. Acridone alkaloids isolated from *Zanthoxylum leprieurii* Guill. The isolated

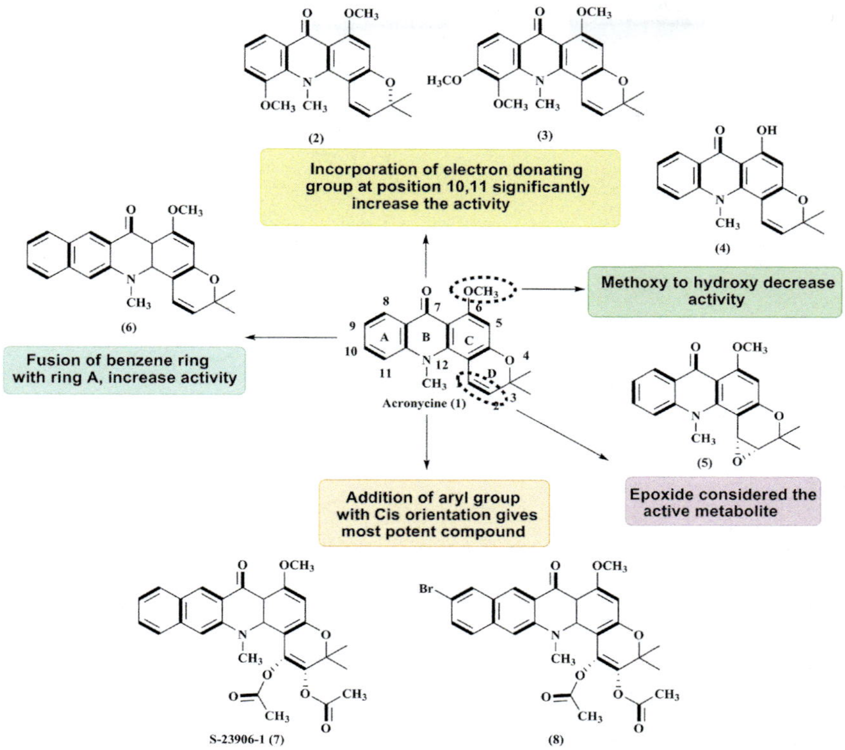

FIGURE 3.6   Structure–activity relationship of acronycine derivatives as anticancer agents.

molecules were screening for cytotoxic activity against lung carcinoma cell (A549), colorectal adenocarcinoma cell (DLD-1), and normal skin fibroblast cells (WSI). The compound **11** showed significant activity in DLD-1 cancer cells ($IC_{50} = 27\,\mu M$) as compared to rest of the molecules. They observed that the presence of O-methyl group at the C-4 position of the acridone moiety inhibited the cytotoxic activity [65]. Acridone alkaloids were obtained from the MeOH extract of the fruits of *Zanthoxylum zanthoxyloides* and *Z. leprieurii*. The anticancer screening of these acridones reveals that compound **(12)** has a moderate cytotoxic effect ($IC_{50} = 86\,\mu M$) against the liver cancer cell line (WRL-68). Computational approaches, QSAR, and modeling studies predicted that compound **12** was the inhibitor of both aromatase and glycosyltransferase [66]. Acridone alkaloids isolated from the *Zanthoxylum poggei* and were evaluated for anticancer activity. Compound **13** showed good cytotoxic activity against the human Caucasian prostant adenocarcinoma cell (PC-3) line with IC_{50} value of $16.5\,\mu M$ as compared to doxorubicin. Compound **13** exhibited strong suppressive effects ($IC_{50} = 13.5\,\mu M$) on the phagocytosis response upon activation with serum opsonized zymosan in the in vitro oxidative burst studies

TABLE 3.1 Summary of Anticancer Activity of Acridones

Compound Code	Structure	Cell Line	Result	References
Natural Compounds				
9	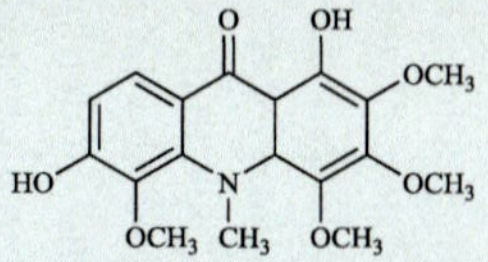	HL-60	$IC_{50} = 2.2\,\mu M$	[63]
10	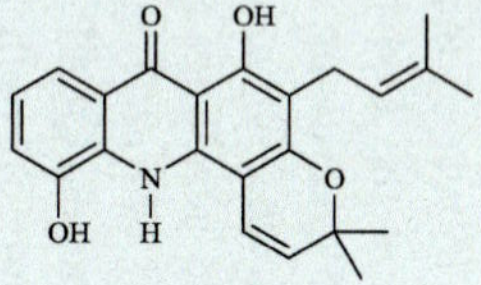	A-549, B-16 melanoma 4A5, CCRFHSB-2, TGBC-11 TKB	$IC_{50} = 1.43–6.26\,\mu M$	[64]
11	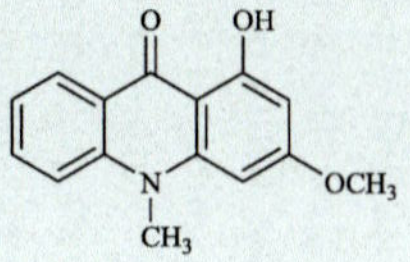	DLD-1	$IC_{50} = 27\,\mu M$	[65]
12	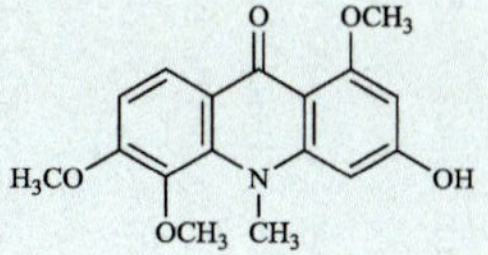	WRL-68	$IC_{50} = 86\,\mu M$	[66]

13	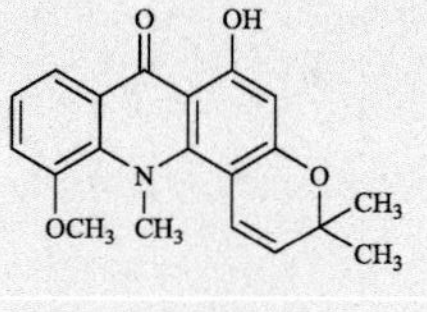	PC-3	$IC_{50} = 16.5\,\mu M$	[67]
14	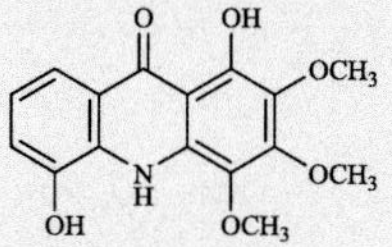	Cathepsin V (catV) inhibitor	$IC_{50} = 1.2\,\mu M$	[68]

Synthetic Compounds (Topoisomerase Inhibitor)

15	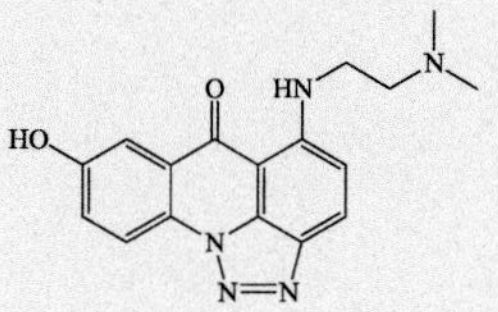	HeLa S_3 NIH 3T3-0 NIH/MDR NIH 3T3-2 NIH/MRP1	$IC_{50} = 8.6\,\mu M$ $IC_{50} = 0.35\,\mu M$ $IC_{50} = 0.82\,\mu M$ $IC_{50} = 0.38\,\mu M$ $IC_{50} = 1.0\,\mu M$	[73,74]
16	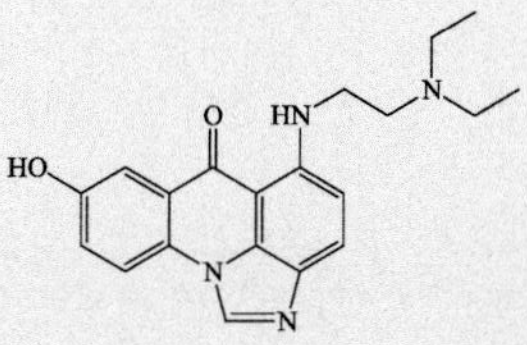	MOLT$_4$	$IC_{50} = 30\,nM$	[77]
17	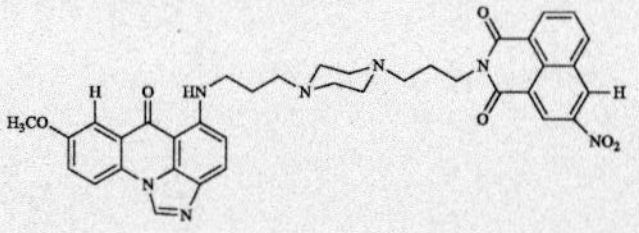	HCT-116 HT-29 HeP3B	$GI_{50} = 1.0\,nM$ $GI_{50} = 1.0\,nM$ $GI_{50} = 2.5\,nM$	[79]

Continued

TABLE 3.1 Summary of Anticancer Activity of Acridones—cont'd

Compound Code	Structure	Cell Line	Result	References
18	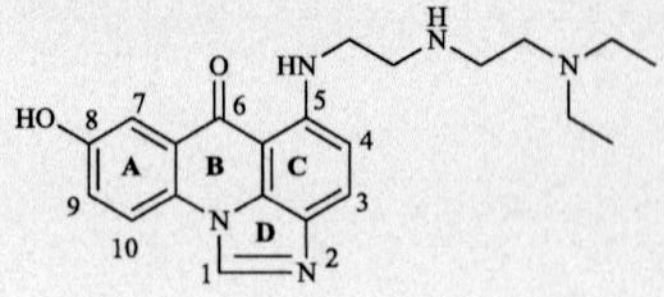	P3888	Optimum dose=100 T/C %=254	[80]
19	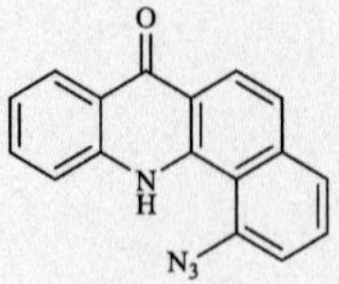	L1210	$IC_{50}=59\,nM$	[81]
20	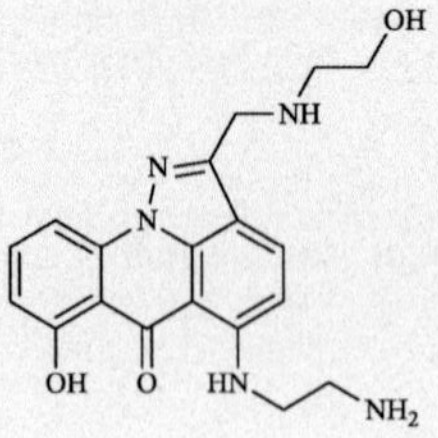	Hela S_3	$IC_{50}=0.0034\,\mu M$	[82]
21	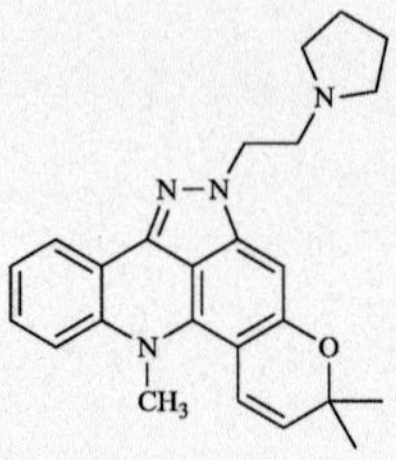	HT-29	$IC_{50}=3.0\,\mu M$	[83]

22	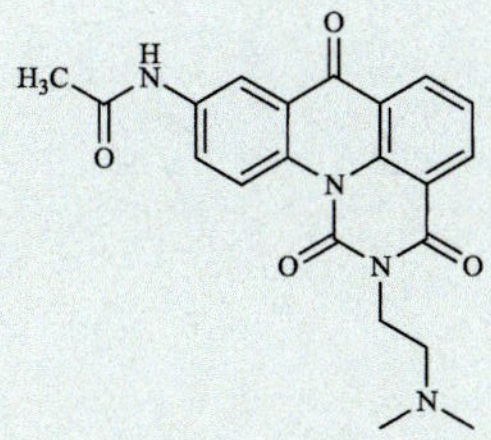	P388 KB	$IC_{50} = 0.86\,\mu M$ $IC_{50} = 1.5\,\mu M$	[84]
23	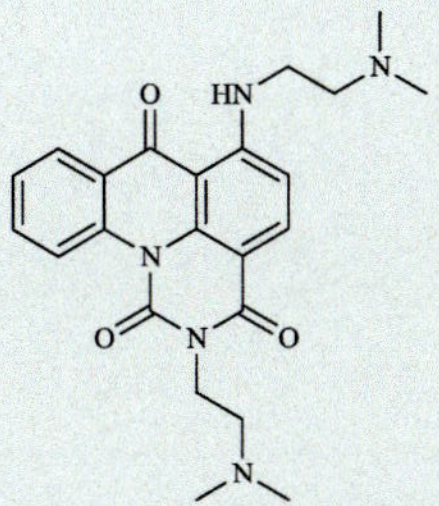	LoVo LoVo/Dx	$EC_{50} = 0.022\,\mu M$ $EC_{50} = 0.029\,\mu M$	[85]
24	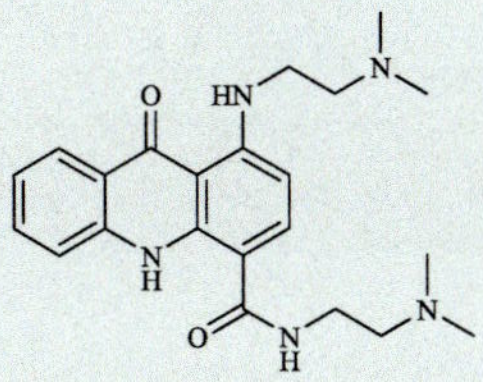	A-2780 G-361 L1210 HT29	$IC_{50} = 0.015\,\mu M$ $IC_{50} = 0.015\,\mu M$ $IC_{50} = 0.044\,\mu M$ $IC_{50} = 0.094\,\mu M$	[86]

Continued

TABLE 3.1 Summary of Anticancer Activity of Acridones—cont'd

Compound Code	Structure	Cell Line	Result	References
25		HT-29 LoVo LoVo/Dx A2780 A2780 cisR CH1 CH1 cisR SKOV-3	$IC_{50} = 0.022\,\mu M$ $IC_{50} = 0.017\,\mu M$ $IC_{50} = 1.0\,\mu M$ $IC_{50} = 0.0041\,\mu M$ $IC_{50} = 0.055\,\mu M$ $IC_{50} = 0.076\,\mu M$ $IC_{50} = 0.015\,\mu M$ $IC_{50} = 0.042\,\mu M$	[87]
26		HT-29	$IC_{50} = 0.0039\,\mu M$	[88]
27		HT-29	$IC_{50} = 0.040\,nM$	[89]

28	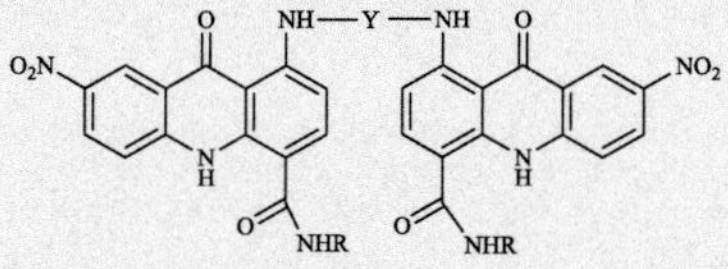 (36) R= $(CH_2)_2N(CH_3)_2$ Y= $(CH_2)_3N(Me)(CH_2)_3$	HT-29	$IC_{50} = 0.00043\,\mu M$	[90]
29	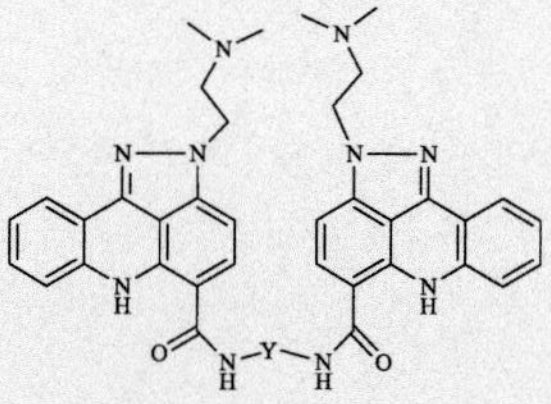 Y= $(CH_2)_3N(Me)(CH_2)_3$	HT-29	$IC_{50} = 3.0\,nM$	[91]
30	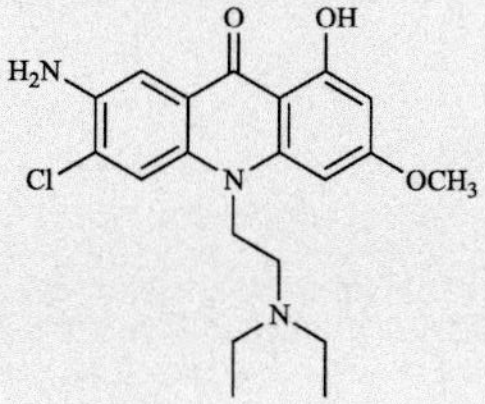	HL-60 P388	$IC_{50} = 3.8\,\mu M$ $IC_{50} = 4.2\,\mu M$	[92]
31	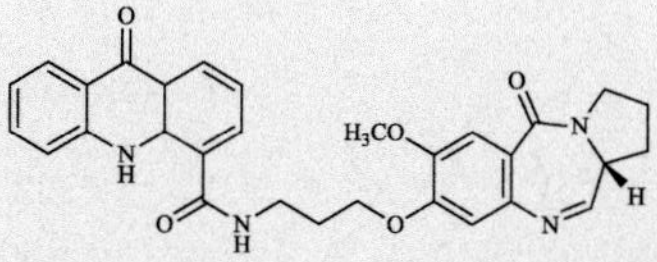	NCI-H23 A498 M14 UACC-62	$LC_{50} = 0.07\,\mu M$ $IC_{50} = 0.05\,\mu M$ $IC_{50} = <0.01\,\mu M$ $IC_{50} = <0.01\,\mu M$	[93]

Continued

TABLE 3.1 Summary of Anticancer Activity of Acridones—cont'd

Compound Code	Structure	Cell Line	Result	References
32		MCF-7 HL-60	$IC_{50}=4.0\,\mu M$ $IC_{50}=11.02\,\mu M$ Binding constant (K_i) $10.32\times10\times M^{-1}$ with CT-DNA	[94]
33		MCF-7 HL-60	$IC_{50}=3.15\,\mu M$ $IC_{50}=1.41\,\mu M$ Binding constant (K_i) $10.38\times10\times M^{-1}$ with CT-DNA	[95]
34		JURKAT M4Beu B160F0 DLD-1	$IC_{50}=2.8\,\mu M$ $IC_{50}=4.7\,\mu M$ $IC_{50}=4.1\,\mu M$ $IC_{50}=5.4\,\mu M$	[96]
35		MCF-7 MDA-MB-435 IGROV1 OVCAR-3 OVCAR-5 SNB-75	$IC_{50}=0.13\,\mu M$ $IC_{50}=1.56\,\mu M$ $IC_{50}=1.29\,\mu M$ $IC_{50}=1.79\,\mu M$ $IC_{50}=1.91\,\mu M$ $IC_{50}=1.46\,\mu M$	[97]

36		HL60 CCRF-CEM K562	$IC_{50}=0.6\,\mu M$ $IC_{50}=0.7\,\mu M$ $IC_{50}=11.4\,\mu M$	[98]
37		Hep-G2	$IC_{50}=1.33\,\mu M$	[99]
38		K562 A375 Hela	$IC_{50}=0.46\,\mu M$ $IC_{50}=0.16\,\mu M$ $IC_{50}=3.79\,\mu M$	[100]
39		MCF-7 T-47D MDA-MB-231	$IC_{50}=11.0\,\mu M$ $IC_{50}=14.5\,\mu M$ $IC_{50}=16.6\,\mu M$	[101]

Continued

TABLE 3.1 Summary of Anticancer Activity of Acridones—cont'd

Compound Code	Structure	Cell Line	Result	References
40	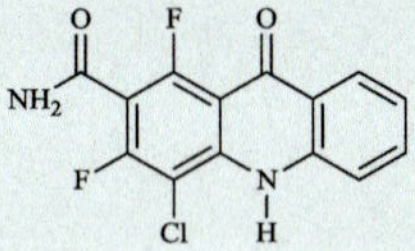	A431 HL-60 K562 Hep42 SKov-3	$IC_{50}=0.08\,\mu M$ $IC_{50}=0.2\,\mu M$ $IC_{50}=0.2\,\mu M$ $IC_{50}=0.2\,\mu M$ $IC_{50}=4.0\,\mu M$	[102]
41	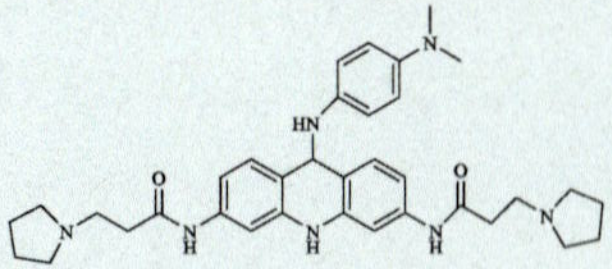	Calu-3 16HBE 14o	$IC_{50}=13.6\,\mu M$ $IC_{50}=3.6\,\mu M$	[107]
Synthetic Compounds (Telomerase Inhibitor)				
42	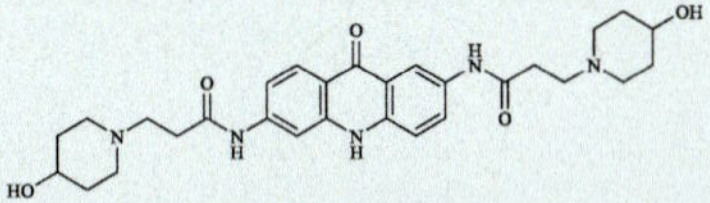	Telomerase inhibitory activity A2780	$EC_{50}=0.2\,\mu M$ $EC_{50}=3.5\,\mu M$	[108]
43	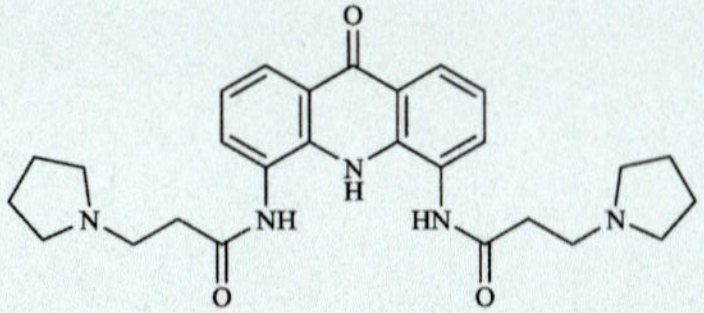	Telomerase inhibitory activity MCF-7 A549	ΔTm (1.0 µM) values 34°C (FRET assays of quadruplex binding) $IC_{50}=23.0\,\mu M$ $IC_{50}=16.0\,\mu M$	[110]
44	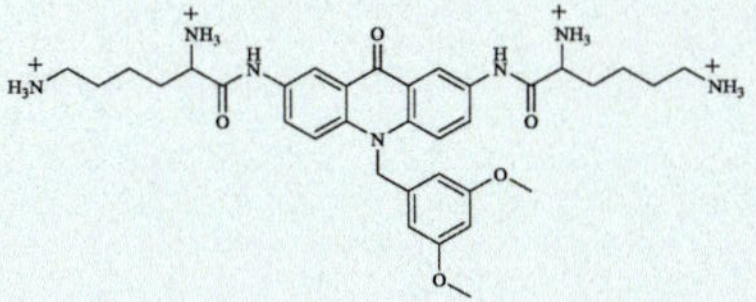	Relative binding affinity with G-quadruplex DNA CCRF-CEM	BAi = 0.70 $IC_{50}=8.0\,\mu M$	[111]

#	Structure	Activity	Result	Ref.
45	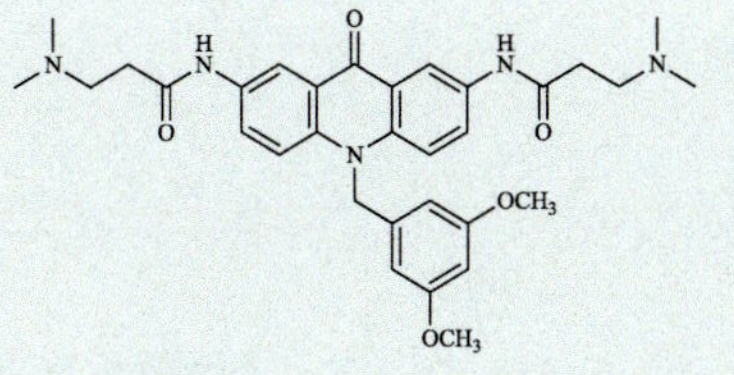 (chemical structure)	G-quadruplex DNA binding ability FRET assays of quadruplex binding CCRF-CEM HCT-116 K562	Binding ability to duplex ctDNA ($K_b = 1.3 \times 10^5\,M^{-1}$) ΔTm ($1.0\,\mu M$) values 10°C $IC_{50} = 0.3\,\mu M$ $IC_{50} = 21.3\,\mu M$ $IC_{50} = 4.7\,\mu M$	[112]
46	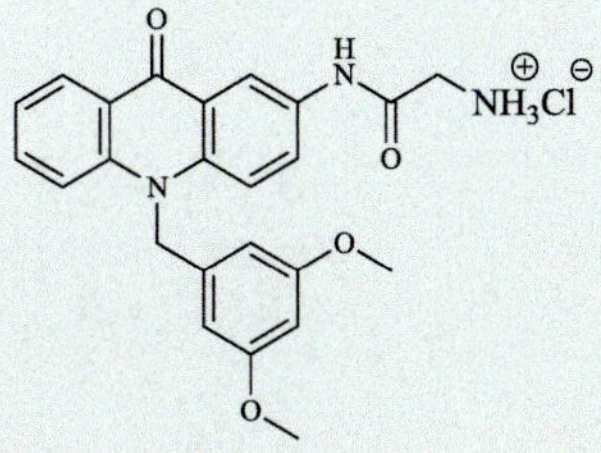 (chemical structure)	G-quadruplex DNA binding ability DNA topo 1 inhibition assay MCF-7 CCRF-CEM HepG-2 A549	Binding ability to duplex ctDNA ($K_b = 1.5 \times 10^5\,M^{-1}$) Moderate activity at $50\,\mu M$ $IC_{50} = 11.7\,\mu M$ $IC_{50} = 0.39\,\mu M$ $IC_{50} = 6.5\,\mu M$ $IC_{50} = 3.4\,\mu M$	[113]

Synthetic Compounds (IMPDH Inhibitor)

#	Structure	Activity	Result	Ref.
47	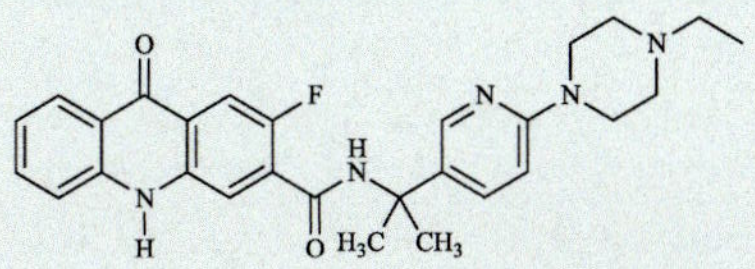 (chemical structure)	IMPDH I IMPDH II CEM PBMC	$IC_{50} = 62.0\,nM$ $IC_{50} = 17.0\,nM$ $IC_{50} = 0.77\,\mu M$ $IC_{50} = 0.22\,\mu M$	[118]
48	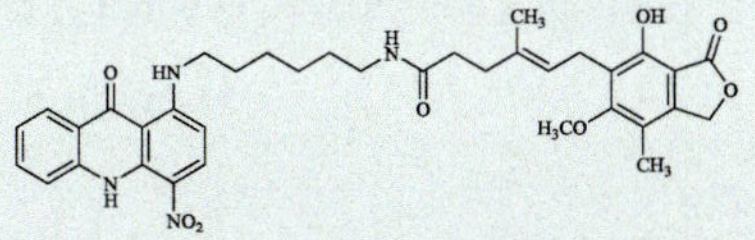 (chemical structure)	IMPDH inhibition MOLT-4 L1210 PBMC	Increase proliferation 70% (highest conc. of GMP in the culture) $IC_{50} = 0.138\,\mu M$ $IC_{50} = 0.03\,\mu M$ $IC_{50} = 1.06\,\mu M$	[119]

Continued

TABLE 3.1 Summary of Anticancer Activity of Acridones—cont'd

Compound Code	Structure	Cell Line	Result	References
Synthetic Compound (Tubulin Polymerization Inhibitor)				
49		Tubulin polymerization assays HT29 HL60 PC3 DU145	$IC_{50} = >10.0\,\mu M$ $IC_{50} = 0.051\,\mu M$ $IC_{50} = 0.1\,\mu M$ $IC_{50} = 0.75\,\mu M$ $IC_{50} = 0.05\,\mu M$	[122]
50		C6 glioblastoma	$IC_{50} = 14.0\,\mu M$	[123]
Miscellaneous Anticancer Agents				
51		NQO1 NQO2 HCT116	$IC_{50} = 111.0\,nM$ $IC_{50} = 1333.0\,nM$ $IC_{50} = 0.333\,\mu M$	[124]

52	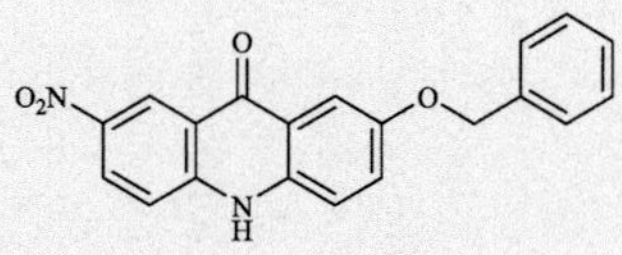	Cat L Cat V	$IC_{50} = 0.5\,\mu M$ $IC_{50} = 1.7\,\mu M$	[125]
53	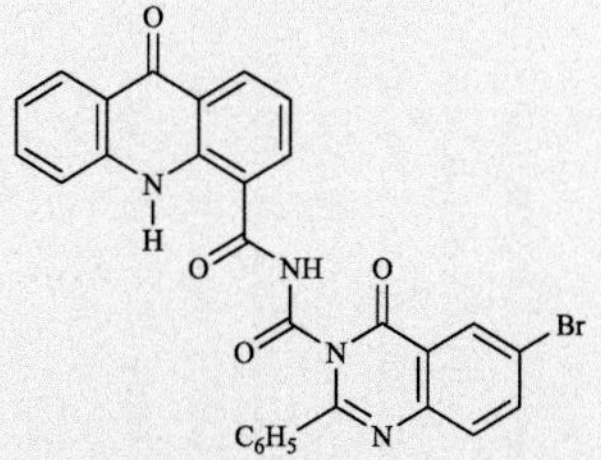	MCF-7 L5174T SW1398 Inhibition of EGFR tyrosine kinase WiDr	$IC_{50} = 2.9\,\mu M$ $IC_{50} = 6.2\,\mu M$ $IC_{50} = 6.0\,\mu M$ 100% $IC_{50} = 11.8\,\mu M$	[126]
54	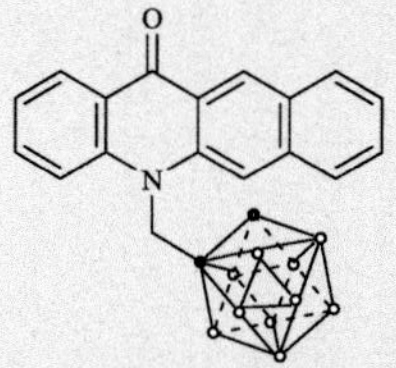	U87 human glioblastoma	10% of cellular viability	[127]
55	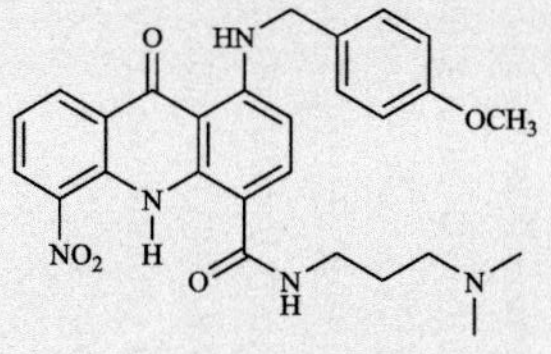	HepG2 MCF-7	$IC_{50} = 3.56\,\mu M$ $IC_{50} = 4.40\,\mu M$	[129]

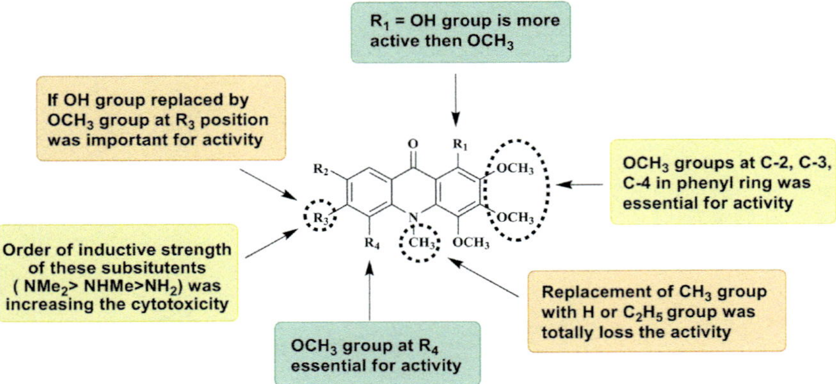

FIGURE 3.7 Structure–activity relationship of glyfoline and its congeners.

using whole blood as compared to ibuprofen [67]. Acridone alkaloids were obtained from *Swinglea glutinosa* as potent and reversible inhibitors of cathepsin V. Cathepsin V (catV) is a lysosomal cyteine peptidase highly expressed in thymus, testis, and corneal epithelium. The potent cathepsin V inhibitor activity was found in compound **14** with an IC_{50} value of $1.2\,\mu M$. Docking study showed that positions C-1 and C-5 hydroxyl group substituent bind to S_1 and S_2 pockets as a hydrogen bond donor to interact with Gly23 and Leu157. Nine-position of carbonyl group bind to S_3 pocket to act as a hydrogen bond from Gly66. The oxygen atom of the 2-methoxy substituent interacts with the NH_2 side chain of Gln19, which is part of the oxyanion hole in catV [68]. Compound **14** $(K_i = 0.2\,\mu M)$ is a promising lead candidate for future medicinal chemistry efforts designed to discover new competitive inhibitors of catV with enhanced affinity and potency.

Acridine/acridone-based derivatives have been studied mostly as anticancer agents that are believed to express their activity through binding or intercalation of DNA. The planar structure of acridones facilitates to act on the nucleotides by intercalating DNA strands, thereby serving as potent anticancer drugs. They perturb the function of cancer cells by decreasing or inhibiting the activity of some enzymes that are crucial for proper DNA functions such as topoisomerase and telomerase [69,70]. Anticancer activities of acridones are summarized in Table 3.1.

4.1.1 Topoisomerase Inhibitor

Acridine and its congener, acridone, which target the covalent enzyme-DNA intermediate (forming a tertiary complex drug-DNA-enzyme), cause DNA fragmentation in dividing cancer cells. The two best-known derivatives for poisoning topoisomerases are based on the structure of a 9-anilinoacridine scaffold amsacrine and DACA. Amsacrine has been used in clinics since 1976, as an antileukemic agent, and was the first drug for which the mode of action proved to be the interaction with topo-II-DNA complex. It is noteworthy that DACA is a member of the few inhibitors that can act against either topo-I or topo-II enzymes [11]. Furthermore, acridones bearing a fused nitrogen ring such as triazoloacridone and imidazoacridone have been designed.

Triazoloacridone (C-1305, **15**) has been synthesized at Gdansk University, and showed potent antitumor activity in vitro and in vivo including HeLa S3, NIH/MDR and NIH/MRP1 cell lines. The surface plasmon resonance studies [13,16,71–74] showed that it binds to DNA by intercalation and possesses higher affinity for GC than AT-DNA. Later on in 1990, Imidazoacridone (C-1311, **16**) was synthesized and is currently in the phase II clinical trial. It also inhibits the cell cycle in the G2 phase in cancer cells [14].

The presence of the —OH group at 8-position in imidazoacridone shows good antitumor activity as compared to the —OCH$_3$ group. Thus, it can be concluded that the activation of the heterocyclic ring is essential for the high anticancer activity of imidazoacridone. Furthermore, it elucidated the sequence of death responses of C-1311 (**16**) in human leukemia MOLT$_4$ cells (IC$_{50}$ = 30 nM). Results showed that C-1311-induced mitotic catastrophe is not an ultimate death event but rather an event that gives rise to apoptosis [14,15,75–77]. Bisimidazoacridones, in which tetracyclic ring systems are held together by either N$_2$-methyl diethylenetriamine or 3,3′-diamino-*N*-methyldipropylamine linker, were synthesized. Compound **17** displayed the good antitumor effect against HCT-116, HT-29, and Hep3B cell line with GI$_{50}$ values of 1.0, 1.0, 2.5 nM, respectively. The presence of methoxy group in 8-position of imidazoacridone ring enhanced the antitumor activity [78,79]. Several new 5-amino-substituted derivatives of 5-amino-6*H*-imidazo[4,5,1-*de*]acridin-6-one were synthesized.

Compound **18** exhibited maximum potency compared with other compounds. The 8-OH substituted compounds or bisubstituted 7-OH and 10-OCH$_3$ compounds showed potent in vivo activity against murine P3888 leukemia cell lines. Most active imidazoacridinones were 8-OH derivatives, which were able to increase the ability for covalent binding to DNA [80]. A new series of pyrido-amsacrine analogues were synthesized and then evaluated in vivo against L1210 leukemia cell lines. Almost all the pyrido-analogues were tighter DNA-binding ligands than amsacrine but none of these compounds were more active than amsacrine. Among all the synthesized compounds pyrido-acridan-7-one analogues **19** appear to be slightly more potent than the others having IC$_{50}$ value 59 nM. The results concluded that a four-ring heterocycle is required to intercalate with DNA whereas the aniline chain is not essential for DNA binding [81].

Pyrazoloacridone derivatives display good antitumor activity against P388 leukemia, P388/ADM leukemia, Hela S$_3$, and sarcoma 180 solid tumor cells. Compound **20** showed potent antiproliferative activity against the Hela S$_3$ cell line (IC$_{50}$ = 0.0034 μM). The SAR studies of these compounds revealed that aminoalkyl-amino and/or hydroxyalky amino side chains at the C-2 and C-5 positions are apparently crucial for the activity. The hydroxy or methoxy substituents at C-7 show only a slight influence on the binding affinity to DNA [82].

A series of heterocyclic compounds were designed and synthesized, and their DNA binding and in vitro cytotoxicities have been investigated. These compounds exhibited interesting cytotoxic activity against the murine leukemia L1210 cell line, being more active than the parent compound. Compound **21** was more active against the human colon HT-29 cell line with IC_{50} value (3.0 µM) as compared to mitoxantrone. The results of this study indicate that the incorporation of an amino-substituted pyrazole ring into the acronycine chromophore, or into its isosteres, results in an improvement of the lead compound's activity [83]. Pyrimidoacridone derivatives with modification at 5- or 9-position with amino or acetylamino substituents were synthesized and evaluated as anticancer agents. Compound **22** displays good activity in murine leukemia P388 (IC_{50}=0.86 µM) and human oral cancer KB (IC_{50} = 1.5 µM) cell lines [84]. Pyrimidoacridine derivatives have been prepared and screened for better anticancer agents. Compound **23** showed potent cytotoxic activity against human colon adenocarcinoma cell lines sensitive (LoVo) and doxorubicin resistant (LoVo/Dx) cell lines with an EC50 value of 0.022–0.029 µM [85].

A series of DNA-intercalating potential antitumor agents, 1-[(ω-aminoalkyl)amino]-4-[N-(ω-aminoalkyl)carbamoyl]-9-oxo-9,10-dihydroacridines, have been prepared and evaluated for anticancer activity against L-1210, A-2780, G-361, and HT-29 cells. Compound **24** from this series has been identified as a worthwhile lead for the development of more potent cytotoxic agents. The IC_{50} value for compound **24** has been found to lie in the range of 0.015–0.094 µM against all the tested cell lines [86]. 2,3-Dihydro-1H,7H-pyrimido[5,6,1-de]acridine-1,3,7-trione derivatives were prepared and evaluated against eight tumor cell lines including human colon adenocarcinoma (HT29, LoVo, and L_OV_O/D_X) and human ovarian carcinoma (A2780, A2780 cisR, CH1, CH1 cisR, and SKOV-3) cells. Among these, compound **25** exhibited potent activity against A2780 (IC_{50}=0.0041 µM) and LoVo/Dx (IC_{50} = 1.0 µM) cell lines as compared with standard drugs. The SAR studies revealed that an amino side chain at 6-position is essential for activity. In these compounds there is a strong intramolecular hydrogen bond between the carbonyl in 7-position and the hydrogen of the amine in 6-position. This hydrogen bond can mimic an additional fifth ring, which may be important for biological activity [87]. Furthermore, these authors have prepared acridine carboxamides and screened for antitumor and DNA-binding activity, and in vitro cytotoxic potency of these derivatives toward the human colon adenocarcinoma cell line (HT-29).

Compound **26** displays potent activity with an IC_{50} value of 0.0039 µM. The substitution in 5-position of the nitro group essential for antitumor activity, with a basic carboxamide side chain that is in position *peri* to an electron withdrawing atom, seems to afford a similar potency in cytotoxicity. They concluded that compound **26** was a highly DNA-affinic and potent cytotoxic compound that led to identification of new lead in the antitumor strategies [88]. On moving to pentacyclic chromophore, they have prepared DNA binding agents with potential antitumor activities bearing two cationic side chains. In vitro cytotoxic properties were determined against three human colon adenocarcinoma cell lines, HT-29, LoVo, and LoVo/Dx. The results indicate that all compounds possess a potent antiproliferative activity with IC_{50} values in the low molecular range. However, compound **27** exhibited the most potent cytotoxic activity ($IC_{50}=0.040\,\mu M$). A SAR study revealed that a methoxy substituent leads to a noticeable improvement in cytotoxic activity [89].

In 2003, Antonini et al. proved marked antitumor activity shown by bis (acridine-4-carboxamides) derivatives in which compound **28** showed potent activity against HT-29 cell line ($IC_{50}=0.00043\,\mu M$). They observed mechanism of these compounds by intercalate with DNA. SAR studies suggested that best results were obtained with $Y=-(CH_2)_3N(CH_3)(CH_2)_3$ and a linker present at $1,1'$ and $4,4'$ position [90]. The various types of previously synthesized potential bis-intercalating agents provided satisfactory results, which prompted them to synthesize a new series of bis (pyrazolo [3,4,5-*kl*]acridine-5-carboxamides) from their corresponding monomers. They exhibited remarkable DNA affinity in which compound **29** shown potent activity against the HT-29 cell line ($IC_{50}=3.0\,nM$). However, such compounds were less efficient in binding than related bis (acridine-4-carboxamides) derivatives [91].

New acridone modified quinolones have been prepared and screened for antitumoral activity against HL-60 and P388 leukemias cell lines. Compound **30** showed a most potent antitumor activity against HL-60 and P388 leukemia cell lines with IC_{50} value (3.8–$4.2\,\mu M$). The SAR studies represent that presence of an intramolecular hydrogen bond between the hydroxyl group at C-1 and the *peri* carbonyl function could play a role in cytotoxicity. The presence of amino group and substituent at N-10 position of the acridone moiety is crucial for cytotoxic activity. In contrast, the substituent at the C-3 position does not greatly influence the activity [92]. C8-linked pyrrolo [2,1-*c*][1,4] benzodiazepine-acridone/acridine hybrid compounds act as potential DNA-binding agents. Their interaction with DNA has been extensively investigated and it is considered unique since they bind within the minor groove of duplex DNA, forming a covalent aminal bond with N_2-amino group of guanine base. Compound **31** showed a high DNA-binding affinity and has been evaluated for in vitro anticancer activity in the standard 60-cancer cell line screen of NCI. It exhibited good activity against NCI-H23, A498, M14, and UACC-62 cell lines [93]. 1,3-dimethyl and 1,3-diacetoxy acridone derivatives were synthesized and evaluated against MCF-7 and HL-60 cell lines. Compounds **32** and **33** showed good cytotoxic activity against these cell lines with highest binding affinity with CT-DNA [94,95].

Iodo-acridone and acridine carboxamides act as potential agents for a targeted chemo-radionuclide and/or chemotherapeutic approach to melanoma. Biodistribution studies carried out in B16F0 murine melanoma tumor bearing mice showed that compound **34** has long-lasting concentration in tumor cells together with an in vivo kinetic profile favorable to targeted radionuclide therapy. This compound showed potent antitumor activity against M4Beu, B16F0, DLD-1, and JURKAT cells with IC_{50} of 2.8–5.4 µM [96]. Eight fluorinated acridones were prepared and evaluated in a panel of 60 cancer cell lines (NCI). Compound **35** exhibited the most potent anticancer activity against a panel of cancer cell lines with IC_{50} value between 0.13 and 26 µM [97]. A novel series of 10-benzyl-9(10H)-acridinones have been synthesized and screened for their activity against CCRF-CEM, K562, and HL60 cell lines. Compound **36** with electron donating dimethoxy groups at C-3 and C-5 positions on a benzene ring showed good inhibition against HL60, CCRF-CEM, and K562 cell lines (IC_{50} = 0.6–11.4 µM). The antitumor effect of compound **36** is believed to be due to the induction of apoptosis, which is further confirmed by propidium iodide (PI) staining and Annexin V-FITC/PI staining assay using flow cytometry analysis [98]. *N*-benzyl-9(10*H*)-acridinones were prepared and evaluated against human liver carcinoma cell line HepG2. Compound **37** exhibited the good inhibitory activity (IC_{50} = 1.33 µM), which is believed to be due to the induction of apoptosis, which was further confirmed by Hoechst 33258 fluorescence staining, agarose gel electrophoresis, and Annexin V-FITC/PI staining assay. SAR studies demonstrated that the substitution at C-4 position of the acridone ring play a very important role in the cytotoxic potency of these acridone derivatives [99]. Novel pyridyl acridones as potent DNA-binding and apoptosis-inducing agents were prepared and evaluated against K562 cells. In the presence of methyl group at C-5 position of acridone compound **38** showed a good ability to inhibit K562 cells (IC_{50} = 0.46 µM). Further, this compound **38** exhibited good results in NCI-H520, U251, A375, A172, HeLa, CNE-2, U118-MG, HepG-2, and MCF-7 cells (IC_{50} = 0.16–3.79 µM). Mechanism studies revealed that this compound **38** interacts with DNA, inhibits topo I activity, and induces apoptosis through the mitochondrial pathway. SAR showed that compounds containing two methylene units between the N,N-dimethyl amino group and 4-carboxamide group displayed better antiproliferative activity than derivatives containing three methylene units [100]. 9(10H)-acridinone-1,2,3-triazole derivatives have been prepared and evaluated against human breast cancer cell lines. Compound **39** with a methoxy group at the 2-position of acridone moiety showed good activity against MCF-7 (IC_{50} = 11.0 µM), T-47D (IC_{50} = 14.5 µM) and MDA-MB-231 (IC_{50} = 16.6 µM) cell lines. SAR study revealed that compounds with mono-methoxy substituents on the pendant benzyl group of the 1,2,3-triazole ring displayed better antitumor activity than derivatives with tri-methoxy groups, which showed no cytotoxicity. Apoptosis induced by compound **39** was confirmed via acridine orange/ethidium bromide and Annexin V-FITC/ (PI) double-staining method [101]. A series of polyhalo acridone heterocyclic compounds were synthesized and evaluated for their in vitro antitumor activity. Among them, compound **40** shown outstanding activity in verity of cell lines such as A431, HL-60, K562, Hep42, and Skov-3 with IC_{50} values in the range of 0.08–4.0 µM. The presence of fluorine atoms in the acridone moiety reduced the Gibbs energy and C log P values of molecules, and may result in an enhancement in cytotoxic activity [102].

4.1.2 Telomerase Inhibitor

Another essential target for anticancer agents is telomerase. Telomeres are repetitive DNA sequences at the ends of chromosomes that protect their termini from being recognized as double-strand breaks. In absence of telomeres, there would be continued repair of ends of the chromosomes leading to genomic instability. Almost all the eukaryotic cells depend on enzyme telomerase for the de novo synthesis of telomeres. Telomerase activity has been found in ~85%–90% of all human tumors. A majority of the tumor cells exhibit telomerase activity in order to evade the telomere checkpoint and to obtain unlimited growth potential, which makes it an attractive target for selective cancer therapy [103,104].

Acridine derivatives have been found to inhibit telomerase activity by π-π interactions with guanine tetrads of quadruplex DNA. G-quadruplexes are the 3′ overhang of telomeres, which are rich in guanine units and have been shown in vitro to produce tetra-stranded DNA structures. These G-quadruplex conformations inhibit telomerase activity and, therefore, drugs that stabilize these tetraplexes could conceivably be effective chemotherapeutic agents [105]. Acridine-based structures inhibiting telomerase can be divided into three subfamilies such as trisubstituted acridines, pyridoacridines, and dibenzophenanthrolines.

Neidle's group studies on disubstituted anthraquinones or disubstituted acridines/acridones and trisubstituted acridines designed a potent and selective telomerase inhibitor having a 3,6,9-trisubstituted acridine structure, named BRACO-19 (**41**) [106]. Biopharmaceutical properties of BRACO-19 showed that it has good solubility (2 mg/mL) in water at pH 7.4 and below, protein binding to human serum albumin at 38%, and interaction with membrane lipids was absent. Cytotoxicity and transport studies were performed with the bronchial cell lines (16HBE14o⁻ and Calu-3), primary human alveolar epithelial cells, and the intestinal cell line (Caco-2). The cytotoxicity in 16HBE 14o⁻, Calu-3 and human alveolar epithelial cells was in the range of IC_{50} value 3.6–13.5 µM while Caco-2 cells were not affected up to 50 µM concentration [107]. Series of 2,6-, 2,7- and 3,6-bis-aminoalkylamido acridones show a similar level of activity against telomerase as compared to disubstituted acridine derivatives. Moreover, computer modeling and calculations of relative binding energies suggest an equivalent binding mode to human intramolecular G-quadruplex DNA, but with significantly reduced affinity, as a result of the limited delocalization of the acridone chromophore compared to the acridine system. Compound **42** displays good telomerase inhibitory activity with $^{tel}EC_{50} = 0.2$ µM [108,109]. A series of 4,5-di-substituted acridones have been designed and synthesized. Several compounds show high affinity for telomeric G-quadruplex DNA in classical and competition FRET assays, together with low duplex DNA affinity, although they do not show activity in a telomerase assay or evidence of telomere shortening. Compound **43** showed high levels of stabilization ability for the telomeric G4 sequence with ΔTm (1 μM) values 34°C. These are comparable to the 32°C ΔTm (1 μM) for the trisubstituted acridine compound BRACO-19 [110]. A novel class of 9(10H)-acridone derivatives with terminal ammonium substituents at C-2 (and C-7) position on the acridone ring were successfully synthesized. The relative affinities of the acridone compounds to G-quadruplex DNA have been investigated and the results showed that these compounds had a binding specificity for G-quadruplex over duplex sequences. In compound **44** the presence of more positive charge to the side chain can improve the formation and stabilization of the G-quadruplex [111]. A novel series of 10-(3,5-dimethoxy)benzyl-9(10H)-acridone derivatives with terminal ammonium substituents at C-2 and C-7 positions on the acridone ring were successfully synthesized as antiproliferation agents. Compound **45** containing dimethylamine substituents at the terminal C-2 and C-7 positions exhibited the highest cytotoxicity in CCRF-CEM cell line (IC$_{50}$ = 0.3 μM). Further study indicated that compound **45** had strong binding activity to human telomeric G-quadruplex DNA, as detected by mass spectrometry, CD spectroscopy, UV absorption, FRET, and fluorescence quenching assays. They suggested that the activity of **45** might be associated with its stabilization of G-quadruplex DNA, which can be developed as potent antitumor agent. The SAR of these compounds revealed that the changing of dimethylamine group to diethylamine and diethanol amine group decreased the activity. The replacement of piperidine to 4-hydroxyl piperidine reduced about 10-fold cytotoxicity. Tertiary amines display more activity as compared to pyrrolidine and morpholine groups, which were essentially inactive [112]. In continued work on acridone moiety Gao and coworkers synthesized 2-amino-10-(3,5-dimethoxy)benzyl-9(10H)-acridinone derivatives as potent DNA-binding antiproliferative agents. These compounds demonstrated promising cytotoxicity to leukemia cells CCRF-CEM, displaying IC$_{50}$ values in the low micromolar range. Compound **46** exhibited good activity against CCRF-CEM, A549, HepG-2, and MCF-7 cells (IC$_{50}$ = 0.39–11.7 μM). Furthermore, compound **45** had strong binding activity to calf thymus DNA (ct DNA), as detected by UV absorption and fluorescence quenching assays, but limited inhibitory activity to human topoisomerase I (topo I) [113]. In addition, UPLC/Q-TOF MS based metabolic assay was performed in order to understand the mechanism of action, and it was proposed to provoke oxidative stress, which resulted in lipid peroxidation and mitochondrial lesions. This mitochondrial alteration leads to cytochrome-C release, which further triggers caspase-3 mediated apoptosis [113,114].

4.1.3 IMPDH Inhibitor

Inosine 5′-monophosphate dehydrogenase (IMPDHa), a key enzyme in the de novo synthesis of guanosine nucleotides, catalyzes the irreversible nicotinamide-adenine dinucleotide (NAD)-dependent oxidation of inosine 5′-monophosphate (IMP) to xanthosine 5′-monophosphate. There are two isoforms of IMPDH-type I and II. IMPDH type II is expressed at very low levels in most resting cell types, but is markedly up-regulated in actively proliferating cell types, including cancer cells and activated peripheral blood lymphocytes. As a result, IMPDH type II has emerged as an attractive target for selectively modulating the immune response without inhibiting the proliferation of other cells [115,116]. Mycophenolic acid (MPA) has been shown to be a potent, uncompetitive, reversible inhibitor of human IMPDH type I and type II. Mycophenolate mofetil (MMF), a prodrug of mycophenolic acid, has clinical utility for the treatment of transplant rejection based on its inhibition of IMPDH. But clinical benefit of MMF is limited due to compound-based, dose-limiting gastrointestinal (GI) toxicity that is related to its specific pharmacokinetic characteristics. Thus, development of an IMPDH inhibitor with a novel structure and a different pharmacokinetic profile may reduce the likelihood of GI toxicity and allow for increased efficacy [117]. In view of these facts, a novel series of acridone-based inhibitors of IMPDH II have been designed and prepared. Based on in vitro potency, in vivo activity, and pharmacokinetic and safety profiles, compound **47** (BMS-566419) was selected as a lead drug candidate. It was a reversible and uncompetitive inhibitor of IMPDH II with a K_i of 25 ± 3 nM with respect to IMP and 20 ± 4 nM with respect to NAD. SAR study suggested that a small group substitution at C-2 position in acridone moiety is essential for activity and also provide enhancement activity and physiochemical properties through restricted rotation of the acridone amide carbonyl bond. Interestingly, the effects were more pronounced when the amide was substituted with a 3-pryidyl at C-3 position of acridone resulting in a 4-5-fold improvement of inhibition activity. In the des-fluoro-acridone pyridyl analogues, the positioning of the nitrogen had very little impact on either the inhibition of the enzyme or on T-cell proliferation. However, in the fluoro-acridone series, the 3-pryridyl analogue was 2- to 3-fold more active than the 2-pyridyl analogue [118]. Hybrid pharmacophore antiproliferative compounds, composed of MPA and 1-nitroacridine/4-nitroacridone derivative, have been synthesized and evaluated as inhibitors of five different leukemia cell lines (Jurkat, Molt-4, HL-60, CCRF-CEM, L1210). These conjugates possess different lengths of the linker between MPA and heterocyclic units. Compound **48** showed IMPDH inhibitions (highest conc. of GMP in the culture) increased proliferation 70% and good cytotoxicity in MOLT-4, L1210, and PBMC cells with IC_{50} value from 0.03 to 1.06 μM [119].

4.1.4 Tubulin Polymerization Inhibitor

Antimitotic agents are one of the major classes of cytotoxic drugs for cancer treatment, and tubulin is the target for numerous natural and synthetic molecules that inhibit the formation of the mitotic spindle [120,121].

New acridone derivatives have been synthesized and evaluated for their tubulin polymerization inhibitory and antiproliferative activities. 4-methoxyphenylethyl acridone derivative **49** provided significant antitubulin activity ($IC_{50} = >10.0$ μM) and good cytotoxicity against HT29 cells ($C_{50} = 0.051$ μM).

Topoisomerase inhibition assays, cell cycle analysis, and modeling studies have been performed to better investigate the mechanism of action of such compounds. The results from topo I or II assay formally established that compounds were not topoisomerase poisons and did not interact with DNA. From cell cycle analysis, it has been indicated that the induction of cellular apoptosis by **49** could be by virtue of their antimitotic activities as there was an increase in cell cycle arrest at G2 phase with parallel increase in apoptotic cells at sub-G1 phase [122]. Docking studies have been performed on the colchicine binding site of tubulin (PDB code: 1SA0) to explore the interaction mechanism and understand the structural features responsible for their activity. Compound **49** showed a single conformation occupying the binding site and engaging in the same hydrogen bonds with Ser 178, Thr 179, and Ala 250. The tricyclic core of acridone was oriented near the residues Ala 250 and Lys 254 and interacted via hydrogen bonding with the carbonyl C-9 group while the 2-(3-hydroxy-4-methoxyphenyl)ethyl adopted a folded up conformation stabilized by hydrogen bonds formed between the methoxy group and Thr 179 [122]. Amino acid derivatives of acridone were synthesized and tested for antiproliferative activity against the C6 glioblastoma cell line. Flow cytometry based detailed cell cycle analysis showed a significant decrease in the cell number in G2/M and S phase after treatment with compound **50** in comparison to the control, thus showing reduction in their proliferation. On the other hand, a significant increase was observed in G0/G1 population of cells after exposing to this compound **50**. These observations suggest that the compound **50** arrested the cancer cell population in the resting phase—that is, G0/G1 phase of cell cycle—like the normal cells, and thus may prove to be a potential antiproliferative compound [123].

4.1.5 Miscellaneous Anticancer Agents

Triazoloacridin-6-ones were prepared and evaluated for NQO1 and NQO2 inhibition. These compounds were computationally docked into the active site of NQO1 and NQO2 and their binding affinities calculated. The oxidoreductases, NAD(P)H:quinone oxidoreductase 1 (NQO1), and NRH:quinone oxidoreductase 2 (NQO2) are homodimeric, cytosolic flavoproteins that catalyze the two electron reduction of quinones to hydroquinones. Although both enzymes are expressed in many of the same tissues throughout the body the highest expression of NQO1 is found in kidney, gut epithelium, and importantly, in many types of solid tumor including cancers of the colon, breast, lung, and liver. In contrast to NQO1, the NQO2 level in normal and malignant tissues is generally lower. According to the molecular design, compound **51** exhibited more selectivity and potency toward inhibition of NQO1 enzyme (IC_{50} of 111.0 nM) over NQO2 enzyme (IC50 of 1333.0 nM). This compound revealed potent antitumor activity against the HCT116 colon cancer cell line with IC_{50} value of 0.333 μM. In addition to the lipophilic effect of the methoxy group with Phe 178, this may account for the 17-fold increased inhibitory potency over unsubstituted C-8 compounds [124].

Cathepsin V (catV) has been associated with the MHC class II complex in humans and the enzyme has been considered a potential diagnostic marker for colon tumors. This enzyme also is expressed in colorectal and breast carcinomas but not in normal colon or mammary tissue. In addition, cathepsin V together with the cathepsins L, K, and S have been implicated in atherosclerosis and are considered potential drug targets. In 2012 Marques et al. synthesized and evaluated the acridones and 4-quinolinones as potent inhibitors of cathepsins L and V. The acridone derivative **52** was the most potent inhibitor for catL and catV ($IC_{50} = 0.5–1.7\,\mu M$). The kinetics revealed that compounds of the classes of acridones are reversible competitive inhibitors of the target enzyme with affinities in the low micromolar range. They represent promising lead candidates for the discovery of novel competitive cathepsin inhibitors with enhanced selectivity and potency [125].

The epidermal growth factor receptor (EGFR) is an important tyrosine kinase receptor involved in many critical processes of cancer development, progression, and proliferation. Due to advances in chemical and structural biology and signaling transduction studies, EGFR was found to be a highly promising target in oncology that is overexpressed in a significant number of human tumors (e.g., breast, ovarian, colon, and prostate); their expression levels often correlate with vascularity and are associated with poor prognosis in patients. Quinazolinone-linked acridones were designed and synthesized as anticancer agents and their effect on inhibition of EGFR was determined. Compound **53**, a phenyl substituted bromoquinazolinone linked to acridone nucleus with an amide linkage, has shown good cytotoxic potentials than the other compounds against MCF7, LS174T, SW1398, and WiDr cell lines. Further, this compound **53** was tested for EGFR tyrosine kinase inhibition at a concentration of $10\,\mu M$ and displayed the highest activity among the selected set of ligands. In silico molecular docking of the quinazolinone tagged acridones into the auto-phosphorylation site of the kinase domain of EGFR (PDB ID: 1M17) revealed the possible interacted residues. Nitrogen and oxygen atoms were positioned appropriately in the 3D spatial arrangement of every ligand that enhances the possibilities of better hydrogen bonding with protein residues. Compound **53** exhibited one hydrogen bond interaction (nitrogen of quinazoline nucleus) with Lys721 (2.004 Å). It also displayed potential hydrophobic interactions with the residues present in the binding pocket, which might be the reason for its higher dock score (7.015) [126].

Benzo[*b*]acridin-12(7H)-ones bearing carbonyl moieties were synthesized and evaluated for their effectiveness as boron neutron capture therapy agents in cancer treatment. As a result, compound **54** revealed considerable activity in the U87 human cells when combined with neutron irradiation with cellular viability of 10%. The cellular uptake of these novel compounds into the U87 human glioblastoma cells was evaluated by boron analysis (ICP-MS) and by fluorescence imaging (confocal microscopy) [127].

Survivin, the smallest member of the inhibitor of apoptosis protein family, is overexpressed in cells of almost all cancers but not in most normal tissues in adults. Survivin expression is required for cancer cell survival, and knocking down its expression or inhibiting its function using molecular approaches results in spontaneous apoptosis. Thus, survivin is an attractive and perhaps ideal target for cancer drug discovery [128]. Zhang et al. designed and synthesized acridone derivatives as survivin inhibitors for the treatment of hepatocellular carcinoma. Compound **55** showed good activity against the HepG2 cell with IC_{50} value of 3.56 µm. The apoptosis mechanism of hepatocellular carcinoma cells further showed that compound **55** could induce the apoptosis of HepG2 cells [129]. SAR of these compounds revealed that compounds with good activity generally had methyl or nitro groups at the C-5 position. The results indicated that electron-negativity and steric effect on C-5 position of acridone scaffold might have little change in the cytotoxic profile. The presence of the benzylamino group in the target compounds was important for the anticancer activity. Molecular docking studies revealed that compound **55** can effectively target the dimerization hydrophobic domain of surviving (PDB ID: 1F3H) and the hydrophobic effect between the ligand and receptor might be due to π-conjugated planar acridone scaffold of **55**. Moreover, three hydrogen bonds of compound **55** were also formed in the hydrophobic domain (residues Phe93, Glu94, and Leu96). Therefore, it was confirmed that compound **55** can effectively bind to the critical hydrophobic core, which suggested that survivin protein may be the main drug target of compound **55**. The dramatic increase in fluorescence of compound **55** in presence of survivin showed that compound **55** binds with survivin. The degradation in survivin was determined by western blot analysis in which HepG2 cells were treated with compound **55** at different concentration gradients, and it was observed that the content of survivin protein in HepG2 cells significantly decreased with the increase of the concentration of compound **55** from 1 to 10 µM [129].

4.2 Multidrug Resistance Inhibitor

During chemotherapy of cancer, cells by some physiological mechanism become able to acquire resistance to toxic chemotherapy. Of these mechanisms the most commonly observed is increased efflux of cytotoxic drugs from cells mediated by energy-dependent transporters known as ATP binding cassette (ABC) transporters. ABC transporters are the proteins that transport the materials across the cell membranes. About 48 transporters have been identified that are grouped in seven subgroups (A–G). Among all ABC transporters, overexpression of three major transporters (i.e., P-glycoprotein (ABCB1), MRP1 (ABCC1), and BCRP (ABCG2)) have been found to be responsible for resistance in tumor cell. Overexpression of P-glycoprotein (P-gp) alone is responsible for efflux of ~50% anticancer agents. Hence it is found to be a major obstacle in cancer treatment. P-gp confers resistance to anticancer drugs including vinca alkaloids, anthracyclines, taxanes, and epipodophyllotoxins. Moreover, there are two major factors that make P-gp the most critical efflux transporter including its broader substrate specificity and prominent presence of P-gp in most excretory and barrier tissues, hence it is considered to be a major cause of MDR in cancer chemotherapy. Thus inhibition of P-gp efflux pump is necessary to restore the sensitivity of anticancer drugs [130,131]. MDR modulators (MDR reversal agents, MDR inhibitors, and chemosensitizers) can be defined as compounds that permit the antitumor drug to enter the cell by occupying the protein active or allosteric site(s), or by altering the physicochemical properties of the biomembranes. Acridone derivatives used as MDR inhibitors are depicted in Table 3.2.

In an attempt to find an effective MDR modulator, Dodic et al. synthesized a novel series of acridone-4-carboxamides and evaluated them to reverse MDR in the CH^{R}C/5 cell line [132]. Among all the derivatives, GF120918 (elacridar, **56**) was selected on the basis of activity, toxicity, and pharmacokinetics as it binds with P-gp by inhibition of [^{3}H] azidopine photoaffinity cross-linking. In phase I trial topotecan coadministration with elacridar increased the bioavailability of topotecan and were was used for the treatment of small cell lung cancer and metastatic carcinoma of the ovary [133]. Acridone derivatives have been prepared and evaluated as breast cancer resistance protein (ABCG2) inhibitors. Compound **57** was even more potent than the reference inhibitor, GF120918, as shown by its ability to inhibit mitoxantrone efflux [134]. Bis-acridone derivatives, in which two acridone moieties were connected by propyl or butyl spacer, were evaluated for their P-gp modulation efficacy [135]. Propyl derivative, 1, 3-bis (9-oxoacridin)-propane (PBA, **58**), was found to potentiate the vinblastine (VLB) cytotoxicity at 1 μM concentration in the MDR KB8-5 cell line. Continuous exposure of PBA at a concentration of 3 μM completely reversed the VLB resistance in this cell line.

2-methoxy-N10-substituted acridones have been prepared and evaluated to modulate the cytotoxicity of VLB in MDR cells (KBCh^R-8-5). Compound **59** at 0.50 μM (IC$_{50}$ = 2.0 nM) caused complete reversal of VLB in KBCh^R-8-5 [136]. 4-methoxy-N10-substituted acridone derivatives were synthesized and evaluated as potential MDR reversing agents. Compound **60** displayed good activity in MDR cell lines such as KBCh^R-8-5 cells (IC$_{50}$ = 6.0 μM) and GC3/c1cell (IC$_{50}$ = 7.53 μM) [31]. 2-methylacridone derivatives were screened for anticancer and reversal of P-gp mediated MDR in breast cancer cell lines (MCF-7 and MCF-7/Adr). Compound **61** displayed good activity in MCF-7 (IC$_{50}$ = 6.10 μM) with completely reversed Dox resistance in MCF-7/Adr cell lines (IC$_{50}$ = 6.20 nM) [137]. Furthermore, substitution at the C-4 position of acridone moiety with —Cl and —F groups led to the development of novel 4-chloro and 4-fluoro N10-substituted acridone derivatives that have been found to increase in vitro cytotoxicity in drug resistant HL-60/VINC and HL-60/DOX cancer cell lines [138,139]. Maximum anti-MDR activity was reported for compounds **62** and **63** with butyl side chain and a hydroxyethyl piperazino side chain at the terminal end.

TABLE 3.2 Summary of Acridones as MDR Inhibitors

Compound Code	Structure	Cell Line	Result	References
56	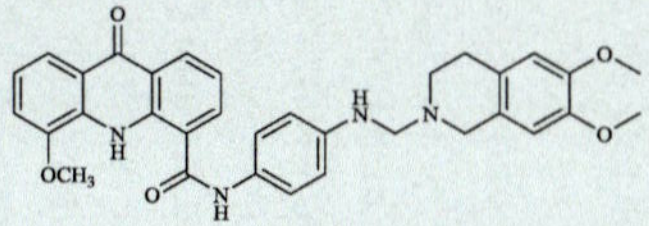	$CH^RC/5$	$EIC_{50}=21\,nM$	[132,133]
57	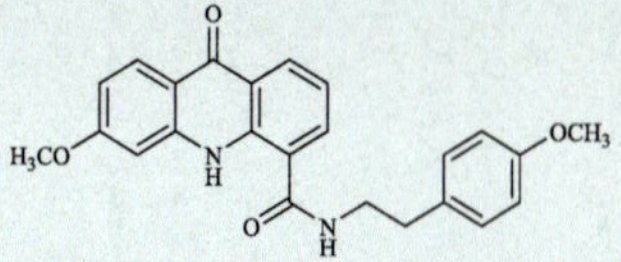	R482	$IC_{50}=0.56\,\mu M$	[134]
58	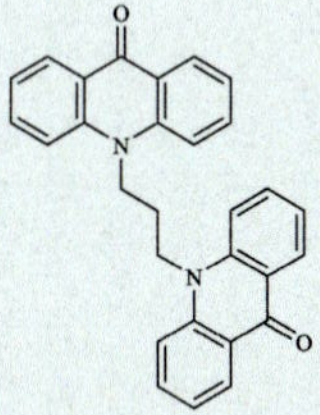	KB8-5	At 1.0 μM conc. potentiate the effect of VLB 50.9%	[135]
59	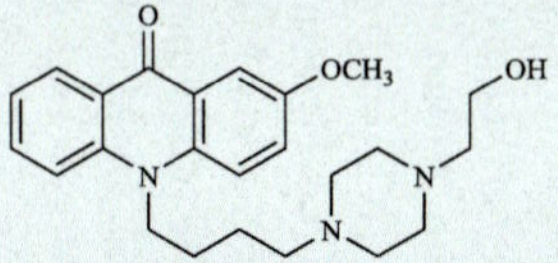	$KBCh^R$-8-5 $KBCh^R$-8-5	$IC_{50}=4.0\,\mu M$ At 0.50 μM conc. ($IC_{50}=2.0\,nM$) complete reversal of VLB resistance	[136]
60	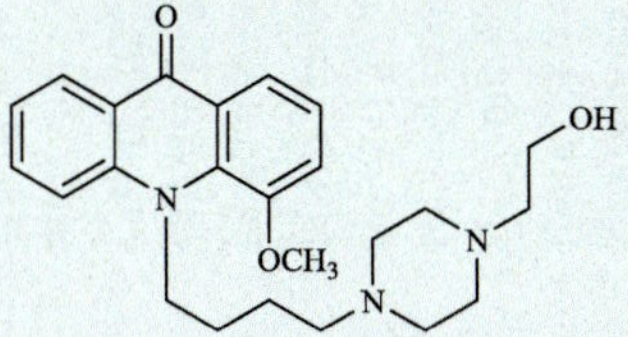	$KBCh^R$-8-5 $KBCh^R$-8-5 GC3/c1 GC3/c1	$IC_{50}=6.0\,\mu M$ At 0.55 μM conc. ($IC_{50}=1.9\,nM$) complete reversal of VLB resistance $IC_{50}=7.53\,\mu M$ At 1.05 μM conc. ($IC_{50}=0.7\,nM$) 8.7-fold potentiating of VLB resistance	[31]

No.	Structure	Cell line / assay	Activity	Ref.
61		MCF-7 MCF-7/Adr	$IC_{50} = 6.10\,\mu M$ $IC_{50} = 6.20\,nM$ complete reversal of Dox resistance	[137]
62		HL-60/VINC HL-60/DOX	497 nM 1074 nM	[138]
63		HL-60/VINC HL-60/DOX PDE1c inhibition	$IC_{50} = 0.9\,\mu M$ $IC_{50} = 3.8\,\mu M$ $IC_{50} = 1.2\,\mu M$	[139]
64		HL-60/DX PDE1c inhibition	$IC_{50} = 5.56\,\mu M$ $IC_{50} = 1.9\,\mu M$	[140]

Continued

TABLE 3.2 Summary of Acridones as MDR Inhibitors—cont'd

Compound Code	Structure	Cell Line	Result	References
65		SW 1573 2R 160 2008MRP1 2008MRP2 2008MRP3 HEK 293 MRP4 HEK 293 MRP5i MCF-7/MR	$IC_{50}=21.3\,\mu M$ $IC_{50}=22.0\,\mu M$ $IC_{50}=21.5\,\mu M$ $IC_{50}=25.5\,\mu M$ $IC_{50}=9.3\,\mu M$ $IC_{50}=9.3\,\mu M$ $IC_{50}=16.0\,\mu M$	[141]
66		DC-3F/ADX	Increase basal activity 33% at 0.05 µM	[142]
67		DC-3F/ADX	Increase basal activity 33% at 0.05 µM	[143]

No.	Structure	Cell line	Activity	Ref.
68		A2780Adr	$IC_{50}=5.70\,\mu M$	[144,145]
69		MCF-7 HeLa A-549	$IC_{50}=6.07\,\mu M$ $IC_{50}=8.80\,\mu M$ $IC_{50}=7.39\,\mu M$	[146]
70		HeLa A-549 MCF-7	$IC_{50}=39.5\,\mu M$ $IC_{50}=43.1\,\mu M$ $IC_{50}=57.42\,\mu M$	[147]
71		KBCh[R]-8-5	Increased uptake of VLB RF=1.25- to 1.9-fold	[148]

Continued

TABLE 3.2 Summary of Acridones as MDR Inhibitors—cont'd

Compound Code	Structure	Cell Line	Result	References
72	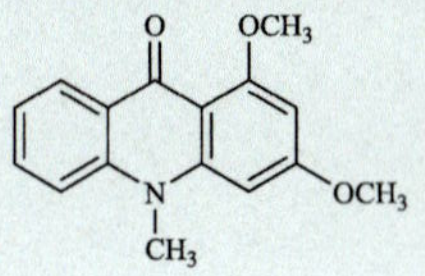	CCRF-CEM CEM/ADR 5000 MDA-MB-231-pcDNA MDA-MB-231-BCRP	$IC_{50}=0.20\,\mu M$ $IC_{50}=195.12\,\mu M$ $IC_{50}=10.30\,\mu M$ $IC_{50}=3.38\,\mu M$	[149]
73	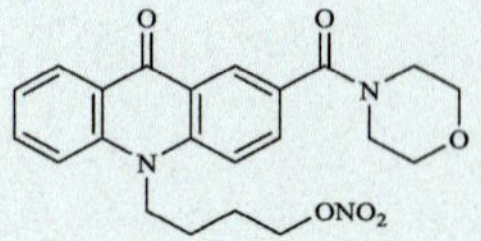	MCF7/Wt MCF7/Dx MCF7/Mr	$IC_{50}=0.8\,\mu M$ $IC_{50}=1.9\,\mu M$ $IC_{50}=2.7\,\mu M$	[150]
74	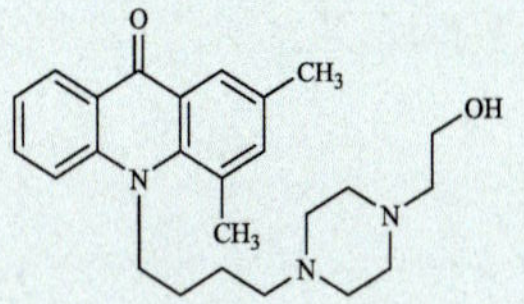	MCF7 MCF7/ADR MCF7/ADR	$IC_{50}=2.0\,\mu M$ $IC_{50}=2.56\,\mu M$ $(IC_{50}=1.25\,nM)$ complete reversal of VLB resistance	[151]
75	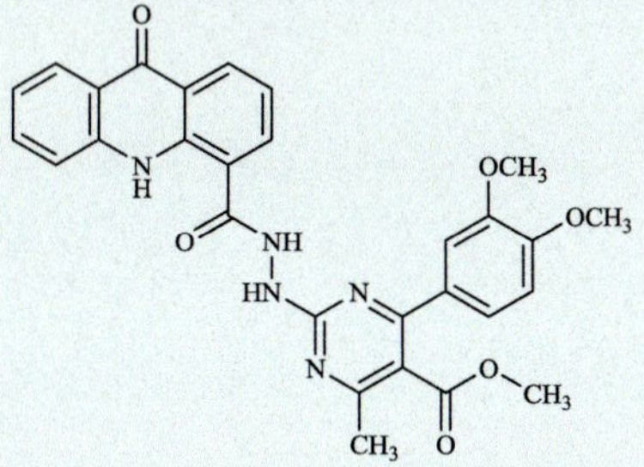	SW 1573 WT SW 1573 2R 160	$IC_{50}=4.0\,\mu M$ $IC_{50}=12.0\,\mu M$	[152]

The results of mechanistic studies suggest that acridone derivatives interact with DNA duplex by intercalation, possess higher affinity to GC than AT base pairs of the DNA, and could not interact noncovalently with the minor grooves of the DNA in solution free gas phase [138,139]. Calmodulin inhibitors have proved to play a significant role in sensitizing MDR cancer cells by interfering with cellular drug accumulation. Prasad et al. investigated the in vitro inhibitory efficacy of chloro acridones against calmodulin-dependent cAMP phosphodi-esterase (PDE1c) [140]. Moreover, molecular docking of acridones was performed with PDE1c in order to identify the possible protein ligand interactions. Results thus obtained were compared with in vitro data. Compound **64** displayed good activity against PDE1c (IC$_{50}$ = 1.9 μM). Docking studies indicated that nitrogen-containing substitution at N-10 position plays a vital role in the hydrophilic interactions. Other groups like carbonyl (C=O) at the C-9 position of acridone moiety showed a significant role in ligand fitness at the binding site of the enzyme. A result of ligand-based 3D-QSAR provides possible structural modifications for the strategic design of more potent chemosensitizing agents in cancer [140]. 2-fluoro N10-substituted acridone derivatives with varying alkyl side chain length, with propyl, butyl, and tertiary amine groups at the terminal end of alkyl side chain were synthesized and screened against different cell lines. These compounds displayed good cytotoxicity potency against 60 human cancer cell line panels of the National Cancer Institute, USA.

During the evaluation of compounds for DNA-binding properties, compound **65** with planar tricyclic ring linked with a butyl β–hydroxyethylpiperazino side chain has shown the highest binding affinity with a binding constant (K$_i$) of 8.04 × 10M^{-1} [141]. N10-acridone derivatives were synthesized and evaluated for their ability to interact with P-gp and Mg^{2+}. Compound **66** bearing a piperidine ring at N10 chain enhanced the basal activity of P-gp by 33% at 0.05 μM concentration. Similarly, it was found to exhibit its strongest interaction with Mg^{2+} having an association constant (K$_a$ value) 1.08 × 10^5 M^{-1}. Anti-MDR activity of compound **66** was attributed to its interaction with P-gp along with Mg^{2+} sequestering behavior of this compound [142].

To search for better MDR modulators acridones interacted with three components of efflux pump, viz. P-gp, ATP, and Mg^{2+}. Biological activity results shows that the compounds bearing COOH group at C-4 position interact with P-gp and Mg^{2+} while compounds bearing Cl at C-4 position interact with ATP and Mg^{2+}. This compound **67** enhanced the basal activity of P-gp by 33% at 0.05 μM concentration [143]. Aryl propyl acridone-4-carboxamide derivatives were developed by pharmacophore drug designed and evaluated for MDR inhibitor in doxorubicin overexpressed A2780Adr cell line by using the standard Hoechst 33342 assay method. Compound **68** exhibited better activity (IC$_{50}$ = 5.70 μM) as compared to the standard modulator verapamil [144,145]. 2-(9-oxoacridin-10(9H)-yl)-*N*-phenyl acetamides has been prepared and evaluated for anticancer activity against three human cancer cell lines such as MCF-7, HeLa, and A-549. Compound **69** showed good inhibition ability against MCF-7

($IC_{50} = 6.07\,\mu M$), HeLa ($IC_{50} = 8.80\,\mu M$), and A-549 ($IC_{50} = 7.39\,\mu M$). The docking simulations revealed better interactions within the ATP binding pocket as compared to the transmembrane binding pocket. The carbonyl group at C-9 position of acridone ring displayed an essential hydrogen bonding interaction with hydroxyl group of Tyr1393. The benzyl ring substituted at the 10th position of the acridone moiety display π–π stacking interaction with Tyr2352. Additionally, the carboxylate group substituted at the benzene ring displayed Mg^{2+} coordination [146]. N10-substituted acridone derivatives were synthesized and screened for cytotoxic activity. Compound **70** showed significant activity against HeLa ($IC_{50} = 39.5\,\mu M$), A-549 ($IC_{50} = 43.1\,\mu M$), and MCF-7 ($IC_{50} = 57.42\,\mu M$) cancer cell lines. The analogues containing higher alkyl, unsaturated alkyl, ester, and oxime groups were not active.

Computational study of compound **70** showed hydrogen bonding between Tyr1393 and a π–π stacking interaction with Tyr2352. Further nitro group showed weak noncovalent interactions with positively charged magnesium ion (Mg^{2+}) [147]. Thimmaiah and coworkers designed acridone-based derivative as a DNA drug intercalation agent. DNA intercalations are important for antitumor activity and inhibit the DNA replication and transcription process responsible for rapidly growing cancer cells. The SAR studies showed that butyl side chain acridone derivative has good affinity to DNA binding. The molecular modeling and docking studies of these compound has identified compound **71** as good affinity ($K = 0.39768 \times 10^5$) with Calf Thymus DNA (CT DNA). All the synthesized compounds have increased the uptake of VLB in MDR KBChR-8-5 cells to an extent of 1.25- to1.9-fold than standard modulator verapamil of similar concentration [148].

Kuete et al. investigated the cytotoxicity of furoquinoline and acridone alkaloids toward multifactorial drug resistant cancer cells. Compound **72** displayed good activity against CCRF-CEM ($IC_{50} = 0.20\,\mu M$) and CEM/ADR5000 ($IC_{50} = 195.12\,\mu M$) cell lines. Biological screening of compound **72** revealed that the mode of action is induced by apoptosis up to a 48.8% increase in reactive oxygen species level.

SAR studies of acridone alkaloids concluded that the presence of N- methyl and two methoxy groups substituted in C-1 and C-3 was more active as compared to two methoxy groups substituted in C-3 and C-10 position. A compound with three cycle rings was more active than the other acridone alkaloids bearing four cycle rings [149]. Nitric oxide donating acridone derivatives were synthesized and evaluated for cytotoxic activity against different sensitive and resistant cancer cell lines, MCF7/Wt, MCF7/Mr (BCRP overexpression) and MCF7/Dx (P-gp expression). Compound **73** fused with NO donor revealed the most potent cytotoxic activity against MCF7/Wt, MCF7/Dx, and MCF7/Mr cells with IC_{50} values varying from 0.8 to 2.7 μM, respectively. Moreover, this compound also showed good inhibition for the colorectal cancer cell line (SW1398, WiDr, and LS174T) with IC_{50} value 1.7–2.8 μM. Exogenous release of nitric oxide by NO donating acridones enhanced the accumulation of doxorubicin in

MCF7/Dx cell lines when it was coadministered with doxorubicin, which inhibited the efflux process of doxorubicin [150]. Murahari and coworkers designed and synthesized acridone derivatives as anticancer agents against breast cancer sensitive MCF7 and resistant MCF7/ADR cell line. Compound **74** showed potent cytotoxicity activity against MCF7 ($IC_{50}=2.0\,\mu M$) and MCF7/ADR ($IC_{50}=2.56\,\mu M$). The SAR studies revealed that compound **74** containing butyl side chains with terminal end of β-hydroxyethyl piperazine moiety was better for anticancer activity than propyl side chains. The binding interactions of compound **74** with P-gp (PDB ID: 1MV5) shows that a hydroxyl group of secondary amine forms a hydrogen bond with Ser355, nitrogen of secondary amine with Asp353, and oxygen of acridone ring with Thr384 [151]. Acridone substituted with pyrimidine hybrids were synthesized and evaluated against cancer sensitive and resistant cell lines. Compound **75** displayed good activity against SW 1573 WT cells ($IC_{50}=4.0\,\mu M$) and SW 1573 2R 160 P-gp cells ($IC_{50}=12.0\,\mu M$). Further studies found that hybrid compounds have the ability to inhibit the Akt kinase activity and induce apoptosis [152].

4.3 Antiviral Activity

The antiviral properties of natural and synthetic acridone show the inhibitory action against diverse viruses both with double-stranded DNA as well as RNA genomes. The degree of inhibitory activity varies with the compound and the virus. The initial and most extensive studies of both natural and synthetic analogues regarding antiviral activity is against herpes simplex virus (HSV), human cytomegalovirus (HCMV), human immunodeficiency virus (HIV), bovine viral diarrhea virus (BVDV), hepatitis C virus (HCV), dengue virus (DENV), and hemorrhagic fever virus (HFV) [153].

Citrusinine-1 (**76**) isolated from the citrus plant (*Rutaceae*) exhibited the potent antiviral activity against HSV-1 ($ED_{50}=0.56\,\mu g/mL$), HSV-2 ($ED_{50}=0.74\,\mu g/mL$), and HCMV ($EC_{50}=1.5\,\mu g/mL$). Citrusinine-1 combined with acyclovir or ganciclovir, synergistically potentiated the antiherpetic activity of these drugs. SAR study of citrusinine-1 and other closely related compounds revealed that the presence of N-alkyl and hydroxyl group either at the C-5 or C-6 position of the acridone moiety is crucial for antiherpetic activity. The presence of the *O*-alkyl moiety at C-3 (CH_3 or pyran ring) instead of an OH group increased the antiviral activity [154] (Fig. 3.8).

Glycofolinine (**77**) isolated from *Glycosmis parva* showed moderate inhibitor activities against both HSV-1 and HSV-2 ($EC_{50}=151\,\mu M$). They observed that hydroxyl groups at the C-5 or C-6 position of acridone could act as additional potential for anti-HSV activity [155]. Interestingly, a different series of acridone alkaloids isolated from citrus plants exhibited a marked inhibitory effect on activation of Epstein–Barr virus leading to tumor-promoting activity. The compounds **78** (5-hydroxynoracronycine) and **79** (acrimarine-F) may be found valuable for antitumor promoter activity [156] (Fig. 3.8). Acridone

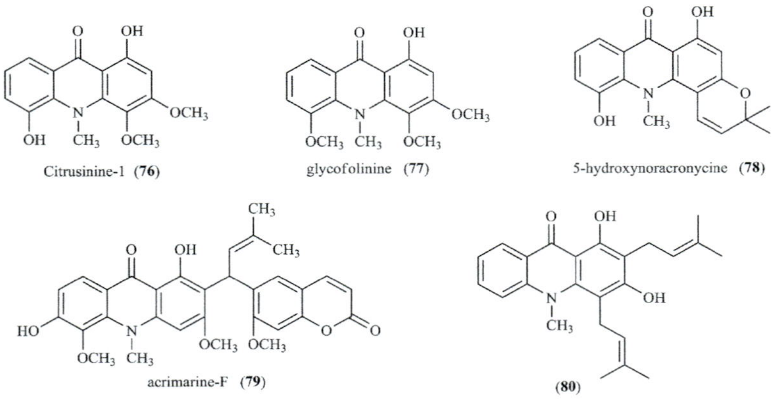

Citrusinine-1 (**76**)

glycofolinine (**77**)

5-hydroxynoracronycine (**78**)

acrimarine-F (**79**)

(**80**)

FIGURE 3.8 Natural acridones as antiviral agents.

alkaloids obtained from the *Rutaceous* plants were tested for their inhibitory effects on 12-O-tetradecanoylphorbol-13-acetate (TPA)-induced Epstein–Barr virus early antigen (EBV-EA) activation. A prenylated acridone 1,3-dihydroxy-10-methyl-2,4-diprenylacridone (**80**) exhibited remarkable inhibitory effect on mouse skin tumor promotion in an in vivo two-stage carcinogenesis test. They were found to be valuable cancer chemopreventive agents [157].

1,3-hydroxyacridone derivatives were synthesized and evaluated as inhibitors of HSV [158]. Halogen-substituted 7-chloro-1,3-dihydroxyacridone (**81**) and its 7-hydroxy derivative 1,3,7-trihydroxyacridone (**82**) showed good antiviral activity. Moreover, both had been shown to inhibit topoisomerase II catalytic activity in vitro and cell-based (SV40-infection) assays, yet neither agent affected the activity of topoisomerase I [159]. Interestingly, it has been found that shifting the chloro group from position C-7 to position C-5 gave compound **83** less DNA topoisomerase II inhibitory activity but more selective antiviral activity. Initially its antiviral action was believed to occur when HSV replication was blocked at a stage after DNA and late protein synthesis. Furthermore, studies have revealed that compound **83** inhibits one or more steps of HSV assembly since treatment results in reduced levels of capsids (particularly B-type) and reduced levels of encapsidated DNA [160]. Using compound **83** as a lead, a series of new compounds were synthesized. Compound **84** (5-methoxy-1,3-dihydroxyacridone), which inhibits replication of HSV-1 ($ED_{50}=2.2\,\mu M$) and HSV-2 ($ED_{50}=0.6\,\mu M$), showed better activity profile than lead compound **83** [161] (Fig. 3.9). SAR studies on substituted 1,3-hydroxyacridone with varied substituents at C-1 and N-10 position of acridone ring suggest that 1-hydroxyl and secondary amine is an important feature of pharmacophore. Replacement of the C-1 hydroxyl group with methoxy and methyl groups and methylation of secondary amine eliminated the antiviral activity. These two important requirements may correspond to hydrogen bond acceptor and donating sites at

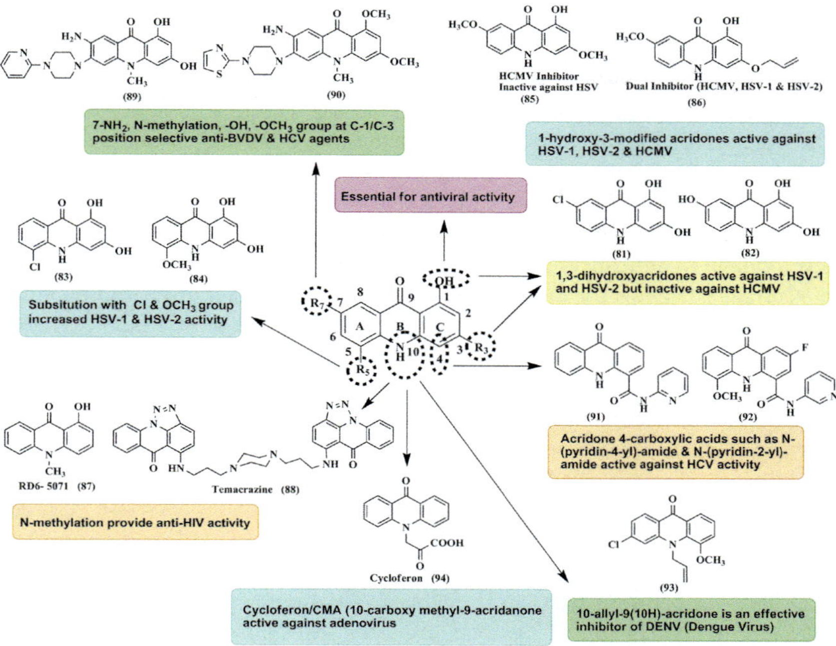

FIGURE 3.9 Structure–activity relationship of antiviral acridones.

a hypothetical binding domain. Substitution of a ring with electron releasing or electron withdrawing group at the C-5 position is preferred for anti-HSV-1 activity. Disubstituted chloro group at position C-5 and C-7 attenuates cell growth inhibition with no change in its anti-HSV activity. 1-hydroxyacridone scaffold generated the 3,7-dimethoxy-1-hydroxyacridone (**85**), which was inactive against HSV replication, yet it eliminated HCMV plaques produced in human embryonic lung fibroblasts with ED_{50} value of 1.4 µg/mL; greater than 35-fold selectivity [162]. Modification in compound **85** at position C-3 led to discovery of a series of compounds that act as dual inhibitors of HCMV and HSV. Compound **86** displays dual activity against HSV-1 ($ED_{50} = 2.2$ µM), HSV-2 ($ED_{50} = 3.9$ µM), and HCMV ($ED_{50} < 2.5$ µM). On the basis of this structure-activity information, it appears that methoxy substituent at C-7 position is important for anti-HCMV activity. For the HSV/HCMV inhibitors the difference between the active C-3 variable alkoxylated series and the inactive C-7 variable compounds could reflect the importance of the hydroxyl group at C-1 for target interaction. However, it is interesting that the C-1 hydroxyl is critical for the anti-HSV activity of the 1,3-dihydroxyacridone class and possibly it is important for several other synthetic and natural antiviral acridones [153,162] (Fig. 3.9). Goodell et al. synthesized the new antiviral compounds such as acridones, xanthones, and acridines, gaining interest for treating herpes virus infections. These compounds, which showed the activity against herpes simplex-1 and/or herpes simplex-2, were further assayed for inhibition of topoisomerase activity to gain insight into the mechanism of action. The results indicate that the acridine analogs bearing substituted carboxamides and bulky 9-amino functionalities are able to inhibit herpes infections as well as inhibit topoisomerase II relaxation of supercoiled DNA. This not only suggests a unique mechanism of action in treating herpes virus infections, but also may be of great interest in the development of anticancer agents that target topoisomerase II activity [163].

Current strategies for the therapeutic treatment of AIDS are based upon the use of combination therapy with one or more inhibitors of the HIV type 1 (HIV-1) reverse transcriptase and protease enzymes. Various acridone derivatives have been synthesized as selective inhibitors of HIV-1 transcription, among which compound **87** (RD6-5071, 1-hydroxy-10-methyl-9,10 dihydroacrid-9-ones) was found to be most potent HIV inhibitor [164]. This compound suppressed tumor necrosis factor (TNF-α), induced HIV-1 expression as well as phorbol 12-myristate 13-acetate (PMA), induced HIV-expression in OM-10.1, U1 and ACH-2 cells. Its 50% effective concentration for HIV-1 p24 antigen production was 2.0 µg/mL in OM-10.1 cells, while its 50% cytotoxic concentration was 18 µg/mL. Furthermore, RD6-5071 inhibited HIV-1 replication in acutely infected U937 and peripheral blood mononuclear cells. An inhibition assay for protein kinase C (PKC) revealed that RD6-5071 could reduce the enzyme activity. These results suggest that the acridone derivatives suppress HIV-1 replication at the transcriptional level primarily through a mechanism of PKC inhibition [165].

An analogue in the bistriazolozoacridones series, temacrazine (NSC 687025, **88**), was found to inhibit HIV-1 replication in cells when used in nanomolar concentrations. Further, it inhibits the transcriptional activation of the HIV-1 LTR by TNF-α inducible cellular transcriptional elements such as NFκB. As an antiviral it was a potent inhibitor of virus production from TNF-α induced latently infected U1 or ACH-2 cell lines or chronically infected H9 cells. Temacrazine-mediated inhibition was found to be independent of transcriptional pathways controlled by TNF-α, Tat-TAR, or Rev-RRE interactions [166].

Some newly synthesized heterocyclic compounds such as pyrimidobenzimidazole, thiazolinoacridine, and acridone for HIV agents. All the compounds synthesized were screened for in vitro inhibition of HIV-1 replication in MT-4 blood lymphocytes. They were found to be inactive at concentrations below 100 µM as compared to the standard drug AZT, which has an IC_{50} value for anti-HIV activity around 0.1 µM. These compounds were also toxic at 100 µM concentrations against uninfected MT-4 cells [167].

BVDV is another pathogen causing a range of clinical manifestations including abortion, early embryonic death, respiratory problems, and immune system disorder that result in mortality rates of 17%–32% [168]. Tabarrani et al. initiated the development of a series of acridone derivatives as potential inhibitors of BVDV through elaboration of nitroacridone [169]. Such series of compounds showed significant antiviral activity, but compound **89** elicited superior anti-BVDV activity with EC_{50} value (0.5 μg/mL), $CC_{50} > 100$ μg/mL, and therapeutic index ($CC_{50}/EC_{50} > 200$ μg/mL). SAR around acridone derivatives suggest that substitution with —NH_2 group at 6-position resulted in compound **89** having higher anti-BVDV activity than the —NO_2 group. Replacement of hydroxyl group at C-1 position with methoxy group abolishes the activity indicating that presence of a hydroxyl group at C-1 position is required for antiviral activity, which is in good agreement with previously reported anti-HSV acridones [170].

About 180 million people worldwide affected with HCV are at high risk for developing chronic hepatitis, liver cirrhosis, and hepatocellular carcinoma [171]. Besides anti-BVDV activity, compound **89** exhibited good anti-HCV activity ($EC_{50} = 1.7$ μg/mL) [169]. 1,3-dimethoxy acridone analogues of compound **89** having no activity against BVDV proved to be equally potent in replication assay with more selectivity and no toxicity, which made it a valid hit compound. Thus structural modifications made on the hit compound obtained a series of new analogues. Compound **90** inhibited HCV NS3 helicase activity with IC_{50} of 3.0 μg/mL a very promising target for the development of anti-HCV drugs. They observed that an amino group when replaced by a fluorine atom at C-6 position resulted in loss of activity while its N,N-alkylation greatly increased the cytostatic effect [172]. Acridone-4-carboxylic acid derivatives were tested using the direct fluorometric helicase activity assay to determine the inhibitory properties of the acridone derivatives toward the NS3 helicase of HCV.

Intercalation of the acridone derivatives into dsRNA may play an important role in inhibition of HCV replication in the replicon system. Pyridylamide compound **91** is efficient HCV replication inhibitors with an IC_{50} value of 3.8 μM (SI = 40.5). Therefore in the replicon system they can act via inhibition of both NS3 and NS5B activities. They even determined that the presence of methyl substituent in the pyridine ring slightly decreased the activity. The addition of an alkyl group that abolishes its inhibitory activity suggested that pyridine bearing amide fragments with no spacer is essential for antihelicase activity [173]. Some acridone derivatives designed and evaluated the inhibitory activity toward NS3 helicase of HCV. Compound **92** constitutes the best inhibitor of HCV replication among all the helicase inhibitors with a TI > 1000 and $EC_{50} < 1$ μM. Although the mechanism of action of molecule **92** in the replicon system remains unclear, it may involve inhibition of various viral and host factors. SAR study of acridone derivatives shows that the presence of the pyridyl ring and the position of the nitrogen atom in the ring determined the strength of the antihelicase activity. On the other hand, addition of a 3,4,5-trimethoxyphenyl group instead of the pyridyl ring led to the weakest inhibitor. Moreover, the substitution of halogens (fluoride) at the C-2 position produced more potent helicase inhibitors (**92**), while the substitution by a chloride atom decreased the anti-HCV activity of these derivatives. This molecule **92** is an excellent lead for further development of anti-HCV drugs (Fig. 3.9) [174].

HFVs belonging to family arenaviridae, bunyaviridae, filoviridae, and flaviviridae can cause severe forms of hemorrhagic disease associated with extremely high morbidity and health problems [175]. Several N-substituted acridones have been prepared and screened for their antiviral activity against JUNV (strain IV4454) and DENV-2 (strain NGC) [176]. They found that 6-Cl substituted *N*-allyl acridone derivative **93** possess potent and selective antiviral activity against JUNV ($EC_{50} = 7.3\,\mu M$) and DENV-2 ($EC_{50} = 7.4\,\mu M$), respectively. Moreover, compound **93** was found to be effective against a wide spectrum of arenaviruses like TCRV ($EC_{50} = 13.4\,\mu M$) and LCMV ($EC_{50} = 7.3\,\mu M$) and flaviviruses like DENV-1 to DENV-4 with an EC_{50} value between 3.1 and 19.9 μM. Mechanistic studies demonstrated that compound **93** did not affect the initial steps of viral adsorption and internalization but the subsequent process of virus RNA synthesis appeared to be the main target of antiviral action. It causes about a 2.5–3 log reduction in the amount of infective plaque forming unit and viral RNA at 48 h postinfection in compound-treated cells relative to nontreated cells [177]. They concluded that N-substitution with carbohydrate groups by cycloaddition generated very cytotoxic derivatives, whereas N-allyl derivatives were so far the more selective group of virus inhibitors. The presence of halogen substituents in *N*-allyl acridones, in general, reduced toxicity and increased antiviral effect. In particular, the 6-chloro substitution seemed to be an important element, whereas the halogen substitution in another position of ring A leads to a minor improvement of antiviral activity (Fig. 3.9).

Adenovirus represents a broad range of human pathogens in numerous diseases like respiratory infections and keratoconjuctivitis, and in posttransplantational infections like hemorrhagic cystitis, enteritis, and hepatitis. Cycloferon/CMA (10-carboxy methyl-9-acridanone, **94**) non-IFN-mediated antiviral agent was evaluated as a potent interferon inducer against adenovirus type-6 (Ad 6) in hep-2 cells. They concluded that CMA does not inhibit Ad 6 hexon protein synthesis but it strongly reduces the yield of infectious virus. Additionally, it suppresses viruses associated with RNA type 1 (VAI) induced repression of IFN synthesis involved in adenoviral infection [178].

4.4 Antimalarial Activity

Malaria infects half a billion, causes serious disease in 100–200 million, and kills roughly 2 million people every year. Its etiological agents are protozoa of the genus *Plasmodium falciparum* and *Plasmodium vivax*. Death and morbidity caused by malaria are on the increase, largely as a result of parasite drug resistance [179]. Several acridine derivatives have demonstrated promising antimalarial activity. Mepacrine/quinacrine (QA), was the first synthetic antimalarial blood schizontocide (Fig. 3.10) used clinically [180]. This 9-aminoacridine was discovered in 1933 following an intensive search for synthetic quinine substitutes and was used when quinine became practically unavailable during World War II. Another acridine-based antimalarial drug, pyronaridine, has been used

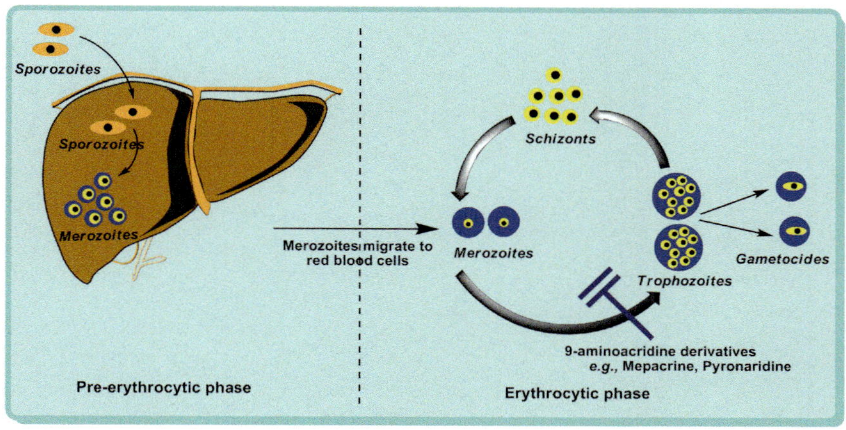

FIGURE 3.10 Site of action of acridine derivatives in life cycle of malarial parasite.

for nearly 20 years as a monotherapy to treat malaria in China. It is being used as fixed-dose combination with artesunate (PYRAMAX) for uncomplicated *P. falciparum* and *P. vivax* malaria as a once-a-day for 3 days treatment [181].

Acronycine (**1**) was first isolated from the Australian scrub ash *A. baueri* and it showed antineoplastic [182] and antiparasitic activity [183]. Furthermore, many derivatives of this alkaloid were evaluated; only 2-nitroacronycine (**95**) (Fig. 3.11) improved the antimalarial activity, reaching IC_{50} values around 2 µg/ mL for susceptible and resistant strains [184]. Atalaphyllinine (**10**) inhibited completely the development of malaria parasites in vivo after prophylactic administration in a daily dose of 50 mg/kg for 3 days with no obvious toxic effect [185]. Acridone alkaloids obtained from *Teclea trichocarpa* and evaluated for antimalarial activity in *P. falciparum* chloroquinine sensitive (HB3) and chloroquinine resistance (K1) strains. Arborinine (**96**) has shown potent activity against the sensitive HB3 cell line ($IC_{50} = 3.85$ µM) and resistance cell line ($IC_{50} = 9.34$ µM). It was observed that the presence of 1-hydroxyl group is essential for activity [186].

In addition, acridone-based derivatives including dihydroacridinediones, haloalcoxyacridinones, and N10-substituted acridones have shown good antimalarial activity. Floxacrine (**97**), a dihydroacridinediones, when originally synthesized was used in of the search for novel and effective antibacterial drugs [187,188]. A QA-floxacrine hybrid, WR 243251 (**98**) was designed and synthesized as a potential antimalarial agent in order to decrease the vascular cardiotoxicity of floxacrine. WR 243251 (**98**) was a potent inhibitor of hematin polymerization ($IC_{50} = 90$ µM, chloroquine (CQ) $IC_{50} = 80$ µM), however, this inhibition did not completely account for the increased antimalarial potency versus CQ for resistant strain K1 ($IC_{50} = 9.5$ nM, CQ $IC_{50} = 450$ nM) [189]. Floxacrine (**97**) was two-fold less potent than CQ for

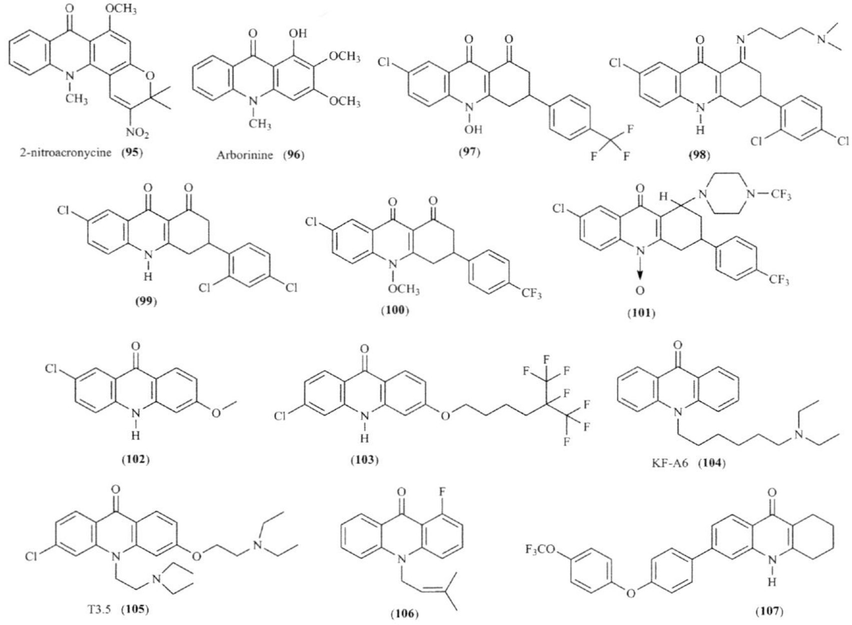

FIGURE 3.11 Antimalarial activity of acridone derivatives.

hematin polymerization inhibition indicating that it exerts antimalarial activity by other mechanisms [189]. In 2008, WR 249685 (**99**) showed stronger antimalarial activity ($IC_{50} \sim 15\,nM$) than floxacrine. It has been investigated that compound **97** kills parasites via a heme-mediated process, and compound **99** acted as a highly selective inhibitor of the quinol oxidation (Qo) of the *P. falciparum* bc1 complex [190]. Docking of WR 249685 (**99**) with *P. falciparum* bc1 model was energetically favorable; the model demonstrated selectivity in the docking of traditional Qo and quinone inhibitors. The most striking feature of the model for dihydroacridine-dione was the putative association between the inhibitor and the E-ef linker region of the cytochrome and later is a region of low sequence identity between *P. falciparum* and mammalian cytochrome b [190]. Deoxyfloxacrine and floxacrine derivatives were evaluated for blood schizontocidal activities against drug-resistant lines of *Plasmodium berghei* and different chloroquine-resistant strains of *P. falciparum*. All the tested compounds showed high activity against drug-resistant lines of *P. berghei* and were superior in their antimalarial potency to floxacrine. Compounds **100** and **101,** however, proved to be highly active against *P. falciparum* strain in vitro with IC_{50} value 0.73–1.78 nM [191]. Haloalkoxyacridones as therapeutic agents are the compounds having potential to treat or prevent malaria in humans and were discovered using an empirical design strategy [192]. First, the authors obtained 2-methoxy-6-chloroacridinone (**102**) by mild hydrolysis of quinacrine. The compound **102** showed strong in vitro antimalarial activity with IC_{50} value of 45 nM for susceptible strain (D6) and 65 nM for multidrug resistant strain (Dd2), with higher selective index. Thirty acridone derivatives were synthesized using compound **102** as a lead to develop more SAR. Among them, 11 acridinones showed IC_{50} values lower than CQ against susceptible (D6) and resistant (Dd2) *P. falciparum* strains. The most potent compounds share a unique and defining structural element in the C ring composed of an extended alkyl group terminated by one or more trifluoromethyl groups at C-3 position of the tricyclic acridone. In this series, compound **103** showed an IC_{50} value of 0.0015 μM against the two strains and an in vitro therapeutic index higher than 1×10^5. Based on a structural similarity to aforementioned antimalarial agents it is proposed that the haloalkoxyacridones exert their antimalarial effects through inhibition of the *Plasmodium* cytochrome bc_1 complex (Fig. 3.11) [192].

N10-substituted acridones bearing alkyl side chains with tertiary amine groups at the terminal position were synthesized. The most potent CQ-chemosensitizer compound **KF-A6 (104)** potentiated the antimalarial activity of CQ 80-fold more than in the Dd2 strain (IC_{50} = 19 nM) similar to that observed for the CQ sensitive strain D6 (13.1 nM) [193]. Later, the same authors reported the discovery of the dual function acridinones, which included both haem targeting structure as well as a chemosensitization moiety at the N10 position. The most potent compound **T3.5 (105)** proved synergistic with chloroquine, amodiaquine, quinine, or piperaquine, but not with mefloquine against the Dd2 strain. It demonstrated in vivo curative action of 256 mg/kg/day orally or 200 mg/kg/day intraperitoneally using murine malaria parasites such as *Plasmodium yoelii* and *P. berghei*. No overt toxicity or behavior change was observed in animals [194]. Furthermore, new derivatives of 10-allyl-, 10-(3-methyl-2-butenyl), and 10-(1,2-propadienyl)-9(10*H*)-acridinone have been evaluated for their antimalarial activity against *P. falciparum* [195]. Compound **106** can be classified as a hit for antimalarial drug development exhibiting IC_{50} less than 0.2 μg/mL (SI=>100). SAR showed that presence of fluorine at the C-1 position is an important structural feature for antimalarial activity. The IC_{50} value of potent antimalarial acridinones that include fluorine can be further reduced by increasing the number of fluorine atoms or the number of CF_3 substituents [192,195,196] (Fig. 3.11).

1,2,3,4-tetrahydroacridin-9(10H)-ones (THAs) were synthesized and evaluated for their antimalarial activity. Of all THA compounds, biaryl ether compound (**107**) with EC_{50} of 12.2 nM for W2 and 9.1 nM for TM90-C2B was the most potent compound, possessing an acceptable resistance index (RI=0.75). SAR study demonstrated that most of the THA compounds possess RI values in the acceptable range and thus lack, in contrast to atovaquone or chloroquine, any cross-resistance. The discovery that the 6- or 7-position of the THA scaffold tolerates aryl substituents provides opportunities for next-generation designs (Fig. 3.12) [197].

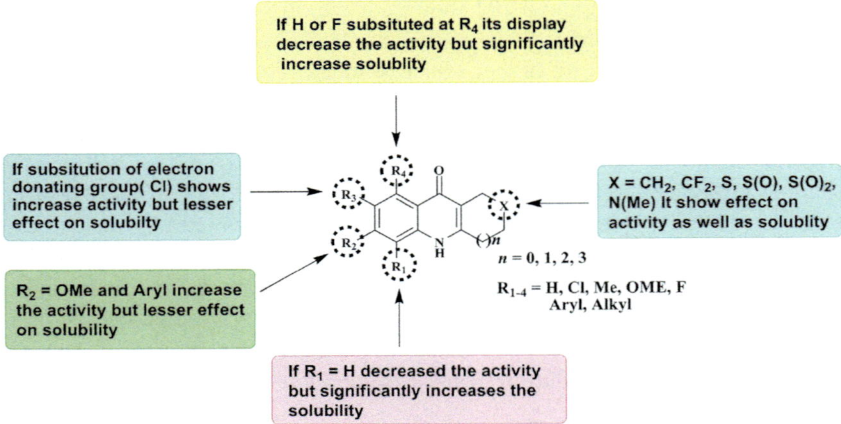

FIGURE 3.12 Structure–activity relationship of 1,2,3,4-tetrahydroacridin-9(10H)-ones (THAs) as antimalarial agent.

4.5 Antiparasitic Activity

The major antiprotozoal agents include antitrichromonarial, antitrypanosomals, and antileishmanial. Trypanosomes and leishmania are the causative agents of African sleeping sickness, Chagas disease, nagana cattle disease, kala-azar, and oriental sore. All these parasitic protozoa lack the nearly ubiquitous enzyme glutathione reductase (GR). Trypanothione reductase (TR) is a flavoenzyme that has been found so far exclusively in trypanosomatid parasites. TR is structurally and mechanistically similar to GR. The important difference between parasite and host enzyme is their mutual exclusive substrate specificity, which is based on the respective charge distributions of their active sites. The absence of the TR enzyme in the mammalian host renders it an attractive target molecule for the development of new antiparasitic drugs. The structures of TR in the free and complex form with its substrates as well as the reversible inhibitor mepacrine have been elucidated [198,199].

Acridone derivatives were isolated from the roots of *Thamnosoma rhodesica*. These compounds were examined then for antiparasitic activity against *Leishmania major* and *Cladsosporium curcumerinum*. They observed that rhodesiacridone (**108**) was showing good activity against *L. major amastigote* (IC$_{50}$ = 6.2 µM) and *L. major promastigote* (IC$_{50}$ = 30.7 µM) as compared to *C. curcumerinum* [200]. Eleven more acridone alkaloids were isolated from *S. glutinosa* (Bl.) Merr. and examined for in vitro activity against chloroquine-sensitive *P. falciparum* 3D7, *Trypanosoma brucei rhodesiense* STIB900, and *Leishmania donovani* L82. Nine of these compounds had IC$_{50}$ values ranging from 0.3 to 11.6 µM against *P. falciparum*. In contrast, a small number of compounds showed significant activity against *T. brucei rhodesiense* and none was active against *L. donovani*. 5-hydroxynoracronycine (**109**) showed

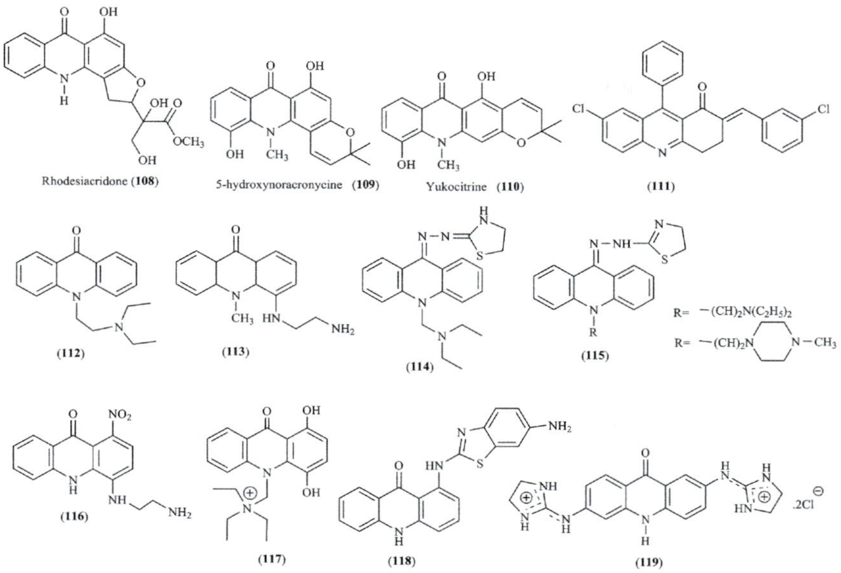

FIGURE 3.13 Antiparasitic activity of acridone derivatives.

potent activity against *P. falciparum* and *T. b. rhodesiense* with IC_{50} value (0.6–1.0 μM). They determined that the presence of a pyrano ring is essential for activity [201,202]. Few acridone alkaloids were isolated from *Citropsis articulata*. The EtOAc extract of its root bark showed in vitro antiplasmodial activity with a 70% growth inhibition of *P. falciparum* at a 10 μg/mL concentration. The antiparasitic activities of the compounds purified from above were evaluated for their inhibition against *P. falciparum*, *Trypanosoma brucei brucei* and *L. donovani*. The isolated compound **109** (5-hydroxynoracronycine) showed good activity in *P. falciparum* ($IC_{50}=0.9$ μM) and *L. donovani* ($EC_{50}=11.2$ μg/mL). All the isolated alkaloids were inactive against *T. brucei* [203]. Acridones alkaloid yukocitrine (**110**) were obtained from *Glycosmis trichanthera*. This compound **110** displayed good antileishmanial activities with EC_{50} of 29.76 μM after 24 h of exposure and EC_{50} of 0.88 μM after 48 h of exposure (Fig. 3.13) [204].

Mosquito-borne diseases represent the remarkable cause for morbidity and mortality in the developing countries. *Aedes aegypti* and *Culex quinquefasciatus* are the two major vectors that cause dreadful diseases such as dengue, yellow fever, and lymphatic flariasis. Acridone analogues were synthesized in which compound **111** displayed good larvicidal activity against *A. aegypti* ($LC_{50}=43.24$ ppm) and *C. quinquefasciatus* ($LC_{50}=59.12$ ppm) [205]. Acridone alkaloids isolated from the plant *Esenbeckia febrifuga* and its antiplasmodial activity were determined against both chloroquine-sensitive

(CQS) (3D7) and chloroquine-resistant (CQR) (W-2) strains of *P. falciparum*. *E. febrifuga* is a plant traditionally used to treat malaria in the Brazilian Amazon region and ethanol extract of stems displayed a good antiplasmodial activity against *P. falciparum* strains W-2 ($IC_{50} = 15.5\,\mu g/mL$) and 3D7 ($IC_{50} = 21.07\,\mu g/mL$). The isolated acridone showed significant antiplasmodial activity against both *P. falciparum* strains with IC_{50} value $>100\,\mu M$ as compared to other isolated alkaloids [206].

9-acridanones and 9-iminoacridines were screened for antiparasitic activity against limax amoeba *Naegleria fawleri*. In addition, their DNA binding affinity was also assessed by various spectroscopic techniques, considering it to be a possible mechanism for their antiparasitic activity. However, 9-acridanone derivatives displayed less DNA intercalating ability as compared to 9-iminoacridines. With regard to antiparasitic activity, 9-acridanone derivatives compound **112** showed good antiparasitic activity (IC_{50} value $50\,\mu g/mL$) [207]. In the same year in vitro antiamoebic activity of some 9-oxo, 9-thio, and 9-imino substituted acridinic compounds were determined against *N. fawleri* and *Acanthamoeba* species. It has been reported that 9-acridanone derivatives displayed the least activity, having IC_{50} value $50–100\,\mu g/mL$ against *N. fawleri*, whereas 9-iminoacridines were more active than acridanone derivatives (IC_{50} value $25\,\mu g/mL$). It has been concluded that double bonds at position C-9 do not favor any antiamoebic activity, but on the other hand, the diethyl-aminoethylthio group is very effective for this activity [208]. In the next year, a series of 9-oxo, 9-thio, and 9-imino substituted acridine derivatives were synthesized and screened for antitrypanosomal activity against *Trypanosoma cruzi*. The most active compounds were 9-thioacridanones and 9-thio-1,2,3,4-tetrahydroacridanones, especially when sulfur is substituted with dialkyl-aminoalkyl group. 9-acridanones were found to be devoid of antiparasitic activity, however, due to the lack of parasitic DNA binding ability of these compounds [209]. Furthermore, 9-acridinones substituted at position C-1 with nitro, amino and acetamido groups were synthesized [210]. Among them, C-1 nitro substituted acridinones, compound **113**, have been screened for antiparasitic activity against three pathogenic strains of *T. cruzi*, *L. donovani*, and *Acanthamoeba culbertsoni*, which elicited significant activity against *A. culbertsoni*. However, such results are in agreement with those obtained either with unsubstituted or 1,4 substituted hydroxyl, methoxy, or carboxyl groups showing that a nitro group is not essentially required for antiparasitic activity.

9-acridanone hydrazones showed high schitosomicidal activity at the post-postural phase of experimental schitosomiasis in primates [211]. Among these derivatives, compound **114** was highly active in cebus monkey infected with *Schistoma mansoni*. Furthermore, the activity of this compound at the prepostural phase of infection was investigated using albino mice and cebus monkeys as models [212]. Results revealed that compound **114** was also effective against 100 *S. manosini cercariae* (LE strain) when administrated at the dose of 100 mg/kg. On the basis of this observation, compounds **114** and **115** were screened for antischitosomal activity [213]. These compounds were tested against 200 cercariae of *S. manosini* (SJ strain) via transcutaneous route using cebus monkey as a model. Both compounds exhibited marked activity even at the lower dose (12.5 mg/kg) [212,213]. In next year two acridone derivatives, **116** and **117** (Fig. 3.13), evaluated both the in vitro and in vivo activity against *L. donovani*. These derivatives inhibited the in vitro growth of the promastigote forms of *L. donovani* only at the high concentration (100 mg/mL). The in vivo results indicated that both compounds reduced the number of amastigotes and decreased parasitism by more than 40%. With regard to the mechanism of action, it has been observed that both of these compounds inhibit the incorporation of [3H] thymidine by inducing alterations at the ultrastructural level in the DNA and mitochondria [214].

1,3 benzothiazole-2-yl amino group substituted acridones were found to possess antileshmanial activity, suggesting that the presence of a benzothiazole group on the parent aminoacridone ring enhances activity. Compound **118** exhibited good antileshmanial activity against promastigotes ($IC_{50}=20.1\,\mu M$) and amastigotes ($IC_{50}=4.3\,\mu M$) leishmania. SAR study revealed that substitution of a 6-nitro or 6-amino group on a benzothiazole ring resulted in formation of compounds having reduced cytotoxicity toward mammalian cells and high antiparasitic activity [215]. Dicationic acridones were synthesized and evaluated for antiprotozoal activity against *T. brucci rhodeisense* STIB900. These dicationic compounds belong to a similar class of diamidine used in parasitic disease. These are selective DNA minor groove binders and their specific binding to kinetoplast DNA (k-DNA), which are responsible for antitrypanosomal activity either by inhibition topoisomerase II or inhibition of transcription. The target acridones displayed low nanomolar activities against bloodstream trypanomastigotes of *T. brucci rhodesiense* and excellent selectivities for the parasite (SI > 2000). Among these, the most active was the 2,6-disubstituted compound **119** ($IC_{50}=7\,nM$), which was equipotent to the reference drug melarsoprol and approximately three times more potent than the *N*-phenylbenzamide-linked compounds. The antiprotozoal activity depends on the relative position of both imidazolinium cations on the fused-heterocyclic scaffold, the 2,6-disubstitution pattern giving the best results against *T. brucei* (Fig. 3.13) [216].

4.6 Antimicrobial Activity

At the end of the 20th century the occurrence and spreading of the agents of new dangerous infectious diseases as well as new strains of the known microorganisms and viruses developing resistance to modern clinical preparations triggered the search for new biologically active compounds. From this standpoint, the heteroaromatic tricycle compounds (e.g., acridine) attract pharmacologists and medical chemists who consider them as the promising basis to create the databases of biologically active compounds. Acridine derivatives are one of the oldest and most successful classes of bioactive agents. Acridone derivatives acting as antimicrobial agents are depicted in Fig. 3.14.

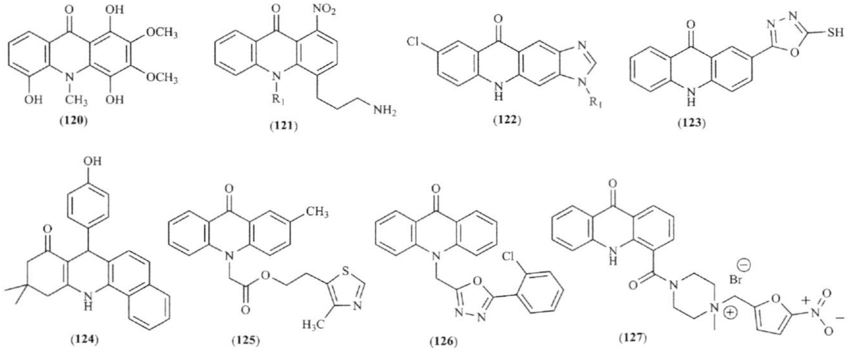

FIGURE 3.14 Acridine derivatives as antimicrobial agents.

Acridone alkaloids are obtained from the ethanol extract of the branch of *Atalantia buxifolia*. These acridone-based alkaloids were evaluated for their antibacterial and AchE inhibitory activity. Compound **120** has shown the significant activity on *Staphylococcus aureus* and weak inhibitory effect on acetylcholinesterase [217]. Nitroacridone derivatives were prepared and evaluated for antibacterial agents against gram-positive, gram-negative, and mycobacterium [218]. Substitution at position C-4 in nitroacridone with propylamine group (**121**) increasing the activity. Thus nitro-substituted acridone derivatives can be good prospectives for antimicrobial activity. Imidazo acridones have been synthesized and screened as antimicrobial agents against *Escherichia coli*, *S. aureus*, *Pseudomonas aeuroginosa*, and *Bacillus subtilis* [219]. Halogen-substituted acridone derivative (**122**) has shown promising antimicrobial activity. 9(10*H*)-acridone bearing 1,3,4-oxadiazole at C-2 position (**123**) derivative has been developed as an antimicrobial agent. Antibacterial and antifungal activity of this compound is due to the presence of an oxadiazole group, which may increase the lipophilicity of the molecules and help cross through the biological membrane of the microorganism, thereby inhibiting their growth [220]. Tetrahydroacridin-8-one derivatives were synthesized and evaluated for antimicrobial activity. The results suggested that these compounds exhibited good inhibitory effect against most of the tested organisms. Compound **124** was shown to be most effective against *Rhodotorula rubra* (MIC = 3.9 μg/mL) and *Aspergillus parasiticus* (MIC = 7.8 μg/mL) [221]. 2-(4-methyl-1,3-thiazol-5-yl) ethyl acridone carboxylates were synthesized by the transesterification of the corresponding butyl esters. These compounds showed good inhibition against strains of *E. coli*, *P. aeruginosa*, *Proteus vulgaris*, *S. aureus*, *B. subtilis*, and *Candida albicans* microorganisms. Moreover, compound **125** with a methyl group at the 2-position of the acridone moiety showed slightly higher activity against *C. albicans* in comparison to standard agent rivanol [222].

Later, acridone acetic acid derivatives bearing 1,3,4-oxadiazoles were prepared and evaluated for antimicrobial activity against bacterial strains *E. coli*,

P. aeruginosa, P. vulgaris, S. aureus, B. subtilis, and *C. albicans.* Compound **126** bearing the chlorine atom at 2-position of the benzene ring is more active against all tested bacterial strains than chlorine atom at 4-position. Introduction of the fluorine atom at 2-position of the benzene ring resulted in a noticeable increase in activity against all tested strains. Thus, the antimicrobial activity of the synthesized compounds depends on the nature of the halogen atom in the benzene ring bonded to oxadiazole moiety [223]. Acridine carboxylic acid derivatives containing a piperazine moiety were synthesized and evaluated for antimicrobial activity. Compound **127** show good antimicrobial activity against multiple test strains such as *E. coli, P. aeruginosa, P. vulgaris, S. aureus, B. subtilis,* and *C. albicans* with zone inhibition (11.5–20 mm) as compared to a standard drug rivanol (Fig. 3.14) [224].

4.7 Antiinflammatory Activity

Control of inflammation has become of prime importance due to its association with numerous diseased states like Alzheimer's disease, asthma, atherosclerosis, Crohn's disease, gout, multiple sclerosis, osteoarthritis, psoriasis, rheumatoid arthritis, diabetes mellitus, carcinoma, bacterial or viral infections, and so on, which result in chronic inflammation [225,226].

Toddaliopsins A-D, four novel 1,2,3-trioxygenated acridone alkaloids, have been isolated from the leaves of *Toddaliopsis bremekampii.* These alkaloids were tested for their antiinflammatory activity by chemiluminescence assay. Toddaliopsin C (**128**) had the greatest activity ($IC_{50} = 4.21 \, \mu M$), suggesting that the presence of a hydroxyl group at C-1 position may enhance the antiinflammatory properties of these compounds [227]. Acridine-alkanoic acids were synthesized and evaluated for antiinflammatory activity using rat hindpaw carrageenan-induced edema assay. In addition, some selected compounds were tested for inhibition of soybean lipoxygenase by the UV absorbance-based enzyme assay. In carrageenan-induced edema assay, only compounds containing the carboxylic acid side chain at the C-2 position of acridone moiety (**129**) showed good activity [228,229]. Cyclooxygenase-2 (COX-2) isozyme is overexpressed in multiple types of cancer relative to that in adjacent noncancerous tissue, which prompted this investigation to prepare a group of hybrid fluorescent conjugates. Acridone conjugates were prepared coupled via a linker group of antiinflammatory drugs such as ibuprofen, (S)-naproxen, acetyl salicylic acid, and chlororofecoxib analogues. All synthesized molecules showed more potent inhibition of COX-2 ($IC_{50} = 0.59–21.2 \, \mu M$) than COX-1 ($IC_{50} = 18.0–88.1 \, \mu M$). Similarly, these derivatives showed a higher COX-2 selectivity index (SI) than antiinflammatory drugs. This may be caused by the fact that these molecules have larger molecular volumes ($353–412 \, Å^3$ range) than antiinflammatory drugs ($155–214 \, Å^3$ range), which may hinder their entry into the smaller COX-1 binding site ($316 \, Å^3$). The most potent and selective COX-2 inhibitor was ibuprofen-acridone conjugate **130** ($IC_{50} = 0.67 \, \mu M$; $SI = 10.6$). Additionally, studies have shown that fluorescence emission (λ_{em} = 417, 440 nm) was not suitable for fluorescent imaging of cancer cells that overexpress the COX-2 isozyme [230].

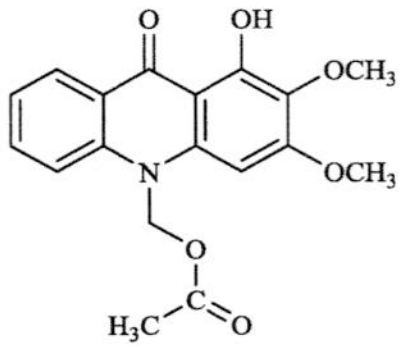

Toddaliopsin C **(128)**

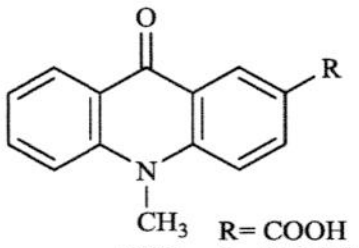

(129) R = COOH
R = CH$_2$COOH

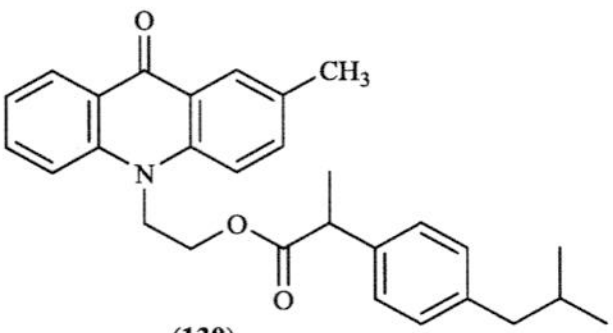

(130)

4.8 Antipsoriatic Activity

Psoriasis is an immune-mediated inflammatory and scaling skin disease characterized by excessive growth of keratinocytes [231]. Owing to the proinflammatory and staining effects of antipsoriatic anthrones such as anthralin (dithranol, 1,8-dihydroxy-10*H*-anthracen-9-one, **131**), this drug is best used in hospital or day treatment centers [232]. But due to side effects associated with use of this topical antipsoriatic agent, Putic et al. have focused on acridones, 10-aza analogue of anthrone, in order to explore their potential as antipsoriatic agents. They have reported the synthesis and evaluation of a series of N-unsubstituted [233] as well as N-substituted hydroxyl-10*H*-acridone [234] against HaCaT keratinocyte growth. SAR for N-unsubstituted acridone derivatives did not follow those of the antipsoriatic anthrones. It has been found that in the series of N-unsubstituted acridones, a 1,3-dihydroxy substitution pattern was most beneficial for potency. However, introduction of a methyl group at C-8 position of 1,3-dihydroxy-substituted skeleton resulted in the most potent acridone **(132)** within this series with keratinocyte growth inhibitory potency (IC$_{50}$ = 0.8 μM) comparable to anthralin. In contrast to anthralin, acridones were devoid of radical generating ability and did not disrupt membrane integrity [233]. In the series of N-substituted hydroxyl-10*H*-acridone, the most potent inhibitor of keratinocyte hyperproliferation was compound **133** having an N-methyl group and a 1,3-dihydroxy arrangement at the acridone scaffold, with an IC$_{50}$ value of 0.6 μM comparable to that of anthralin. They concluded that hydroxyl group at C-1 and C-3 positions are crucial for antipsoriatic activity where N-alkylation increased the potency of the molecule. Benzyl substitution at the 10-position yielded keratinocyte growth inhibitory activity in the low micromolar range [234].

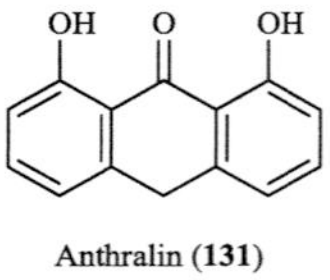

Anthralin **(131)**

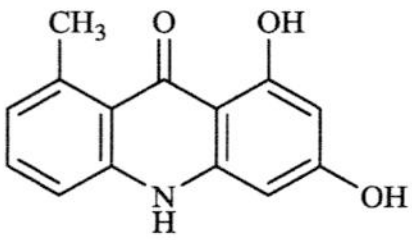

(132)

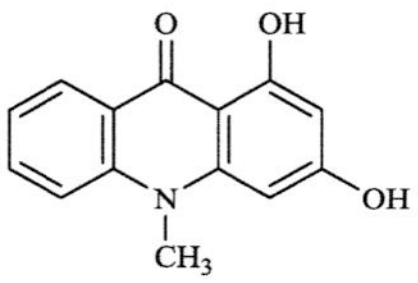

(133)

4.9 Fluorescent Probes

The use of fluorescent molecules in biological research is important in many applications, and their use is increasing due to their versatility, sensitivity, and quantitative capabilities. Among their myriad of uses, fluorescent probes are employed to detect protein location and activation, identify complex formation and conformational changes in protein, and monitor biological processes in vivo. Acridone is highly fluorescent and stable against photodegradation, oxidation, and heat [19]. Several acridone derivatives have been used as fluorescent labels for peptides [235], amino acids [236], antibodies [237], and substrates for catalysis [238]. DNA tagged with a fluorescent agent is widely used as DNA probes for the detection of special DNA, and studies of protein interactions and single-nucleotide polymorphism typing. Shoji et al. have prepared acridone-tagged thymidine nucleotide (**134**) (Fig. 3.15), which can be incorporated into DNA during the polymerase chain reaction, enzymatically forming multiacridone-labeled DNA [239]. The acridone derivative has a high quantum yield of fluorescence at 420–480 nm by excitation at 360–400 nm. These wavelengths of the emissions indicate that acridone could be useful as a donor in a fluorescent energy transfer (FRET) system in combination with dabcyl (4-[4-(dimethylamino)phenylazo]benzoyl) as an acceptor molecule, which has absorption at 420–520 nm. Acridone-labeled DNA by postmodification compound **135** examined its use as a donor for the FRET system in combination with a 3′-terminal dabcyl-tagged DNA as an acceptor, in which a special target DNA can be detected by FRET [240]. The FRET with an acridone and dabcyl pair has been found to be somewhat superior and complements that with fluorescence-quencher pairs. Also significant quenching of the acridone fluorescence by guanine in the DNA was observed depending on the DNA sequence.

Acridone-tagged DNA can be used as a base-discriminating fluorescent probe for single-nucleotide polymorphism, although it has some sequence limitations [241]. N-4-butylamine acridone (BAA, **136**) was synthesized and its use as the fluorescent probe for detection of calf thymus DNA (ctDNA) was investigated and used for determination of ctDNA [242]. The corresponding linear response range was found to be from 1.0 to 20.0 mg/L of DNA concentration with BAA as the fluorescent probe under optimal conditions. The interaction between BAA and ctDNA was investigated by various methods such as fluorescence, absorption, and viscosity measurements. The mechanistic studies revealed that it consists of three binding modes of DNA such as surface binding, groove binding, and intercalation binding. By the quenching phenomenon they have concluded that the interaction of the probe with ctDNA mostly follows the groove binding mode. Furthermore, polyhalo acridones were prepared by the reaction of polyhalo isophthalonitriles with substituted anilines and subsequently cyclocondensation in the presence of sulfuric acid [243]. These polyhalo acridones were proven to be useful as pH-sensitive fluorescent probes for a wide range of acidic and basic conditions. Among these polyhalo-substituted

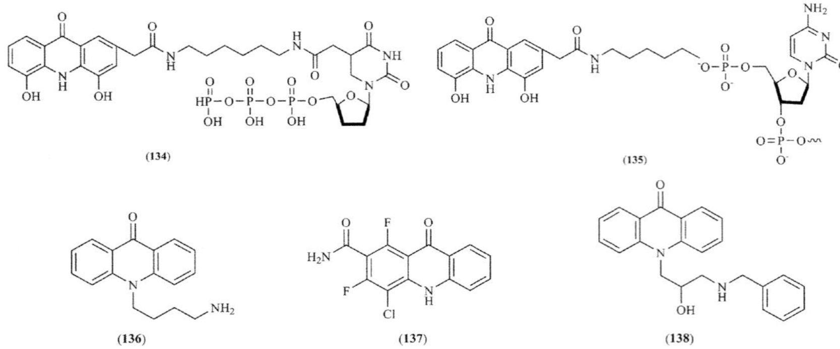

FIGURE 3.15 Fluorescent probes application of acridones.

derivatives, compound **137** (Fig. 3.15) gave the highest fluorescence quantum yield (up to 0.98). Moreover, acridone carrying an appropriate substituent at N10 position showed significant fluorescence changes on interacting with ATP in HEPES buffer at pH 7.2. These compounds in combination with ATP exhibited quenching of fluorescence. The fluorescence spectrum of a solution of compound **138** (0.1 mM, $\Phi = 0.67$) in HEPES buffer (pH 7.2) showed emission bands at 417 nm and 440 nm when excited at 253 nm. The selectivity and sufficient binding of these probes with ATP could be useful for monitoring of metabolic processes (Fig. 3.15) [244].

A Mobil Composition of Matter (MCM)-41 type mesoporous silica material containing *N*-propylacridone groups has been successfully prepared by cocondensation of an appropriate organic precursor with tetraethyl orthosilicate (TEOS) under alkaline sol-gel conditions. The fluorescence properties of the organic chromophore can be tuned via complexation of its carbonyl group with scandium triflate, which makes the material a good candidate for solid state sensors and optics [245]. Swist et al. designed and synthesized new acridone-based branched blocks as highly fluorescent material. The luminescence studies represent acridones to be a good type of chromophores and confirmed the applicable potential of these novel aryl-based π-conjugated polymers for the development of various emitters in electro-and chemiluminescence [246].

4.10 Central Nervous System Activity

Acridone alkaloids isolated from *Glycosmis chlorosperma* as DYRK1A inhibitors. Dual-specificity tyrosine phosphorylation-regulated kinase 1A (DYRK1A) is a protein kinase with diverse functions and is implicated in neuronal development and adult brain physiology. Higher levels of DYRK1A are associated with the pathology of neurodegenerative diseases and have been implicated in some neurobiological alterations of Down syndrome, such as mental retardation. The isolated acridone molecules were also tested on the closely related

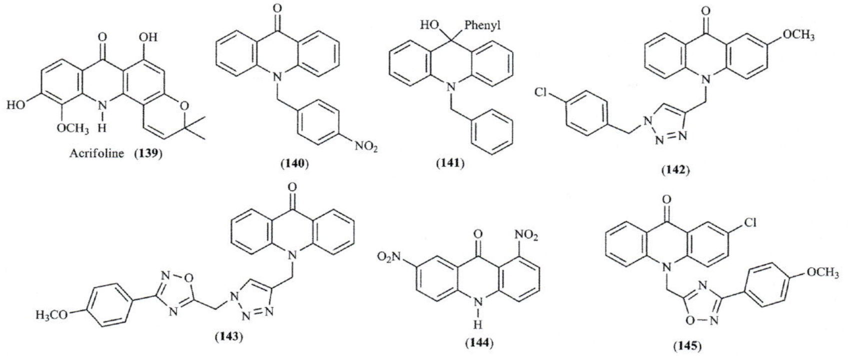

FIGURE 3.16 Central nervous system activity of acridones.

cdc2-like kinase CLK1 and the selectivity profile of these compounds were finally evaluated by including cyclin-dependent kinases 1 and 5 (CDK1 and CDK5), glycogen synthase kinase-3 (GSK3), and casein kinase 1 (CK1). In order to better understand the influence of the substitution pattern of the acridone tricyclic system on their specific binding to DYRK1A, a molecular docking study of the compounds isolated into the DYRK1A ATP binding domain was performed and the results obtained with the β-carboline alkaloid harmine were compared [247]. Acrifoline (**139**) (Fig. 3.16) showed inhibition activities of DYRK1A, CLK1, GSK3, CDK1, and CDK5 with IC_{50} values of 0.075, 0.17, 2, 5.3, and 9 µM, respectively. Moreover, molecular docking studies of compound **139** form hydrogen bonds between the hydroxy group (at C-6 and Glu203) and the conserved Lys188 and interact with backbone atoms of Glu239 and Leu241 through hydrogen bonds involving the C-1 hydroxy group. Docking studies predict a binding mode for acrifoline, which showed the most potent DYRK1A-inhibiting activity, similar to that of harmine, a β-carboline alkaloid, and leucettines, currently considered the most potent bioavailable inhibitors of this enzyme [247]. P2X4 receptor antagonists as drugs have potential for the treatment of neuropathic pain and neurodegenerative diseases. N-substituted phenoxazine and related acridone and benzoxazine derivatives were synthesized and optimized with regard to their potency to inhibit ATP-induced calcium influx in 1321N1 astrocytoma cells stably transfected with the human P2X4 receptor. They determined that *para*-nitro-substitution resulted in the most potent compound of this series (**140**, $IC_{50} = 2.49$ µM). Virtually the same potency was observed for the N1-benzyl substituted 6-hydroxy-6-phenyldihydroacridine (**141**, $IC_{50} = 2.44$ µM). SAR study revealed moderate potency for *N*-benzoylacridone whereas *N*-benzylacridone appeared to be more potent than *N,N*-diethyl-substituted urea derivative (Fig. 3.16) [248].

A novel series of acridone linked to 1,2,3-triazole derivatives has been synthesized and evaluated in vitro for acetylcholinesterase (AChE) and butyrylcholinesterase (BChE) inhibitory activities. The best ability to inhibit AChE was shown by compound **142** (IC_{50} = 7.31 µM) with a 4-substituted chlorine on the pendant benzyl group, whereas introduction of strong electron-withdrawing nitro group decreases the activity. Additionally, the presence of groups in the order of OMe > Me ≥ Cl > Br instead of hydrogen on the 2-substituted acridone moieties enhanced inhibitory activity. Docking studies of compound **142** showed π–π interactions between Phe330 and the 1,2,3-triazole moiety and between Trp84 and the acridone moiety. Additionally, the compound forms hydrogen bonds between the oxygen of OMe and Ser122. Furthermore, a weak interaction between chlorine and the backbone carbonyl group of amino acid is beneficial [249]. Acridone-1,2,4-oxadiazole-1,2,3-triazole hybrid molecules were synthesized and evaluated as inhibitors of AChE and BChE. Among these compound **143** showed an inhibitory effect against AChE with an IC_{50} value of 11.55 µM. The more activity was shown by compound **161** containing an unsubstituted acridone ring and 4-methoxyphenyl-1,2,4-oxadiazole moieties. SAR studies of these molecules suggested that a methoxy group into the pendant benzyl group imparted a higher activity than chlorine, methyl, and bromine substituent. Docking studies showed that compound **161** forms π–π interaction between the acridone moiety and Phe331 and Trp84 in the catalytic anionic site. Additionally, compound **143** is proposed to interact with the hydroxyl group of Ser200 in the catalytic triad site through hydrogen bonds involving the carbonyl group of acridone. Further, the 1,2,4-oxadiazole moiety exhibited π–π interaction with a phenyl group of Tyr121 in the peripheral anionic site. It should be noted that there was an interaction via hydrogen bond between a methoxy group of aryl and backbone NH atoms [250]. Nitro acridones were synthesized and evaluated for acetylcholinestrase inhibitors. Compound **144** (Fig. 3.16) has shown potent activity against acetylcholinesterase inhibitors with IC_{50} value 0.22 µM. The activity of these compounds is explained on the basis of hydrogen bonding, secondary interaction, and orientation and electronic features of the substituent against the active site of the target enzyme. The hydrogen bond interacts with amino acids of the enzyme and showed good binding affinities. The two nitro groups at 1- and 7-position of acridone moiety could interact with THR31, LYS89, and SER88 amino acids through hydrogen bonds. Accordingly, compound **144** is an effective blocker for acetylcholinesterase and it also plays an important role in stabilizing the ligand receptor complex by π-π interaction. Further, compound **144** showed the highest antioxidant activity (IC_{50} = 27.80 µ/mL) by DPPH method. This molecule also displayed good antimicrobial activity in gram positive bacteria except *Streptococcus pyrogenes* [251].

A series of acridone-based oxadiazoles were synthesized and evaluated for their anticonvulsant activity against pentylenetetrazole (PTZ)- and maximal electroshock (MES)-induced seizures in mice. Most of the compounds exhibited better anticonvulsant activity and higher safety with respect to the standard drug, phenobarbital. Among the tested derivatives, compounds **145** with ED_{50} value of 2.08 mg/kg was the most potent compound in the

PTZ test. The anticonvulsant effect of compound **145** was blocked by flumazenil, suggesting the involvement of benzodiazepine (BZD) receptors in the anticonvulsant activity. Docking study of compound **145** revealed that acridone ring is responsible for two important interactions with BZD-binding pocket of $GABA_A$: π-cation interaction with the positively charged nitrogen of α1 His101 and π-π interaction with γ2 Phe77. It is obvious that there is a hydrogen bond between N-2 of 1,2,4-oxadiazole ring and α1Thr206. Also, α1Tyr209 possessing aromatic moiety is involved in a π-π interaction whit 3-phenyl moiety. Therefore, the results showed that binding mode of compound **145** in the BZD-binding pocket of $GABA_A$ receptor resembled that of diazepam (Fig. 3.16) [252].

4.11 Miscellaneous

Acridone alkaloids isolated from *Atalantia monophylla* and evaluated their antiallergic activity in an RBL-2H3 cell model. All these compounds possessed stronger antiallergic activity than ketotifen fumarate, a clinically used drug (IC_{50}=47.5 μM). They observed that buxifoliadine- E (**146**) showed a potent antiallergic effect (IC_{50}=6.1 μM) as compared to another isolated compounds. These compounds were also tested on the enzyme activity of β-hexosaminidase and showed weak inhibition at 100 μM, whose results indicated the inhibition of the antigen-induced degranulation but not the activity of β-hexosaminidase. Hence, buxifoliadine-E (**146**) could provide a platform for structure-based design of acridone as an antiallergic agent [253].

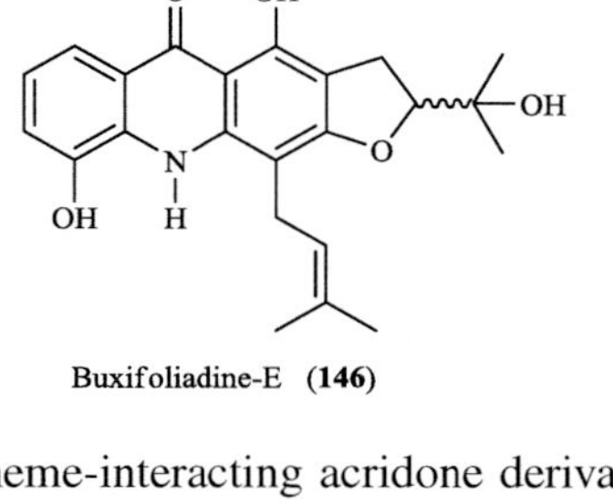

Buxifoliadine-E (**146**)

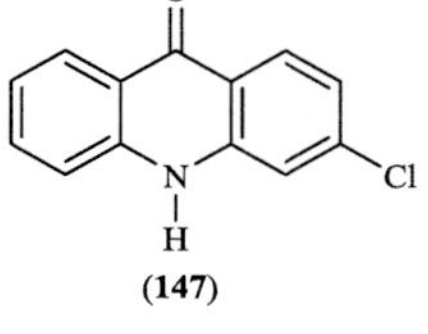

(**147**)

Novel heme-interacting acridone derivatives were prepared and screened to prevent free heme-mediated protein oxidation and degradation. Heme is an important molecule for living aerobic organisms, and plays an essential role in drug detoxification, oxygen transport, respiration, and signal transduction. Free heme is responsible for oxidative stress, hemolysis, and inflammation. It also affects the function of kidneys, neurones, cardiac cells, hepatocytes, and peripheral leukocytes. These acridones in vitro block heme-mediated protein oxidation and degradation and are used as markers for heme-induced oxidative stress. Compound **147** showed greater protective activity against heme-H_2O_2 systems and oxidative protein degradation with the highest heme binding capacity [K_a=(4.6±0.17)$10^4 M^{-1}$] and inhibiting protein carbonyl formation (IC_{50}=36 μM) [254].

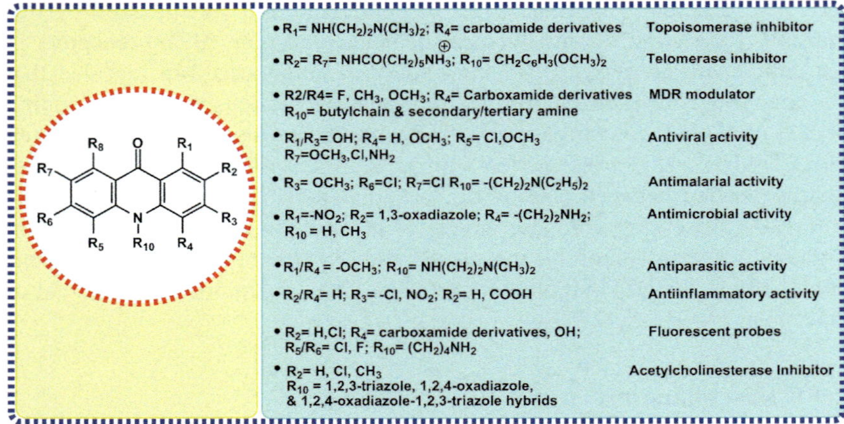

FIGURE 3.17 Enormous activities displayed by acridone and its derivatives.

5. CONCLUSION

The heterocyclic aromatic structure of acridone has a lot of potential for biological utilization. Despite the active, exhaustive, and target-based research on development of many compounds of acridone, no molecule has made its way to the market. Acronycine derivatives such as compound **7** underwent phase I clinical trials against various solid tumors. However, triazoloacridone (C-1305, **15**) has been selected for extended preclinical trials and imidazoacridone (C-1311, **16**) is currently undergoing phase II clinical trials as a drug under the name of symadex TM. In patients with solid tumor, elacridar (**56**) in combination with topotecan and doxorubicin exhibited as MDR inhibitor activity in Phase I clinical trials.

This chapter is expected to provide a comprehensive view of acridone-based compounds to a drug designer and medicinal chemist. The compiled data of SAR analysis (Fig. 3.17) of this chapter may be utilized for designing potent, selective, and multitargeted compounds as the novel therapeutic intervention of various diseased conditions. The future discovery of the acridone core lies in the design of new molecules with more selectivity and less inherent toxicity. It has been found that N10 substituted acridones (with oxadiazole and triazole derivatives) are effective against acetylcholinesterase inhibitors. So these molecules can be explored further for treatment of neurodegenerative disease.

REFERENCES

[1] M. Wainwright, Acridine – a neglected antibacterial chromophores, Journal of Antimicrobial Chemotherapy 47 (1) (2001) 1–13.

[2] G.K. Hughes, F.N. Lahey, J.R. Price, Alkaloids of the Australian, Rutaceae, Nature 162 (4110) (1948) 223–224.

[3] F.N. Lahey, W.C. Thomas, Alkaloids of the Australian Rutaceae: *Acronychia baueri* I the isolation of the alkaloids, The Australian Journal of Scientific Research 2 (1949) 423–426.

[4] M. Gniazdowski, L. Szmigiero, Nitracrine and its congeners-an overview, General Pharmacology 26 (1995) 473–481.

[5] G. Cholewiński, K. Dzierzbicka, A.M. Kolodziejczyk, Natural and synthetic acridines/acridones as antitumor agents: their biological activities and methods of synthesis, Pharmacological Reports 63 (2) (2011) 305–336.

[6] G.J. Atwell, B.F. Cain, R.N. Seelye, Potential antitumor agents. 12. 9-anilinoacridines, Journal of Medicinal Chemistry 15 (6) (1972) 611–615.

[7] W.A. Denny, Acridine derivatives as chemotherapeutic agents, Current Medicinal Chemistry 9 (18) (2002) 1655–1665.

[8] G.J. Finlay, G.J. Atwell, B.C. Baguley, Inhibition of the action of the topoisomerase II poison amsacrine by simple aniline derivatives: evidence for drug-protein interactions, Oncology Research Featuring Preclinical and Clinical Cancer Therapeutics 11 (6) (1999) 255–264.

[9] P. Belmont, J. Bosson, T. Godet, M. Tiano, Acridine and acridone derivatives, anticancer properties and synthetic methods: where are we now? Anti-Cancer Agents in Medical Chemistry 7 (2) (2007) 139–169.

[10] V.J. Harvey, J.R. Hardy, S. Smith, W. Grove, B.C. Baguley, Phase II study of the amsacrine analogue CI-921 (NSC 343499) in non-small cell lung cancer, European Journal of Cancer 27 (12) (1991) 1617–1620.

[11] W.A. Denny, B.C. Baguley, Dual topoisomerase I/II inhibitors in cancer therapy, Current Topics in Medicinal Chemistry 3 (3) (2003) 339–353.

[12] www.clinicaltrial.gov [NCT00004151], [NCT00004937].

[13] W.M. Cholody, S. Martelli, J. Konopa, 8-substituted 5-[(aminoalkyl)amino]-6H-v-triazolo[4,5,1-de]acridin-6-ones as potential antineoplastic agents. Synthesis and biological activity, Journal of Medicinal Chemistry 33 (10) (1990) 2852–2856.

[14] W.M. Cholody, S. Martelli, J. Paradziej-Lukowicz, J. Konopa, 5-[(aminoalkyl) amino]-imidazolo[4,5,1-de]acridin-6-ones as a novel class of antineoplastic agents. Synthesis and biological activity, Journal of Medicinal Chemistry 33 (1) (1990) 49–52.

[15] W.M. Cholody, S. Martelli, J. Konopa, Chromophore-modified antineoplastic imidazoloacridinones. Synthesis and activity against murine leukemias, Journal of Medicinal Chemistry 35 (2) (1992) 378–382.

[16] K. Lemke, M. Wojciechowski, W. Laine, C. Bailly, P. Colson, M. Baginski, A.K. Larsen, A. Skladanowsk, Induction of unique structural changes in guanine-richDNA regions by the triazoloacridone C-1305, a topoisomerase II inhibitor with antitumor activities, Nucleic Acids Research 33 (18) (2005) 6034–6047.

[17] N. Isambert, M. Campone, E. Bourbouloux, M. Drouin, A. Major, W. Yin, P. Loadman, R. Capizzi, C. Grieshaber, P. Fumoleau, C-1311 (SYMADEX) administered in a phase 1 dose escalation trial as a weekly infusion for 3 consecutive weeks in patients with advanced solid tumours, European Journal of Cancer 46 (4) (2010) 729–734.

[18] R. Kumar, M. Kaur, O. Silakari, Chemistry and biological activities of thioacridines/thioacridones, Mini-Reviews in Medicinal Chemistry 13 (2013) 1220–1230.

[19] D. Karak, A. Banerjee, A. Sahana, S. Guha, S. Lohar, S.S. Adhikari, D. Das, 9-Acridone-4-carboxylic acid as an efficient Cr (III) fluorescent sensor: trace level detection, estimation and speciation studies, Journal of Hazardous Materials 188 (1–3) (2011) 274–280.

[20] M. Mitsui, Y. Ohshima, Structure and dynamics of 9(10H)-acridone and its hydrated clustures. I. Electronic spectroscopy, Journal of Advances in Physical Chemistry A 104 (38) (2000) 8638–8648.

[21] M. Mitsui, Y. Ohshima, Structure and dynamics of 9(10H)-acridone and its hydrated clustures. II. Structural characterization of hydrogen-bonding networks, The Journal of Physical Chemistry A 104 (38) (2000) 8649–8659.

[22] M. Mitsui, Y. Ohshima, Structure and dynamics of 9(10H)-acridone and its hydrated clustures. III. Microscopic solvation effects on non radiative dynamics, The Journal of Physical Chemistry A 104 (38) (2000) 8660–8670.

[23] M. Gensicka-Kowalewska, G. Cholewinski, K. Dzierzbicka, Recent development in the synthesis and biological activityof acridone/acridine analogues, RSC Advances 7 (2017) 15776–15804.

[24] G.D. Potis, W. Jones, 9(10H)-acridone, Acta Crystallographica Section C 51 (1995) 267–268.

[25] A. Bouzyk, L. Jozwiak, A. Wroblewska, J. Rak, J. Blazejowski, Structure, properties thermodynamics and isomerisation ability of 9-acridinones, The Journal of Physical Chemistry A 106 (15) (2002) 3957–3963.

[26] I.D. Postescu, D. Suciu, A method for N-alkylation of acridones, Journal für praktische Chemie 318 (3) (1976) 515–518.

[27] C.F.H. Allen, G.H.W. McKee, Acridone, Organic Syntheses 2 (1943) 15–17.

[28] K. Matsumura, Sulfonation of acridone, Journal of the American Chemical Society 57 (1935) 1533–1536.

[29] R.F. Pellon, R. Carrasco, T. Marquez, T. Mamposo, Use of N,N-dimethylformamide as solvent in the synthesis of N-phenylanthranillic acids, Tetrahedron Letters 38 (1997) 5107–5110.

[30] J.P. Hanoun, J.P. Galy, A. Tenaglia, A convenient synthesis of N-arylanthranillic acids using ultrasonics in the Ullmann-Goldberg condensation, Synthetic Communications 25 (1995) 2443–2448.

[31] R. Hegde, P. Thimmaiah, M.C. Yerigeri, G. Krishnegowda, K.N. Thimmaiah, P.J. Houghton, Anti-calmodulin acridone derivatives modulate vinblastine resistance in multidrug resistant (MDR) cancer cells, European Journal of Medicinal Chemistry 39 (2) (2004) 161–177.

[32] J.H. Adams, P.M. Brown, P. Gupta, M.S. Khan, J.R. Lewis, Rutaceous constituents-131:A biomimetic synthesis of acronycine 2, Tetrahedron 37 (1) (1981) 209–217.

[33] S. Horne, R. Rodrigo, A short efficient route to acronycine and other acridones, Journal of the Chemical Society, Chemical Communications (1991) 1046–1048.

[34] K. Lehmstedt, Eine einfache synthese des acridons und 3-substituierter acridone (IX. Mitteil, Über Acridin), Chemische Berichte 65 (5) (1932) 834–839.

[35] K. Matsumura, The synthesis of certain acridine compounds, Journal of the American Chemical Society 51 (3) (1929) 816–820.

[36] G.S.J. Chen, M.S. Gibson, Synthesis of 10-benzoyl-9-acridone from 2,2'- and 2,3'-disubstituted benzophenones, Journal of the Chemical Society Perkin Transactions I (1975) 1138–1139.

[37] J.H. Adams, P. Gupta, M.S. Khan, J.R. Lewis, R.A. Watt, Intramolecular cyclisation of 2-aminobenzophenones; a new 9-acridone synthesis, Journal of the Chemical Society Perkin Transactions I (1976) 2089–2093.

[38] R. Krishan, S.A. Lang, Y.I. Lin, R.G. Wilkinion, Reactions of hydroxybenzophenone with hydrazines, Journal of Heterocyclic Chemistry 25 (1988) 447–452.

[39] Y. Suzuki, T. Toyota, A. Miyashita, M. Sato, Synthesis of Heterocyclic compounds via nucleophilic aroylation catalyzed by imidazolidenyl carbine, Chemical and Pharmaceutical Bulletin 54 (12) (2006) 1653–1658.

[40] V.Y. Orlov, V.V. Ganzha, A.D. Kotov, V.G. Sokolov, Synthesis of nitroacridones from 2,1-benzisoxazole, Russian Journal of Organic Chemistry 43 (10) (2007) 1502–1507.

[41] T.L. Ho, D.G. Jou, Generation of 9(10H)-acridone from anthranilic acid, Journal of the Chinese Chemical Society 48 (2001) 81–82.

[42] A.F. Pozharkii, K. Konstantinchenko, Behaviour of acridine towards sodium amide & alkali, Chemistry of Heterocyclic Compounds 8 (12) (1972) 1518–1520.

[43] G.M. Coppola, The chemistry of 2H-3,1-benzoxazine-2,4(1H)-dione (isatoicanhydride). 14. A facile entry into the acronycine ring system, Journal of Heterocyclic Chemistry 21 (3) (1984) 913–914.

[44] G.M. Coppola, Chemistry of 2H-3,1-benzoxazine-2,4(1H)-dione (isatoicanhydride). 16. Facile construction of the furacridone ring system, Journal of Heterocyclic Chemistry 21 (5) (1984) 1569–1570.

[45] G.M. Coppola, H.F. Schuster, The chemistry of 2H-3,1-benzoxazine-2,4(1H)-dione (isatoic anhydride). 21. A mild process for the preparation of 10-alkyl-9- acridanones and it's application to the synthesis of acridone alkaloids, Journal of Heterocyclic Chemistry 26 (4) (1989) 957–964.

[46] A.A. Aly, M.A. Gomaa, New cycloaddition of diarylazines with 1,2- dehydrobenzene, 1,1,2,2-tetracyanoethylene, and dibenzoylacetylene-facile synthesis of acridinones, pyrazolidine, and pyridazine derivatives, Canadian Journal of Chemistry 83 (2005) 57–62.

[47] S.J. Tu, B. Jiang, R.H. Jia, J.Y. Zhang, Y. Zhang, C.S. Yao, F. Shi, An efficient one pot, three component synthesis of indeno[1,2-b] quinoline-9,11(6H,10H)-dione, acridine-1,8(2H,5H)-dione and quinoline-3-carbonitrile derivatives from enaminones, Organic and Biomolecular Chemistry 4 (2006) 3664–3668.

[48] A.F. Morgera, P. Zanirato, BF$_3$.OEt$_2$-promoted synthesis of acridines via N-aryl nitrenium-BF$_3$ions generated by dissociation of 2-oxo azidoarenes in benzene, Arkivoc 12 (2006) 111–120.

[49] J. Zhao, R.C. Larock, Synthesis of xanthones, thioxanthones, and acridones by the coupling of arynes and substitutedbenzoates, Journal of Organic Chemistry 72 (2007) 583–588.

[50] D.G. Pintori, M.F. Greaney, Insertion of benzene rings into the amide bond: one-step synthesis of acridones and acridones from aryl amides, Organic Letters 12 (1) (2010) 168–171.

[51] A.V. Dubrovskiy, R.C. Larock, Synthesis of O-(dimethylamino)aryl ketones and acridones by the reaction of 1,1-dialkylhydrazones and arynes, Organic Letters 13 (15) (2011) 4136–4139.

[52] P.B. Bach, J.R. Jett, U. Pastorino, M.S. Tockman, S.J. Swensen, C.B. Begg, Computed tomography screening and lung cancer outcomes, Journal of the American Medical Association 297 (9) (2007) 953–961.

[53] R.L. Siegel, K.D. Miller, A. Jemal, Cancer statistics, CA: A Cancer Journal for Clinicians 68 (2018) 7–30.

[54] Q.C. Nguyen, T.T. Nguyen, R. Yougnia, T. Gaslonde, H. Dufat, S. Michel, F. Tillequin, Acronycine derivatives: a promising series of anti-cancer agents, Anti-Cancer Agents in Medicinal Chemistry 9 (7) (2009) 804–815.

[55] F. Tillequin, S. Michel, E. Seguin, Structure activity relationships in the acronycine series, Current Medicinal Chemistry 9 (2002) 1241–1253.

[56] T.C. Chou, C.C. Tzeng, T.S. Wu, K.A. Watanabe, T.L. Su, Inhibition of cell growth and macromolecule biosynthesis of humanpromyelocytic leukemic cells by acridone alkaloids, Phytotherapy Research 3 (1989) 237–242.

[57] M. Brum-Bousquet, S. Mitaku, A.L. Skaltsounis, F. Tillequin, M. Koch, Acronycine epoxide a new acridone alkaloid from several *Sarcomelicope* species, Planta Medica 54 (5) (1988) 470–471.

[58] A.F.M. Motiur Rhman, J.L. Liang, S.H. Lee, J.K. Son, M.J. Jung, Y. Kwon, Y. Jahng, 2,2-Dimethyl-2H-pyran-derived alkaloids practical synthesis of acronycine and benzo[b] acronycine and their biological properties, Archives of Pharmacal Research 31 (2008) 1087–1093.

[59] E. Seguin, F. Tillequin, Structure activity relationships and mechanism of action of antitumor benzo[b]acronycine antitumor agents, Annales Pharmaceutiques Françaises 63 (1) (2005) 44–52.

[60] N. Costes, H. Le Deit, S. Michel, F. Tillequin, M. Koch, B. Pfeiffer, P. Renard, S. Léonce, N. Guilbaud, L. Kraus-Berthier, A. Pierré, G. Atassi, Synthesis and cytotoxic and antitumor activity of benzo[b]pyrano[3, 2-h]acridin-7-one analogues of acronycine, Journal of Medicinal Chemistry 43 (12) (2000) 2395–2402.

[61] N. Guilbaud, L. Kraus-Berthier, F. Meyer-Losic, V. Malivet, C. Chacun, M. Jan, F. Tillequin, S. Michel, M. Koch, B. Pfeiffer, G. Atassi, J. Hickman, A. Pierre, Marked antitumor activity of a new potent acronycine derivative in orthotopic models of human solid tumors, Clinical Cancer Research 7 (2001) 2573–2580.

[62] T. Gaslonde, F. Covello, L. Velazquez-Alonso, S. Léonce, A. Pierré, B. Pfeiffer, S. Michel, F. Tillequin, Synthesis and cytotoxic activity of benzo[a]acronycine and benzo[b] acronycine substituted on the A ring, European Journal of Medicinal Chemistry 46 (5) (2011) 1861–1873.

[63] T.L. Su, B. Kohler, T.C. Chou, M.W. Chun, K.A. Watanabe, Synthesis of the acridone alkaloids glyfoline and congeners structure-activity relationship studies of cytotoxic acridones, Journal of Medicinal Chemistry 35 (14) (1992) 2703–2710.

[64] S. Kawaii, Y. Tomono, E. Katase, K. Ogawa, M. Yano, Y. Takemura, M.C. Ito, H. Furukawa, The antiproliferative effect of acridone alkaloids on several cancer cell lines, Journal of Natural Products 62 (4) (1999) 587–589.

[65] R.M. Ngoumfo, J.B. Jouda, F.T. Mouafo, J. Komguem, C.D. Mbazoa, T.C. Shiao, M.I. Choudhary, H. Laatsch, J. Legault, A. Pichette, R. Roy, In vitro cytotoxic activity of isolatedacridones alkaloids from *Zanthoxylum leprieurii* Guill. et Perr, Bioorganic and Medicinal Chemistry 18 (10) (2010) 3601–3605.

[66] V.N. Wouatsa, L. Misra, S. Kumar, O. Prakash, F. Khan, F. Tchoumbougnang, R.K. Venkatesh, Aromatase and glycosyl transferase inhibiting acridone alkaloids from fruits of Cameroonian *Zanthoxylum* species, Chemistry Central Journal 7 (2013) 125.

[67] J.D. Wansi, A.T. Tchoa, F.A.A.Z. Tozea, L. Naharb, C. Martind, S.D. Sarkerb, Cytotoxic acridone and indoloquinazoline alkaloids from *Zanthoxylum poggei*, Phytochemistry Letters 17 (2016) 293–298.

[68] R.P. Severino, R.V.C. Guido, E.F. Marques, D. Brömme, M.F. da Silva, J.B. Fernandes, A.D. Andricopulo, P.C. Vieira, Acridone alkaloids as potent inhibitors of cathepsin V, Bioorganic and Medicinal Chemistry 19 (4) (2011) 1477–1481.

[69] P. Belmont, I. Dorange, Acridine/acridone: a simple scaffold with a wide range of application in oncology, Expert Opinion on Therapeutic Patents 18 (11) (2008) 1211–1224.

[70] M. Kukowska, Amino acid or peptide conjugates of acridine/acridone and quinoline/ quinolone-containing drugs. A critical examination of their clinical effectiveness within a twenty-year timeframe in antitumor chemotherapy and treatment of infectious diseases, European Journal of Pharmaceutical Sciences 109 (2017) 587–615.

[71] E. Augustin, A. Moś-Rompa, A. Skwarska, J.M. Witkowski, J. Konopa, Induction of G2/M phase arrest and apoptosis of human leukemia cells by potent antitumor triazoloacridinone C-1305, Biochemical Pharmacology 72 (12) (2006) 1668–1679.

[72] J. Wesierska-Gadek, D. Schloffer, M. Gueorguieva, M. Uhl, A. Skladanowski, Increased susceptibility of poly(ADP-ribose) polymerase-1 knockout cells to antitumor triazoloacridone C-1305 is associated with permanent G2 cell cycle arrest, Cancer Research 64 (13) (2004) 4487–4497.

[73] K. Lemke, V. Poindessous, A. Skladanowski, A.K. Larsen, The antitumor triazoloacridone C-1305 is a topoisomerase II poison with unusual properties, Molecular Pharmacology 66 (4) (2004) 1035–1042.

[74] M. Koba, J. Konopa, Interactions of antitumor triazoloacridinones with DNA, Acta Biochemica Polonica 54 (2) (2007) 297–306.

[75] Z. Mazerska, P. Sowinski, J. Konopa, Molecular mechanism of the enzymatic oxidation investigated for imidazoacridinone antitumor drug C-1311, Biochemical Pharmacology 66 (9) (2003) 1727–1736.

[76] A. Wisniewska, A. Chrapkowska, A. Kot-Wasik, J. Konopa, Z. Mazerska, Metabolic transformations of antitumor imidazoacridinone, C-1311 with microsomal fractions of rat and human liver, Acta Biochimica Polonica 54 (4) (2007) 831–838.

[77] A. Skwarska, E. Augustin, J. Konopa, Sequential induction of mitotic catastrophe followed by apoptosis in human leukemia MOLT4 cells by imidazoacridinone C-1311, Apoptosis 12 (12) (2007) 2245–2257.

[78] W.M. Cholody, L. Hernandez, L. Hassner, D.A. Scudiero, D.B. Djurickovic, C.A. Michejda, Bisimidazoacridones and related compounds: new antineoplastic agents with high selectivity against colon tumors, Journal of Medicinal Chemistry 38 (16) (1995) 3043–3052.

[79] H.K. Haripraksha, T. Kosakowska-Cholody, C. Meyer, W.M. Cholody, S.F. Stinson, N.I. Tarasova, C.J. Michejda, Optimization naphthalimide-imidazoacridone with potent antitumor activity leading to clinical candidate (HKH40A, RTA.502), Journal of Medicinal Chemistry 50 (23) (2007) 5557–5560.

[80] W.M. Cholody, B. Horowska, J. Paradziez-Lukowicg, S. Martelli, J. Konopa, Structure-activity relationship for antineoplastic imidazoacridinones: synthesis and antileukemic activity in vivo, Journal of Medicinal Chemistry 39 (5) (1996) 1028–1032.

[81] E. Llama, C.D. Campo, M. Capo, M. Anadon, Synthesis and antitumour activity of pyrido-amsacrine analogue and related compounds, Journal of the Pharmaceutical Society 82 (1993) 262–265.

[82] T. Sugaya, Y. Mimura, Y. Shida, Y. Osawa, I. Matsukuma, S. Ikeda, S. Akinaga, M. Morimoto, T. Ashizawa, M. Okabe, H. Ohno, M. Gomi, M. Kasai, 6H-pyrazolo[4,5,1-de] acridin-6-ones as a novel class of antitumor agents, Journal of Medicinal Chemistry 37 (7) (1994) 1028–1032.

[83] I.K. Kostakis, P. Magiatis, N. Pouli, P. Marakos, A.L. Skaltsounis, H. Pratsinis, S. Léonce, A. Pierré, Design, synthesis, and antiproliferative activity of some new pyrazole-fused amino derivatives of the pyranoxanthenone, pyranothioxanthenone, and pyranoacridone ring systems: a new class of cytotoxic agents, Journal of Medicinal Chemistry 45 (12) (2002) 2599–2609.

[84] J. Kamata, T. Okada, Y. Kotake, J. Niijima, K. Nakamura, T. Uenaka, A. Yamaguchi, K. Tsukahara, T. Nagasu, N. Koyanagi, K. Kitoh, K. Yoshimatsu, H. Yoshino, H. Sugumi, Synthesis and evaluation of novel pyrimido-acridone, -phenoxadine, and -carbazole as topoisomerase II inhibitors, Chemical and Pharmaceutical Bulletin 52 (9) (2004) 1071–1081.

[85] I. Antonini, D. Cola, P. Polucci, M. Bontemps-Gracz, E. Borowski, S. Martelli, Synthesis of (dialkylamino)alkyl-disubstituted pyrimido[5,6,1-de]acridines, a novel group of anticancer agents active on a multidrug resistant cell line, Journal of Medicinal Chemistry 38 (17) (1995) 3282–3286.

[86] I. Antonini, P. Polucci, T.C. Jenkins, L.R. Kelland, E. Menta, N. Pescalli, B. Stefanska, J. Mazerski, S. Martelli, 1-[(ω-aminoalkyl)amino]-4- [N-(ω-aminoalky)carba moyl] 9-oxo-9,10-dihydroacridines as intercalating cytotoxic agents: synthesis, DNA binding and biological evaluation, Journal of Medicinal Chemistry 40 (23) (1997) 3749–3755.

[87] I. Antonini, P. Polucci, L.R. Kelland, E. Menta, N. Pescalli, S. Martelli, 2,3-Dihydro-1H,7H-pyrimido[5,6,1-de]acridine-1,3,7-trione derivatives, a class of cytoxic agents active on multidrug-resistant cell lines: synthesis, biological evaluation and structure activity relationships, Journal of Medicinal Chemistry 42 (14) (1999) 2535–2541.

[88] I. Antonini, P. Polucci, A. Magnano, S. Martelli, Synthesis, antitumour, cytotoxicity and DNA-binding of novel N-5,2-Di(ω-aminoalkyl)-2,6-dihydropyrazolo[3,4,5-kl] acridine-5-carboxamides, Journal of Medicinal Chemistry 44 (20) (2001) 3329–3333.

[89] I. Antonini, P. Polucci, A. Magnano, B. Gatto, M. Palumbo, E. Menta, N. Pescalli, S. Martelli, 2,6-Di[ω-aminoalkyl]-2,5,6,7-tetrahydropyrazolo[3,4,5-mn]pyrimido [5,6,1-de]-acridine-5,7-diones: novel, potent, cytotoxic and DNA-binding agents, Journal of Medicinal Chemistry 45 (3) (2002) 696–702.

[90] I. Antonini, P. Polucci, A.B. Gatto, M. Palumbo, E. Menta, N. Pescalli, S. Martelli, Design, synthesis and biological properties of new bis-(acridine-4-carboxamides) as anticancer agents, Journal of Medicinal Chemistry 46 (14) (2003) 3109–3115.

[91] I. Antonini, P. Polucci, A. Magnano, S. Sparapani, S. Martelli, Rational design, synthesis and biological evaluation of bis(pyrazolo[3,4,5-k,l] acridine-5-carboxamides as new anticancer agents, Journal of Medicinal Chemistry 47 (21) (2004) 5244–5250.

[92] O. Tabarrini, V. Cecchetti, A. Fravolini, G. Nocentini, A. Barzi, S. Sabatini, H. Miao, C. Sissi, Design and synthesis of modified quinolones as antitumoral acridones, Journal of Medicinal Chemistry 42 (12) (1999) 2136–2144.

[93] A. Kamal, O. Srinivas, P. Ramulu, G. Ramesh, P.P. Kumar, Synthesis of C8-linked pyrrolo[2,1-c][1,4]benzodiazepine-acridone/acridine hybrids as potential DNA-binding agents, Bioorganic and Medicinal Chemistry Letters 14 (15) (2004) 4107–4111.

[94] N.K. Sathish, P. GopKumar, V.V. Rajendra Prasad, S.M. Shanta kumar, Y.C. Mayur, Synthesis, chemical characterisation of novel 1,3-dimethyl acridones as cytotoxic agents, and their DNA-binding studies, Medicinal Chemistry Research 19 (2010) 674–689.

[95] N.K. Sathish, V.V. Rajendra Prasad, N.M. Raghavendra, S.M. Shanta Kumar, Y.C. Mayur, Synthesis of novel 1,3-diacetoxy-acridones as cytotoxic agents and their DNA-binding studies, Scientia Pharmaceutica 77 (2009) 19–32.

[96] N. Desbois, M. Gardette, J. Papon, P. Labarre, A. Maisonial, P. Auzeloux, C. Lartigue, B. Bouchon, E. Debiton, Y. Blache, O. Chavignon, J.C. Teulade, J. Maublant, J.C. Madelmont, N. Moins, J.M. Chezal, Design, synthesis and preliminary biological evaluation of acridine compounds as potential agents for a combined targeted chemo-radionuclide therapy approach to melanoma, Bioorganic and Medicinal Chemistry 16 (2008) 7671–7690.

[97] O.O. Fadeyi, S.T. Adamson, E.L. Myles, C.O. Okoro, Novel fluorinated acridone derivatives. Part 1: synthesis and evaluation as potential anticancer agents, Bioorganic and Medicinal Chemistry Letters 18 (14) (2008) 4172–4176.

[98] C. Gao, Y. Jiang, C. Tan, X. Zu, H. Liu, D. Cao, Synthesis and potent antileukemic activities of 10-benzyl-9(10H)-acridinones, Bioorganic and Medicinal Chemistry 16 (18) (2008) 8670–8675.

[99] X.F. Huang, Y. Zhua, H.L. Zhu, Synthesis and evaluation of N-benzyl-acridinone derivatives induced apoptosis in human liver cancer cell-lines, Letters in Drug Design and Discovery 8 (2011) 606–611.

[100] B. Zhang, K. Chen, N. Wang, C. Gao, Q. Sun, L. Li, Y. Chen, C.H. Liu, Y. Jiang, Molecular design, synthesis and biological research of novel pyridyl acridones as potent DNA-binding and apoptosis-inducing agents, European Journal of Medicinal Chemistry 77 (2015) 214–226.

[101] M. Mohammadi-Khanaposhtani, M. Safavi, R. Sabourian, M. Mahdavi, M. Pordeli, M. Saeedi, S.K. Ardestani, A. Foroumadi, A.T. Akbarzadeh, Design, synthesis, invitro cytotoxic activity evaluation,and apoptosis-induction study of new9(10H)-acridinone-1,2,3-triazoles, Molecular Diversity 19 (4) (2015) 787–795.

[102] H. Chao, Y. Jiao-Sheng, Z. Xiang-Hui, S. Bo, L. Min-Bo, L. Jun, Synthesis and evaluation of the antitumor activity of polyhalo acridone derivatives, RSC Advances 5 (2015) 17444–17450.

[103] S. Zimmermann, U.M. Martens, Telomeres and telomerase as targets for cancer therapy, Cellular and Molecular Life Sciences 64 (7–8) (2007) 906–921.

[104] C.K. Kwok, C.J. Merrick, G-Quadruplexes: prediction, characterization, and biological application, Trends in Biotechnology 35 (10) (2017) 997–1013.

[105] R. Riqo, M. Palumbo, C. Sissi, G-quadruplexes in human promoters: a challenge for therapeutic applications, Biochimica et Biophysica Acta 1861 (5 Pt B) (2017) 1399–1413.

[106] S. Neidle, M.A. Read, G-quadruplexes as therapeutic targets, Biopolymers 56 (3) (2001) 195–208.

[107] S. Taetz, C. Baldes, T.E. Murdter, E. Kleideiter, K. Piotrowska, U. Bock, E. Haltner-Ukomadu, J. Huwer, U.F. Schaefer, U. Klitz, C.M. Lehr, Biopharmaceutical characterization of telomerase inhibitor BRACO19, The Pharmaceutical Research 23 (5) (2006) 1031–1037.

[108] R.J. Harrison, A.P. Reszka, S.M. Haider, B. Romagnoli, J. Morrell, M.A. Read, S.M. Gowan, C.M. Incles, L.R. Kelland, S. Neidle, Evaluation of by substituted acridone derivatives as telomerase inhibitors: the importance of G-quadruplex binding, Bioorganic and Medicinal Chemistry Letters 14 (23) (2004) 5845–5849.

[109] V.P. Zambre, P.R. Murumkar, R. Giridhar, M.R. Yadav, Development of highly predictive 3D-QSAR CoMSIA models for anthraquinone and acridone derivatives as telomerase inhibitors targeting G-quadruplex DNA telomere, Journal of Molecular Graphics and Modelling 29 (2) (2010) 229–239.

[110] F. Cuenca, M.J. Moore, K. Johnson, B. Guyen, A. De Cian, S. Neidle, Design, synthesis and evaluation of 4,5-disubstituted acridone ligands with high G-quadruplex affinity and selectivity, together with low toxicity to normal cell, Bioorganic and Medicinal Chemistry Letters 19 (17) (2009) 5109–5113.

[111] C. Gao, S. Li, X. Lang, H. Liu, F. Liu, C. Tan, Y. Jiang, Synthesis and evaluation of 10-(3,5-dimethoxy) benzyl-9(10H)-acridone derivatives as selective telomeric G-quadruplex DNA ligands, Tetrahedron 68 (38) (2012) 7920–7925.

[112] C. Gao, W. Zhang, S. He, S. Li, F. Liu, Y. jiang, Synthesis and antiproliferative activity of 2,7-diamino10-(3,5-dimethoxy)benzyl-9(10H)-acridone derivatives as potent telomeric G-quadruplex DNA ligands, Bioorganic Chemistry 60 (2015) 30–36.

[113] C. Gao, F. Liu, X. Luan, C. Tan, H. Liu, Y. Xie, Y. Jin, Y. Jiang, Novel synthetic 2-amino-10-(3,5-dimethoxy)benzyl-9(10H)-acridinone derivatives as potent DNA-binding antiproliferative agents, Bioorganic and Medicinal Chemistry 18 (21) (2010) 7507–7514.

[114] Y. Wang, D. Gao, Z. Chen, S. Li, C. Gao, D. Cao, F. Liu, H. Liu, Y. Jiang, Acridone derivative 8a induces oxidativestress-mediated apoptosis in CCRF CEM leukemia cells: application of metabolomics in mechanistic studies of antitumor agents, PLoS One 8 (5) (2013) e63572.

[115] R.C. Jackson, G. Weber, H.P. Morris, IMP dehydrogenase, an enzyme linked with proliferation and malignancy, Nature 256 (5515) (1975) 331–333.

[116] H.N. Jayaram, M. Grusch, D.A. Cooney, G. Krupitza, Consequences of IMP dehydrogenase inhibition and its relationship to cancer and apoptosis, Current Medicinal Chemistry 6 (7) (1999) 561–574.

[117] G. Cholewinski, M. Malachowska-Ugarte, K. Dzierzbicka, The chemistry of mycophenolic acid-synthesis and modifications towards desired biological activity, Current Medicinal Chemistry 17 (18) (2010) 1926–1941.

[118] S.H. Watterson, P. Chen, Y. Zhao, H.H. Gu, T.G. Murali Dhar, Z. Xiao, S.H. Ballentine, Z. Shen, C.A. Fleener, K.A. Rouleau, M. Obermeier, Z. Yang, K.W. McIntyre, D.J. Shuster, M. Witmer, D. Dambach, S. Chao, A. Mathur, B.C. Chen, J.C. Barrish Robl, R. Townsend, E.J. Iwanowicz, Acridone-based inhibitors of inosine $5'$-monophosphate dehydrogenase: discovery and SAR leading to the identification of n-(2-(6-(4-ethylpiperazin-1-yl)pyridin-3-yl) propan-2-yl)-2-fluoro-9-oxo-9,10-dihydro acridine-3-carboxamide (BMS-566419), Journal of Medicinal Chemistry 50 (15) (2007) 3730–3742.

[119] M. Malachowska-Ugarte, G. Cholewinski, K. Dzierzbicka, P. Trzonkowski, Synthesis and biological activity of novel mycophenolic acid conjugates containing nitro-acridine/acridone derivatives, European Journal of Medicinal Chemistry 54 (2012) 197–201.

[120] A.L. Risinger, F.J. Giles, S.L. Mooberry, Microtubule dynamics as a target in oncology, Cancer Treatment Reviews 35 (3) (2009) 255–261.

[121] S. Honore, E. Pasquier, D. Braguer, Understanding microtubule dynamics for improved cancer therapy, Cellular and Molecular Life Sciences 62 (24) (2005) 3039–3056.

[122] V. Verones, N. Flouque, M. Lecoeur, A. Lemoine, A. Farce, B. Baldeyrou, C. Mahieu, N. Wattez, A. Lansiaux, J.F. Goossens, P. Berthelot, N. Lebegue, Synthesis, antiproliferative activity and tubulin targeting effect of acridinone and dioxophenothiazine derivatives, European Journal of Medicinal Chemistry 59 (2013) 39–47.

[123] P. Singh, A. Kumar, A. Sharma, G. Kaur, Identification of amino acid appended acridinesas potential leads to anti-cancer drugs, Bioorganic and Medicinal Chemistry Letters 25 (18) (2015) 3854–3858.

[124] K.A. Nolan, M.P. Humphries, J. Barnes, J.R. Doncaster, M.C. Caraher, N. Tirelli, R.A. Bryce, R.C. Whitehead, I.J. Stratford, Triazoloacridin-6-ones as novel inhibitors of the quinone oxidoreductases NQO1 and NQO2, Bioorganic and Medicinal Chemistry 18 (2) (2010) 696–706.

[125] E.F. Marques, M.A. Bueno, P.D. Duarte, L.R. Silva, A.M. Martinelli, C.Y. dos Santos, R.P. Severino, D. Brömme, P.C. Vieira, A.G. Corrêa, Evaluation of synthetic acridones and 4-quinolinones as potent inhibitors of cathepsins L and V, European Journal of Medicinal Chemistry 54 (2012) 10–21.

[126] Y.R. Babu, M. Bhagavanraju, G.D. Reddy, G.J. Peters, V.V. Prasad, Design and synthesis of quinazolinone tagged acridones as cytotoxic agents and their effects on EGFR tyrosine kinase, Archiv der Pharmazie Chemistry in Life Sciences 347 (9) (2014) 624–634.

[127] A.F. da Silva, R.S. Seixas, A.M. Silva, J. Coimbra, A.C. Fernandes, J.P. Santos, A. Matos, J. Rino, I. Santos, F. Marques, Synthesis, characterization and biological evaluation of carboranylmethyl benzo[b]acridones as novel agents for boron neutron capture therapy, Organic and Biomolecular Chemistry 12 (28) (2014) 5201–5211.

[128] R.C. Peery, J.Y. Liu, J.T. Zhang, Targeting surviving for therapeutic discovery: past, present, and future promises, Drug Discovery Today 22 (10) (2017) 1466–1477.

[129] B. Zhang, N. Wang, C. Zhang, C. Gao, W. Zhang, K. Chen, W. Wu, Y. Chen, C. Tan, F.F. Liu, Y. Jiang, Novel multi-substituted benzyl acridone derivatives as survivin inhibitors for hepatocellular carcinoma treatment, European Journal of Medicinal Chemistry 129 (2017) 337–348.

[130] F.J. Sharom, ABC multidrug transporters: structures, function and role in chemoresistance, Pharmacogenomics 9 (1) (2008) 105–127.

[131] R. Kumar, M. Kaur, O. Silakari, Modulation approaches to improve cancer chemotherapy: a review, Anti-Cancer Agents in Medical Chemistry 14 (5) (2014) 713–749.

[132] N. Dodic, B. Dumaitre, A. Daugan, P. Pianetti, Synthesis and activity against multidrug resistance in Chinese hamster ovary cells of new acridone-4-carboxamides, Journal of Medicinal Chemistry 38 (13) (1995) 2418–2426.

[133] I.E. Kuppens, E.O. Witteveen, R.C. Jewell, S.A. Radema, E.M. Paul, S.G. Mangum, J.H. Beijnen, E.E. Voest, J.H. Schellens, A phase I, randomized, open-label, parallel-cohort, dose-finding study of elacridar (GF120918) and oral topotecan in cancer patients, Clinical Cancer Research 13 (11) (2007) 3276–3285.

[134] A. Boumendjel, S.A. Ahmed-Belkacem, M. Blanc, A. Di Pietro, Acridone derivatives: design, synthesis, and inhibition of breast cancer resistance protein ABCG2, Bioorganic and Medicinal Chemistry 15 (8) (2007) 2892–2897.

[135] J.K. Horton, K.N. Thimmaiah, G.A. Altenberg, A.F. Castro, G.S. Germain, G.K. Gowda, P.J. Houghton, Characterization of a novel bisacridone and comparison with psc 833 as a potent and poorly reversible modulator of P-glycoprotein, Molecular Pharmacology 52 (6) (1997) 948–957.

[136] G. Krishnegowda, P. Thimmaiah, R. Hegde, C. Dass, P.J. Houghton, K.N. Thimmaiah, Synthesis and chemical characterization of 2 methoxy-N(10)-substituted acridones needed to reverse vinblastine resistance in multidrug resistant (MDR) cancer cells, Bioorganic and Medicinal Chemistry 10 (7) (2002) 2367–2380.

[137] Y.C. Mayur, O. Ahmad, V.V. Rajendra Prasad, M.N. Purohit, N. Srinivasulu, S.M. Shanta Kumar, Synthesis of 2-methyl N10-substituted acridones as selective inhibitors of multidrug resistance (MDR) associated protein in cancer cells, Medicinal Chemistry 4 (5) (2008) 457–465.

[138] V.V. Rajendra Prasad, J. VenkatRao, R.S. Giri, N.K. Sathish, S.M. Shanta Kumar, Y.C. Mayur, Chloroacridone derivatives as cytotoxic agents active on multidrug-resistant cell lines and their duplex DNA complex studies by electrospray ionization mass spectrometry, Chemico-Biological Interactions 176 (2–3) (2008) 212–219.

[139] V.V. Rajendra Prasad, G.J. Peters, C. Lemos, I. Kathmann, Y.C. Mayur, Cytotoxicity studies of some novel fluoro acridone derivatives against sensitive and resistant cancer cell lines and their mechanistic studies, European Journal of Pharmaceutical Sciences 43 (4) (2011) 217–224.

[140] V.V. Rajendra Prasad, G.D. Reddy, D. Appaji, G.J. Peters, Y.C. Mayur, Chemosensitizing acridones: in vitro calmodulin dependent cAMP phosphodiesterase inhibition, docking, pharmacophore modeling and 3D QSAR studies, Journal of Molecular Graphics and Modelling 40 (2013) 116–124.

[141] Y.C. Mayur, Zaheeruddin, G.J. Peters, C. Lemos, L. Kathmann, V.V. Rajendra Prasad, Synthesis of 2-fluoro n10-substituted acridones and their cytotoxicity studies in sensitive and resistant cancer cell lines and their DNA intercalation studies, Archiv der Pharmazie 342 (11) (2009) 640–650.

[142] P. Singh, J. Kaur, P. Kaur, S. Kaur, Search for MDR modulators: design, syntheses and evaluations of N-substituted acridones for interactions with P-gp and Mg^{2+}, Bioorganic and Medicinal Chemistry 17 (6) (2009) 2423–2427.

[143] P. Singh, J. Kaur, B. Yadav, S.S. Komath, Design, synthesis and evaluations of acridone derivatives using *Candida albicans*-search for MDR modulators led to the identification of an anti-candidiasis agent, Bioorganic and Medicinal Chemistry 17 (11) (2009) 3973–3979.

[144] V.S. Velingkar, V.D. Dandekar, Microwave-assisted synthesis and evaluation of substituted aryl propyl acridone-4-carboxamides as potential chemosensitizing agents for cancer, Letters in Drug Design and Discovery 8 (3) (2011) 268–275.

[145] V.S. Velingkar, V.D. Dandekar, Design, synthesis and evaluation of substituted N-(3-arylpropyl)-9,10-dihydro-9-oxoacridine-4-carboxamides as potent MDR reversal agents in cancer, Chinese Journal of Chemistry 29 (2011) 504–510.

[146] R. Kumar, M. Kaur, M.S. Bahia, O. Silakari, Synthesis, cytotoxic study and docking based multidrug resistance modulator potential analysis of 2-(9-oxoacridin-10(9H)-yl)-N-phenyl acetamides, European Journal of Medicinal Chemistry 80 (2014) 83–91.

[147] R. Kumar, M.S. Bahia, O. Silakari, Synthesis, cytotoxic activity, and computational analysis of N10-substituted acridone analogs, Medicinal Chemistry Research 24 (3) (2015) 921–933.

[148] K. Thimmaiah, A.G. Ugarkar, E.F. Martis, M.S. Shaikh, E.C. Coutinho, M.C. Yergeri, Drug-DNA interaction studies of acridone-based derivatives, Nucleosides, Nucleotides and Nucleic Acids 34 (5) (2015) 309–331.

[149] V. Kuete, H. Fouotsa, A.T. Mbaveng, B. Wiench, A.E. Nkengfack, T. Efferth, Cytotoxicity of a naturally occurring furoquinoline alkaloid and four acridone alkaloids towards multifactorial drug-resistant cancer cells, Phytomedicine 22 (10) (2015) 946–951.

[150] V.V. Rajendra Prasad, G. Deepak Reddy, I. Kathmann, M. Amareswararao, G.J. Peters, Nitric oxide releasing acridone carboxamide derivatives as reverters of doxorubicin resistance in MCF7/Dx cancer cells, Bioorganic Chemistry 64 (2016) 51–58.

[151] M. Murahari, P.S. Kharkar, N. Lonikar, Y.C. Mayur, Design, synthesis, biological evaluation, molecular docking and QSAR studies of 2,4-dimethylacridones as anticancer agents, European Journal of Medicinal Chemistry 130 (2017) 154–170.

[152] M. Murahari, K.V. Prakash, G.J. Peters, Y.C. Mayur, Acridone-pyrimidine hybrids-design, synthesis, cytotoxicity studies in resistant and sensitive cancer cells and molecular docking studies, European Journal of Medicinal Chemistry 139 (2017) 961–981.

[153] C.S. Sepulveda, M.L. Fascio, C.C. Garcia, N.B.D. Accorso, E.B. Damonate, Acridones as antiviral agents: synthesis, chemicals and biological properties, Current Medicinal Chemistry 20 (2013) 2402–2414.

[154] N. Yamamoto, H. Furukawa, Y. Ito, S. Yoshida, K. Maeno, Y. Nishiyama, Anti-herpes virus activity of citrusinine-I, a new acridone alkaloid, and related compounds, Antiviral Research 12 (1989) 21–36.

[155] C. Chansriniyom, N. Ruangrungsi, V. Lipipun, T. Kumamoto, T. Ishikawac, Isolation of acridone alkaloids and N-[(4-monoterpenyloxy)phenylethyl]-substituted sulfur-containing propanamide derivatives from *Glycosmis parva* and their anti-herpes simplex virus activity, Chemical and Pharmaceutical Bulletin 57 (11) (2009) 1246–1250.

[156] Y. Takemura, M. Ju-ichi, C. Ito, H. Furukawa, H. Tokuda, Studies on the inhibitory effects of some acridone alkaloids on Epstein-Barr virus activation, Planta Medica 61 (4) (1995) 366–368.

[157] M. Itoigawa, C. Ito, T.S. Wu, F. Enjo, H. Tokuda, H. Nishino, H. Furukawa, Cancer chemopreventive activity of acridone alkaloids on EpsteinBarr virus activation and two-stage mouse skin carcinogenesis, Cancer Letters 193 (2003) 133–138.

[158] P. Akanitapichat, C.T. Lowden, K.F. Bastow, 1, 3-Dihydroxyacridone derivatives as inhibitors of herpes virus replication, Antiviral Research 45 (2) (2000) 123–134.

[159] J.R. Vance, K.F. Bastow, Inhibition of DNA topoisomerase II catalytic activity by the antiviral agents 7-chloro-1, 3-dihydroxyacridone and 1,3,7-trihydroxyacridone, Biochemical Pharmacology 58 (4) (1999) 703–708.

[160] P. Akanitapichat, K.F. Bastow, The antiviral agent 5-chloro-1,3-dihydroxyacridone interferes with assembly and maturation of herpes simplex virus, Antiviral Research 53 (2) (2002) 113–126.

[161] C.T. Lowden, K.F. Bastow, Anti-herpes simplex virus activity of substituted 1-hydroxyacridones, Journal of Medicinal Chemistry 46 (23) (2003) 5015–5020.

[162] C.T. Lowden, K.F. Bastow, Cell culture replication of herpes simplex virus and, or human cytomegalovirus is inhibited by 3, 7-dialkoxylated, 1-hydroxyacridone derivatives, Antiviral Research 59 (3) (2003) 143–154.

[163] J.R. Goodell, A.A. Madhok, H. Hiasa, D.M. Ferguson, Synthesis and evaluation of acridine -and acridone-based anti-herpes agents with topoisomerase activity, Bioorganic and Medicinal Chemistry 14 (2006) 5467–5480.

[164] M. Fujiwara, M. Okamoto, M. Okamoto, M. Watanabe, H. Machida, S. Shigeta, K. Konno, T. Yokota, M. Baba, Acridone derivatives are selective inhibitors of HIV-1 replication in chronically infected cells, Antiviral Research 43 (3) (1999) 179–189.

[165] S. Schutze, S. Nottrott, K. Pfizenmaier, M. Kronke, Tumor necrosis factor signal transduction. Cell-type-specific activation and translocation of protein kinase C, The Journal of Immunology 144 (7) (1990) 2604–2608.

[166] J.A. Turpin, R.W. Jr Buckheit, D. Derse, M. Hollingshead, K. Williamson, C. Palamone, M.C. Osterling, S.A. Hill, L. Graham, C.A. Schaeffer, M. Bu, M. Huang, W.M. Cholody, C.J. Michejda, W.G. Rice, Inhibition of acute-, latent-, and chronic-phase human immunodeficiency virus type 1 (HIV-1) replication by a bistriazoloacridone analog that selectively inhibits HIV-1 transcription, Antimicrobial Agents and Chemotherapy 42 (3) (1998) 487–494.

[167] S.M. Sondhi, R.P. Verma, V.K. Sharma, S.N. Singhal, J.L. Kraus, M. Camplo, J.C. Chermann, Synthesis and HIV screening of some heterocyclic compound, Phosphorus sulfur silicon 122 (1997) 215–225.

[168] C. Pellerin, J. van den Hurk, J. Lecomte, P. Tijssen, Identification of a new group of bovine viral diarrhea virus stari associate with severe outbreaks and high mortalities, Virology 203 (2) (1994) 260–268.

[169] O. Tabarrini, G. Manfroni, A. Fravolini, V. Cecchetti, S. Sabatini, E.D. Clercq, J. Rozenski, C. Bruno, H. Dutartre, J. Paeshuyse, J. Neyts, Synthesis and anti-BVDV activity of acridones as new potential antiviral agents, Journal of Medicinal Chemistry 49 (8) (2006) 2621–2627.

[170] K.F. Bastow, New acridone inhibitors of human herpes virus replication, Current Drug Targets – Infectious Disorders 4 (4) (2004) 323–330.

[171] K. Mohd Hanafiah, J. Groeger, A.D. Flaxman, S.T. Wiersma, Global epidemiology of hepatitis C virus infection: new estimates of age-specific antibody to HCV seroprevalence, Hepatology 57 (4) (2013) 1333–1342.

[172] G. Manfroni, J. Paeshuyse, S. Massari, S. Zanoli, B. Gatto, G. Maga, O. Tabarrini, V. Cecchetti, A. Fravolini, J. Neyts, Inhibition of subgenomic hepatitis C virus RNA replication by acridone derivatives: identification of NS3 helicase inhibitor, Journal of Medicinal Chemistry 52 (10) (2009) 3354–3365.

[173] A. Stankiewicz-Drogon, L.G. Palchykovska, V.G. Kostina, I.V. Alexeeva, A.D. Shved, A.M. Boguszewska-Chachulska, New acridone-4-carboxylic acid derivatives as potential inhibitors of hepatitis C virus infection, Bioorganic and Medicinal Chemistry 16 (19) (2008) 8846–8852.

[174] A. Stankiewicz-Drogoń, B. Dorner, T. Erker, A.M. Boguszewska-Chachulska, Synthesisofnewacridone derivatives, inhibitors of NS3 helicase,which efficiently and specifically inhibit subgenomic HCV replication, Journal of Medicinal Chemistry 53 (8) (2010) 3117–3126.

[175] E.B. Damonte, C.E. Coto, Treatment of arenavirus infections: from basic studies to the challenge of antiviral therapy, Advances in Virus Research 58 (2002) 125–155.

[176] C.S. Sepúlveda, M.L. Fascio, M.B. Mazzucco, M.L. Palacios, R.F. Pellón, C.C. García, N.B. D'Accorso, E.B. Damonte, Synthesis and evaluation of N substituted acridones as antiviral agents against haemorrhagic fever viruses, Antiviral Chemistry and Chemotherapy 19 (1) (2008) 41–47.

[177] C.S. Sepúlveda, C.C. García, M.L. Fascio, N.B. D'Accorso, M.L. Docampo Palacios, R.F. Pellón, E.B. Damonte, Inhibition of Junin virus RNA synthesis by an antiviral acridone derivative, Antiviral Research 93 (1) (2012) 16–22.

[178] V.V. Zarubaev, A.V. Slita, V.Z. Krivitskaya, A.K. Sirotkin, A.L. Kovalenko, N.K. Chatterjee, Direct antiviral effect of cycloferon (10-carboxymethyl-9-acridanone) against adenovirus type 6 in vitro, Antiviral Research 58 (2) (2003) 131–137.

[179] R.W. Snow, J.F. Trape, K. Marsh, The past, present and future of childhood malaria mortality in Africa, Trends in Parasitology 17 (12) (2001) 593–597.

[180] W.H. Wernsdorfer, D. Payne, The dynamics of drug resistance in *Plasmodium falciparum*, Pharmacology and Therapeutics 50 (1) (1991) 95–121.

[181] F. Kurth, P. Pongratz, S. Bélard, B. Mordmuller, P.G. Kremsner, M. Ramharter, In vitro activity of pyronaridine against *Plasmodium falciparum* and comparative evaluation of antimalarial drug susceptibility assays, Malaria Journal 8 (2009) 79–84.

[182] G.H. Svoboda, G.A. Poore, P.J. Simpson, G.B. Boder, Alkaloids of *Acronychia baueri* Schott. I. Isolation of the alkaloids and a study of the antitumor and other biological properties of acronycine, Journal of Pharmaceutical Sciences 55 (1996) 758–768.

[183] J. Schneider, E.L. Evans, E. Grunberg, R.I. Fryer, Synthesis and biological activity of acronycine analogs, Journal of Medicinal Chemistry 15 (1972) 266–270.

[184] L.K. Basco, S. Mitaku, A.L. Skaltsounis, N. Ravelomanantsoa, F. Tillequin, M. Koch, J. Le Bras, In vitro activities of furoquinoline and acridone alkaloids against *Plasmodium falciparum*, Antimicrobial Agents and Chemotherapy 38 (1994) 1169–1171.

[185] H. Fujioka, Y. Nishiyama, H. Furukawa, N. Kumada, In vitro and in vivo activities of atalaphillinine and related acridone alkaloids against rodent malaria, Antimicrobial Agents and Chemotherapy 33 (1989) 6–9.

[186] M.W. Muriithi, W.R. Abraham, J. Addae-Kyereme, I. Scowen, S.L. Croft, P.M. Gitu, H. Kendrick, E.N. Njagi, C.W. Wright, Isolation and in vitro antiplasmodial activities of alkaloids from *Teclea trichocarpa*: in vivo antimalarial activity and x-ray crystal structure of normelicopicine, Journal of Natural Products 65 (2002) 956–959.

[187] A.F. Valdés, Acridine and acridinones: old and new structures with antimalarial activity, The Open Medicinal Chemistry Journal 5 (2011) 11–20.

[188] L.H. Schmidt, Antimalarial properties of floxacrine, a dihydroacridinedione derivative, Antimicrobial Agents and Chemotherapy 16 (4) (1979) 475–485.

[189] A. Dorn, J.P. Scovill, W.Y. Ellis, H. Matile, R.G. Ridley, J.L. Vennerstrom, Short report: floxacrine analog WR243251 inhibits hematin polymerization, The American Journal of Tropical Medicine and Hygiene 65 (2001) 19–20.

[190] G.A. Biagini, N. Fisher, N. Berry, P.A. Stocks, B. Meunier, D.P. Williams, R. Bonar- Law, P.G. Bray, A. Owen, P.M. Neill, S.A. Ward, Acridinediones: selective and potent inhibitors of the malaria parasite mitochondrial bc1 complex, Molecular Pharmacology 73 (5) (2008) 1347–1355.

[191] W. Raether, B. Enders, J. Hofmann, U. Schwannecke, H. Seidenath, H. Hänel, M. Uphoff, Anti- malarial activity of new floxacrine- related acridinedione derivatives studies on blood schizontocidal action of potential candidates against *P. berghei* in mice and *P. falciparum* in vivo and in vitro, Parasitology Research 75 (8) (1989) 619–626.

[192] R. Winter, J.X. Kelly, M.J. Smilkstein, R. Dodean, G.C. Bagby, R.K. Rathbun, J.I. Levin, D. Hinrichs, M.K. Ricoe, Evaluation and lead optimization of anti-malarial acridones, Experimental Parasitology 114 (1) (2006) 47–56.

[193] J.X. Kelly, M.J. Smilkstein, R.A. Cooper, K.D. Lane, R.A. Johnson, A. Janowsky, R.A. Dodean, D.J. Hinrichs, R. Winter, M.K. Ricoe, Design, synthesis, and evaluation of 10-N-substituted acridones as novel chemosensitizers in *Plasmodium falciparum*, Antimicrobial Agents and Chemotherapy 51 (11) (2007) 4133–4140.

[194] J.X. Kelly, M.J. Smilkstein, R. Brun, S. Wittlin, R.A. Cooper, K.D. Lane, A. Janowsky, R.A. Johnson, R.A. Dodean, R. Winter, D.J. Hinrichs, M.K. Ricoe, Discovery of dual function acridones as a new antimalarial chemotype, Nature 459 (7244) (2009) 270–273.

[195] A. Fernández-Calienes, R. Pello'n, M. Docampo, M. Fascio, N. D'Accorso, L. Maes, J. Mendiola, L. Monzote, L. Gille, L. Rojas, Antimalarial activity of new acridinone derivatives, Biomedicine and Pharmacotherapy 65 (3) (2011) 210–214.

[196] A.A. Walker, A new approach for developing anti-malarial agents, Drug Discovery Today 14 (2009) 19–20.

[197] R.M. Cross, J.R. Maignan, T.S. Mutka, L. Luong, J. Sargent, D.E. Kyle, R. Manetsch, Optimization of 1,2,3,4-tetrahydroacridin-9(10H)-ones as antimalarials utilizing structure activity and structure property relationships, Journal of Medicinal Chemistry 54 (13) (2011) 4399–4426.

[198] S. Bailey, K. Smith, A.H. Fairlamb, W.N. Hunter, Substrate interactions between trypanothione reductase and N1-glutathionylspermidine disulphide at 0.28-nm resolution, European Journal of Biochemistry 213 (1) (1993) 67–75.

[199] C.S. Bond, Y. Zhang, M. Berriman, M.L. Cunningham, A.H. Fairlamb, W.N. Hunter, Crystal structure of *Trypanosoma cruzi* trypanothione reductase in complex with trypanothione, and the structure-based discovery of new natural product inhibitors, Structure 7 (1) (1999) 81–89.

[200] K.M. Ahua, J.R. Ioset, A. Ransijn, J. Mauël, S. Mavi, K. Hostettmann, Antileishmanial and antifungal acridone derivatives from the roots of *Thamnosma rhodesica*, Phytochemistry 65 (7) (2004) 963–968.

[201] A.F. Waffo, P.H. Coombes, N.R. Crouch, D.A. Mulholland, S.M. El Amin, P.J. Smith, Acridone and furoquinoline alkaloids from *Teclea gerrardii* (rutaceae: Toddalioideae) of southern Africa, Phytochemistry 68 (5) (2007) 663–667.

[202] D.A.P. dos Santos, P.C. Vieira, F.G.F. da Silva, J.B. Fernandas, L. Rattray, S.L. Croft, Antiparasitic activity of acridone alkaloids from *Swinglea glutinosa* Merr, Journal of the Brazilian Chemical Society 20 (4) (2009) 644–651.

[203] D. Lacroix, S. Prado, D. Kamoga, J. Kasenene, B. Bodo, Structure and in vitro antiparasitic activity of constituents of *Citropsis articulata* root bark, Journal of Natural Products 74 (2011) 2286–2289.

[204] F. Astelbauer, A. Obwaller, A. Raninger, B. Brem, H. Greger, M. Duchêne, W. Wernsdorfer, J. Walochnik, Anti-leishmanial activity of plant-derived acridones, flavaglines, and sulfur-containing amides, Vector Borne and Zoonotic Diseases 11 (7) (2011) 793–798.

[205] S.M. Roopan, A. Bharathi, N.A. Al-Dhabi, M.V. Arasu, G. Madhumitha, Synthesis and insecticidal activity of acridone derivatives to *Aedes aegypti* and *Culex quinquefasciatus* larvae and non-target aquatic species, Scientific Reports 7 (2017) 39753.

[206] F. Dolabela, S.G. Oliveira, J.M. Nascimento, J.M. Peres, H. Wagner, M.M. Póvoa, A.B. de Oliveira, In vitro antiplasmodial activity of extract and constituents from *Esenbeckia febrifuga*, a plant traditionally used to treat malaria in the Brazilian Amazon, Phytochemistry 15 (5) (2008) 367–372.

[207] D. Sharples, J. Barbe, A.M. Galy, J.P. Galy, New antiamebic acridines II. Synthesis and DNA binding of a series of 9-acridanones and 9-Iminoacridines, Chemotherapy 33 (5) (1987) 347–354.

[208] A. Osuna, J.I. Rodriguez-Santiago, L.M. Ruiz-Perez, F. Gamarro, S. Castanys, G. Giovannangeli, A.M. Galy, J.P. Galy, J.C. Soyfer, J. Barbe, Antiamebic activity of new acridinic derivatives against *Naegleria* and *Acanthamoeba* species in vitro, Chemotherapy 33 (1) (1987) 18–21.

[209] A. Osuna, L.M. Ruiz-Perez, F. Gamarro, J.I. Rodriguez-Santiago, D. Sharples, S. Castanys, A.M. Galy, G. Giovannangeli, J.P. Galy, J.C. Soyfer, J. Barbe, New antiparasitic agents: III. Comparison between trypanocidal activities of some acridine derivatives against *Trypanosoma cruzi* in vitro, Chemotherapy 34 (2) (1988) 127–133.

[210] L. Ngadi, N.G. Bisiri, A. Mahamoud, A.M. Galy, J.P. Galy, J.C. Soyfer, J. Barbe, M. Placidi, R.S. Rodriguez-Santiago, C. Mesa-Valle, R. Lombardo, C. Mascaro, A. Osuna, Synthesis and antiparasitic activity of new 1-nitro, 1-amino and 1-acetamido 9-acridinones, Arzneimittel-Forschung Drug Research 43 (1) (1993) 480–483.

[211] P.M. Coelho, L.H. Pereira, *Schitosoma mansoni*: preclinical studies with 9-acridanone-hydrazones in Cebus monkeys experimentally infected, Revista do Instituto de Medicina Tropical de Sao Paulo 33 (1) (1991) 50–57.

[212] L.H. Pereira, P.M. Coelho, J.O. Costa, R.T. de Mello, Activity of 9-acridanone-hydrazone drugs detected at the pre-postural phase, in the experimental chistosomiasis mansoni, Memorias do Instituto Oswaldo Cruz 90 (3) (1995) 425–428.

[213] P.M. Coelho, L.H. Pereira, R.T. de Mello, Antischistosomal activity of acridanone-hydrazonesin cebus monkeys experimentally infected with the SJ strain of schisto- soma mansoni, Revista da Sociedade Brasileira de Medicina Tropical 28 (3) (1995) 179–183.

[214] C.M. Mesa-Valle, J. Castilla-Calvente, M. Sanchez-Moreno, V. Moraleda-Lindez, J. Barbe, A. Osuna, Activity and mode of action of acridine compounds against *Leishmania donovani*, Antimicrobial Agents and Chemotherapy 40 (3) (1996) 684–690.

[215] F. Delmas, A. Avellaned, C. Di Giorgio, M. Robin, E. De Clercq, P. Timon-David, J.P. Galy, Synthesis and antileishmanial activity of (1,3-benzothiazol-2-yl) amino-9-(10H)-acridinone derivatives, European Journal of Medicinal Chemistry 39 (8) (2004) 685–690.

[216] S. Montalvo-Quiros, A. Taladriz-Sender, M. Kaiser, C. Dardonville, Antiprotozoal activity and DNA binding of dicationic acridone, Journal of Medicinal Chemistry 58 (4) (2015) 1940–1949.

[217] Y.Y. Yang, W. Yang, W.J. Zuo, Y.B. Zeng, S.B. Liu, W.L. Mei, H.F. Dai, Two new acridone alkaloids from the branch of *Atalantia buxifolia* and their biological activity, Journal of Asian Natural Products Research 15 (2013) 899–904.

[218] L. Ngadi, A.M. Galy, J.P. Galy, J. Barbe, Some new 1-nitroacridine derivatives as antimicrobial agents, European Journal of Medicinal Chemistry 25 (1990) 67–70.

[219] M. Rahimizadeh, M. Pordel, M. Bakavoli, Z. Bakhtiarpoor, A. Orafaie, Synthesis of imidazo [4,5-a] acridones and imidazo[4,5-a]acridines as potential antibacterial agents, Monatshefte für Chemie 140 (2009) 633–638.

[220] J. Salimon, N. Salih, E. Yousif, A. Hameed, A. Kareem, Synthesis and biological evaluation of 9(10-H)-acridone bearing 1,3,4-oxadiazole derivatives as antimicrobial agents, Arabian Journal of Chemistry 3 (4) (2010) 205–210.

[221] V. Nadaraj, S.T. Selvi, S. Mohan, Microwave-induced synthesis and anti-microbial activity of 7,10,11,12-tetrahydrobenzo[c]acridin-8(9H)-one derivative, European Journal of Medicinal Chemistry 44 (3) (2009) 976–980.

[222] Y.D. Markovich, T.N. Kudryavtseva, K.V. Bogatyrev, P.I. Sysoev, L.G. Klimova, G.V. Nazarov, Synthesis of 2-(4-methyl-1,3-thiazol-5-yl)ethyl esters of acridone carboxylic acids and evaluation of their antibacterial activity, Russian Chemical Bulletin 63 (5) (2014) 1153–1158.

[223] T.N. Kudryavtseva, P.I. Sysoev, S.V. Popkov, G.V. Nazarov, L.G. Klimovac, Synthesis and antimicrobial activity of some acridone derivatives bearing 1,3,4-oxadiazole moiety, Russian Chemical Bulletin 64 (6) (2015) 1341–1344.

[224] T.N. Kudryavtseva, A.Yu. Lamanov, L.G. Klimova, G.V. Nazarova, Synthesis and antimicrobial activity of acridine carboxylic acid derivatives containing a piperazine moiety, Russian Chemical Bulletin 66 (1) (2017) 123–128.

[225] R. Medzhitov, Inflammation 2010: new adventures of an old flame, Cell 140 (6) (2010) 771–776.

[226] I. Grivennikov, F.R. Greten, M. Karin, Immunity, inflammation, and cancer, Cell 140 (6) (2010) 883–899.

[227] D. Naidoo, P.H. Coombes, D.A. Mulholland, N.R. Crouch, A.J. van den Bergh, N-substituted acridone alkaloids from *Toddaliopsis bremekampii* (Rutaceae: Toddalioideae) of south-central Africa, Phytochemistry 66 (14) (2005) 1724–1728.

[228] I.B. Taraporewala, J.M. Kauffman, Synthesis and structure activity relationships of anti-inflammatory 9,10-Di hydro-9-oxo-2-acridine-alkanoic acids and 4-(2-carboxyphenyl) aminobenzene alkanoic acids, Journal of Pharmaceutical Sciences 79 (2) (1990) 173–178.

[229] J.C. Sircar, C.F. Schwender, E.A. Johnson, Soybean lipoxygenase inhibition by nonsteroidal antiinflammatory drugs, Prostaglandins 25 (3) (1983) 393–396.

[230] A. Bhardwaj, J. Kaur, S.K. Sharma, Z. Huang, F. Wuest, E.E. Knaus, Hybrid fluorescent conjugates of COX-2 inhibitors: search fora COX2 isozyme imaging cancer biomarker, Bioorganic and Medicinal Chemistry Letters 23 (1) (2013) 163–168.

[231] M.A. Lowes, A.M. Bowcock, J.G. Krueger, Pathogenesis and therapy of psoriasis, Nature 445 (7130) (2007) 866–873.

[232] A. Menter, C.E. Griffiths, Current and future management of Psoriasis, Lancet 370 (9853) (2007) 272–284.

[233] A. Putic, L. Stecher, H. Prinz, K. Müller, Structure activity relationship studies of acridones as potential antipsoriatic agents. 1. Synthesis and antiproliferative activity of simple N-unsubstituted 10H-acridin-9-ones against human keratinocyte growth, European Journal of Medicinal Chemistry 45 (8) (2010) 3299–3310.

[234] A. Putic, L. Stecher, H. Prinz, K. Müller, Structure activity relationship studies of acridones as potential antipsoriatic agents. 2. Synthesis and antiproliferative activity of 10-substituted hydroxy-10H-acridin-9-ones against human keratinocyte growth, European Journal of Medicinal Chemistry 45 (11) (2010) 5345–5352.

[235] T. Faller, K. Hutton, G. Okafo, A. Gribble, P. Camilleri, D.E. Games, A novel acridone derivative for the fluorescence tagging and mass spectrometric sequencing of peptides, Chemical Communications 16 (1997) 1529–1530.

[236] A. Szymanska, K. Wegner, L. Lankiewitz, Synthesis of N-[(tert-butoxy) carbonyl]-3-(9,10-dihydro-9-oxoacridin-2-yl)L-alanine, a new fluorescent amino acid derivative, Helvetica Chimica Acta 86 (10) (2003) 3326–3331.

[237] N. Bahr, E. Tierney, J.L. Reymond, Highly photoresistant chemosensor using acridone as fluorescent label, Tetrahedron Letters 38 (9) (1997) 1489–1492.

[238] J.L. Reymond, T. Koch, J. Schroder, E. Tierney, A general assay for antibody catalysis using acridone as a fluorescent tag, Proceedings of the National Academy of Sciences of the United States of America 93 (9) (1996) 4251–4256.

[239] A. Shoji, T. Hasegawa, M. Kuwahara, H. Ozaki, H. Sawai, Chemico-enzymatic synthesis of a new fluorescent-labeled DNA by PCR with a thymidine nucleotide analogue bearing an acridone derivative, Bioorganic and Medicinal Chemistry Letters 17 (3) (2007) 776–779.

[240] Y. Hagiwara, T. Hasegawa, A. Shoji, M. Kuwahara, H. Ozaki, H. Sawai, Acridone-tagged DNA as a new probe for DNA detection by fluorescence resonance energy transfer and for mismatch DNA recognition, Bioorganic and Medicinal Chemistry 16 (14) (2008) 7013–7020.

[241] Y. Saito, K. Hanawa, N. Kawasaki, S.S. Bag, I. Saito, Acridone-labeled base-discriminating fluorescence (BDF) nucleoside: synthesis and their photophysical properties, Chemistry Letters 35 (2006) 1182–1183.

[242] B. Qiu, L. Guo, Z. Chen, Y. Chi, L. Zhang, G. Chen, Synthesis of N-4-butylamine acridone and its use as fluorescent probe for ctDNA, Biosensors and Bioelectronics 24 (5) (2009) 1281–1285.

[243] C. Huang, S.J. Yan, Y.-M. Li, R. Huang, J. Lin, Synthesis of polyhalo acridones as pH-sensitive fluorescence probes, Bioorganic and Medicinal Chemistry Letters 20 (15) (2010) 4665–4669.

[244] J. Kaur, P. Singh, ATP selective acridone based fluorescent probes for monitoring of metabolic events, Chemical Communications 47 (15) (2011) 4472–4474.

[245] M. Hemgesberg, G. Dorr, Y. Schmitt, A. Seifert, Z. Zhou, R.K. Taylor, S. Bay, S. Ernst, M. Gerhards, T.J. Muller, W.R. Thiel, Novel acridone-modified MCM-41 type silica: synthesis, characterization and fluorescence tuning, Beilstein Journal of Nanotechnology 2 (2011) 284–292.

[246] A. Swist, J. Cabaj, J. Soloducho, P. Data, M. Łapkowski, Novel acridone-based branched blocks as highly fluorescent materials, Synthetic Metals 180 (2013) 1–8.

[247] M.A. Beniddir, E. Le Borgne, B.I. Iorga, N. Loaëc, O. Lozach, L. Meijer, K. Awang, M. Litaudon, Acridone alkaloids from *Glycosmis chlorosperma* as DYRK1A inhibitors, Journal of Natural Products 77 (5) (2014) 1117–1122.

[248] V. Herandez-Olmos, A. Abdelrahman, A. El-Tayeb, D. Freudendahl, S. Weinhausen, C.E. Müller, N-substituted phenoxazine and acridone derivatives: structure-activity relationships of potent P2X4 receptor antagonists, Journal of Medicinal Chemistry 55 (22) (2012) 9576–9588.

[249] M. Mohammadi-Khanaposhtani, M. Saeedi, N.S. Zafarghandi, M. Mahdavi, R. Sabourian, E.K. Razkenari, H. Alinezhad, M. Khanavi, A. Foroumadi, A. Shafiee, T. Akbarzadeh, Potent acetylcholinesterase inhibitors: design, synthesis, biological evaluation, and docking study of acridone linked to 1,2,3-triazole derivatives, European Journal of Medicinal Chemistry 92 (2015) 799–806.

[250] M. Mohammadi-Khanaposhtani, M. Mahdavi, M. Saeedi, R. Sabourian, M. Safavi, M. Khanavi, A. Foroumadi, A. Shafiee, T. Akbarzadeh, Design, synthesis, biological evaluation, and docking study of acetylcholinesterase inhibitors: new acridone-1,2,4-oxadiazole-1,2,3-triazole hybrids, Chemical Biology and Drug Design 86 (6) (2015) 1425–1432.

[251] M. Parveen, A. Aslam, S.A.A. Nami, A.M. Malla, M. Alam, D.U. Lee, S. Rehman, P.S. Silva, M.R. Silva, Potent acetylcholinesterase inhibitors synthesis, biological assay and docking study of nitro acridone derivatives, Journal of Photochemistry and Photobiology B 161 (2016) 304–311.

[252] M. Mohammadi-Khanaposhtani, M. Shabani, M. Faizi, I. Aghaei, R. Jahani, Z. Sharafi, N. Shamsaei Zafarghandi, M. Mahdavi, T. Akbarzadeh, S. Emami, A. Shafiee, A. Foroumadi, Design, synthesis, pharmacological evaluation,and docking study of new acridone-based 1,2,4-oxadiazoles as potential anticonvulsant agents, European Journal of Medicinal Chemistry 112 (2016) 91–98.

[253] A. Chukaew, C. Ponglimanont, C. Karalai, S. Tewtrakul, Potential anti-allergic acridone alkaloids from the roots of *Atalantia monophylla*, Phytochemistry 69 (2008) 2616–2620.

[254] C. Pal, M.K. Kundu, U. Bandyopadhyay, S. Adhikari, Synthesis of novel heme-interacting acridone derivatives to prevent free heme-mediated protein oxidation and degradation, Bioorganic and Medicinal Chemistry Letters 21 (12) (2011) 3563–3567.

Flavone: An Important Scaffold for Medicinal Chemistry

Manjinder Singh[1], Om Silakari[2]

[1]*Chitkara university, Patiala, India;* [2]*Punjabi University, Patiala, India*

Chapter Outline

1. INTRODUCTION

Flavonoids are low molecular weight polyphenolic phytochemicals derived from secondary metabolism of plants, and play an important role in several biological processes. They exhibit diverse properties that are beneficial for human health via interacting with a number of cellular targets involved in critical cell signaling pathways in the body [1].

Plant flavonoids have been shown in recent years to be of vital significance to mankind as well as to plants. They have been strongly implicated as active

Key Heterocycle Cores for Designing Multitargeting Molecules. https://doi.org/10.1016/B978-0-08-102083-8.00004-2

contributors to the health benefits of beverages such as tea and wine, foods such as fruit and vegetables, and even, recently, chocolate. The widely lauded Mediterranean diet, for example, is thought to owe much of its benefits to the presence of flavonoids in food and beverages. In the early 1990s, the epidemiological correlation between high food flavonoid intake and a lowering in the risk of coronary heart disease has been reported [2].

Flavonoids can be classified into various classes: flavonols (quercetin, kaempferol, myricetin, fisetin), flavones (luteolin, apigenin), flavanones (hesperetin, naringenin), flavonoid glycosides (astragalin, rutin), flavonolignans (silibinin), flavans (catechin, epicatechin), isoflavones (genistein, daidzein), anthocyanidins (cyanidin, delphinidin), aurones (leptosidin, aureusidin), leucoanthocyanidins (teracacidin), neoflavonoids (coutareagenin, dalbergin), and chalcones (Fig. 4.1). All the classes exhibit a variety of biological activities but among them, the flavones have been considerably explored. Various natural, semisynthetic, and synthetic derivatives of flavones have been synthesized and evaluated for several therapeutic activities like antiinflammatory, antiestrogenic, antimicrobial [3], antiallergic, antioxidant [4], antitumor, and cytotoxic activities [5]. The majority of metabolic diseases are speculated to originate from oxidative stress, and it is therefore significant that recent studies have

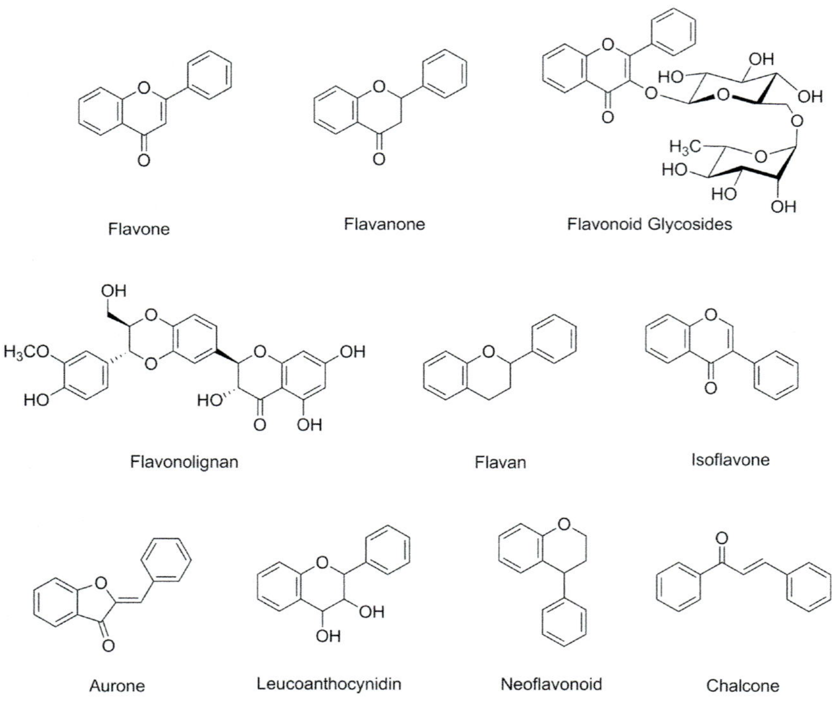

FIGURE 4.1   Various types of flavonoids.

shown the positive effect of flavones on diseases related to oxidative stress such as atherosclerosis, diabetes, cancer, Alzheimer's disease, and others. Some of the flavones of natural origin like naringenin (Natural Female Support), gingko flavone glycosides (Gingko Smart), and synthetic origin like Flavopiridol are currently available in the market.

In order to explore diverse roles of flavones, investigating various methods for their synthesis and structural modification of the flavone ring have now become important goals of several research groups. This chapter is an effort to provide some understanding about this nucleus for such tasks.

2. CHEMISTRY

Flavone is a class of flavonoids based on the backbone of 2-phenylchromen-4-one(2-phenyl-1-benzopyran-4-one). The molecular formula of the flavone molecule is $C_{15}H_{10}O_2$. It has three-ring skeletons, C6-C3-C6, and the rings are referred to as A-, C-, and B-rings, respectively (Fig. 4.2). Flavones have three functional groups including hydroxy, carbonyl, and conjugated double bond; consequently they give typical reactions of all three functional groups. Flavones are colorless-to-yellow crystalline substances, soluble in water and ethanol. They give a yellow color solution when dissolved in alkali. Flavones are moderate-to-strong oxygen bases, and are soluble in acids due to the formation of oxonium salts having pKa values ranging from 0.8 to 2.45 [6]. Flavones have a planar structure with its C—O—C bond angle 120.9 degrees. Its bond length between C—O is 1.376 Å and its dihedral angle is around 179.2 degrees.

Synonyms of flavone are 2-phenyl-4*H*-chromen-4-one; 2-phenyl-1-benzo-pyran-4-one. Flavones can react in several ways including reduction reactions, degradation in the presence of base, oxidation, rearrangement, substitution, addition, condensation, and reaction with organometallic reagents.

Several synthetic methods have been developed and modified to get products of high yield, purity, and desired quality. Flavones can be synthesized by various synthetic schemes like Claisen–Schmidt condensation [7], Baker-Venkataraman-rearrangement [8], Ionic Liquid Promoted synthesis [9],

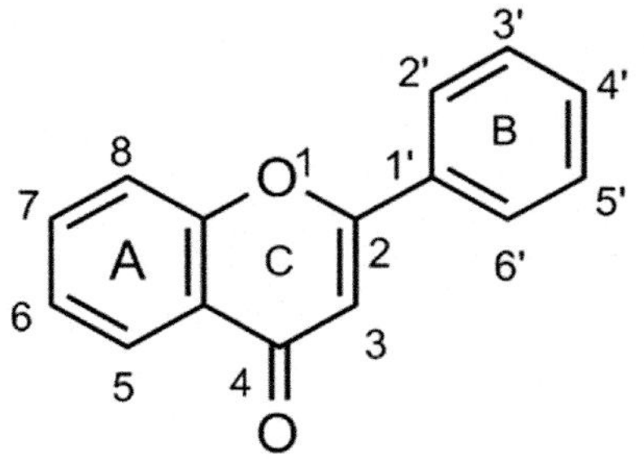

FIGURE 4.2 Basic scaffold of flavone.

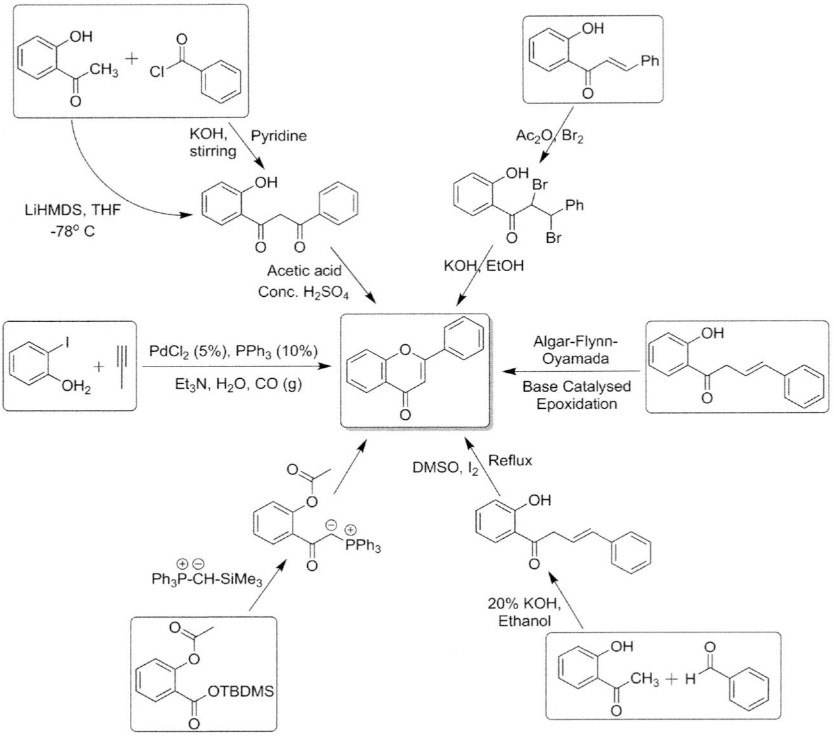

FIGURE 4.3 Synthetic schemes for flavone synthesis.

Allan-Robinson [10], Vilsmeier-Haack reaction [11], Wittig reaction, Fries rearrangement, and modified Schotten-Baumann reaction. Currently, most of the flavones are synthesized based on the Baker-Venkataraman method.

Traditionally, flavones were synthesized with the Baker-Venkataraman-rearrangement but these reactions consequently undergo the use of strong bases, acids, long reaction time, and low yields. The synthesis of a flavone nucleus has been carried out using different starting materials as displayed in Fig. 4.3.

3. FLAVONE AS A PRIVILEGED SUBSTRUCTURE

A flavone scaffold can be termed "skeleton key," and considered a privileged substructure in many compounds, acting at different targets to elicit varied pharmacological properties with various substitution patterns (Fig. 4.4). It is the diversity of this structure that gives flavones a wide range of biological activity. Due to this wide range of biological activities, their structure–activity relationships (SARs) have generated interest among medicinal chemists, which has culminated in the discovery of several lead molecules in numerous disease conditions. This information gives a comprehensive account of SARs or structural requirements of flavone derivatives necessary for wide biological activity spectrum.

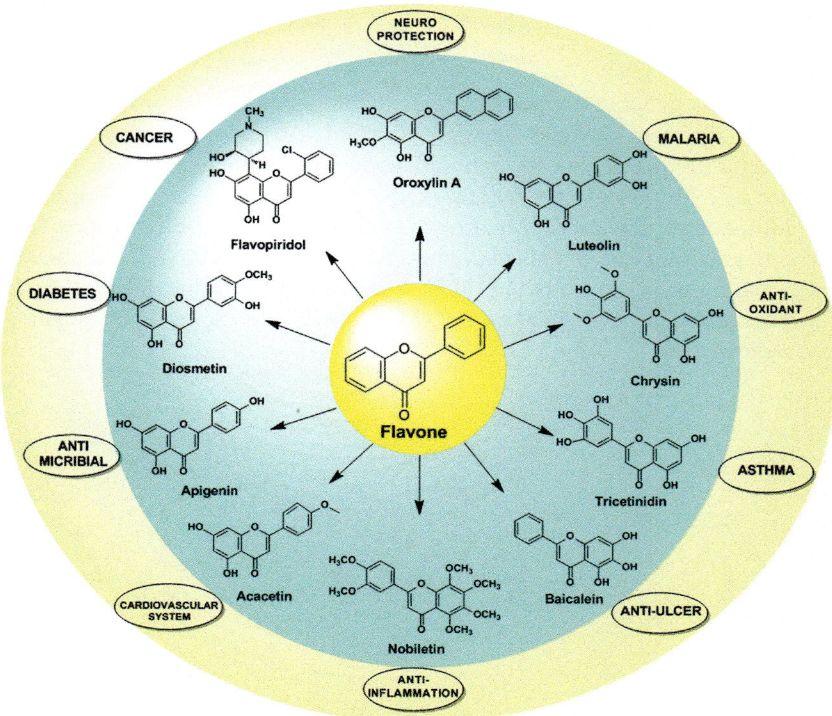

FIGURE 4.4   Multitargeted activities of flavone.

3.1 Antioxidants

The high levels of free radicals in living systems are able to oxidize biomolecules, leading to tissue damage, cell death, or various diseases such as cancer, cardiovascular diseases, arteriosclerosis, neural disorders, and skin irritations and inflammations [12]. Oxidative species and free radicals are involved in the pathophysiology of numerous diseases like neurodegenerated disorders, cardiovascular, and cerebrovascular; and autoimmune disorders like diabetes, rheumatoid arthritis, and psoriasis. Therefore, various natural as well as synthetic antioxidants are used to scavenge free radicals.

It was reported that flavones have well-known antioxidant activity, and can act through several pathways. Therefore, flavones are significantly used in pharmaceutical and food industries [13]. Flavones like chrysin (**1**), luteolin (**2**), and apigenin (**3**), which contain two or three free hydroxyl groups in A- or B-rings, show antioxidant properties at low concentrations [14].

Hyun et al. generated a quantitative structure–activity relationship (QSAR) model for different flavones having a number of hydroxyl groups to evaluate structure requirements for antioxidant activity and concluded that as the number of hydroxyl groups increases, the scavenging effect also increases. In addition, two neighboring hydroxyl groups show better effects too. They found that among all flavone analogues, 6,7,3'-trihydroxyflavone (4) showed a scavenging effect up to 87.8% [15]. It was also found that substitution of 3,4-dihydroxybenzoyl at position C-3 (5) increases scavenging activity against free radicals [16].

4   5   6

A natural flavone, Chrysoeriol (6), extracted from the tropical plant *Coronopus didymus* was tested for its ability to inhibit lipid peroxidation induced by γ-radiation, Fe (III) and Fe (II), and showed better protecting effects with DPPH radicals at millimolar concentrations [17].

SAR models using various flavones were developed for their antioxidant effects. On the basis of SAR and QSAR studies of flavone-based antioxidants, some salient features were identified for compounds to be antioxidants. Catechol group (3', 4'-OH) in the B-ring and presence of C2=C3 double bond in the C-ring enables the conjugation of the B-ring to the 4-oxo group. Increase in the number of hydroxyl groups usually increases the scavenging effect. Two neighboring hydroxyl groups show better effects too. Further extension of conjugation in α and β carbonyl groups increase the scavenging activity against free radicals (Fig. 4.5).

3.2 Antitumor Agents

Various antitumor agents (also referred as antiproliferative and antineoplastics) reported for treatment of various kinds of cancers act through different mechanisms. However, the major side effect associated with these agents is cytotoxicity toward normal cells due to lack of selectivity for the abnormal cells. From 1967 to 1991, a large group study was organized to explore the protective role of flavonoids against lung cancer and other malignant neoplasms and the risk of ovarian cancer [18].

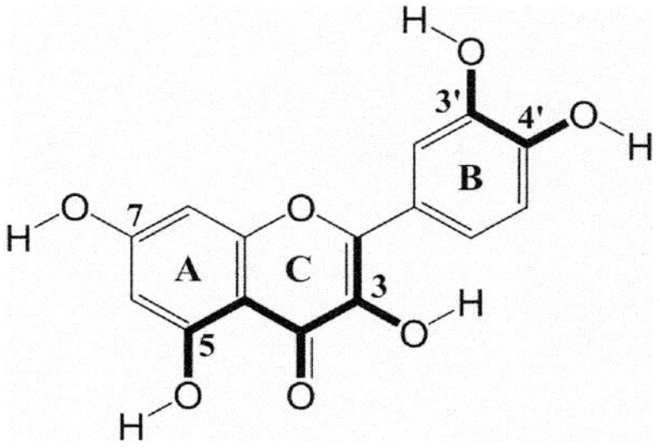

FIGURE 4.5 Structure features essential for antioxidant activity.

Nguyen et al. developed 8-substituted regioisomer; among all compound (**7**) was more potent than capitavine (**8**) and chrysin against various kinases. In cyclic amine analogues, hydroxyl group at C-5 is essential for activity. The structures of these molecules are reminiscent of flavopiridol (**9**), the first synthetic cCDK inhibitor entered in clinical trials as an anticancer drug. Similar to flavopiridol, another new synthetic flavonoidal alkaloid, P276–00 (**10**), entered into clinical studies as a small-molecule CDK inhibitor [19].

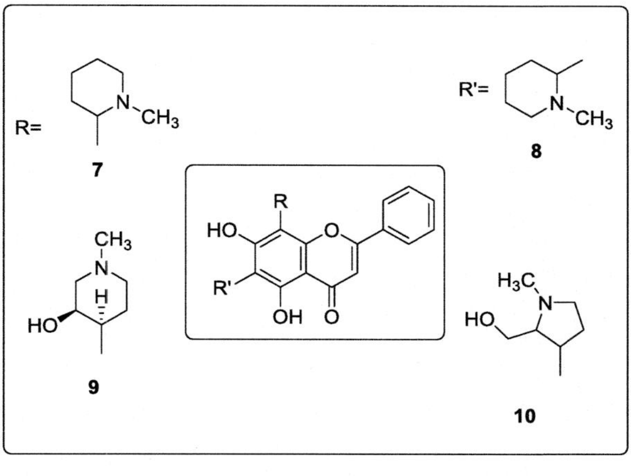

Compounds (**11**), (**12**), and (**13**), in which an imidazole ring is fused with the A-ring of the flavone, exhibit modest activity against the target protein eukaryotic elongation factor 2A (eEF1A1) responsible for breast cancer cells.

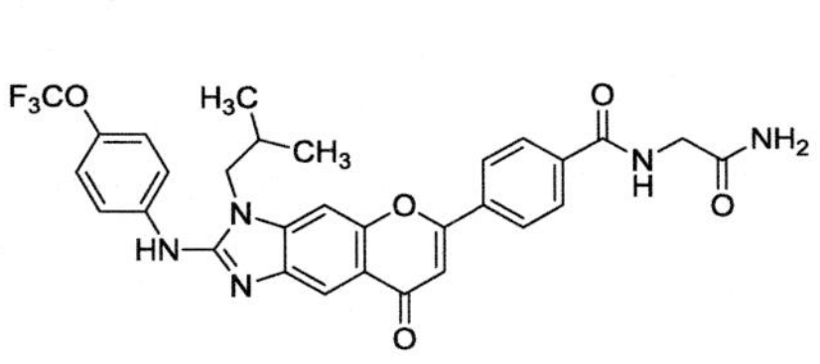

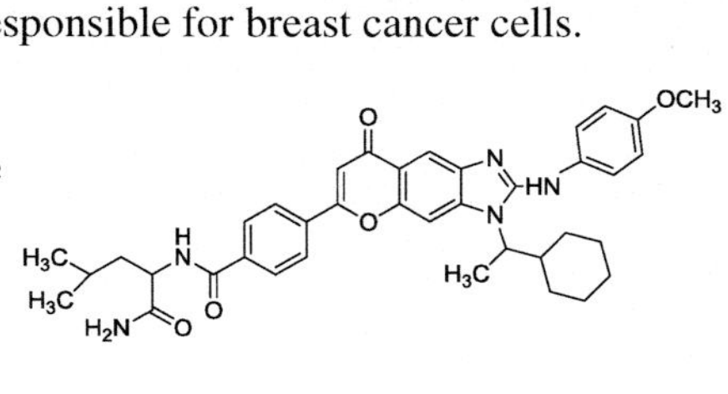

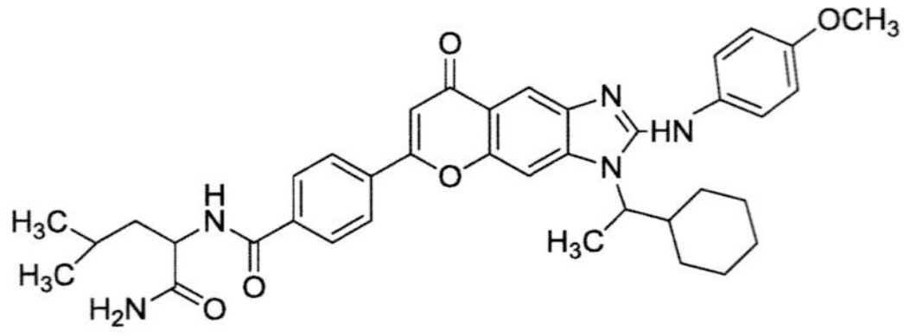

13

Polymethoxy flavones are also reported for various targets of cancer. Comparative in vitro antitumor cytotoxicity screening and subsequent tublin polymerization studies lead to identification of 5,3′-dihydroxy-3,6,7,8,4′-pentamethoxy-flavone **(14)**, and its 5-amino semisynthetic analogues aminoflavone **(15)** as tublin polymerization inhibitors [20].

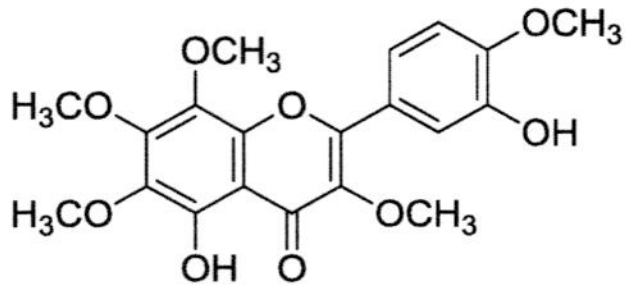

14 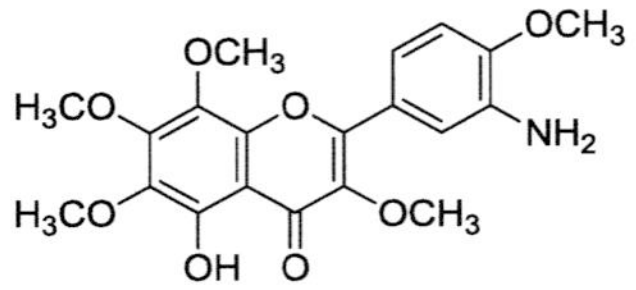   **15**

Flavones after oxidation to quinones via autooxidation, enzymatic oxidation, or reactions with reactive oxygen species inhibit human DNA topoisomerase-I [21]. Various synthetic and semisynthetic analogues have been synthesized by introducing different kinds of cyclic amines in the A-ring and subsequently evaluated against various types of cancers.

The antioxidant effect of apigenin **(3)** was studied on hepatoma cell growth through alteration of gene expression patterns. It also exhibited strong cytotoxic activity in various types of cancer including hepatocarcinogenesis, neuroblastoma, breast cancer, esophageal squamous cell carcinoma, colon cancer, lung, prostate cancer cells, cell mitosis impairment, and cell apoptosis promotion [22]. A number of substituents—chloride, isopropyl, methoxy, and nitro groups on flavone scaffold—have been tried to develop effective anti-cancer agents. The compound **(16)** ($IC_{50} = 1.1\ \mu M$) with chloride at A-ring and dimethoxy modifications at B-ring are active against HepG-2, nasopharyngeal carcinoma cells (CNE-2 and CNE-1), breast adeno-carcinoma cells (MCF-7), epithelial carcinoma cells (Hela), and mitochondrial-dependent apoptosis via triggering the caspase cascade [23].

16   **17**

Telomere and telomerase are closely related to the occurrence and development of gastric cancer. 2-chloro-pyridine derivatives of flavone moiety **(17)** ($IC_{50} = 18.45 \pm 2.79\,\mu g/mL$) is a potential telomerase inhibitor having marked effects against gastric cancer cells [24].

The SAR depicts that substitution of *p*-hydroxyl, 3, 5′-dimethoxy, 5′-amino, 2′-chloro at B-ring is important for anticancer activity by diverse mechanisms as described earlier. Substituted aliphatic or aromatic amino moiety at C-6 or C-8 position, chloro at C-6, and hydroxyl at C-5 on A-ring, nitrogen at 1-position, and 3-methoxy on C-ring are important for anticancer activity. Imidazole ring fused with flavone via C-6 and C-7 position and nitro at C-6 is active against the target protein responsible for cytotoxicity in breast cancer cells. Overall SAR patterns of flavone derivatives for anticancer activities via diverse mechanisms as described earlier are shown in Fig. 4.6.

3.3 Antiinflammatory Agents

Inflammation has foremost role in several disease conditions like asthma, atherosclerosis, Alzheimer's disease, rheumatoid arthritis, diabetes mellitus, carcinoma, Crohn's disease, gout, multiple sclerosis, osteoarthritis,

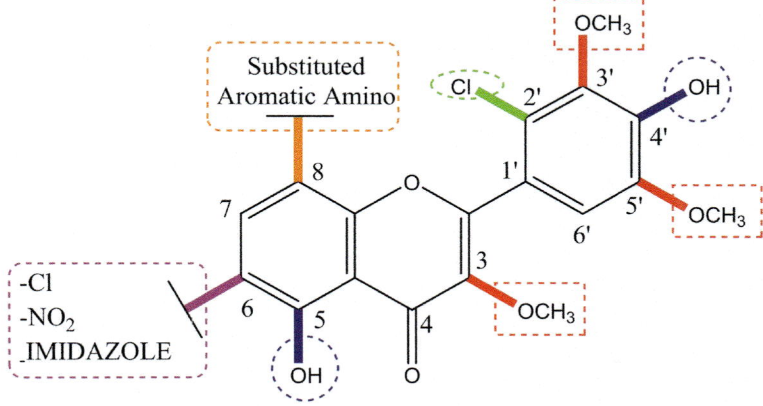

FIGURE 4.6 Structure–activity relationship for anticancer activity of flavones.

psoriasis, bacterial or viral infections, among others. Different inflammatory mediators involved in these conditions are plasma proteases, prostaglandins, leukotrienes, histamine, serotonin, nitric oxide, interleukins (IL-1 to IL-16), iNOS production, tumor necrosis factor-α (TNF-α), NF-κB, and chemokines [25]. These mediators are produced through diverse signaling pathways involving cyclooxygenases, caspases and kinases like cyclin-dependent kinases (CDK1 and CDK5), mitogen-activated protein kinase 38 (MAPK38), c-Jun N-terminal kinase (JNK), serine threonine kinases (IKK1 and IKK2), interleukin receptor associated kinase 4 (IRAK-4), and Janus kinases (JAK1- JAK3) [26].

Kim et al. synthesized a number of flavones that have been identified and further semisynthetically optimized for various inflammatory conditions. 5,6,7-trimethoxy- and 5,6,7-trihydroxyflavones were synthesized from cinnamic acid derivatives; among these derivatives, 4'-bromo-5,6,7-trimethoxyflavone (18) was the most potent compound that inhibited the production of nitric oxide (NO), a free radical, and PGE2 in LPS-treated RAW 264.7 cells. It also reduced the LPS-induced expressions of iNOS and COX-2 at the protein and mRNA levels and release of TNF-α, IL-6, and IL-1β [27].

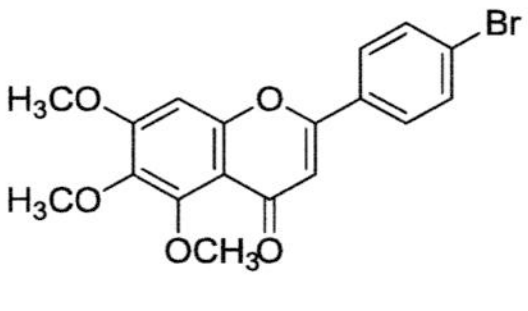

18

Some potent biflavones, among theses (19) ($IC_{50} = 3.0 \pm 0.9\,\mu M$) showed higher inhibitory activity than natural biflavonoid, ochnaflavone (20, $IC_{50} = 3.5 \pm 0.6\,\mu M$) against phospholipase A2 (PLA2) [28].

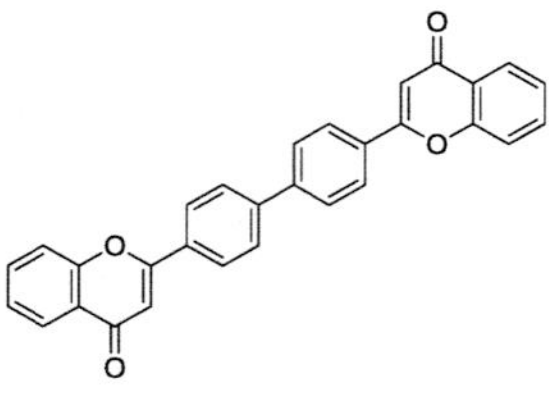

19 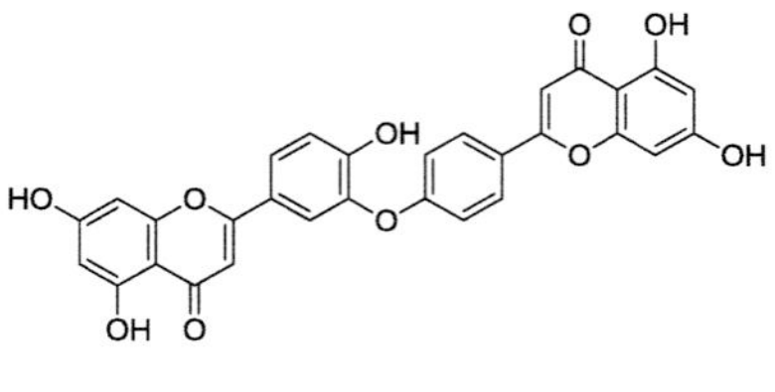 **20**

Also, the flavones without a hydroxyl residue on B-ring were more effective than flavone analogues with two or more than two hydroxyl residues on B-ring. Two new furanoflavonoids, (21) and (22), were found to be analgesic and antiinflammatory and were synthesized by a new methodology of benzyl deprotection [29].

21   **22**

Overall structural feature analysis of natural and semisynthetic flavones revealed that having a hydroxyl group at C-2′,C-3′,C-4′ of B-ring and a methoxy group at C-5, C-6, C-7 positions of A-ring have shown good antiinflammatory activity [30].

3.4 Neuroprotective Agents

The central nervous system (CNS) is balanced by enormous intracellular signaling pathways, neurotransmitters, and diverse type of enzymes like kinases, esterases, and so on. A minor alteration in these pathways or enzymes causes CNS disturbances, which leads to various neurodegenerative disorders like dementia, depression, anxiety, Alzheimer's disease (AD), Parkinson's disease, convulsions, and so on.

Flavonoids have been evaluated to treat various neurodegenerative disorders, and flavones, a class among all subclasses of flavonoids, can protect the brain by their ability to modulate intracellular signals to promote cellular survival. Natural flavones Oroxylin A (**23**) and its analogues have a marked effect on the CNS to improve memory and cognition due to their binding affinity with benzodiazepine or GABA receptors [31].

23

Tropomyosin-receptor-kinase B (TrkB) agonist 7,8-dihydroxyflavones (**24**) have been isolated and found to have significant antidepressant effects. Later, 4′-Dimethylamino-7,8-dihydroxyflavone (**25**) was synthesized, and displayed higher TrkB agonistic activity than the original hit (**24**) [32]. SAR analysis of these compounds demonstrated that the 7,8-dihydroxy groups on the A-ring and the middle heteroatomic chromen-4-one, C-ring are essential for the TrkB stimulatory effect. On the other hand, the electron-withdrawing group, such as F, or an electron-donating OH at 4′-position suppresses the activity whereas dimethylamino or pyrrolidino group at 4′-position yield the good activity (**26**).

Replacement of 7, 8-dihydroxy groups with an imidazole ring also showed increase in activity. Additionally, replacement of dimethylamino group with a pyrrolidino group in compound **27** not only displayed higher agonistic activity than compound **26** but also attenuated the locomotor enhancement and showed better antidepressant effect [33].

24

25

26

27

Flavones have also been reported to have the ability to delay the initiation or progression of Alzheimer's disease like pathology and related neurodegenerative disorders by disrupting the β-amyloid aggregation through the inhibition of β-secretase (BACE-1) and/or activation of α-secretase (ADAM10) [34]. Compounds bearing pyrrolidine-1-yl methyl (**28**) or piperidine-1-yl methyl (**29**), in C-4′ and double bond at C2=C3 showed the highest AChE inhibitory activity in nanomolar range, $IC_{50} < 100\,nM$ [35].

28

29

Several natural flavones including jaceosidine, eupafolin, luteolin, and apigenin have been reported in the literature. The flavones analogues with C2=C3 double bond, along with flouro group at A-ring (as in **30**) or sulfur group at C-4 of C-ring in the thioflavones (**31**) are selective MAO-B inhibitors [36].

30

31

Most synthetic flavone derivatives with electronegative groups (halogens, nitro, etc.) at C-6 and/or C-3′ of the flavones possess a high affinity for central benzodiazepine receptors. Representative compounds such as 6,3′-dinitroflavone (**32**) and 6-bromo-3′-nitroflavone (**33**) exhibited a high affinity for the BDZ-Rs ($K_i = 1.5$–30 nM) and possess anxiolytic effects [37]. Familial amyloid polyneuropathy and senile systemic amyloidosis are associated with transthyretin amyloid fibrils. It was reported that apigenin (**3**) completely inhibited transthyretin amyloid fibril at a concentration of 36 μM [38].

32

33

3.5 Antihyperglycemic Agents

Various flavone-based hybrids of 6- and 7-hydroxy flavones with aminopropanol as novel antidiabetic agents were synthesized. Among all synthesized hybrids **34** and **35** were found to be the most potent. These active compounds possess bulky lipophilic substitution on B-ring of flavone and smaller substituents at the nitrogen atom. The hybrids with *N*-tert-butyl and isopropyl-2-propanol are superior in activity [39].

34

35

Flavone scaffold is also explored for designing aldose reductase (AR) inhibitors for the treatment of diabetic complications. New flavonyl-2,4-thiazolidinediones are also reported to be insulinotropic with additional AR inhibitory activity. Compounds **36** and **37** ($IC_{50} < 1\,\mu M$) were obtained as aldose reductase inhibitors [40].

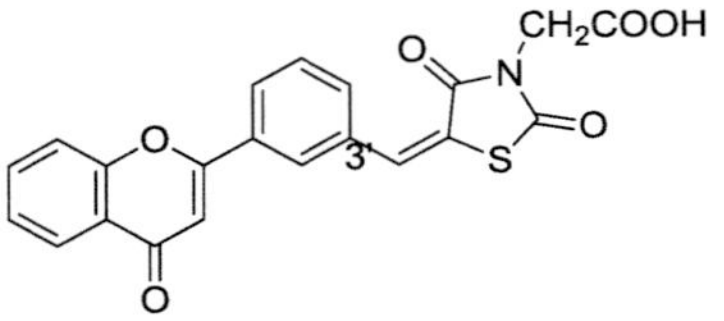

36 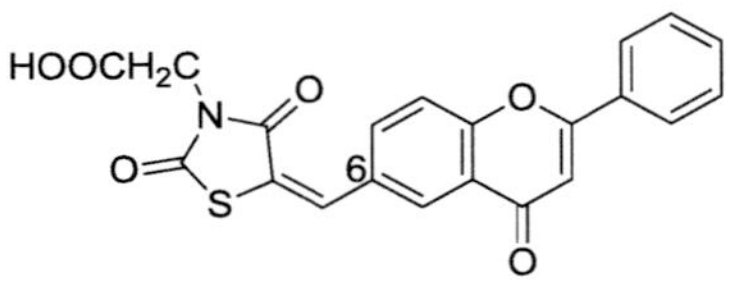   **37**

Matin et al. designed and synthesized novel pharmacophore having 7-hydroxy-benzopyran-4-one moiety as potential dual peroxisome proliferator-activated receptor (PPAR)-α and -γ agonists. Among all compounds, (**38**) 5, 4′-dihydroxy-flavone and diosmetin (**39**) were identified as novel and potent dual PPAR-α and -γ agonists. It has been proposed that dual agonists may exhibit enhanced effectiveness in treatment of type II diabetes and metabolic syndrome [41]. Some natural flavones and isoflavones have dual PPAR-α and -γ agonistic properties. SARs of semisynthetic derivatives of flavones identified 7-hydroxybenzopyran-4-one moiety as a pharmacophore for dual PPAR-α and γ agonistic activity.

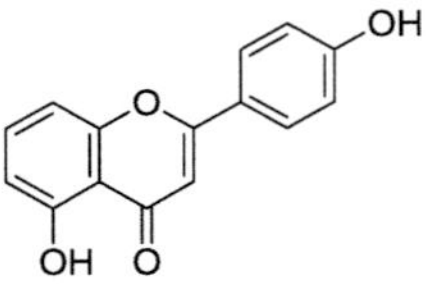

38 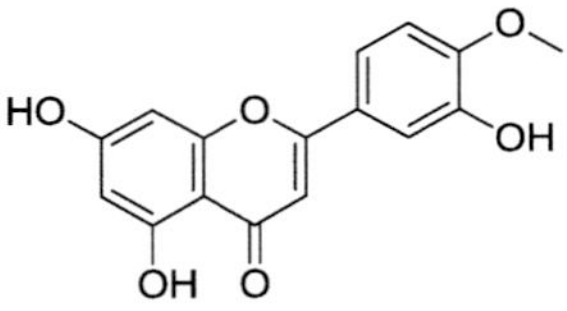   **39**

Several flavonoids have been reported to inhibit advanced glycation end-products (AGEs) formation because of their radical scavenging activities. Baicalein ($IC_{50} = 93\,\mu M$) and luteolin (**2**, $IC_{50} = 99\,\mu M$) substantially exhibited AGEs formation inhibitory activity with DPPH radical scavenging activity. Hydroxyl groups at the C-3′, C-4′, C-3, C-5, and C-7 positions were found to be important for the AGEs formation inhibitory activity whereas methylation or glucosylation of the 4′and 7-hydroxyl reduced the activity [42].

3.6 Antiulcer Agents

Several polyphenols, such as flavones and isoflavones, constitute one of the most important classes used in the treatment of peptic ulcers. Rutin, a natural flavone, has been evaluated for ulcer-protecting effects against gastric lesions induced by 50% ethanol and also for gastritis and peptic ulcer provoked by oxidative stress and inflammation [43]. DA-6034 (7-carboxymethyloxy-3′,4′,5-trimethoxy flavone) **40**, is a synthetic flavone that suppresses the *Helicobacter pylori*-induced inflammatory bowel disease in animal models by targeting NF-κB and extracellular signal-regulated kinase (ERK), a representative MAPK [44].

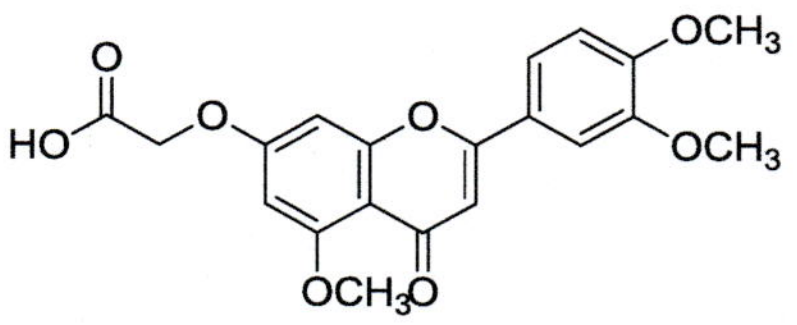

40

Several substituted flavones were also found to show good gastroprotective activity. Flavone glycosides **(41)** along with the other compounds 7-*O*-methylchrysin **(42)**, 5-hydroxy-4′, 7,4′-dimethoxyflavone **(43)**, oroxylin A, chrysin, and baicalein (constituents from *Oroxylum indicum*) showed gastroprotective properties in rats under various ulcer-inducing conditions [45].

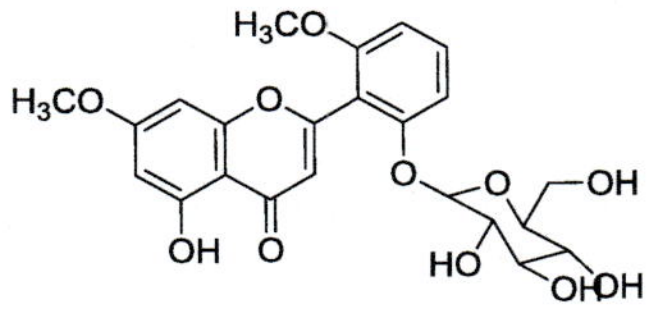

41

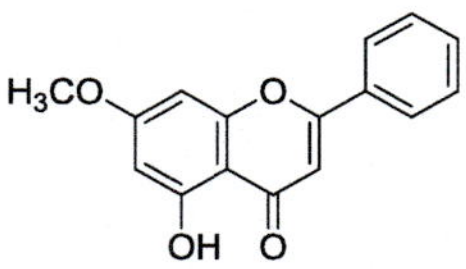

42

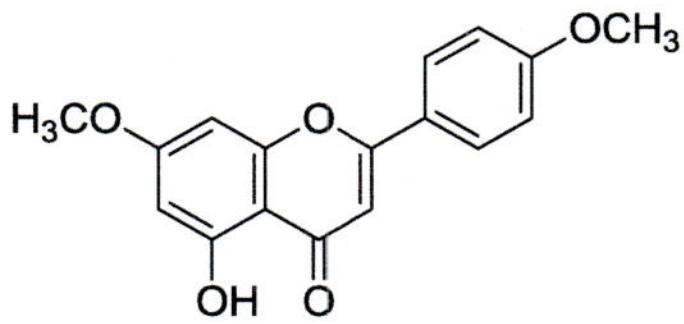

43

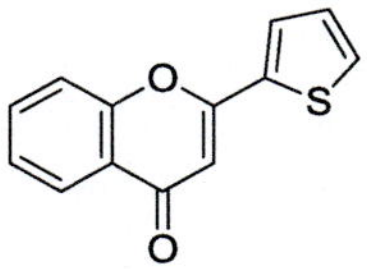

44

Replacement of the aromatic B-ring with a small alkyl group or heterocyclic rings of similar size (i.e., thiophene, **44** and pyridine, **45**) or even of much larger size (i.e., indole, **46**) retain gastroprotective properties. Compound

5-methoxy-4′-fluoroflavone (**47**, $ED_{50} = 5.5$ (4.0–7.0)) was found to be the most potent compound to protect against indomethacin-induced gastric damage [46].

45   **46**   **47**

SAR depicts that the presence of hydroxyl groups at C-5 and C-7 in chrysin cannot be ruled out, as its absence in compounds **41** and **42** drastically decreased gastroprotective activities. However, the methoxy group at the C-7 position appears to improve gastroprotection. Glycosidation at the 2′-position on the B-ring in compound **41** was observed to enhance the activity as compared to **43**. A C2=C3 double bond and an intact C-ring appear necessary for optimum activity [47]. It was reported that substitution with methoxy at different positions of the flavone nucleus—for example, at 5-position (**48**), at C-7 (**49**), or methyl substitution at the 7-position (**50**)—retained the potency of gastroprotective effect whereas substitution with methoxy or hydroxy groups at the C-3, C-6, or C-8 positions led to reduction in gastroprotective activity.

48   **49**   **50**

3.7 Antimicrobial Agents

Resistance to antimicrobial agents has become a progressively more important and vital global problem; for example, a major cause for concern in the United Kingdom is methicillin-resistant *Staphylococcus aureus*. In this context major pharmaceutical industries have concentrated their efforts to improve antimicrobial agents. Flavones have been reported for their protective role against microbial invasion [3]. This class of natural products is becoming the subject of antiinfective research and many groups have isolated and identified different flavones possessing antifungal, antiviral, and antibacterial activity. Nitrogen-containing flavones have been reported to have considerable antimicrobial activity. The compounds, bearing amino alkyl, cyano, or alkenylalkyl group on piperazine are found to be potent antibacterial and antifungal agents.

3.7.1 Antifungals

Numerous flavones like 6,7,4′-trihydroxy-3′,5′-dimethoxyflavone (**51**), 5,5′-dihydroxy-8,2′,4′-trimethoxyflavone (**52**), and 5,7,4′-trihydroxy-3′,5′-dimethoxyflavone (**53**), 6-methoxy-2-(piperazin-1-yl)-4H-chromen-4-one (**54**) derivatives have been reported to exhibit activity against *Aspergillus flavus*, a species of fungi that causes invasive disease in immunosuppressed patients. Among all, **52** were found to possess the most potent antifungal activity against *A. flavus* [48].

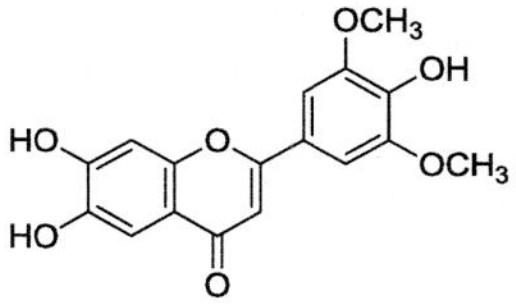

51

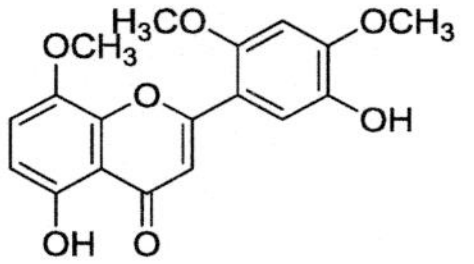

52

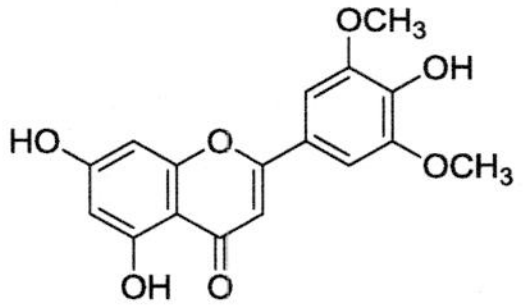

53

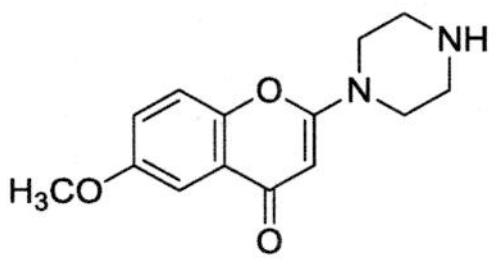

54

Various synthetic biflavones showed antibacterial, antifungal, and antiviral activity. Considering the antifungal properties, amentoflavone, isocryptomerin, ginkgetin, and bilobetin were shown to have good antifungal action [49]. Dimerization of flavones resulted in increased fungi toxicity. 3′-3‴biflavone (**55**) was found to be active against *Aspergillus niger* [50]. The compound **56**, 2-(2-chloroquinolin-3-yl)-6-methoxy-4H-chromen-4-one also showed excellent antifungal activity against *A. niger* [51]. 4′-Bromo derivatives (**57**, $IC_{50} = 28.2\,\mu g/mL$) have displayed significant antifungal activity against *A. niger* and *Fusarium oxysporium* [52].

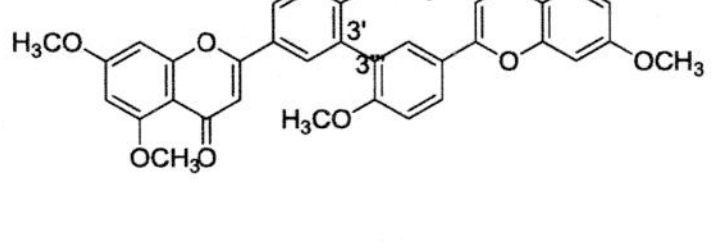

55

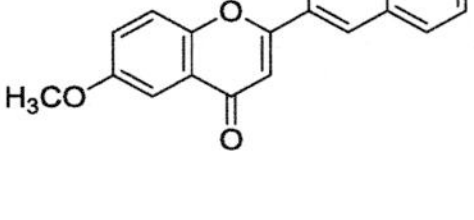

56

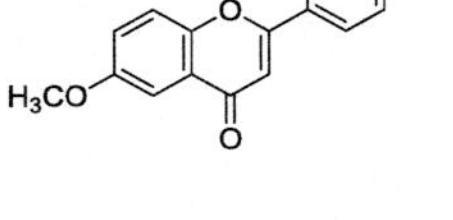

57

3.7.2 Antibacterials

Several flavone and flavonolignan derivatives were synthesized and evaluated for inhibitory activity against MDR efflux pump NorA of *S. aureus*. Compound (**58**, MIC = 0.08 µg/mL) was found to be the most active flavonolignan derivative [53]. Stermitz et al. reported that a flavonolignan, namely (G)-5′-methoxyhydnocarpin-D (**59**), acts as a potent NorA multidrug resistant (MDR) pump inhibitor. Compounds free from the hydroxyl group at A-ring was active and substitution with hydroxyl group gradually decreased activity.

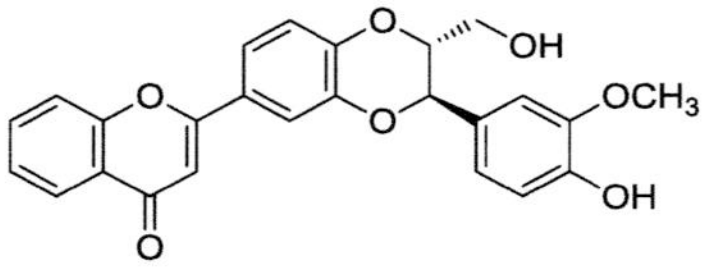

58

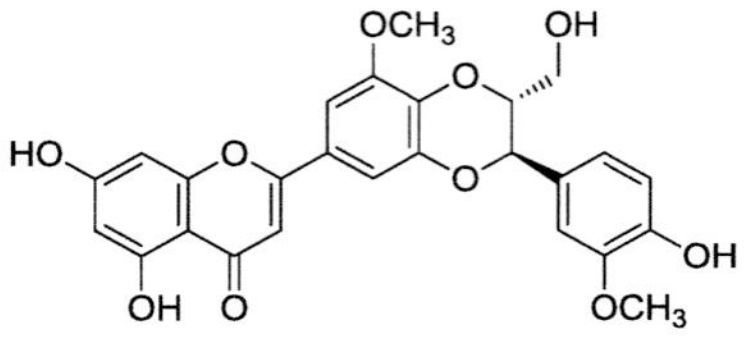

59

It was proposed that the B-ring of flavones is involved in intercalation or hydrogen bonding with the stacking of nucleic acid bases, thus imparting inhibitory action on DNA and RNA synthesis [54]. SAR depicted that 5-hydroxyflavones with additional hydroxyl groups at the 7- and 4′-positions did not exhibit antibacterial activity [55]. Methoxy groups substituted on B-ring drastically decrease the antibacterial activity of flavones.

3.7.3 Antivirals

Different natural and semisynthetic flavones reported effective against antiviral properties of various flavone derivatives have been evaluated using different virus strains such as influenza virus, human herpes simplex virus (HSV-2), measles virus, poliovirus, rotavirus, Epstein–Barr virus, human immunodeficiency virus (HIV), and hepatitis B virus (HBV).

Various scientists explored the structure requisites of flavone derivatives for influenza H1N1 neuraminidase (NA) inhibition by various computational approaches like 3D-QSAR and molecular docking. Natural flavones like scutellarin (**60**), dinatin (**61**), the hydroxyl group substituted at different positions of B-ring and A-ring, C=O at C-4, C2=C3, and different substituents at 8-position of A-ring, possess the NA inhibitory activity [56]. It was suggested that hydrogen bonds and hydrophobic and electrostatic interactions were closely related to the NA inhibitory activity. 5-OH and 7-OH might be essential for NA inhibitory activity; electron withdrawing group at 8-position decreased the activity, while hydrophobic groups increased the activity. Among all the compounds, **62** is the most active compound that displayed six hydrogen bonding interactions (including Glu225, Glu116, Arg115, Arg146, and Asp148) with key residues of the NA active site [57].

60   **61**   **62**

Ryu et al. reported that the parent flavone, **63** ($IC_{50} = 1.1\,\mu M$), was more effective at NA inhibition than alkylated flavones, **64** ($IC_{50} = 10.14 \pm 3.0\,\mu M$). New ginkgetin-sialic acid conjugates (**65**, $IC_{50} = 5.50\,\mu g/mL$) remarkably have more potent antiinfluenza virus activity against A/PR/8/34 (H1N1) than F36 (**77**, $IC_{50} = 9.78\,\mu g/mL$) [58].

63   **64**

65

It was also reported that flavones have an important role in inhibiting the poliovirus by interfering with RNA synthesis of poliovirus [59]. Kwon et al. showed that polyphenolic flavones isolated from the roots of *Glycyrrhiza uralensis* relieve acute gastroenteritis with fever, vomiting, abdominal pain, diarrhea, dehydration, and rhinitis by inactivating rotaviruses via interfering viral absorption and replication and found that compound (**66**) was the most potent inhibitor of bovine rotavirus [60].

66   **67**

Robustaflavone, **67** ($EC_{50} = 0.25\,\mu M$), a naturally occurring biflavone isolated from the seed kernel extract of *Rhus succedanea*, was found to be a potent inhibitor of HBV acting via DNA polymerase inhibition [61].

3.8 Anti-HIV Agents

Numerous flavones having anti-HIV activity like Quercetagetin (3,3′,4′,5, 6,7-Hexahydroxy flavone) and Morin (2′,3,4′,5,7-Pentahydroxy flavone), **68** and **69,** have been isolated from the leaves of *Nelumbo nucifera* [62]. Flavones react chemically by forming a complex with essential groups of integrase enzymes. High electron density on a carbonyl oxygen atom increases the integrase inhibitory activity [63].

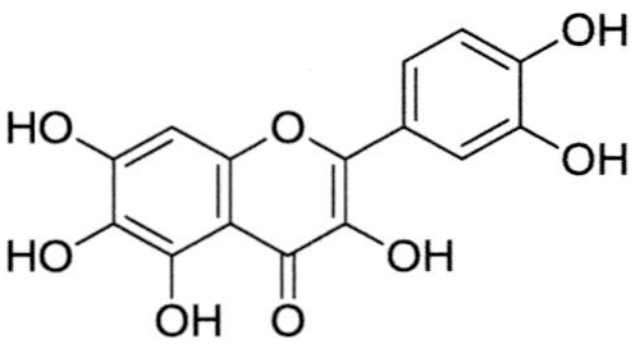

68

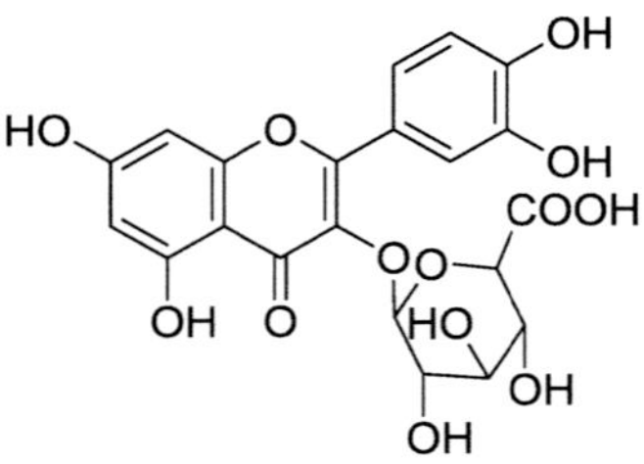

69

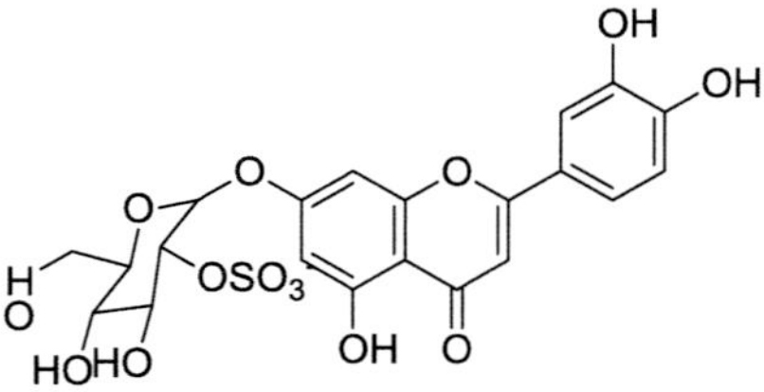

70

Another natural flavone, thalassiolins, with the HIV-integrase inhibitory property, has been isolated from the Caribbean sea grass *Thalassia testudinum*. It contains β-D-glucopyranosyl-2″-sulfate at hydroxyl group of C-7 of flavone, which increased potency against HIV-integrase by interacting with catalytic core domain of HIV-1 integrase. Thalassiolin A (**70**, luteolin 7-β-D-glucopyranosyl-2″-sulfate) was the most active, inhibiting the integrase terminal cleavage and strand transfer activities with IC_{50} values of 2.1 and 0.4 mM, respectively [64].

Various flavones and chalcones have been synthesized for inhibition of HIV multiplication. Among all, 3-methoxyflavone **(71)** was found to be active against HIV-1 and HIV-2. The *para* substitution on B-ring **(72)** seemed to increase HIV-2 potency. Compounds **72**, **73**, and **74** were found to be specific inhibitors of the HIV-2 multiplication [65].

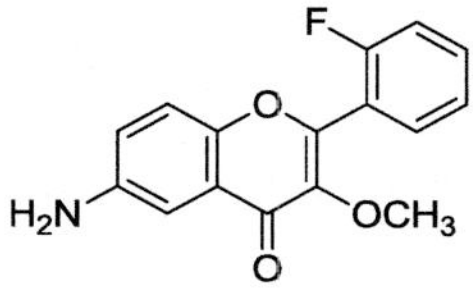

71

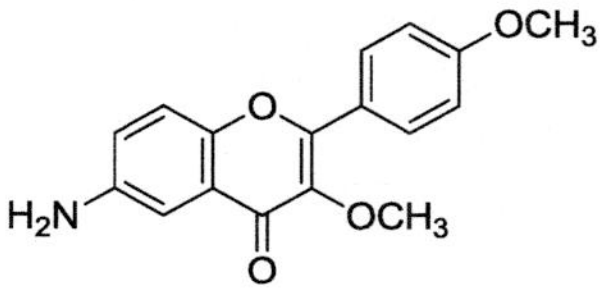

72

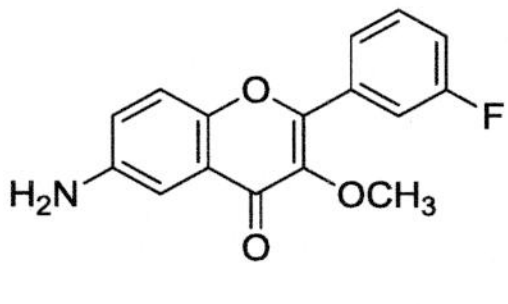

73

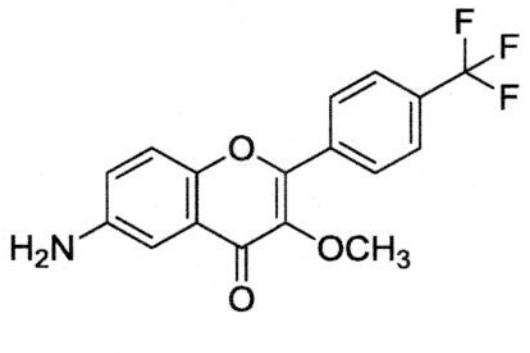

74

SAR studies have revealed that A-ring bearing an amino group at 6-position, C-ring with H, OH, and OCH_3 at 3-position, and mono substituents (H, OCH_3, F, CF_3) in various positions of B-ring are needed to increase HIV-2 potency. Inhibitory activity is reduced or eliminated by methoxy or glycosidic substitutions or by saturation of the 2, 3 double bond.

3.9 Cardiovascular Diseases

Since ancient times, numerous flavonoids, including flavones (Apigenin, **3**), have been reported to reduce occurrence of various heart diseases like coronary diseases, arrhythmias, atherosclerosis, hypertension, ischemic stroke, peripheral arteriopathy, congestive heart failure, and so on. Flavones attributed to an increase in HDL-cholesterol levels, antioxidant capacity, lipid regulation, inhibition of platelet aggregation, improving endothelial function, and the antiinflammatory effects [66]. Apart from all these properties, flavones are involved in the direct inhibition of some radical-forming enzymes (xanthine oxidase, NADPH oxidase, and lipoxygenases) for the management of several cardiovascular disorders.

A natural flavone luteolin-7-O-β-D-glucopyranoside exhibited cardioprotective effects against doxorubicin-induced toxicity in H9c2 cells, reduced lactate dehydrogenase and creatine kinase levels, and decreased the elevated intracellular concentration of reactive oxygen species and [Ca^{2+}]. SAR analysis revealed that the C2=C3 double bond on C-ring and 3′,4′-dihydroxyl on B-ring are necessary for cardioprotective effects. 4-keto group is a requisite for vasodilation [67,68].

Budriesi et al. have synthesized a series of 1,4-dihydropyridines bearing a 3-methoxyflavone ring at 4-position, and evaluated them for cardiovascular activity in isolated right atria of the guinea pig. Among all synthesized compounds, **75** was found to be a potent calcium channel modulator, thus it could be useful in the treatment of myocardial ischemia, where negative inotropic and hypotensive effects could be potentially deleterious [69]. Various flavone derivatives have been evaluated for lipid peroxidation inhibitory activity on isolated rat hearts. Compound (**76**) reduced ischemia/reperfusion-induced cardiac dysfunction [70].

75   **76**

Oxypropanolamine substituted flavones were shown to exhibit antihypertensive activity in spontaneously hypertensive rats. The position of the oxypropanolamine side chain, hydroxy group of the side chain, steric bulkiness and length of N-substituents, and degree of the N-substitution and substitution on the B-ring of the flavones scaffold play a significant role for antihypertensive effects. Among all tested analogues, the most effective one was flavodilol (**77**). Compounds with the functionality at the C-7 displayed better activity than the corresponding congeners at the 5-, 6-, and 8-position. The optimal chain length of three carbon atoms in substituents of the secondary amine was found to be important for drug-receptor interactions. Substitution of the para position of the B-ring with a variety of substituents (i.e., NO_2, NH_2, OCH_3, and CH_3) causes a loss of activity [71].

77   **78**

In cardiac disorders, abnormal rhythm of the heart increases the risk of death, congestive heart failure, and heart strokes. Atrial fibrillation is the most common form of cardiac dysarrhythmia. Several antiarrhythmic agents like amiodarone and sotalol are effective in treating atrial fibrillation but have major limitations, such as inducing severe ventricular arrhythmia [72]. Acacetin, (5,7-dihydroxy-4′-methoxyflavone, **78**), a natural flavone, selectively inhibits human atrial repolarization potassium currents and prevents atrial fibrillation. Acacetin is an orally effective atrium-selective agent [73].

3.10 Antiplatelet Agents

O-sulfated oligosaccharide-derived flavones have been synthesized in order to improve the anticoagulant potency by increasing both molecular size and number of sulfate groups. The flavonoside **79** was the most potent in prolonging activated partial thromboplastin time (APTT) (APTT2 = 66 μM). Compound **79** acts by interfering with coagulation cascade factors or platelet function. Their influence on platelets, thrombin, FXa, and ATIII was studied. The anticoagulant effects are increased by increasing the number of sulfate groups in the compound. Decasulfated compound **79** was more potent than all other compounds that increase clotting time, clot formation time, and decrease deceleration progress of the clot. The possibility of a dual anticoagulant and antiplatelet activity of compounds could enhance their potential as antithrombotic agents [74]. Wang et al. reported that oxime- and methyloxime-containing isoflavone-7-yl derivatives have better antiplatelet activity with respect to flavones [75].

79

3.11 Xanthine Oxidase Inhibitors

Xanthine oxidase (XO) is concerned in the pathogenesis of viral infections, inflammation, brain tumors, and postischemic reperfusion injury. Thus, its inhibition is important to decrease the production of excessive uric acid and to prevent the formation of superoxide radicals, and it provides protection against postischemic reperfusion injuries [76].

The planar structure and C-5 and C-7 hydroxyl group of flavones are important for the inhibition of XO enzyme at low concentrations. The C-5 and C-7 hydroxyl groups of flavones may contribute to the inhibition by mimicking C-2 and C-6 binding interaction sites of xanthine that forms hydrogen bonds in the active site of XO [77]. Apigenin (**3**, $K_i = 0.52\,\mu M$) and luteolin (**2**, $K_i = 2.9\,\mu M$) exhibited strong inhibitory effect on the XO as compared to a synthetic XO inhibitor allopurinol (K_i of $7.3\,\mu M$) [78].

3.12 Lipid-Lowering Agents

The flavone derivative CM108 (**80**) has also been reported for lipid regulation. It was observed that this compound increases the high-density lipoprotein (HDL) level and reduced triglyceride, cholesterol, and low-density lipoproteins in serum and liver [79]. The hybrid congeners 6- or 7-hydroxy flavones with aminopropanol have been synthesized and evaluated for antidyslipidemic activities. The compounds **81** and **82** lowered cholesterol and triglyceride profiles and improved HDL cholesterol in db/db mice.

80   **81**   **82**

From the SAR point of view, bulky lipophilic substitution like tert-butyl- on B-ring of flavones, the oxygen atoms at the $3',5'$-positions in the B-ring seem necessary for antidyslipidemic activities [80].

3.13 Spasmolytic Agents

Several flavones have been reported to possess muscle-relaxant properties. Flavones relax smooth muscles via blockade of muscarinic receptors. Various polyoxygenated flavones were isolated and among them, **83** showed competitive antagonism of muscarinic receptor binding. The potency of the methoxyflavones to inhibit muscarinic receptor binding is influenced by the position and number of methoxy group substitutions. All the flavones that bind to the muscarinic receptor possessed C-3, C-5 methoxy groups and C2=C3 double bond in conjugation with 4-oxo group [81].

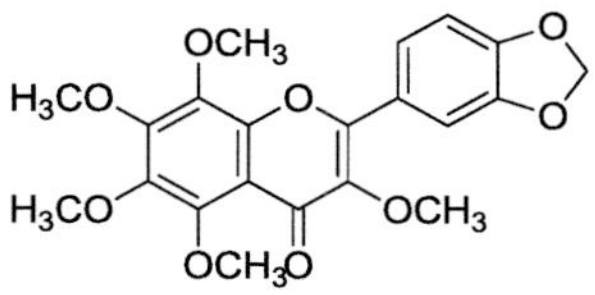

83

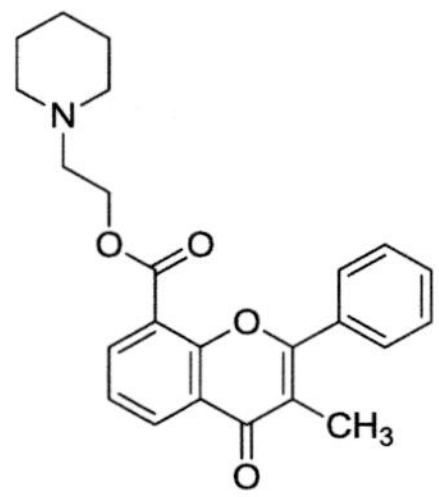

84

Flavoxate, **84**, possesses anticholinergic activity along with antimuscarinic effects. Thus, it can be used to treat urinary bladder spasms, while flavoxate hydrochloride is prescribed for symptomatic relief of interstitial cystitis, dysuria, nocturia, suprapubic pain, and incontinence, which may occur in various other disorders like cystitis, prostatitis, urethritis, and urethrocystitis/urethrotrigonitis [82].

3.14 Vasorelaxants

A series of flavones like 3-hydroxyflavone **(85)**, 6-hydroxyflavone **(86)**, 7-hydroxyflavone **(87)**, and chrysin **(1)** have been evaluated for ex vivo and in vitro vasorelaxant effect, therefore, they can be the drugs of interest as novel antihypertensive agents. All the flavone derivatives possess an endothelium-dependent vasorelaxant effect, with an increased production of NO and prostacyclin PGI_2 in a concentration-dependent manner [83].

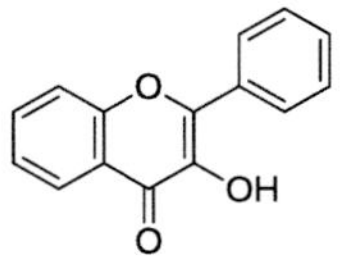

85 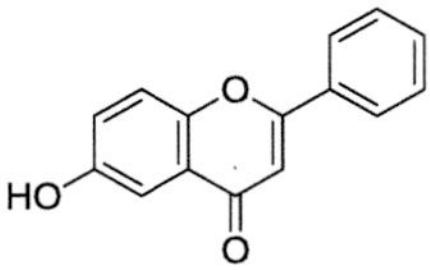   **86** 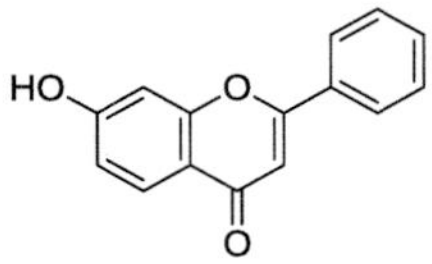   **87**

Dong et al. designed and synthesized several flavone derivatives for vasorelaxant activity. SAR was developed with 3D-QSAR analysis, and carried out by a comparative molecular field analysis method. Hydroxyl groups at C-5 and C-7 positions in the A-ring, C-4 carbonyl group, C2=C3 double bond and hydroxyl groups at different positions in the B-ring are important features for the vasorelaxant activity, while the presence of C-glycosyl group at C-8, hydroxyl group at the C-3 position, greatly reduce the relaxation effect. On the other hand, bulky substituents at ortho- and meta-positions on the B-ring **(88)** of flavone derivatives decrease the potency of vasorelaxant activities. 5, 7-dihydroxy-3′-bromo-flavone **(89)** exhibited the highest vasodilator activity [84].

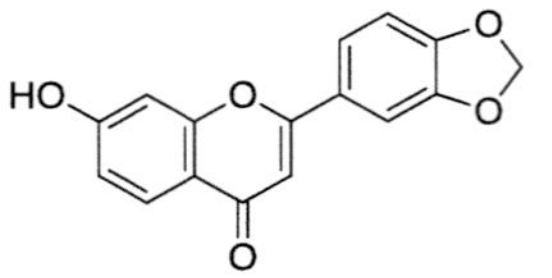

88

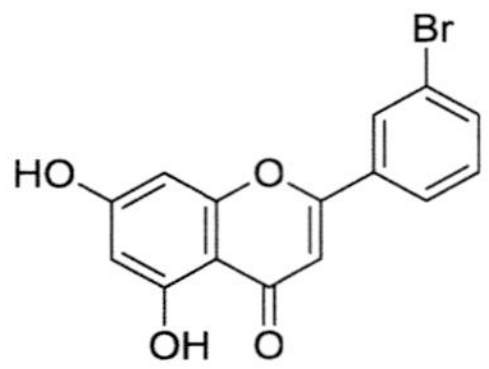

89

3.15 Antileishmanial Agents

Leishmaniasis is one of the major parasitic diseases. Currently, there is no effective vaccine for leishmaniasis. Therefore, there is an urgent need to speed up the development of a new generation of effective and safe antileishmanials. Various natural and synthetic flavones have been reported to have potent leishmanicidal activity [85]. Synthetic flavone dimers with either polyethylene glycol linker or amino ethylene-glycol linker showed marked leishmanicidal activity. Among all compounds, **90** (IC_{50} = 0.13–0.21 µM) showed very reliable and promising leishmanicidal activity with a high therapeutic index without toxicity.

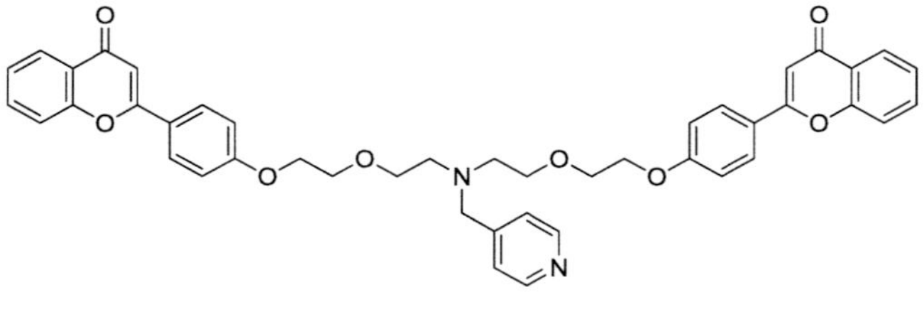

90

SAR revealed that the pyridine ring with nitrogen at 4-position is important for selective antileishmanials activity. Activity depends upon various structural features like (1) length of the linker; shortening the linker length results in loss of activity. (2) Changing the attachment position of the two flavones to the amino polyethylene glycol (PEG) linker, attachment at C-3′ position B-ring showed significant activity as compared to compound having linker attached at C-4′ position. However, attachment at C-2′ position of the B-ring or C-3 position of C-ring resulted in at least a 10-fold decrease in activity. (3) Substituting the two flavones with different groups like methoxy at C-3′ of B-ring caused remarkable cytotoxicity toward both promastigotes and host peritoneal elicited macrophage (PEM) cells, whereas methoxy at C-3 or fluorine at C-6 did not cause any toxic effect toward the PEM cells and retained the activity. (4) Position of attachment of pyridine ring at the amino PEG linker. The activity profile rank order in pyridine is obtained like substitution through para-position > meta-position ≫ ortho-position. Also, substitution with bromo and cyano group at meta-position or ortho-position of the pyridine ring reduced the activity. Replacement of the pyridine ring by pyrimidine ring further lowered activity two-fold [86].

3.16 Antiosteoporotic Agents

Osteoporosis, a silent epidemic, is characterized by decreased bone mineral density, increased risk of fractures, and is associated with microarchitectural deterioration of bone tissue that results in low bone mass. Osteoporosis is especially common in women after menopause as a result of reduced estrogen level.

Flavonoids have been intensively reported on regarding their estrogenlike activities and particularly their ability to affect bone metabolism. Flavones like Baicalin were screened for their osteogenic properties by measuring alkaline phosphatase activity in cultured rat osteoblasts.

Natural dihydroflavonol derivative K 058 (**91**) isolated from *Ulmus wallichiana* increases mRNA levels of various osteoblast-specific genes. It was not osteogenic in the rat uterus, but promotes osteoprogenitor cells in immature rats [87].

91

3.17 Cosmetic Agents

Tyrosinase is a copper-containing enzyme that catalyzes different reactions in melanin synthesis. Undergoing several reactions, it oxidized L-tyrosine to melanin. Melanin synthesis is of considerable importance because its alteration causes many skin diseases. A mutation in the tyrosinase gene resulting in impaired tyrosinase production leads to type I oculocutaneous albinism, a hereditary disorder. The mutated tyrosinase results in increased melanin synthesis. So, tyrosinase inhibitors have gradually become more important in medicinal and cosmetic products. Tyrosinase inhibitors have been reported for the treatment of melanin hyperpigmentation and used in cosmetic materials for whitening skin after sunburn [88].

It was reported that luteolin (**2**) showed tyrosinase inhibitory activity by acting via disruption of the tertiary structure of the enzyme through intermolecular hydrogen bonding as a noncompetitive inhibitor. Luteolin 7-*O*-glucoside also showed tyrosinase inhibitory activity [89]. Baicalein has also been used as a depigmentation agent in cosmetics, having several numbers and relative positions of the hydroxyl groups that enhance its tyrosinase inhibitory activity [90]. Other flavones involve in tyrosinase inhibition activity are 6-hydroxyapigenin, scutellarein (**92**), 6-hydroxygalangin (**93**), and 6-hydroxykaempferol (**94**).

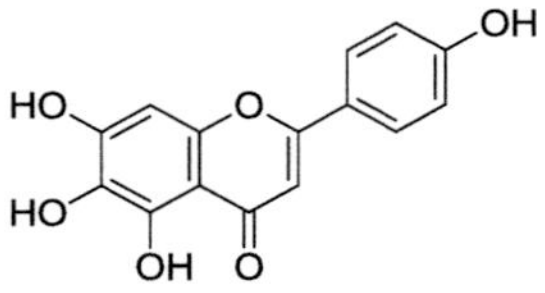

92 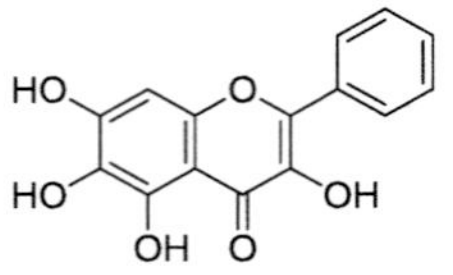 **93** 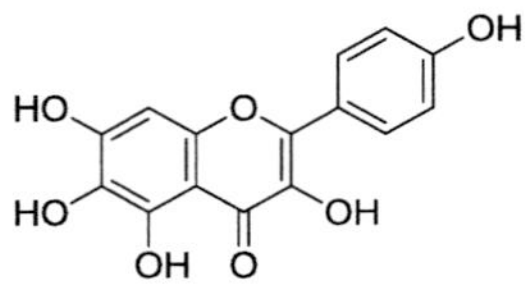 **94**

Venous insufficiency is a condition closely related to varicose veins, affecting larger veins deep within the leg, and is characterized by pain, aching, swelling, and feelings of heaviness and fatigue. The varicose veins condition is considered a cosmetic problem with stuffed, blue, or purple lines visible on the skin of the lower legs. Troxerutin, **95**, has been tested for the management of varicose veins and venous insufficiency and showed substantial relief from swelling, aching, leg pains, and other uncomfortable symptoms [90,91].

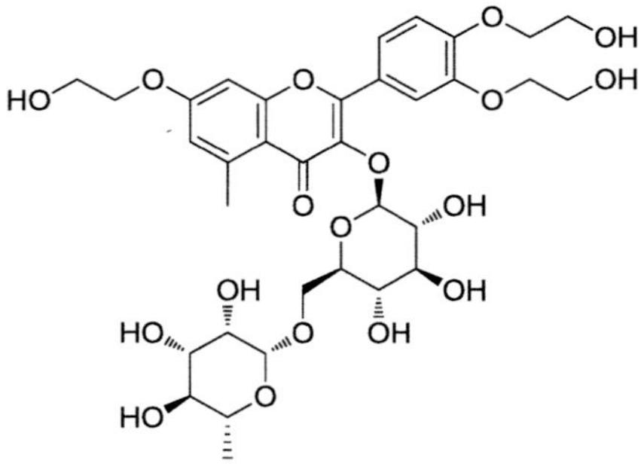

95

3.18 Immunomodulators

The immune system is the body's defense system against various types of diseases and pathogens like viruses, parasites, and so on. Sometimes immune cells fail to recognize the body's own constituents and become hyperactive and attack normal tissues; such anomalous immune response is called an autoimmune disease. Common autoimmune diseases include rheumatoid arthritis, diabetes mellitus type-I, and psoriasis.

Verbeek et al. reported that luteolin (**2**) and apigenin (**3**) were found to be strong inhibitors of both murine and human T-cell responses and inhibited the human autoantigen α-B-crystallin, an autoantigen in multiple sclerosis. Major contributor antigen-specific IFN-γ production was reduced effectively by flavones, and thus manage T-cell mediated multiple sclerosis [92].

SAR revealed that the presence of C2=C3 double bond in C-ring and absence of a hydroxyl group at C-3 are required for immune-suppressive action. Another novel flavone PMF **96** (5,3′-dihydroxy-3,6,7,8,4′-pentamethoxyflavone) induced immune response against tumors by suppressing signal transducers and activators of transcription (STAT3) [93].

96

3.19 Photoprotectants

Ionizing radiation by interaction with living cells through generation of toxic free radicals causes single strand breaks, double strand breaks, oxidative damage, chromosomal abnormality, and mutation leading to the cell death and an increased risk for numerous genetic diseases [94]. To protect the normal cells from radiation, efforts have been increased to search photoprotective agents. Various polyphenols like flavones have been evaluated for photoprotective activity.

Flavones having the UV-absorbing and radical-quenching capability exerted photoprotective properties. Flavones are able to trap peroxyl and related radicals with slight modifications in their structure [95]. Radioprotective effects of quercetin (3-hydroxy-flavone), chrysin (**1**), luteolin (**2**), vicenin (**97**), and orientin (**98**) flavones are well reported in the literature.

97   **98**

3.20 Antiasthmatic Agents

FcεRI receptors, present on the surface of mast cells and basophils, are critically involved in various chronic inflammatory conditions like atopic dermatitis, allergic rhinitis, and asthma. Interaction of allergen-bound IgE with FcεRI receptors triggers the mast cell signaling, resulting in the release of allergic mediators, histamine, proteases, chemotactic factors, leukotrienes, and arachidonic acid metabolites [96]. Antiasthmatic therapy can be targeted at blocking the first step (i.e., FcεRI receptor expression) or antagonizing the effects of the mediators released in the later steps. Flavones have been reported to have an inhibitory effect on FcεRI receptor expression, antihistaminic activity, and leukotriene antagonism.

Tricetinidin (**99**) possesses similar inhibitory activity as compared to epicatechin gallate (**100**), a potent FcεRI expression inhibitor. Both these compounds possess pyrogallol moiety in common, thus, this partial structure can be proposed to be significantly associated with inhibition of FcεRI expression [97].

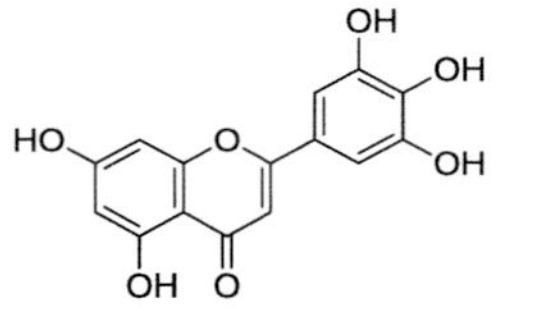

99 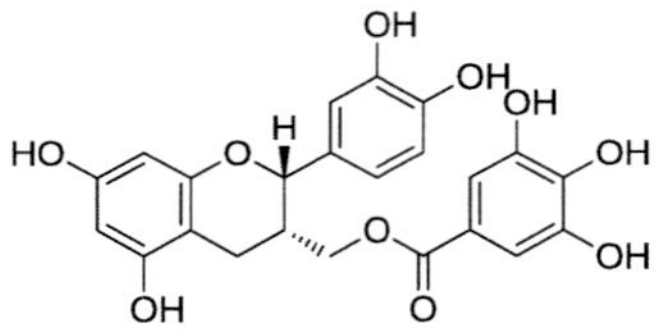   **100**

Flavones also exhibited leukotriene antagonism. The leukotrienes cause constriction of the pulmonary airways and small blood vessels, and therefore are involved in asthma, psoriasis, myocardial infarction, and vasospasmic diseases [98].

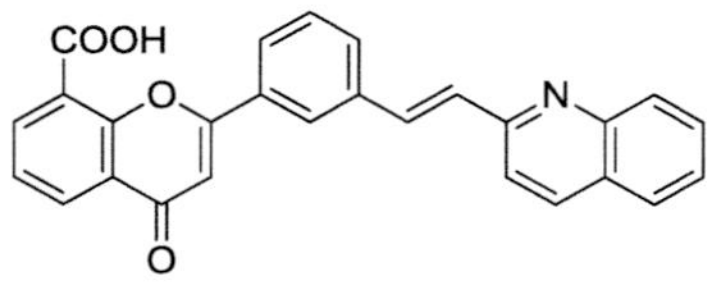

101 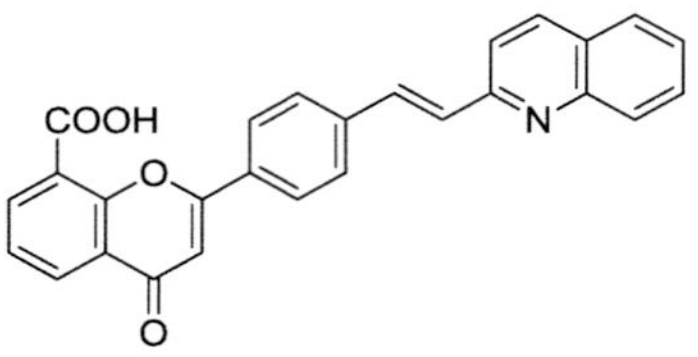   **102**

SAR depicted that 4′-substituted flavones showed 10–20 times less leukotriene antagonism as compared to 3′-substituted analogues (**101** vs. **102**). Also, substitution of carboxylic acid at the C-8 position is favorable for the antagonism.

3.21 GABA Antagonists

γ-amino butyric acid (GABA) is the major inhibitory neurotransmitter in the CNS and plays a crucial role in managing neuronal excitability in various CNS related disorders by binding to GABA$_A$ receptors. GABA$_A$ receptors also have some additional binding sites for compounds like benzodiazepines (BZDs), β-carboline, barbiturates, and such that allosterically modify chloride channel gating [99]. Among these, BZDs form the most important $GABA_A$ receptor modulating drugs with anxiolytic, anticonvulsant, muscle relaxant, and sedative hypnotic effects [100]. It has been found that some naturally occurring and synthetic flavones bind to the BZD binding site with high affinity and exhibit anxiolytic effect selectively [101].

Dekermendjian et al. reported in 1999 that flavones with substitution at 6-position (methyl) and 3′-position (nitro group) (**103**) have high affinity for benzodiazepines [102]. But later in 2002, Kahnberg et al. identified 5′-bromo-2′-hydroxy-6-methylflavone (**104**) as a potent binder of BZD with K_i value of 0.9 nM [103].

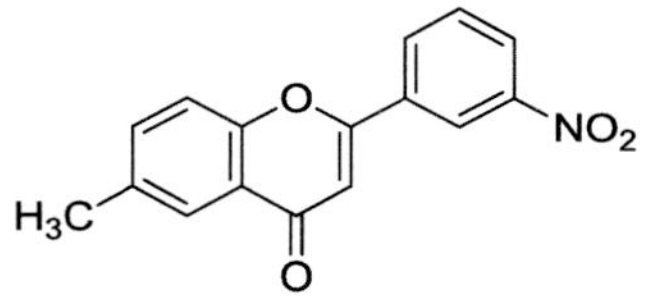

103

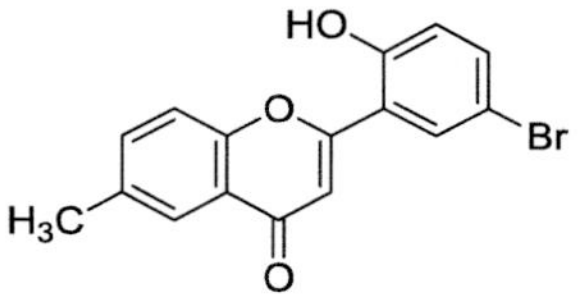

104

3.22 Anthelmintics

Various parasitic diseases caused by helminthes pose a major threat to public health worldwide. Anthelmintics currently in use encounter resistance, thus there is a need for new anthelmintics with a specific mode of action. In 2008, Ayers et al. reported the isolation of three flavones from the whole plant extract of *Struthiola argentea* having anthelmintic activity (**105**, **106**, and **107**) [104].

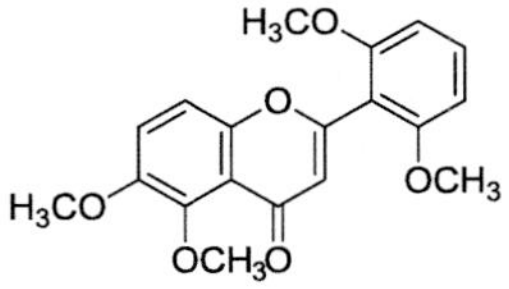

105

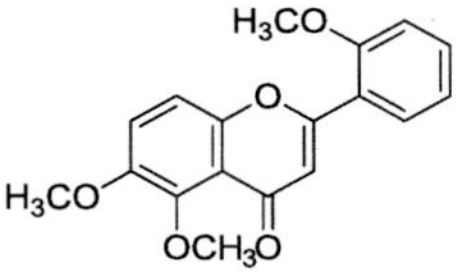

106

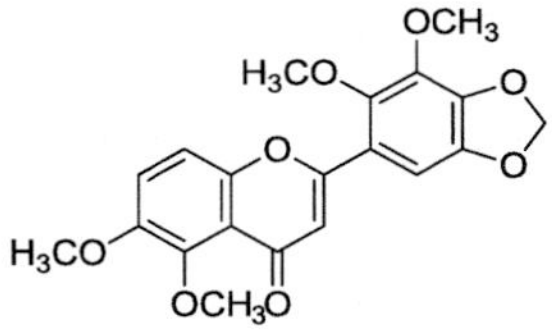

107

4. ROLE OF FLAVONES AS MULTITARGETING AGENTS IN MULTIFACTORIAL DISEASES

As discussed in the previous section, flavone scaffold has been used in numerous drug developments for the management of varied diseases. Moreover, there are other reports that indicate that some flavone derivatives have been designed as multitargeted agents.

For the treatment of AD, various flavones were used as adjuvant with some active moieties like tacrine due to their antioxidant activity. In that context, a new family of tacrine-4-oxo-4*H*-chromene hybrids has been designed, synthesized, and evaluated biologically for AD by Fernandez-Bachiller et al. Considering the tacrine fragment with cholinesterase inhibitory activity and the flavone scaffold with free radical scavenging and β-secretase (BACE-1) inhibitory activities, the hybrids were found to be more potent than the parent inhibitor, tacrine as well as apigenin. Among all the hybrids, compound (Fig. 4.7) showed potent combined inhibition of BACE-1 and ChEs, as well as good antioxidant and CNS-permeable properties [105].

Chiruta et al. in 2011 reported the multitarget neuroprotective compound Fisetin (**108**), which provides effective treatment for the management of several neurodegenerative disorders. Fisetin has direct antioxidant activity and can also increase the intracellular levels of glutathione, the major endogenous antioxidant. In addition, fisetin has both neurotrophic and antiinflammatory activity [106].

108

Selective dual inhibitors of Raf1 and JNK1 kinases, 1-(3-chloro-4-(4-oxo-4H-chromen-2-yl) phenyl)-3-phenylurea derivatives, have been developed for

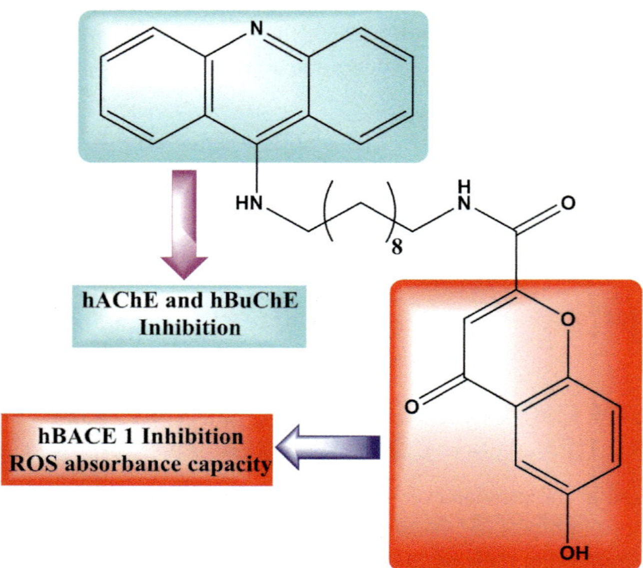

FIGURE 4.7 The modulation of amyloid β (Aβ) pathology by flavones.

antitumor treatment. The Raf protein involved in Ras-Raf-ERK MAPKs pathway regulates cellular processes like cell proliferation, migration, and apoptosis. c-Jun-N-terminal Kinase (JNK), another MAPK member, is activated in some cancers by Ras to cooperatively promote cancer together with the Ras-Raf-ERK signaling. Dysregulation of both Raf-MAPKs and JNK pathways had been well elucidated in various human cancers. Some compounds targeting either Raf or JNK had been approved or in preclinical trials for anticancer treatment, but no multitargeted drugs had entered clinical trials yet. Compound (109) at 50 μM has selectivity against p38-alpha kinase, good antiproliferative activity against the HepG2 cell-line, and relatively low toxicity against normal liver cell lines QSG7701 and HL7702. Due to low toxicity, 2′-chloro-4′-aminoflavones scaffold may be potentially explored for deriving multitarget Raf1 and JNK1 inhibitors [107,108].

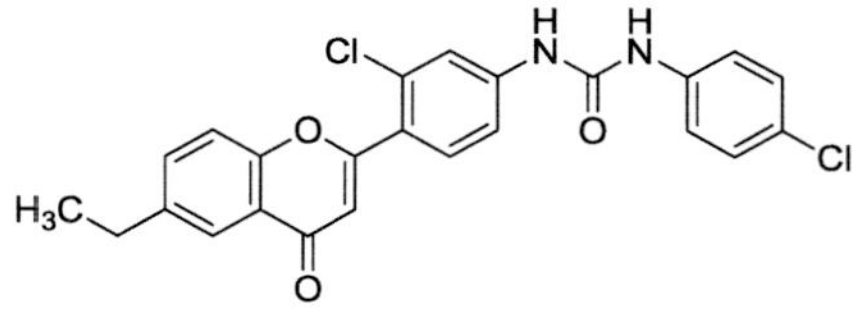

109

Sang et al. developed various rivastigmine-scutellarin hybrids on the basis of multitarget-directed ligand strategy. Among all the synthesized hybrids, **110** exhibited dual inhibitory potency on AChE and BuChE (IC$_{50}$=0.57 and 22.6 μM, respectively) with good antioxidant activity, with a value 1.3-fold of Trolox. Additionally, theses hybrids have the ability to cross the blood–brain barrier in vitro and had significant neuroprotective effects in scopolamine-induced cognitive impairment in mice.

Scutellarin (4′,5,6-trihydroxyflavone-7-glucuronide), itself possesses a broad range of pharmacological properties related to neurological disorders, such as free radical scavenging effects, metal chelating properties, antiinflammatory effects, neuroprotective action, and the inhibition of Aβ fibril formation but, due to low oral absorption and poor blood–brain transport of scutellarin, its clinical use as an anti-AD drug is restricted, which was improved by linking it with rivastigmine. The kinetic characterization recommended that **110** showed a mixed-type inhibition, binding to both PAS and CAS of AChE [109].

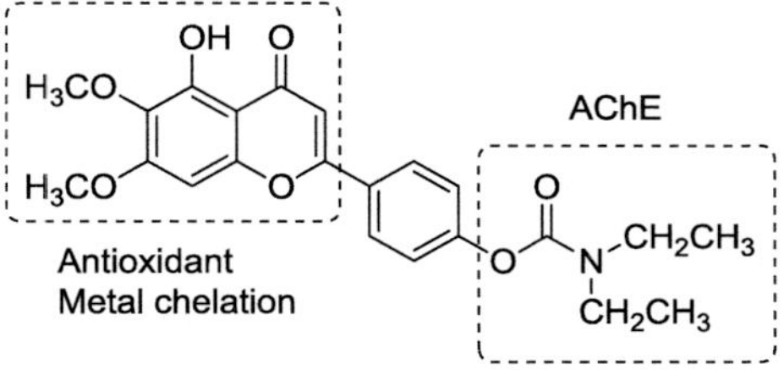

110

Kang et al. examined neuroprotective effects of biflavonoids obtained from natural source on amyloid β-peptide-induced and oxidative stress-induced cell death in neuronal cells and suggested that amentoflavone (**111**) and ginkgetin (**112**), homodimers of flavone exhibit strong neuroprotective effects in ischemic stroke and AD [110].

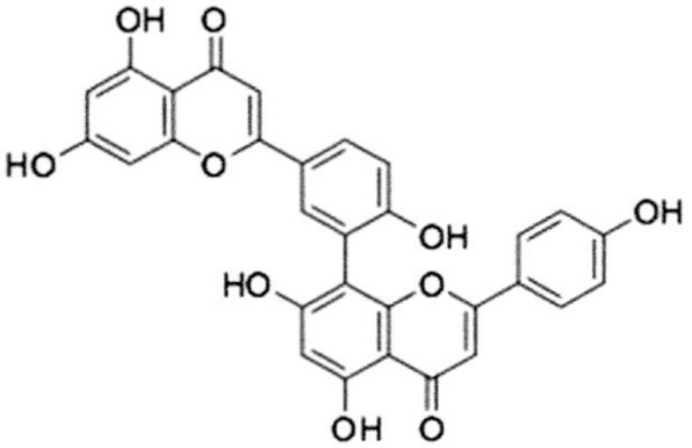

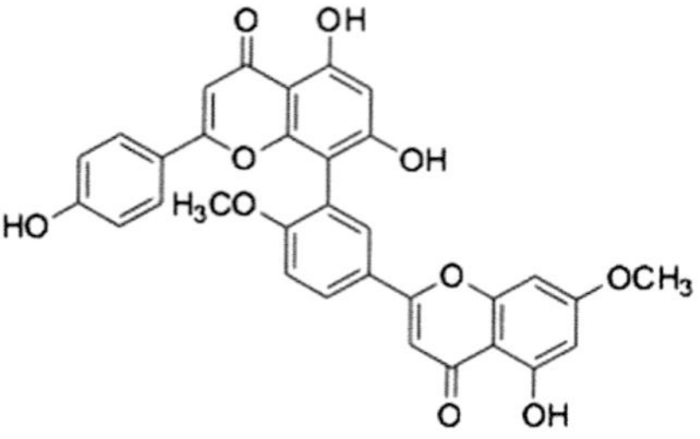

111

112

In 2013, Li et al. designed a series of novel flavone derivatives as potential multifunctional acetylcholinesterase inhibitors against AD. Compounds exhibited potent AChE inhibitory activity, high selectivity for AChE over BuChE, and moderate to good inhibitory potency toward Aβ aggregation. Specifically, compound **113** was the strongest AChE inhibitor, being 20-fold more potent than galanthamine and 2-fold more potent than tacrine, and it also had ability to inhibit Aβ aggregation and to function as a metal chelator. The SAR revealed that the diethylamine group linked to flavonoid scaffold by a four-carbon spacer exhibited the most potent activity [111].

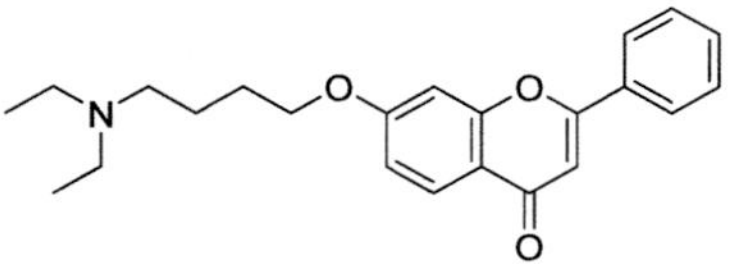

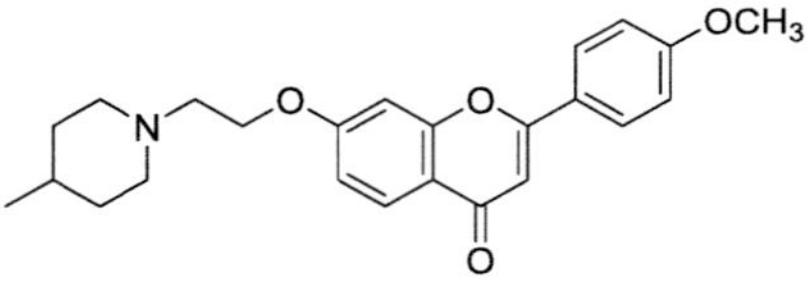

113

114

Similarly in 2016, Singh et al. designed, synthesized, and evaluated a novel series of 2-phenyl-1-benzopyran-4-one derivatives as potential polyfunctional anti-Alzheimer's agents. The in vitro studies showed that the majority of the synthesized derivatives inhibited AChE with IC_{50} values in the low-micromolar range. Compound **114**, strongly inhibited AChE with IC_{50} values of 6.33 nM, was found to be more potent than the reference compound donepezil. The SAR discovered that the extent of AChE inhibition was mainly influenced by the presence of the electron releasing group ($-OCH_3$) on B-ring and substituted cyclic amine linked to flavone scaffold by a two-carbon spacer [112].

5. CONCLUSION/PROSPECTIVES

The flavones are important members of the flavonoid family present in fruits and vegetables, which have received wide interest for their antioxidant potential and their ability to modulate several enzyme systems involved in a number of diseases. Flavones are lipophilic as well as hydrophilic, having polar functionalities in different positions, and the skeleton itself is amenable for the generation of functionalities for selective modulation of different enzymes.

Studies in various disease areas have shown that many diseases, specifically those that are metabolic and multifactorial, are best treated with combinations of drugs acting either with different mechanisms or with a drug exhibiting multiple pharmacological actions. The complex diseases like cancer, diabetes, AD, cardiovascular diseases, and so on can be treated with either a number of drugs each acting on different pathways in the diseases (polypharmacy) or a single agent having polyfunctional potential (multitargeted agents, hybrid molecules, etc.). Research is focused on polyfunctional drugs that may have much improved potency due to synergistic effects with fewer side effects.

As we discussed earlier, flavones with varied substituent's at different positions exhibited a wide spectrum of pharmacological activities. Despite this, the nucleus has been explored to design only single-target drugs. That is, those possessing only one type of pharmacological activity. Various natural and synthetic flavones and its derivatives are present in the market, but act via blocking a single pathway. In spite of exhibiting a wide range of biological activities, flavones are yet to achieve the status of promising drug candidates, and only a few flavones have undergone clinical studies. The reason for this could be the lack of optimization of biological activities.

The scenario has been changed, the research focused toward design and synthesis of polyfunctional flavones. The medicinal chemists have been paying attention to the design and synthesis of the multitargeted flavones endowed with requisite biological activities in a single entity. The investigation of various methods for the synthesis and structural modification of the flavone ring have become important goals of several research groups. The naturally obtained flavone moiety can be taken as lead compound for synthesis of semi- and purely synthetic flavone derivatives with different functional groups at different positions of flavone skeleton against different complex diseases. The newly designed multitargeted flavones would improve efficacy (potency) and reduce side effects in the treatment of pharmacologically interrelated diseases.

The flavone-based multitargeted strategies against multifarious diseases has become a central point in the medicinal chemistry field. The present information gives details about the structural requirement of flavone derivatives for various pharmacological activities. This information may provide an opportunity to scientists of medicinal chemistry discipline to design selective, optimized, and polyfunctional flavone derivatives for the treatment of multifactorial diseases.

REFERENCES

[1] J.L. Richard, F. Cambien, P. Ducimetiere, Epidemiologic characteristics of coronary disease in France, Nouvelle Presse Medicale 10 (1981) 1111–1114.

[2] C.A. Williams, Flavonoids chemistry, biochemistry and applications, in: K.R. Markham (Ed.), Historical Advances in the Flavonoid Field-A Personal Perspective, Talor & Francis, London, 2006.

[3] T.P.T. Cushnie, A.J. Lamb, Antimicrobial activity of flavonoids, International Journal of Antimicrobial Agents 26 (2005) 343–356.

[4] B. Havsteen, Flavonoids, a class of natural products of high pharmacological potency, Biochemical Pharmacology 32 (1983) 1141–1148.

[5] E. Middleton, K. Chithan, The impact of plant flavonoids on mammalian biology: implications for immunity, inflammation and cancer, in: J.B. Harborne (Ed.), The flavonoids: Advances in Research since 1986, Chapman and Hall, London, 1993, pp. 145–166.

[6] C.T. Davis, T.A. Geissman, Basic dissociation constants of some substituted flavones, Journal of the American Chemical Society 76 (1954) 3507–3511.

[7] K.V. Sashidhara, S.R. Avula, G.R. Palnati, S.V. Singh, K. Srivastava, S.K. Puri, J.K. Saxena, Synthesis and in vitro evaluation of new chloroquine-chalcone hybrids against chloroquine-resistant strain of *Plasmodium falciparum*, Bioorganic and Medicinal Chemistry Letters 22 (2012) 5455–5459.

[8] H.S. Mahal, K. Venkataraman, Synthetical experiments in the chromone group. Part XIV. The action of sodamide on 1-acyloxy-2-acetonaphthones, Journal of the Chemical Society 10 (1934) 1767–1769.

[9] S.R. Sarda, M.Y. Pathan, V.V. Paike, P.R. Pachmase, W.N. Jadhav, R.P. Pawar, A facile synthesis of flavones using recyclable ionic liquid under microwave irradiation, Arkivoc 16 (2006) 43–48.

[10] J. Allan, R. Robinson, An accessible derivative of chromonol, Journal of the Chemical Society 125 (1924) 2192–2195.

[11] W.K. Su, X.Y. Zhu, Z.H. Li, First Vilsmeier-Haack synthesis of flavones using bis-(trichloromethyl) carbonate/dimethylformamide, Organic Preparations and Procedures International 41 (2009) 69–75.

[12] I. Gulcin, Antioxidant activity of L-adrenaline: an activity-structure insight, Chemico-Biological Interactions 179 (2009) 71–80.

[13] (a) G. Cao, E. Sofic, R.L. Prior, Antioxidant and prooxidant behavior of flavonoids: structure-activity relationships, Free Radical Biology and Medicine 22 (1997) 749–760.
(b) Z.Y. Chen, P.T. Chan, K.Y. Ho, K.P. Fung, J. Wang, Antioxidant activity of natural flavonoids is governed by number and location of their aromatic hydroxyl groups, Chemistry and Physics of Lipids 76 (1996) 157–163.

[14] J.M. McCord, Superoxide radical: controversies, contradictions, and paradoxes, Proceedings of the Society for Experimental Biology and Medicine 209 (1995) 112–117.

[15] J. Hyun, Y. Woo, D. Hwang, G. Jo, S. Eom, Y. Lee, J.C. Park, Y. Lim, Relationships between structures of hydroxyflavones and their antioxidative effects, Bioorganic and Medicinal Chemistry Letters 20 (2010) 5510–5513.

[16] A. Gomes, O. Neuwirth, M. Freitas, D. Couto, D. Ribeiro, A. Fueiredo, A.M.S. Silva, R. Seixas, D. Pinto, A.C. Tome, J.A.S. Cavaleir, E. Fernandes, J. Lima, Synthesis and antioxidant properties of new chromone derivatives, Bioorganic and Medicinal Chemistry 17 (2009) 7218–7226.

[17] B. Mishra, I. Priyadarsini, S. Kumar, K. Unnikrishnan, H. Mohan, Effect of O-glycosilation on the antioxidant activity and free radical reactions of a plant flavonoid, chrysoeriol, Bioorganic and Medicinal Chemistry 11 (2003) 2677–2685.

[18] C. Nicolas, F. Sebastien, M.C. Martinez, A. Ramaroson, Anticancer properties of flavonoids: roles in various stages of carcinogenesis, Cardiovascular and Hematological Agents in Medicinal Chemistry 9 (2011) 62–77.

[19] T.B. Nguyen, O. Lozach, G. Surpateanu, Q. Wang, P. Retailleau, B.I. Iorga, L. Meijer, F. Gueritte, Synthesis, biological evaluation, and molecular modeling of natural and unnatural flavonoidal alkaloids, inhibitors of kinase, Journal of Medicinal Chemistry 55 (2012) 2811–2819.

[20] A.M. Senderowicz, Flavopiridol: the first cyclin-dependent kinase inhibitor in human clinical trials, Investigational New Drugs 17 (1999) 313–320.

[21] J.J. Lichius, O. Thoison, A. Montagnac, M. Pais, F. Gueritte-Voegelein, T. Sevenet, J.P. Cosson, A.H. Hadi, Antimitotic and cytotoxic flavonols from *Zieridium pseudobtusifolium* and *Acronychia porteri*, Journal of Natural Products 58 (1994) 1012–1016.

[22] H.F. Lu, Y.J. Chie, M.S. Yang, C.S. Lee, J.J. Fu, J.S. Yang, T.W. Tan, S.H. Wu, Y.S. Ma, S.W. Ip, J.G. Chung, Apigenin induces caspase-dependent apoptosis in human lung cancer A549 cells through Bax- and Bcl-2-triggered mitochondrial pathway, International Journal of Oncology 36 (2010) 1477–1484.

[23] X.H. Liu, H.F. Liu, X. Shen, B.A. Song, P.S. Bhadury, H.L. Zhu, J.X. Liu, X.B. Qi, Synthesis and molecular docking studies of novel 2-chloro pyridine derivatives containing flavone moieties as potential antitumor agents, Bioorganic and Medicinal Chemistry Letters 20 (2010) 4163–4167.

[24] A.J. Gillis, A.P. Schuller, E. Skordalakes, Structure of the *Tribolium castaneum* telomerase catalytic subunit TERT, Nature 455 (2008) 633–637.

[25] C.A. Feghali, T.M. Wright, Cytokines in acute and chronic inflammation, Frontiers in Bioscience 2 (1997) 12–26.

[26] C.A. Dinarello, Anti-inflammatory agents: present and future, Cell 140 (2010) 935–950.

[27] D.H. Kim, C.H. Yun, M.H. Kim, C.H. Naveen Kumar, B.H. Yun, J.S. Shin, H.J. An, Y.H. Lee, Y.D. Yun, H.K. Rim, M.S. Yoo, K.T. Lee, Y.S. Lee, 4′-Bromo-5,6,7-trimethoxyflavone represses lipopolysaccharide-induced iNOS and COX-2 expressions by suppressing the NF-jB signaling pathway in RAW 264.7 macrophages, Bioorganic and Medicinal Chemistry Letters 22 (2012) 700–705.

[28] J. Chen, H.W. Chang, H.P. Kim, H. Park, Synthesis of phospholipase A2 inhibitory biflavonoids, Bioorganic and Medicinal Chemistry Letters 16 (2006) 2373–2375.

[29] V. Anuradha, P.V. Srinivas, R.R. Rao, K. Manjulatha, M.G. Purohit, J.M. Rao, Isolation and synthesis of analgesic and anti-inflammatory compounds from *Ochna squarrosa* L, Bioorganic and Medicinal Chemistry 14 (2006) 6820–6826.

[30] R.L. Mower, R. Landolfi, M. Steiner, Inhibition ill vitro of platelet aggregation and arachidonic acid metabolism by flavones, Biochemical Pharmacology 33 (1984) 357–364.

[31] T.A. Pham, H. Che, P.T. Phan, J.W. Lee, S.S. Kim, H. Park, Oroxylin A analogs exhibited strong inhibitory activities against iNOS-mediated nitric oxide (NO) production, Bioorganic and Medicinal Chemistry Letters 22 (2012) 2534–2535.

[32] X. Liu, C.B. Chan, J.W. Sang, S. Pradoldej, J. Huang, K. He, L.H. Phun, S. France, G. Xiao, Y. Jia, H.R. Luo, K. Ye, A synthetic 7,8-dihydroxyflavone derivative promotes neurogenesis and exhibits potent antidepressant effect, Journal of Medicinal Chemistry 53 (2010) 8274–8286.

[33] X. Liu, C.B. Chan, Q. Qi, G. Xiao, H.R. Luo, X. He, K. Ye, Optimization of a small tropomyosin-related kinase B (TrkB) agonist 7,8-dihydroxy fl avone active in mouse models of depression, Journal of Medicinal Chemistry 55 (2012) 8524–8537.

[34] Y. Shimmyo, T. Kihara, A. Akaike, T. Niidome, H. Sugimoto, Flavonols and flavones as BACE-1 inhibitors: structure–activity relationship in cell-free, cell-based and in silico studies reveal novel pharmacophore features, Biochimica et Biophysica Acta 1780 (2012) 819–825.

[35] I. Uriarte-Pueyo, M.I. Calvo, Flavonoids as acetylcholinesterase inhibitors, Current Medicinal Chemistry 18 (2011) 5289–5302.

[36] F. Chimenti, R. Fioravanti, A. Bolasco, P. Chimenti, D. Secci, F. Rossi, M. Yáñez, F. Orallo, F. Ortuso, S. Alcaro, R. Cirilli, R. Ferretti, M.L. Sanna, A new series of flavones, thioflavones, and flavanones as selective monoamine oxidase-B inhibitors, Bioorganic and Medicinal Chemistry 18 (2010) 1273–1279.

[37] J.H. Medina, H. Viola, C. Wolfman, M. Marder, C. Wasowski, D. Calvo, A.C. Paladini, Neuroactive flavonoids: new ligands for the benzodiazepine receptors, Phytomedicine 5 (1998) 235–243.

[38] P.W. Baures, S.A. Peterson, J.W. Kelly, Discovering transthyretin amyloid fibril inhibitors by limited screening, Bioorganic and Medicinal Chemistry 6 (1998) 1389–1401.

[39] A.K. Verma, H. Singh, M. Satyanarayana, S.P. Srivastava, P. Tiwari, A.B. Singh, A.K. Dwivedi, S.K. Singh, M. Srivastava, C. Nath, R. Raghubir, A.K. Srivastava, R. Pratap, Flavone-based novel antidiabetic and antidyslipidemic agents, Journal of Medicinal Chemistry 55 (2012) 4551–4567.

[40] O. Bozdag-Dündar, E.J. Verspohl, N. Das-Evcimen, R.M. Kaup, K. Bauer, M. Sarıkaya, B. Evranos, R. Ertan, Synthesis and biological activity of some new flavonyl-2,4-thiazolidine-diones, Bioorganic and Medicinal Chemistry 16 (2008) 6747–6751.

[41] A. Matin, M.R. Doddareddy, N. Gavande, S. Nammi, P.W. Groundwater, R.H. Roubin, D.E. Hibbs, The discovery of novel isoflavone pan peroxisome proliferator-activated receptor agonists, Bioorganic and Medicinal Chemistry 21 (2013) 766–778.

[42] H. Matsuda, T. Wang, H. Managi, M. Yoshikawa, Structural requirements of flavonoids for inhibition of protein glycation and radical scavenging activities, Bioorganic and Medicinal Chemistry 11 (2003) 5317–5323.

[43] C. La Casa, I. Villegas, C. Alarcón de la Lastra, V. Motilva, M.J. Martín Calero, Evidence for protective and antioxidant properties of rutin, a natural flavone, against ethanol induced gastric lesions, Journal of Ethnopharmacology 71 (2000) 45–53.

[44] Y.S. Kim, M. Son, J.I. Ko, H. Cho, M. Yoo, W.B. Kim, I.S. Song, C.Y. Kim, Effect of DA-6034, a derivative of flavonoid, on experimental animal models of inflammatory bowel disease, Archives of Pharmacal Research 22 (1999) 354–360.

[45] J.J. Ares, P.E. Outt, J.L. Randall, J.N. Johnston, P.D. Murray, L.M. O'Brien, P.S. Weisshaar, B.L. Ems, Synthesis and biological evaluation of flavonoids and related compounds as gastroprotective agents, Bioorganic and Medicinal Chemistry Letters 6 (1996) 995–998.

[46] J.J. Ares, P.E. Outt, J.L. Randall, P.D. Murray, P.S. Weisshaar, L.M. O'Brien, B.L. Ems, S.V. Kakodkar, G.R. Kelm, W.C. Kershaw, Synthesis and biological evaluation of substituted flavones as gastroprotective agents, Journal of Medicinal Chemistry 38 (1996) 4937–4943.

[47] T.H. Babu, K. Manjulatha, G.S. Kumar, A. Hymavathi, A.K. Tiwari, M. Purohit, J.M. Rao, K.S. Babu, Gastroprotective flavonoid constituents from *Oroxylum indicum* vent, Bioorganic and Medicinal Chemistry Letters 20 (2010) 117–120.

[48] W.F. Zheng, R.X. Tan, L. Yang, Z.L. Liu, Two flavones from *Artemisia giraldii* and their antimicrobial activity, Planta Medica 62 (1996) 160–162.

[49] H.J. Jung, W.S. Sung, S.H. Yeo, H.S. Kim, I.S. Lee, E.R. Woo, D.G. Lee, Antifungal effect of amentoflavone derived from *Selaginella tamariscina*, Archives of Pharmacal Research 29 (2006) 746–751.

[50] G. Sagrera, A. Bertucci, A. Vazquez, G. Seoane, Synthesis and antifungal activities of natural and synthetic biflavonoids, Bioorganic and Medicinal Chemistry 19 (2011) 3060–3073.

[51] P. Venkatesan, T. Maruthavanan, Synthesis of substituted flavone derivatives as potent antimicrobial agents, Bulletin of the Chemical Society of Ethiopia 25 (2011) 419–425.

[52] S.B.A. Ghani, L. Weaver, Z.H. Zidan, H.M. Ali, C.W. Keevil, R. Brown, Microwave-assisted synthesis and antimicrobial activities of flavonoid derivatives, Bioorganic and Medicinal Chemistry Letters 18 (2008) 518–522.

[53] N.R. Guz, F.R. Stermitz, J.B. Johnson, T.D. Beeson, S. Willen, J. Hsiang, K. Lewis, Flavonolignan and flavone Inhibitors of a *Staphylococcus aureus* multidrug resistance pump: structure activity relationships, Journal of Medicinal Chemistry 44 (2001) 261–268.

[54] A. Mori, C. Nishino, N. Enoki, S. Tawata, Antibacterial activity and mode of action of plant flavonoids against *Proteus vulgaris* and *Staphylococcus aureus*, Phytochemistry 26 (1987) 2231–2234.

[55] K. Osawa, H. Yasuda, T. Maruyama, H. Morita, K. Takeya, H. Itokawa, Isoflavanones from the heartwood of *Swartzia polyphylla* and their antibacterial activity against cariogenic bacteria, Chemical and Pharmaceutical Bulletin 40 (1992) 2970–2974.

[56] A.L. Liu, H.D. Wang, S.M. Lee, Y.T. Wang, G.H. Du, Structure–activity relationship of flavonoids as influenza virus neuraminidase inhibitors and their in vitro anti-viral activities, Bioorganic and Medicinal Chemistry 16 (2008) 7141–7147.

[57] L. Gao, M. Zu, S. Wu, A.L. Liu, G.H. Du, 3D QSAR and docking study of flavone derivatives as potent inhibitors of influenza H1N1 virus neuraminidase, Bioorganic and Medicinal Chemistry Letters 21 (2011) 5964–5970.

[58] Y.B. Ryu, M.J. Curtis-Long, J.W. Lee, H.W. Ryu, J.Y. Kim, W.S. Lee, K.H. Park, Structural characteristics of flavanones and flavones from *Cudrania tricuspidata* for neuraminidase inhibition, Bioorganic and Medicinal Chemistry Letters 19 (2009) 4912–4915.

[59] H.J. Jeong, Y.B. Ryu, S.J. Park, J.H. Kim, H.J. Kwon, J.H. Kim, K.H. Park, M.C. Rho, W.S. Lee, Neuraminidase inhibitory activities of flavonols isolated from *Rhodiola rosea* roots and their in vitro anti-influenza viral activities, Bioorganic and Medicinal Chemistry 17 (2009) 6816–6823.

[60] P.A. Miller, K.P. Milstrey, P.W. Trown, Specific inhibition of viral ribonucleic acid replication by gliotoxin, Science 159 (1968) 431–432.

[61] H.J. Kwon, H.H. Kim, Y.B. Ryu, J.H. Kim, H.J. Jeong, S.W. Lee, J.S. Chang, K.O. Cho, M.C. Rho, S.J. Park, W.S. Lee, In vitro anti-rotavirus activity of polyphenol compounds isolated from the roots of *Glycyrrhiza uralensis*, Bioorganic and Medicinal Chemistry 18 (2010) 7668–7674.

[62] Y.M. Lin, D.E. Zembower, M.T. Flavin, R.M. Schure, H.M. Anderson, B.E. Korba, F. Chen, Robustaflavone, a naturally occurring biflavanoid, is a potent non-nucleoside inhibitor of hepatitis B virus replication in vitro, Bioorganic and Medicinal Chemistry Letters 7 (1907) 2325–2328.

[63] Y. Kashiwada, A. Aoshima, Y. Ikeshiro, Y.P. Chen, H. Furukawa, M. Itoigawa, T. Fujioka, K. Mihashi, L.M. Cosentino, S.L. Morris-Natschke, K.H. Lee, Bioorganic and Medicinal Chemistry 13 (2005) 443–448.

[64] J. Lameira, I.G. Medeiros, M. Reis, A.S. Santos, C.N. Alves, Structure–activity relationship study of flavone compounds with anti-HIV-1 integrase activity: a density functional theory study, Bioorganic and Medicinal Chemistry 14 (2006) 7105–7112.

[65] D.C. Rowley, M.S. Hansen, D. Rhodes, C.A. Sotriffer, H. Ni, J.A. McCammon, F.D. Bushman, W. Fenical, Thalassiolins A–C: new marine-derived inhibitors of HIV cDNA integrase, Bioorganic and Medicinal Chemistry 10 (2002) 3619–3625.

[66] G. Casano, A. Dumètre, C. Pannecouque, S. Hutter, N. Azas, M. Robin, Anti-HIV and anti-plasmodial activity of original flavonoid derivatives, Bioorganic and Medicinal Chemistry 18 (2010) 6012–6023.

[67] L.A. Friedman, A.W. Kimball, Coronary heart disease mortality and alcohol consumption in Framingham, American Journal of Epidemiology 124 (1986) 481–489.

[68] S.Q. Wang, X.Z. Han, X. Li, D.M. Ren, X.N. Wang, H.X. Lou, Flavonoids from *Dracocephalum tanguticum* and their cardioprotective effects against doxorubicin-induced toxicity in H9c2 cells, Bioorganic and Medicinal Chemistry Letters 20 (2010) 6411–6415.

[69] R. Budriesi, A. Bisi, P. Ioan, A. Rampa, S. Gobbi, F. Belluti, L. Piazzi, P. Valenti, A. Chiarini, 1,4-Dihydropyridine derivatives as calcium channel modulators: the role of 3-methoxy-flavone moiety, Bioorganic and Medicinal Chemistry 13 (2005) 3423–3430.

[70] J. Lebeau, R. Neviere, N. Cotelle, Beneficial effects of different flavonoids, on functional recovery after ischemia and reperfusion in isolated rat heart, Bioorganic and Medicinal Chemistry Letters 11 (2001) 23–27.

[71] E.S. Wu, T.E. Cole, T.A. Davidson, M.A. Dailey, K.G. Doring, M. Fedorchuk, J.T. Loch, T.L. Thomas, J.C. Blosser, A.R. Borrelli, C.R. Kinsolving, R.B. Parker, J.C. Strand, B.E. Watkins, Synthesis and structure-activity relationship of flavodilol and its analogues, a novel class of antihypertensive agents with catecholamine depleting properties, Journal of Medicinal Chemistry 32 (1989) 183–192.

[72] D.M. Roden, Current status of class III antiarrhythmic drug therapy, The American Journal of Cardiology 72 (1993) 44B–49B.

[73] G.R. Li, H.B. Wang, G.W. Qin, M.W. Jin, Q. Tang, H.Y. Sun, X.L. Du, X.L. Deng, X.H. Zhang, J.B. Chen, L. Chen, X.H. Xu, L.C. Cheng, S.W. Chiu, H.F. Tse, P.M. Vanhoutte, C.P. Lau, Acacetin, a natural flavone, selectively inhibits human atrial repolarization potassium currents and prevents atrial fibrillation in dogs, Circulation 11 (2008) 2449–2457.

[74] M. Correia-da-Silva, E. Sousa, B. Duarte, F. Marques, F. Carvalho, L.M. Cunha-Ribeiro, M.M. Pinto, Flavonoids with an oligopolysulfated moiety: a new class of anticoagulant agents, Journal of Medicinal Chemistry 54 (2011) 95–106.

[75] T.C. Wang, I.L. Chen, P.J. Lu, C.H. Wong, C.H. Liao, K.C. Tsiao, K.M. Chang, Y.L. Chen, C.C. Tzeng, Synthesis, antiproliferative, and antiplatelet activities of oxime- and methyloxime-containing flavone and isoflavone derivatives, Bioorganic and Medicinal Chemistry 13 (2005) 6045–6053.

[76] J.M. McCord, Free radicals and myocardial ischemia: overview and outlook, Free Radical Biology and Medicine 4 (1988) 9–14.

[77] U. Takahama, Y. Koga, S. Hirota, R. Yamauchi, Inhibition of xanthine oxidase activity by an oxathiolanone derivative of quercetin, Food Chemistry 126 (2011) 1808–1811.

[78] J. Flemmig, K. Kuchta, J. Arnhold, H.W. Rauwald, Olea europaea leaf (Ph.Eur.) extract as well as several of its isolated phenolics inhibit the gout-related enzyme xanthine oxidase, Phytomedicine 18 (2011) 561–566.

[79] L. Guo, W.R. Hu, J.H. Lian, W. Ji, T. Deng, M. Qian, B.Q. Gong, Anti-hyperlipidemic properties of CM108 (a flavone derivative) in vitro and in vivo, European Journal of Pharmacology 551 (2006) 80–86.

[80] R. Pratap, M. Satyanarayana, C. Nath, R. Raghubir, A. Puri, R. Chander, P. Tiwari, B.K. Tripathi, A.K. Srivastava, Oxy-Substituted Flavones as Antihyperglycemic and Antidyslipidemic Agents, U.S. Patent US7635779B2, 2009.

[81] L.Y. Chung, K.F. Yap, S.H. Goh, M.R. Mustafa, Z. Imiyabir, Muscarinic receptor binding activity of polyoxygenated flavones from *Melicope subunifoliolata*, Phytochemistry 69 (2008) 1548–1554.

[82] C.R. Chapple, H. Parkhouse, C. Gardener, E.J. Milroy, Double-blind, placebo-controlled, cross-over study of flavoxate in the treatment of idiopathic detrusor instability, British Journal of Urology 66 (1990) 491–494.

[83] M. Torres-Piedra, M. Figueroa, O. Hernández-Abreu, M. Ibarra-Barajas, G. Navarrete-Vázquez, S. Estrada-Soto, Vasorelaxant effect of flavonoids through calmodulin inhibition: Ex vivo, in vitro, and in silico approaches, Bioorganic and Medicinal Chemistry 19 (2011) 542–546.

[84] Z. Chen, Y. Hu, H. Wu, H. Jiang, Synthesis and biological evaluation of flavonoids as vaso-relaxant agents, Bioorganic and Medicinal Chemistry Letters 14 (2004) 3949–3952.

[85] J.B. Harborne, C.A. Williams, Advances in flavonoid research since 1992, Phytochemistry 55 (2000) 481–504.

[86] I.L. Wong, K.F. Chan, T.H. Chan, L.M. Chow, Flavonoid dimers as novel, potent antileish-manial agents, Journal of Medicinal Chemistry 55 (2012) 8891–8902.

[87] R. Maurya, P. Rawat, K. Sharan, J.A. Siddiqui, G. Swarnkar, G. Mishra, K.R. Arya, N. Chattopadhyay, World Pat. 110003, 2009.

[88] I. Kubo, I. Kinst-Hori, S.K. Chaudhuri, Y. Kubo, Y. Sánchez, T. Ogura, Flavonols from *Heterotheca inuloides*: tyrosinase inhibitory activity and structural criteria, Bioorganic and Medicinal Chemistry 8 (2000) 1749–1755.

[89] D.E. Wilcox, A.G. Porras, Y.T. Hwang, K. Lerch, M.E. Winkler, E.I. Solomon, Substrate ana-log binding to the coupled binuclear copper active site in tyrosinase, Journal of the American Chemical Society 107 (1985) 4015–4027.

[90] T. Tatsuhiko, Application to cosmetics of oil soluble licorice extract, Fragrance Journal 17 (1989) 122–125.

[91] S. Beutner, B. Bloedorn, S. Frixel, I.H. Blanco, T. Hoffmann, H.D. Martin, B. Mayer, P. Noack, C. Ruck, M. Schmidt, I. Schulke, S. Sell, H. Ernst, S. Haremza, G. Seybold, H. Sies, W. Stahl, R. Walsh, Quantitative assessment of antioxidant properties of natural colorants and phytochemicals: carotenoids, flavonoids, phenols and indigoids. The role of β-carotene in antioxidant functions, Journal of the Science of Food and Agriculture 81 (2001) 559–568.

[92] R. Verbeek, A.C. Plomp, E.A. van Tol, J.M. van Noort, The flavones luteolin and apigenin inhibit in vitro antigen-specific proliferation and interferon-gamma production by murine and human autoimmune T cells, Biochemical Pharmacology 68 (2004) 621–629.

[93] K. Phromnoi, S. Prasad, S.C. Gupta, R. Kannappan, S. Reuter, P. Limtrakul, B.B. Aggarwal, Dihydroxypentamethoxyflavone downregulates constitutive and inducible signal transducers and activators of transcription (STAT)-3 through the induction of tyrosine phosphatase SHP-1, Molecular Pharmacology 80 (2011) 889–899.

[94] K. Sankaranarayanan, Estimation of the genetic risks of exposure to ionizing radiation in humans: current status and emerging perspectives, Journal of Radiation Research 47 (2006) B57–B66.

[95] M.R. Cesarone, G. Belcaro, L. Pellegrini, A. Ledda, G. Vinciguerra, A. Ricci, G. Gizzi, E. Ippolito, F. Fano, M. Dugall, G. Acerbi, M. Cacchio, A. Di Renzo, S. Stuard, M. Corsi, HR, 0-(beta-hydroxyethyl)-rutosides; (venoruton): rapid relief of signs/symptoms in chronic venous insufficiency and microangiopathy: a prospective, controlled study, Angiology 56 (2005) 165–172.

[96] U. Blank, C. Ra, L. Miller, K. White, H. Metzger, J.P. Kinet, Complete structure and expression in transfected cells of high affinity IgE receptor, Nature 337 (1989) 187–189.

[97] S. Tamura, K. Yoshihira, K. Fujiwara, N. Murakami, New inhibitors for expression of IgE receptor on human mast cell, Bioorganic and Medicinal Chemistry Letters 20 (2010) 2299–2302.

[98] E.S. Wu, A. Kover, Synthetic Flavonoids as Inhibitors of Leukotrienes and 5-Lipoxygenase, U.S. Patent 4889941, 1989.

[99] P.J. Whiting, R.M. McKernan, K.A. Wafford, Inter- national review of neurobiology, in: R.J. Bradley, R.A. Harris (Eds.), Structure and Pharmacology of Vertebrate GABA a Receptor Subtypes, Academic Press, New York, 1995, pp. 95–138.

[100] A. Doble, I.L. Martin (Eds.), The GABAA/benzodiazepine Receptor as a Target for Psychoactive Drugs, Springer-Verlag, Heidelberg, Germany, 1996.

[101] J. Ai, K. Dekermendjian, O. Sterner, M. Nielsen, M.R. Witt, Compounds isolated from medicinal plants as ligands for benzodiazepine receptors, Recent Research Development in Phytochemistry 1 (1997) 365–385.

[102] K. Dekermendjian, P. Kahnberg, M.R. Witt, O. Sterner, M. Nielsen, T. Liljefors, Structure-activity relationships and molecular modeling analysis of flavonoids binding to the benzodiazepine site of the rat brain GABA a receptor complex, Journal of Medicinal Chemistry 42 (1999) 4343–4350.

[103] P. Kahnberg, E. Lager, C. Rosenberg, J. Schougaard, L. Camet, O. Sterner, E. Stergaard Nielsen, M. Nielsen, T. Liljefors, Refinement and evaluation of a pharmacophore model for flavone derivatives binding to the benzodiazepine site of the GABA a receptor, Journal of Medicinal Chemistry 45 (2002) 4188–4201.

[104] S. Ayers, D.L. Zink, K. Mohn, J.S. Powell, C.M. Brown, T. Murphy, R. Brand, S. Pretorius, D. Stevenson, D. Thompson, S.B. Singh, Flavones from *Struthiola argentea* with anthelmintic activity in vitro, Phytochemistry 69 (2008) 541–545.

[105] M.I. Fernández-Bachiller, C. Pérez, L. Monjas, J. Rademann, M.I. Rodríguez-Franco, New tacrine-4-oxo-4 H-chromene hybrids as multifunctional agents for the treatment of Alzheimer's disease, with cholinergic, antioxidant, and β-amyloid-reducing properties, Journal of Medicinal Chemistry 55 (2012) 1303–1317.

[106] C. Chandramouli, S. David, D. Richard, M. Pamela, Chemical modification of the multitarget neuroprotective compound fisetin, Journal of Medicinal Chemistry 55 (2012) 378–389.

[107] W.H. Park, MAPK inhibitors differentially affect gallic acid-induced human pulmonary fibroblast cell growth inhibition, Molecular Medicine Reports 4 (2011) 193–204.

[108] Y. Muftuoglu, G. Mustata, Pharmacophore modeling strategies for the development of novel nonsteroidal inhibitors of human aromatase (CYP19), Bioorganic and Medicinal Chemistry Letters 20 (2010) 3050–3054.

[109] Z. Sang, Y. Li, X. Qiang, G. Xiao, Q. Liu, Z. Tan, Y. Deng, Multifunctional scutellarin-rivastigmine hybrids with cholinergic, antioxidant, biometal chelating and neuroprotective properties for the treatment of Alzheimer's disease, Bioorganic and Medicinal Chemistry 23 (2015) 668–680.

[110] S.S. Kang, J.Y. Lee, Y.K. Choi, S.S. Song, J.S. Kim, S.J. Jeon, Y.N. Han, K.H. Son, B.H. Han, Neuroprotective effects of naturally occurring biflavonoids, Bioorganic and Medicinal Chemistry Letters 15 (2005) 3588–3591.

[111] R. Li, X. Wang, X. Hu, L. Kong, Design, synthesis and evaluation of flavonoid derivatives as potential multifunctional acetylcholinesterase inhibitors against Alzheimer's disease, Bioorganic and Medicinal Chemistry Letters 23 (2013) 2636–2641.

[112] M. Singh, O. Silakari, Design, synthesis and biological evaluation of novel 2-phenyl-1-benzopyran-4-one derivatives as potential poly-functional anti-Alzheimer's agents, RSC Advances 6 (2016) 108411–108422.

Thiazolidine-2,4-Dione: A Potential Weapon for Targeting Multiple Pathological Conditions

Navriti Chadha, Om Silakari
Punjabi University, Patiala, India

Chapter Outline

1. INTRODUCTION

Thiazolidine-2,4-dione (TZD) is a biologically important heterocyclic ring system that demonstrates a range of pharmacological activities, including antihyperglycemic [1], anticancer [2], antiarthritic [3], antiinflammatory [4], and antimicrobial [5]. Among these, antihyperglycemic is the widely studied effect of TZD derivatives that has also been extended to the development of clinically used glitazone drugs such as rosiglitazone [6], pioglitazone [7], lobeglitazone [8], and troglitazone [9].

Key Heterocycle Cores for Designing Multitargeting Molecules. https://doi.org/10.1016/B978-0-08-102083-8.00005-4

In the literature, several TZD derivatives have been reported by structural variations at 1- and 5-positions, which in turn led to the development of biologically active molecules against a broad spectrum of protein targets such as peroxisome proliferator-activated receptor (PPAR)-γ, aldose reductase (ALR2), phosphoinositide 3-kinase (PI3K)-γ, PI3K-α/mitogen-activated protein kinase (MEK), Pim kinase, protein tyrosine phosphatase 1B (PTP1B), cyclooxygenase (COX-2), UDP-*N*-acetylmuramoylalanine D-glutamate ligase (MurD ligase), histone deacetylase (HDAC), and tyrosinases (Fig. 5.1). The detailed description related to the TZD activity against the aforementioned protein has already been published [10]. This chapter is an attempt to compile the multitargeting ability of the scaffold as reported in the literature.

The first ever pharmacological evaluation (anti-TB activity) of any TZD derivative was reported by the Italian scientist Vistentini in 1954 [11]. Following this, in the same year, Marshall and Vallance [12] reported the anticonvulsing activity of other TZD derivatives. In the 1960s and 1970s, various other pharmacological and toxicological effects of TZD derivatives were explored by different research groups [13–19]. In 1982, the research on TZD gained high standards when Sohda and coworkers selected ciglitazone for clinical evaluation in hyperglycemia [20]. Although ciglitazone failed to achieve desired

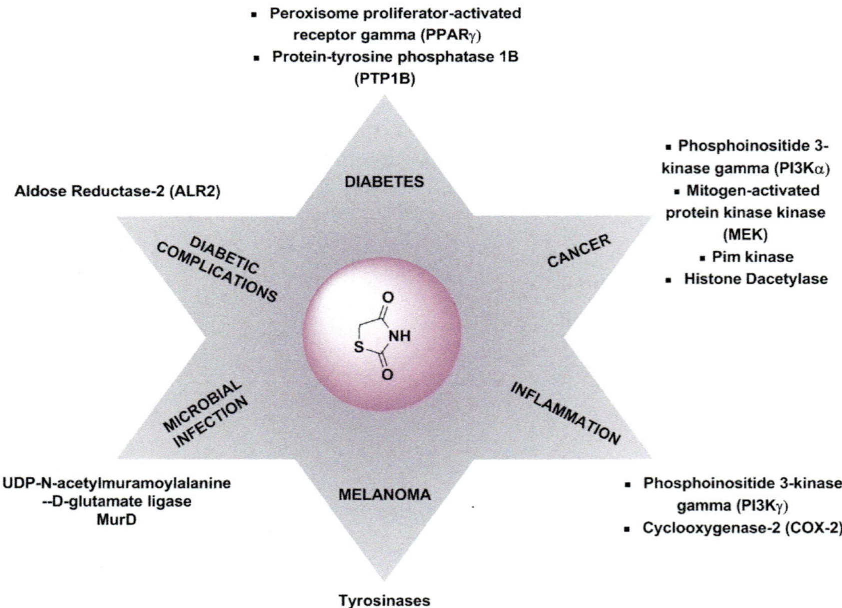

FIGURE 5.1 Multitargeted activity of thiazolidine-2,4-dione. *(Adapted from N. Chadha, M.S. Bahia, M. Kaur, O. Silakari, Thiazolidine-2,4-dione derivatives: programmed chemical weapons for key protein targets of various pathological conditions, Bioorganic and Medicinal Chemistry 23 (13) (2015) 2953–2974.)*

clinical standards, the continuous research on TZD resulted in troglitazone that was approved by FDA in 1997 for hyperglycemic conditions [21]. Thereafter, in 1999, two more TZD derivatives, rosiglitazone and pioglitazone, gained FDA approval, and in 2013, lobeglitazone was approved for use in Korea [8,22,23].

2. CHEMISTRY

Chemically, TZD nucleus is a five-membered ring containing two carbonyl moieties at 2- and 4-position sparing behind –NH and methylene ($-CH_2$) of thiazole ring for structural modifications to develop various analogues (molecule **1**). The unsubstituted TZD nucleus exists as a white crystalline solid with melting point range of 120–122°C [24]. The infrared spectrum of the same depicts an NH stretch at $3145\,cm^{-1}$, CH_2 stretch at $2923\,cm^{-1}$, carbonyl stretch at $1738\,cm^{-1}$, and $1659\,cm^{-1}$ (for C2 and C4 carbonyls, respectively), C-N stretch at $1318\,cm^{-1}$ and $1165\,cm^{-1}$, and C-S stretch at $808\,cm^{-1}$ and $727\,cm^{-1}$ [24]. The ^{1}H-NMR depicts a singlet for CH_2 protons at δ 3.98 (ppm) and a broad singlet for NH protons at δ 12.51 (ppm), which is due to the deshielding effect of two carbonyl groups present on both sides [25]. The mass spectrum of TZD depicts a base peak at m/z 116 (100%) [24].

1

3. SYNTHESIS

The synthesis of TZD nucleus has been carried out using different starting materials including thiocarbamates, thioureas, thiosemicarbazones, and alkali thiocynates as displayed in Fig. 5.2. The in situ preparation of alkyl thioncarbamate is carried out by reacting carbonyl sulfide with primary amine in the presence of potassium hydroxide. These alkyl thioncarbamates are then reacted with α-haloalkanoic acids to produce thiolcarbamates, which cyclize to yield TZD nucleus [26]. The most commonly used synthetic protocol is refluxing of α-chloroacetic acid with thiourea for 12 h, which yields TZD nucleus via 2-imino-4-thiazolidinone intermediate as described in Scheme 5.2 [27]. This reaction can be further accelerated a using microwave assisted technique in which initially α-chloroacetic acid is reacted with thiourea in cold conditions to yield 2-imino-4-thiazolidinone, which is further irradiated with microwave at 250 W for 5 min to obtain white crystals of TZD [25]. The third synthetic protocol is the reaction of thiosemicarbazone of acetone with chloroacetic acid ester, which, in the presence of sodium ethoxide, yields 2-hydrazino-4-thiazolidinone that further, in the presence of dilute

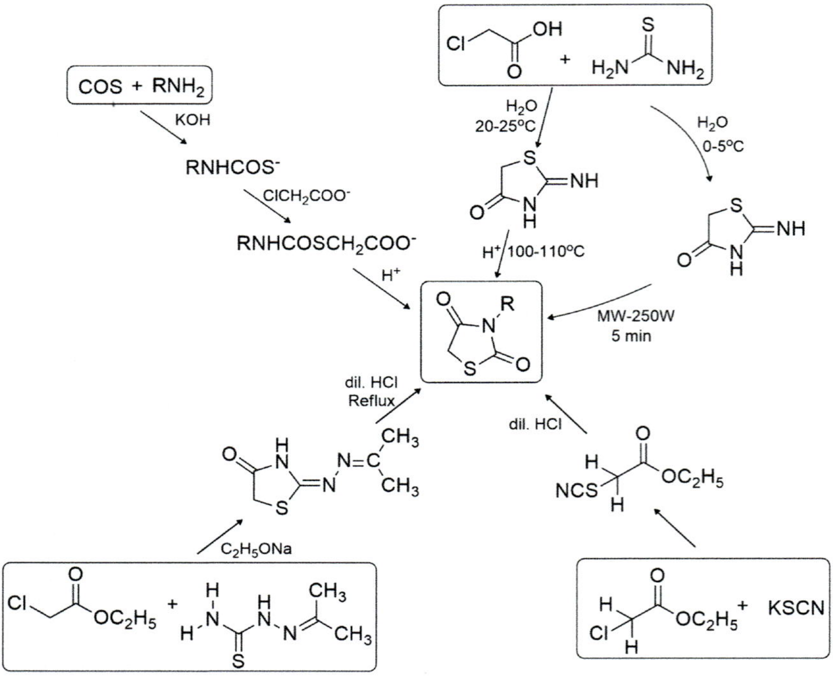

FIGURE 5.2 Synthetic scheme for thiazolidine-2,4-dione synthesis.

hydrochloric acid, yields TZD nucleus [28]. The other way to obtain TZD nucleus includes acidification (i.e., with dilute HCl) of the product obtained from chemical reaction of ethylchloroacetate with potassium thiocyanate [29]. Due to the presence of two carbonyl groups and one α-hydrogen, TZD nucleus undergoes different tautomerism such as amide-imidol and/or keto-enol (Fig. 5.3) [24,25]. The pK_a value of the TZD nucleus has been reported to be 6.82 [30].

The free –NH and –CH$_2$ moieties of TZD core are referred to as substitution positions, which have been explored to develop a wide variety of TZD derivatives.

3.1 Substitutions at –NH Moiety of TZD Core

The free –NH moiety of TZD has primarily been alkylated using alkyl or aryl halides in the presence of alkali including potassium carbonate [30], tetrabutylammonium iodide [31], or sodium hydride [32] using acetone or dimethylformamide (DMF) as solvent (Scheme 5.1). The different derivatives that can be obtained with substituting free –NH of TZD core have been mentioned in Table 5.1.

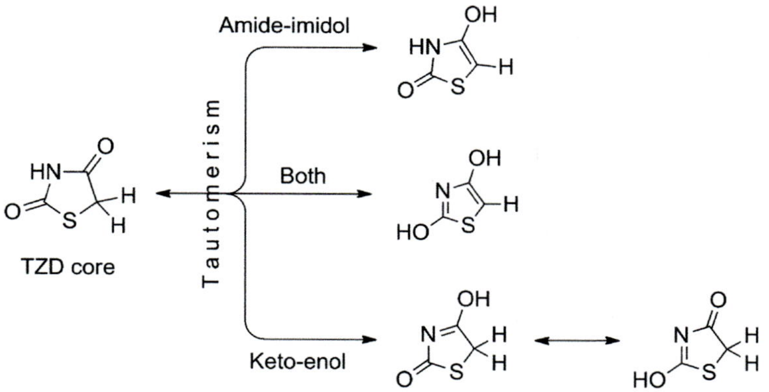

FIGURE 5.3 Tautomeric structures of thiazolidine-2,4-dione.

SCHEME 5.1 Substitution of –NH by alkyl or aryl group.

3.2 Substitution at Free –CH$_2$ Moiety of TZD Core

The methylene moiety has critically been substituted with aldehydes or ketones leading to formation of arylidene derivatives, via Knoevenagel condensation. The condensation of aldehyde and TZD has been carried out under different reaction conditions including a few drops of piperidine using ethanol or methanol as solvents for 7–42 h [43] (Scheme 5.2), or anhydrous sodium acetate in glacial acetic acid [27] while condensation of TZD with ketones has been carried out in the presence of ammonium acetate or piperidinium acetate in toluene or ethyl acetate [44]. Attempts have also been made to develop an eco-friendly reaction condition for Knoevenagel condensation using L-tyrosine in water [45] or β-alanine in acetic acid [46], or baker's yeast in ethanol [47]. These reactions have enabled coupling of TZD with various benzylidene derivatives as well as other heterocyclic ring moieties such as chalcones, flavones, acridines, furfurals, and dibenzocycloheptanone. Various aldehydes and ketones that are condensed with the –CH$_2$ moiety of TZD are discussed in Table 5.2.

4. THIAZOLIDINE-2,4-DIONE AS A PRIVILEGED SUBSTRUCTURE

The Evan's conception of privileged substructure of "being able to bind multiple targets" is fulfilled by thiazalodine-2,4-dione substructure as it binds a broad

TABLE 5.1 Reported Substitutions on NH of the TZD Ring

S. No.	Reactants	Product	References
1	TZD 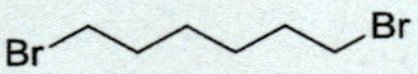 + Br—(CH₂)₆—Br + Triethylamine (TEA)	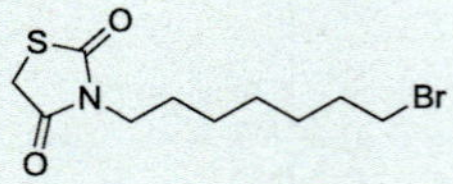 N-(6-bromohexyl)thiazolidine-2,4-dione	[34]
	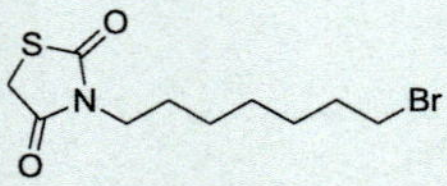 N-(6-bromohexyl)thiazolidine-2,4-dione + Hexamethylene tetramine /hexamine	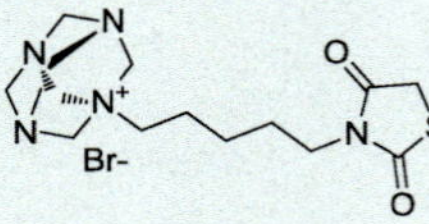 Hexaminium bromide salt of N-hexyl-thiazolidine-2,4-dione (Br⁻)	
	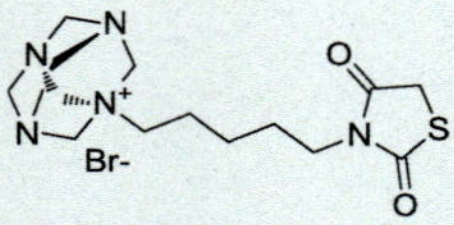 Hexaminium bromide salt (Br⁻) + Conc. HCl	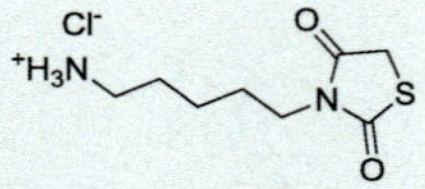 N-(aminoalkyl)thiazolidine-2,4-dione (Cl⁻, ⁺H₃N–)	
	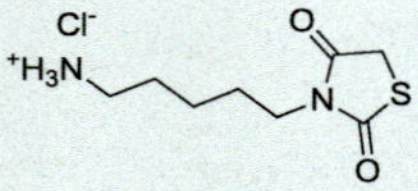 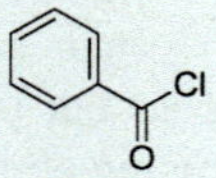 N-(aminoalkyl)thiazolidine-2,4-dione (Cl⁻, ⁺H₃N–) + benzoyl chloride or benzenesulfonyl chloride 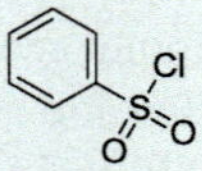	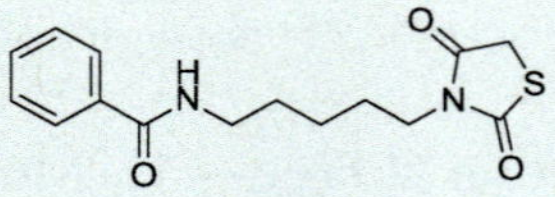 N-benzoylamino derivative of thiazolidine-2,4-dione or N-benzenesulfonylamino derivative of thiazolidine-2,4-dione 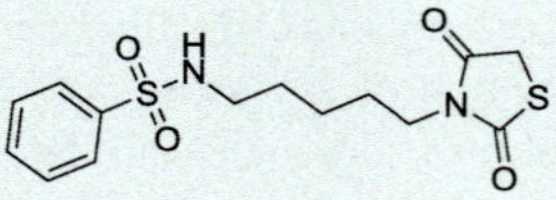	
2	TZD + $BrCH_2COOCH_3$ + K_2CO_3/acetone	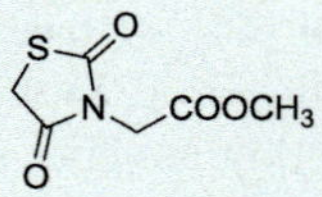 Methyl 2-(2,4-dioxothiazolidin-3-yl)acetate (N–COOCH₃)	[33,35]
	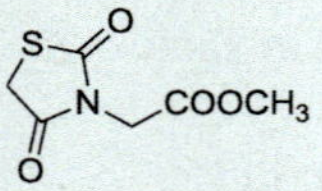 Methyl 2-(2,4-dioxothiazolidin-3-yl)acetate (N–COOCH₃) + Acetic acid/HCl	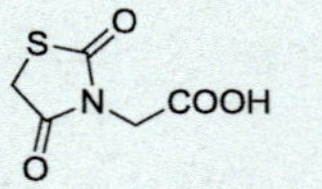 2-(2,4-dioxothiazolidin-3-yl)acetic acid (N–COOH)	

TABLE 5.1 Reported Substitutions on NH of the TZD Ring—cont'd

S. No.	Reactants	Product	References
3	TZD + BocNHCH$_2$CH$_2$Br + TBAI/acetone	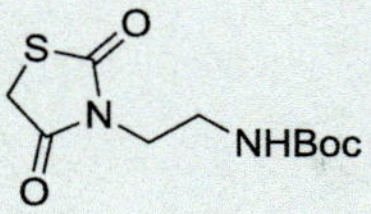	[31,36]
	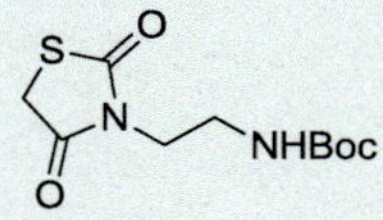 (structure) + H$_2$/Pd-C in MeOH or acetic anhydride/DCM	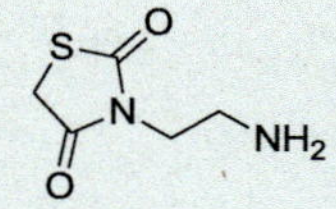 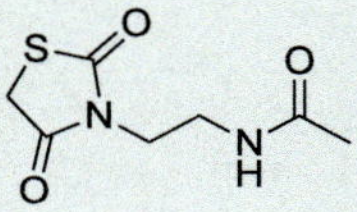	
4	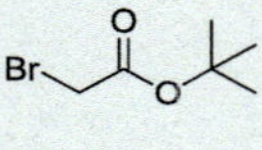 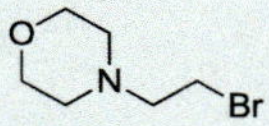 (structure) or (structure) + K$_2$CO$_3$, TBAI/acetone	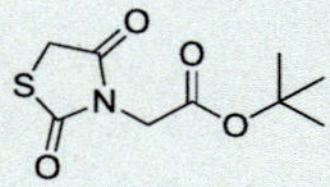 or 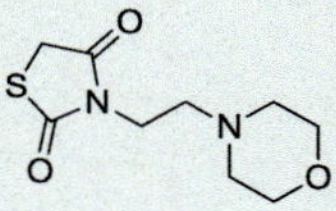	[37]
5	TZD 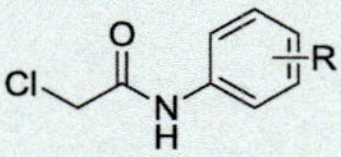 + (structure) R = 4-Me, 4-OMe + KOH/EtOH	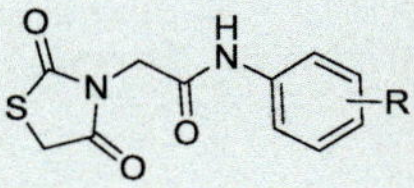	[38]
6	TZD 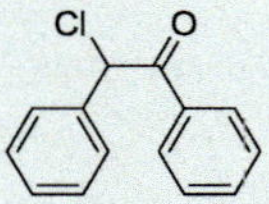 + (structure) + KOH/acetone	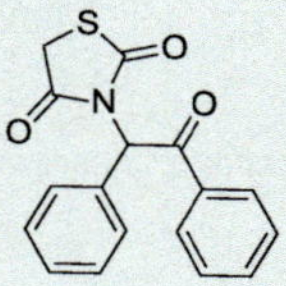	[39]

Continued

TABLE 5.1 Reported Substitutions on NH of the TZD Ring—cont'd

S. No.	Reactants	Product	References
7	TZD 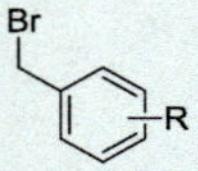 + Br-CH₂-C₆H₄-R R = H, 4-CH₃, 4-OCH₃, 4-Cl, 4-Br + K₂CO₃, TBAI/DMF	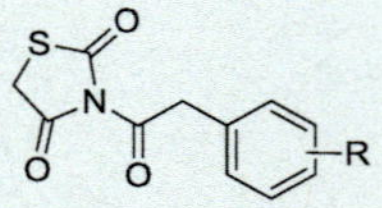 (TZD–N-acyl-benzyl product)	[40]
8	TZD + R-Cl R = methyl, allyl, isopropyl, butyl + K₂CO₃/acetone	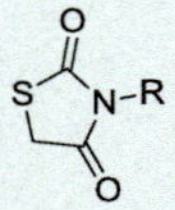 (N-R thiazolidinedione)	[41]
9	TZD 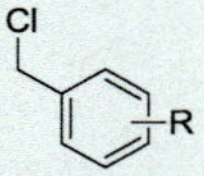 + Cl-CH₂-C₆H₄-R R = 4-Cl, 4-CH₃ + K₂CO₃	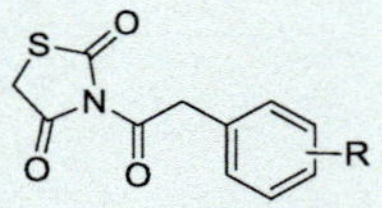 (TZD–N-acyl-benzyl product)	[42]

DMF, dimethylformamide; *HCHO*, formaldehyde; K_2CO_3, potassium carbonate; *KOH*, potassium hydroxide; *TBAI*, tetrabutylammonium iodide; *TEA*, triethylamine.

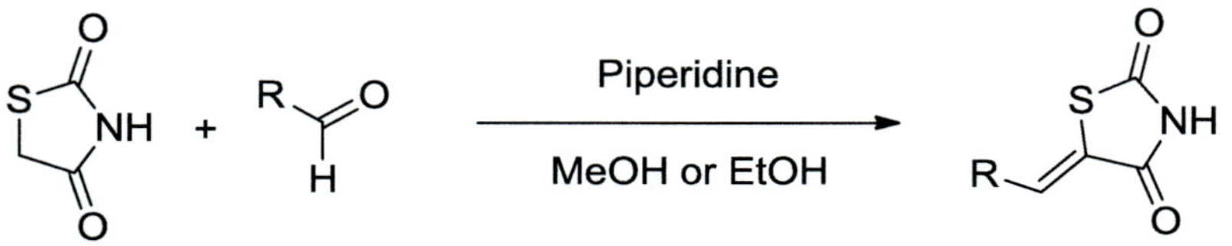

SCHEME 5.2 Knoevenagel condensation of TZD with aldehyde group.

spectrum of protein targets, but not limited to, such as PPAR-γ, ALR2, PI3K-γ, PI3K-α/ MEK, pim kinase, PTP1B, COX-2, MurD ligase, HDAC and tyrosinases, and poly (ADP-ribose) polymerase (PARP). These subsequently have therapeutic implications as antidiabetic (glitazones), anticancer, antiarthritic, antiinflammatory, antioxidant, and antimicrobial agents.

The TZD derivatives came into light in 1982 when AL-321 (**2**) was reported to potentially reduce the levels of cholesterol and triglycerides in genetically obese and diabetic mice [64]; however, the molecular basis

TABLE 5.2 Examples of Some of the Substitutions of Aldehydes and Ketones Reported in Literature

S. No.	Reactants	Product	References
1	TZD 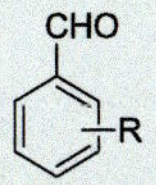 + CHO (benzaldehyde with R substituent) R = OH, OMe, 2-F, 2-Cl, 2-Br, 3-F, 3-Cl, 3-Br, 3-OCH$_3$, 3-F, 4-Cl, 4-Br, 4-OCH$_3$, 4-NO$_2$	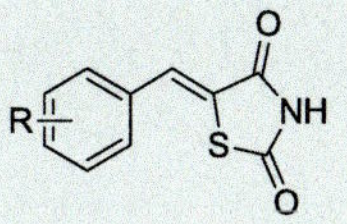	[39,43,48]
2	TZD 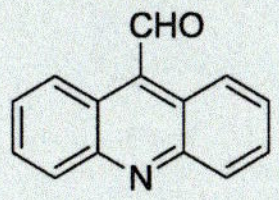 + CHO (acridine-9-carbaldehyde)	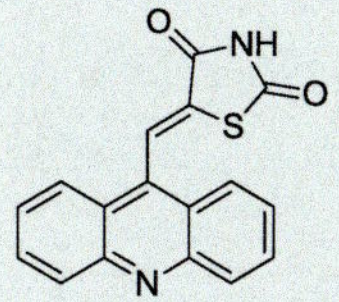	[42]
3	TZD 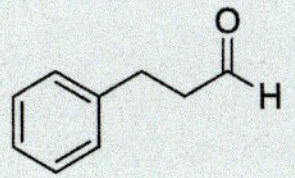 + (3-phenylpropanal)	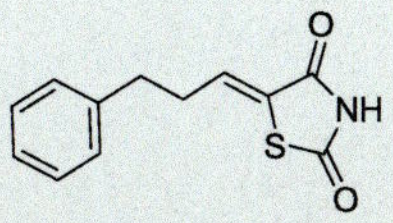	[36,37]
4	TZD 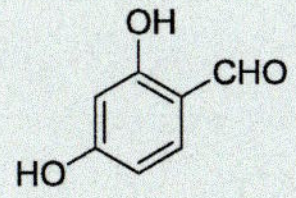 + (2,4-dihydroxybenzaldehyde)	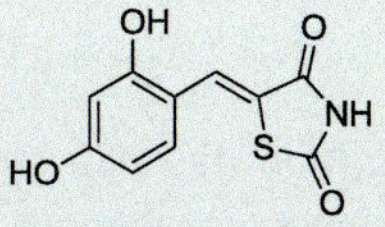	[49]

Continued

TABLE 5.2 Examples of Some of the Substitutions of Aldehydes and Ketones Reported in Literature—cont'd

S. No.	Reactants	Product	References
5	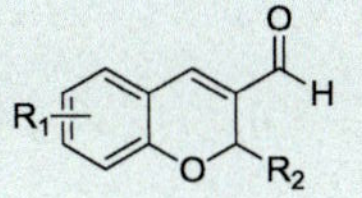 R_1 = H, 6-Br, 8-OMe; R_2 = Me, H	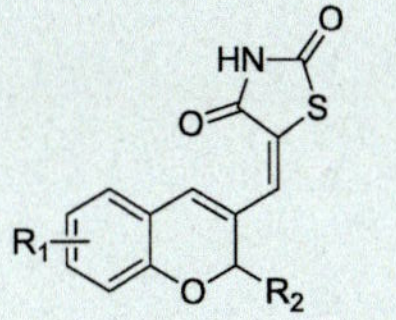	[50]
6	TZD 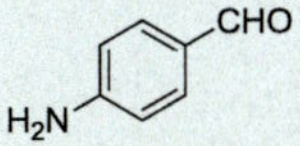 + (4-aminobenzaldehyde)	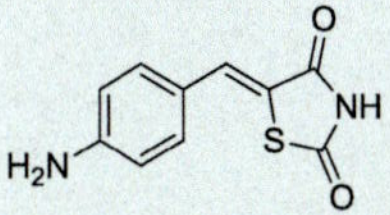	[38]
	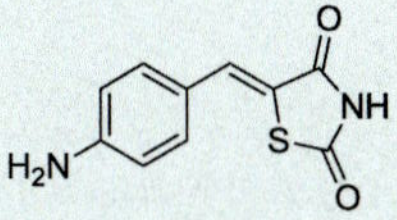 (aminobenzylidene-thiazolidinedione) + α-bromoacrylic acid	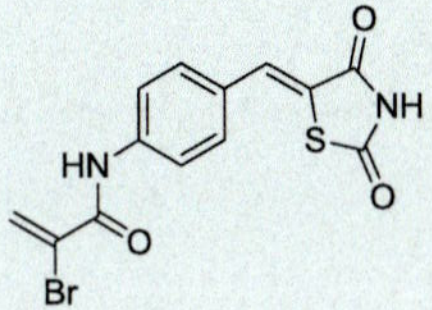	
7	TZD 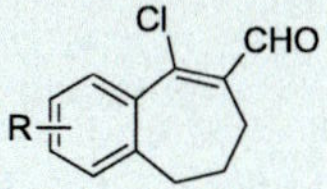 + (chloro-benzosuberone carbaldehyde); R = H, CH_3	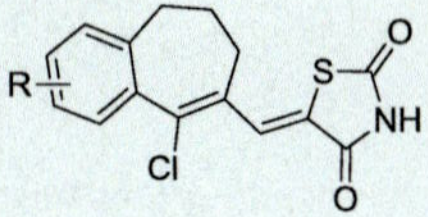	[40]

8	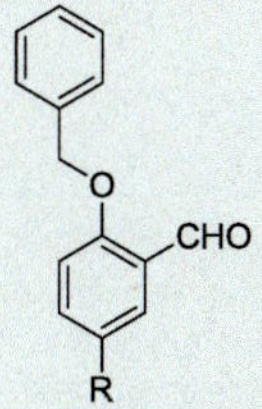 TZD + [structure] R = F, Cl, Br, Me, t-Bu, Phenyl, methoxy, nitro, piperidine, morpholine	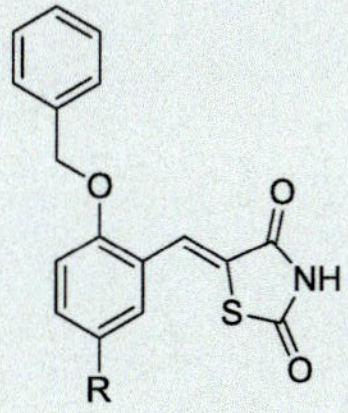 [structure]	[51]
9	TZD + [structure] X = NH, O + D-Glutamic acid dimethyl ester hydrochloride + TBTU/TEA/ DCM	[structures]	[52–54]

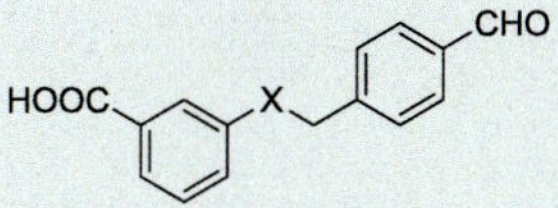

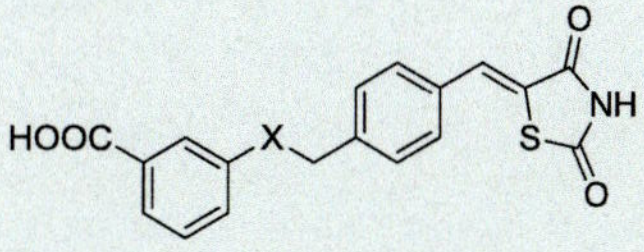

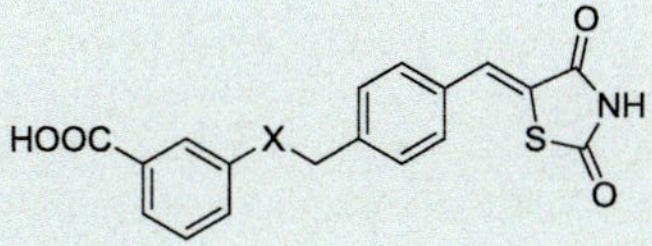

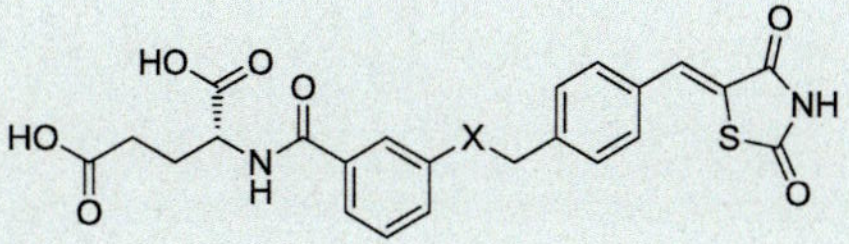

Continued

TABLE 5.2 Examples of Some of the Substitutions of Aldehydes and Ketones Reported in Literature—cont'd

S. No.	Reactants	Product	References
10	TZD 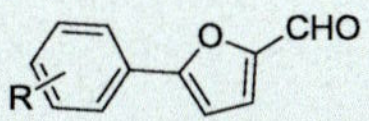 + (furan-2-carbaldehyde with aryl substituent) R = 2-OH, 3-OH, 4-OH, 2-OMe, 2-Cl, 2-Br, 2-OH, 2-F, 3-F, 4-F	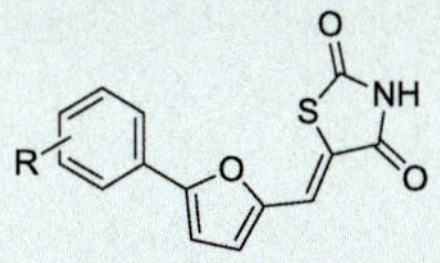	[55]
11	TZD 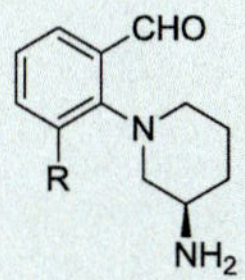 + (benzaldehyde with aminopiperidine substituent) R = H, Cl, Br, OMe, OEt. The aminopiperidine is replaced by aminopyrolidine or piperazine	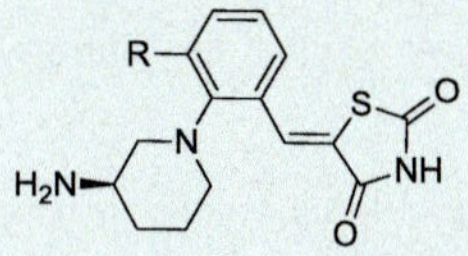	[56]

Continued

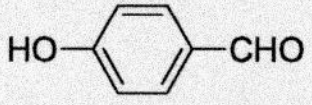

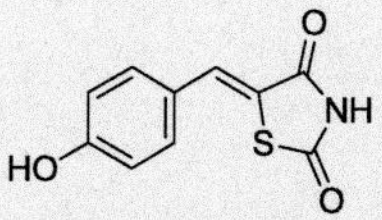

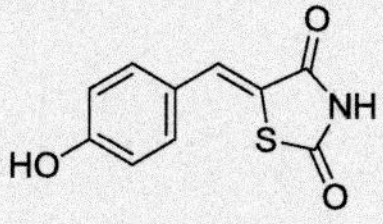

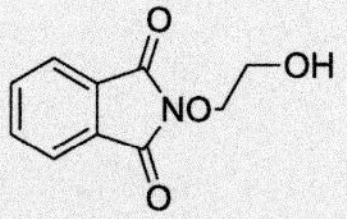

12

TZD + 4-HO-C6H4-CHO

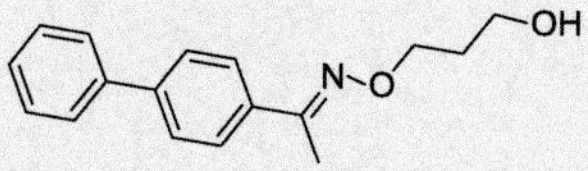

PPh3/DAED/aq. acetic acid

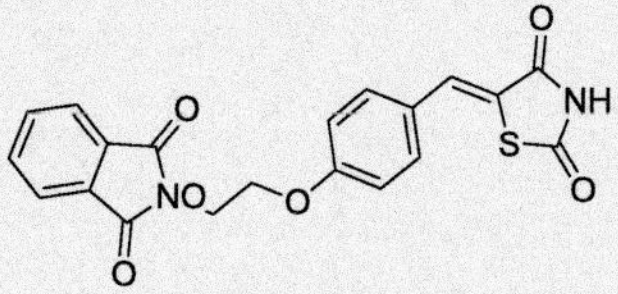

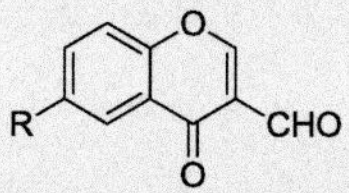

or

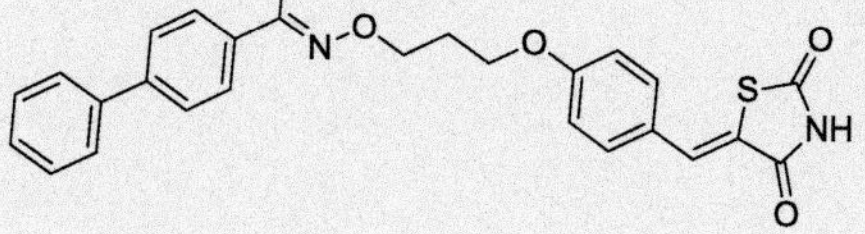

[57]

13

TZD +

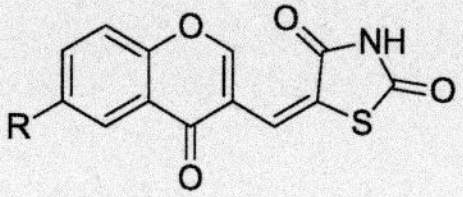

R = Cl, F, Br, OMe

[58]

TABLE 5.2 Examples of Some of the Substitutions of Aldehydes and Ketones Reported in Literature—cont'd

S. No.	Reactants	Product	References
14	TZD + aldehyde. R_1 = 2-OCH$_2$COOH, 4-OCH$_2$COOH. R_2 = H, 3-OMe, 3-OEt, 5-Br	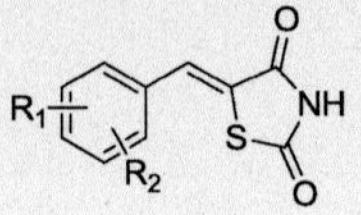	[59]
15	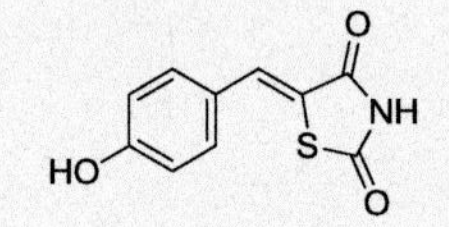 TZD + 4-hydroxybenzaldehyde. 1-Bromo-3-chloropropane	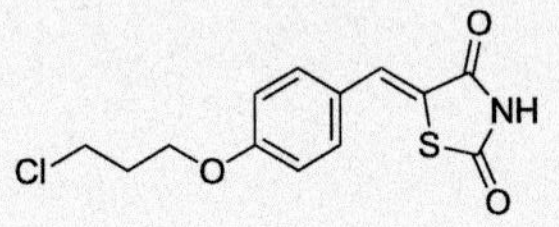	[60]

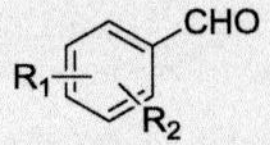

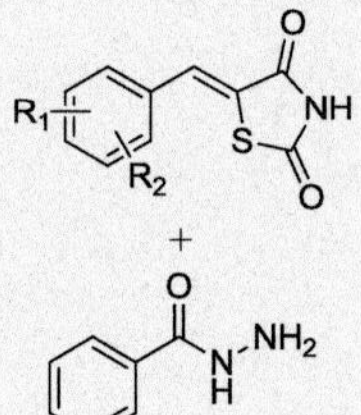

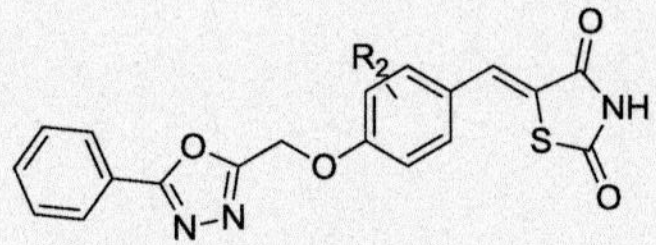

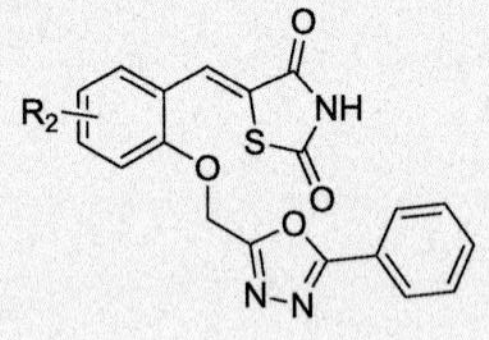

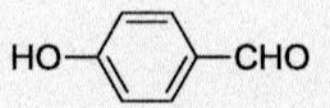

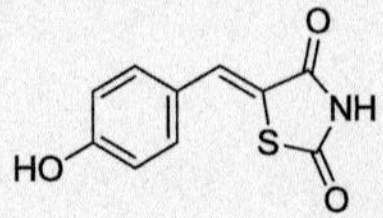

No.	Reactants	Product	Ref.
16	TZD + [aldehyde]		[61]
17	TZD + [aldehyde]		[62]
18	TZD + [ketone]		[63]
19	TZD + [aldehyde]		[27]

of this effect was later described by Lehmann and coworkers [65] who reported TZD derivatives as the potent and selective activators of PPARγ— later termed glitazone class of PPAR-γ agonists. Further, the modifications of lipophilic tail of AL-321 resulted in the identification of ciglitazone (**3**), pioglitazone (**4**), rosiglitazone (**5**), troglitazone (**6**), and englitazone (**7**), which are potent agonists of PPAR-γ and act as insulin sensitizers [66]. Lobeglitazone (**8**), a pyrimidine aryl ether derivative of rosiglitazone discovered by Chon Kun Dang Pharmaceuticals as a potent compound— compared to rosiglitazone—was approved for clinical use in Korea by their drug usage regulatory authority (Ministry of Food and Drug Safety, Korea) [67]. Rivoglitazone (**9**), discovered by Daiichi Sankyo Co., showed 178-fold more efficacy compared with pioglitazone, and also described beneficial effects in Phase II clinical trials [68,69]. Balaglitazone (**10**), a product of Dr. Reddy's Laboratories, has been engaged in Phase III clinical evaluation that is five-fold more potent than rosiglitazone [70].

Although, glitazones achieved clinical success for the treatment of diabetes, troglitazone (Rezulin) and rosiglitazone (Avandia) were withdrawn from the clinical practice in 2000 and 2010 due to severe hepatic and cardiac toxicity, respectively [71]. Moreover, in 2013 in light of new re-evaluation of rosiglitazone for cardiovascular outcomes and regulation of glycaemia in diabetes trial, the prescribing and dispensing restrictions of rosiglitazone were withdrawn by the FDA [72,73]. Among other discovered glitazones, pioglitazone (Actos), rosiglitazone (Avandia), and lobeglitazone (Duvie) are the only marketed drugs at present, since ciglitazone and englitazone failed clinical evaluation and troglitazone was withdrawn from market.

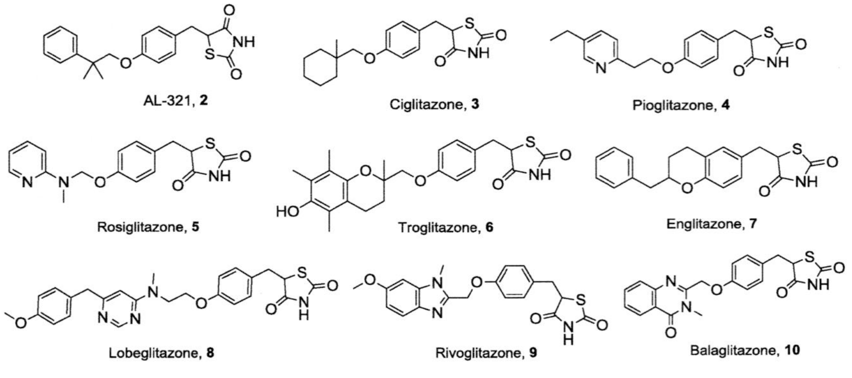

AL-321, 2

Ciglitazone, 3

Pioglitazone, 4

Rosiglitazone, 5

Troglitazone, 6

Englitazone, 7

Lobeglitazone, 8

Rivoglitazone, 9

Balaglitazone, 10

In addition to glitazones, various other TZD derivatives have also demonstrated the PPAR-γ agonism with less side effects and/or improved potency. For instance, oxime derivatives of TZD (**11**, $IC_{50} = 0.19\,\mu M$) [57] where the aromatic group at the meta- or para-position of acetophenone oxime or acetylpyridine oxime, and the ethylene bridge between oxygen atoms of oxime moiety is required for PPAR-γ activation. The hybrids of TZD with quercetin through diether linkage

(**12**) [74] and with substituted chromones through methylene linkage (**13**) showed potent PPAR-γ agonism with reduced hepatotoxicity. The conjugation of TZD core with the oxadiazole ring resulted in compound **14**, which exhibits no hepatic and cardiac toxicity [58]. The other conjugates of TZD include with benzoxazinone (**15**) [61], benzothiathiazole (**16**) [75], and chalcones (**17**) [76], which possess potent PPAR-γ agonism and antihyperglycemic potential.

Some of the TZD derivatives were reported as PTP1B inhibitors by Maccari and coworkers [66]. The reported molecules consisted of *p*-methylbenzoic acid moiety at the –NH of TZD ring whereas different substitutions at the para-position of central arylidene ring afforded micromolar (μM) activity of the resulting molecules (**18**, IC$_{50}$= 1.1 μM). Later, Bhattarai and coworkers reported the structural modifications at O-benzyl group attached to benzylidene ring, and the highest potency was observed with 4-trifluoromethane (**19**, IC$_{50}$=5.0 μM) [77]. The further optimization of substitutions at the central benzylidene ring resulted in the improved potency, for example, molecule **20** with methyl-4-(trifluoromethyl)benzene sulfonate and 1-(methoxymethyl)-4-(trifluoromethyl)benzene group at the para- and ortho-positions, respectively displayed IC$_{50}$ value of 1.3 μM, and demonstrated the benefit of dual substitution for activity improvement [78]. Moreover, the extension of hydrophobic tail at the para-position of benzylidene ring with benzyloxyphenyl group drastically improved the potency, and yielded the most potent compound of TZD class of PTP1B inhibitors (compound **21**, IC$_{50}$=0.69 μM) [79].

Bruno and coworkers exhibited the ALR2 inhibitory potential of TZD derivatives, and reported a series of molecules by varying substituents at the para-position of benzylidene ring. The substitution of phenoxy group at the benzylidene ring resulted in the most active compound of TZD class of ALR2 inhibitors, molecule **22** ($IC_{50}=0.13\,\mu M$) [80]. Further, the structure–activity relationship (SAR) was also studied for similar derivatives, which revealed that methyl carboxylic acid substituted on NH position incurs active derivatives and compound **23** ($IC_{50}=6.14\,\mu M$) was observed to be less potent compared to compound **22** [35]. Thereafter, the SAR analysis of TZD derivatives resulted in flavonyl-substituted TZD derivatives and similar observations were recorded that methyl carboxylic acid substitution give active compounds (**24**, $IC_{50}=0.41\,\mu M$) [81].

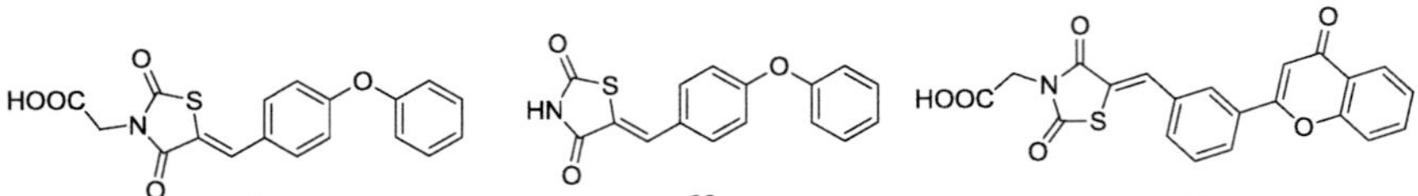

22   **23**   **24**

Initially, Xia and coworkers developed TZD inhibitors of Pim 1 and 2—in an effort to treat prostate cancer and leukemia—by screening ChemBridge library of molecules [82]. Among the synthesized TZD derivatives, molecule **25** (SMI-4a, $IC_{50}=3\,\mu M$) showed the highest potency for Pim1—primarily due to the lipophilic interactions of triflouromethyl group with hydrophobic residues of the active site. At present, SMI-4a is under clinical surveillance to evaluate its therapeutic profile in leukemia [83]. In addition, Dakin and coworkers have also reported different TZD derivatives, and molecule **26** ($IC_{50}=9\,nM$) bearing aminopiperidine at the ortho-position of benzylidene ring showed the highest potency primarily due to the interactions with acidic amino acid residues Asp128 and Glu171 of Pim1 [56]. Based on these observations, it can be interpreted that the substitution of positively charged groups at ortho and small hydrophobic groups at meta-position of the benzylidene ring is required to possess Pim1 inhibitory activity of TZD derivatives.

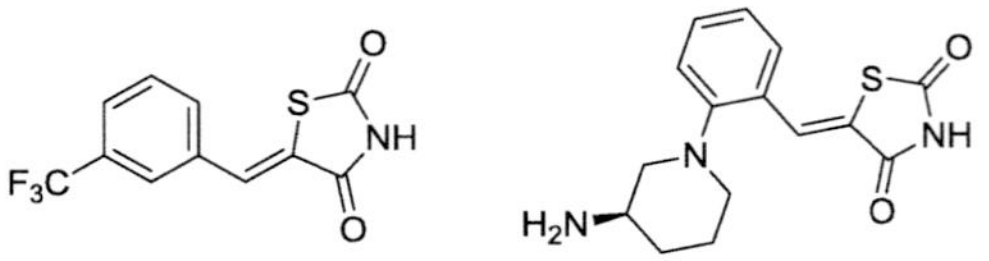

25   **26**

Pomel and coworkers reported the discovery of TZD inhibitors of PI3K-γ with the help of a SAR study [55]. The furan ring present at TZD was substituted with phenyl group, and the variation of hydroxyl group exhibited maximal PI3K-γ inhibitory activity at the ortho-position of phenyl ring (**27**, o>m>p, IC_{50} 30>290>380 nM). Thereafter, the optimization process of molecule **27** exhibited that 1,4-di-substitution of hydroxyl groups at phenyl ring are much favorable (**29**, $IC_{50}=20\,nM$) compared to *o*-hydroxy-*p*-fluoro (**28**, $IC_{50}=33\,nM$) and *o*-hydroxy-*p*-methylester (**30**, $IC_{50}=29\,nM$) substitutions. Furthermore, some TZD inhibitors of PI3K-γ including AS-605240 ($IC_{50}=9\,nM$) and AS-604850 ($IC_{50}=250\,nM$), are the patent molecules of Merk Sereno [84–86].

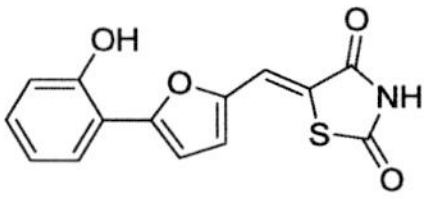

27 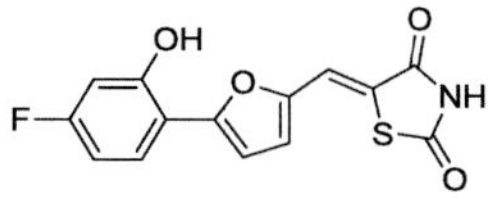   28 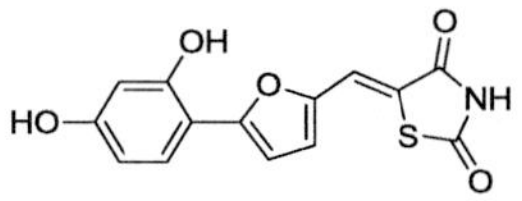   29

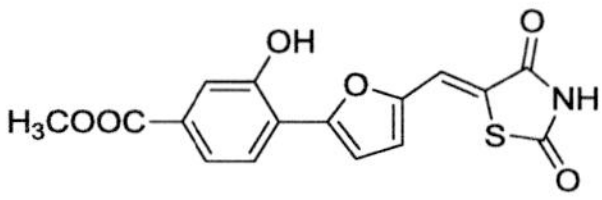

30 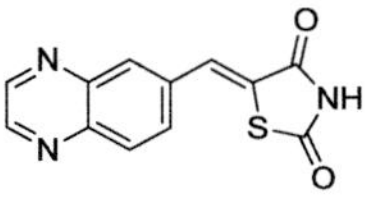   AS-605240 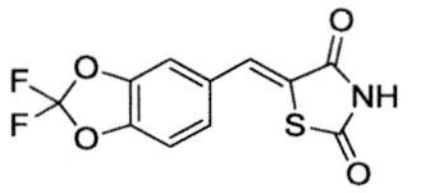   AS-604850

Ha and coworkers developed TZD inhibitors of tyrosinase, and among them, molecule **36** exhibited prominent inhibition of tyrosinase ($IC_{50}=9.87\,\mu M$) [43]. The SAR analysis of these molecules described the importance of meta and para substitutions at the benzylidene ring. For instance, the removal of the meta hydroxyl group from compound **31** compromised the biological activity against tyrosinase (**32** {24.08 %inhibition}31 vs. **31**{82.27%inhibition}), the replacement of para methoxy group with hydroxyl maintains the biological potential (**33** {62.68 %inhibition} vs. **31** {82.27 %inhibition}) while the substitution of hydroxyl group at both meta- and para-positions potentiate the activity of resulting compound **34** (MHY498, $IC_{50}=3.55\,\mu M$) as compared to compound **31** ($IC_{50}=9.87\,\mu M$) [49].

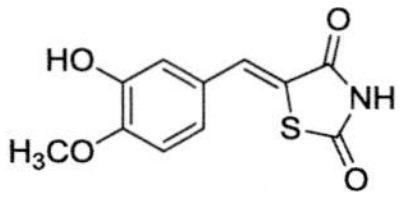

31 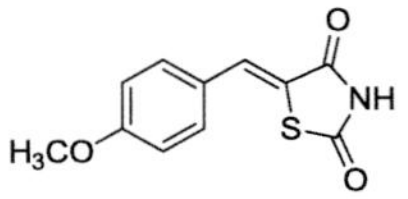   32 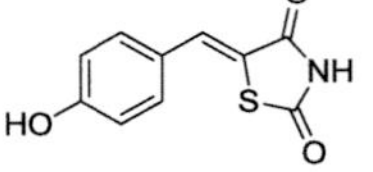   33 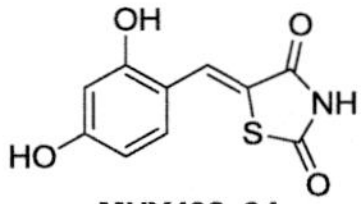   MHY498, 34

Tomasic and coworkers reported TZD inhibitors of MurD as antimicrobial agents [52]. However, the results showed that these compounds are inactive as compared to its rhodanine derivative **35** (S-isomer $IC_{50}=206\,\mu M$, R-isomer $IC_{50}=174\,\mu M$). Their investigations showed that the substitution at 4-position reduced inhibitory activity (**36**, % inhibition$=74$ at $250\,\mu M$). Further, derivatives were synthesized and TZD derivative **37** (R-isomer, $IC_{50}=35\,\mu M$) was observed to be more potent than its rhodanine derivative (R-isomer, $IC_{50}=49\,\mu M$) [53]. Substitution of the benzyl amine linker with alkyl linkers (**38**) gave compounds devoid of inhibitory activity. Modification of compound **37** was carried out by formulating its methyl ether derivatives [54]. However, evaluation of these molecules showed that rhodanine derivatives were more potent than its TZD derivatives.

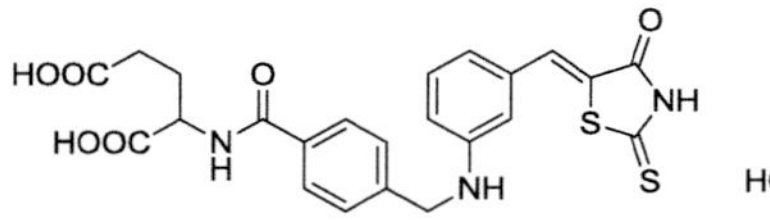

35

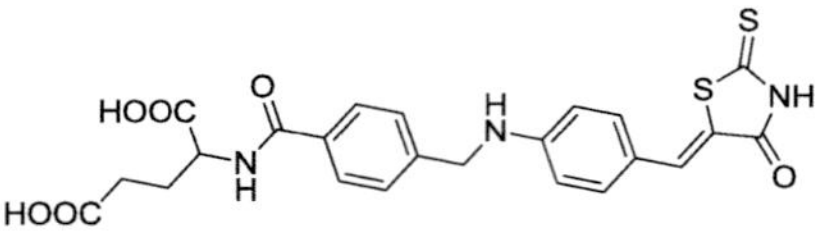

36

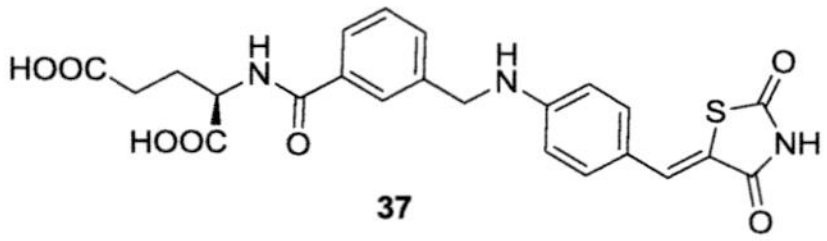

37

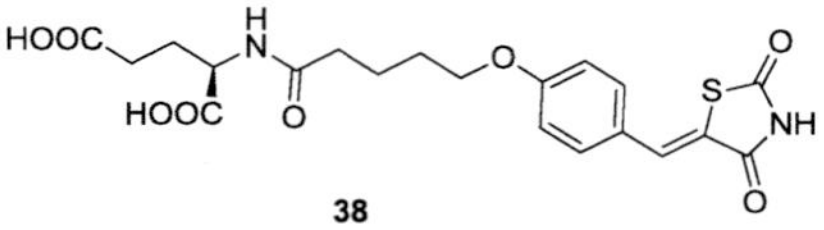

38

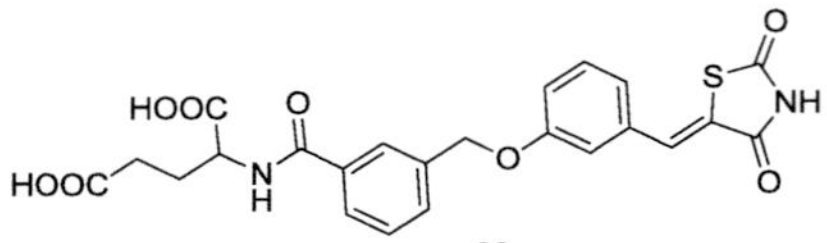

39

Ali and coworkers reported TZD derivatives (**40** and **41**) as potent antiinflammatory and analgesic compounds, and thereafter suggest that the biological activity of these molecules is conferred through COX-2 inhibition [39].

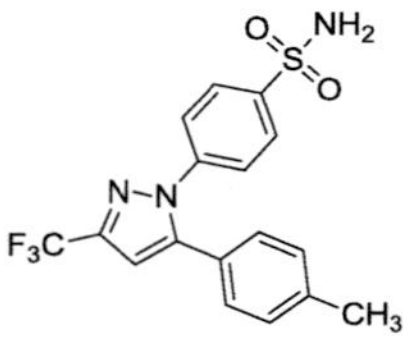

Celecoxib

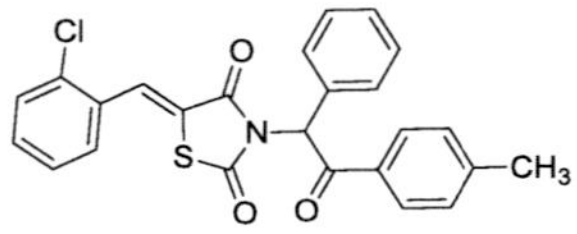

40

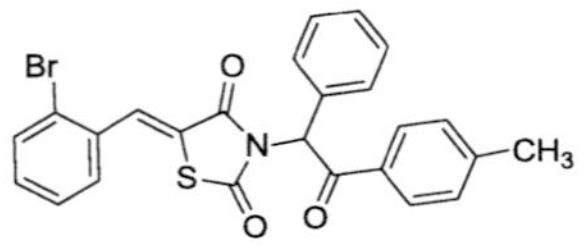

41

Mohan and coresearchers developed TZD inhibitors of HDAC1 (**42** and **43**), and subsequently evaluated their cancer cell line activity [34]. These molecules exhibited 43.31% and 57.27% inhibition in HepG2 cell lines, and 42.11% and 56.85% inhibition of HDAC1 at a dose of 100 μM.

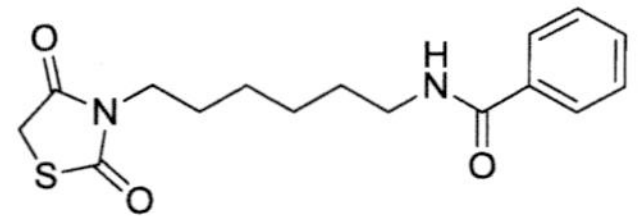

42

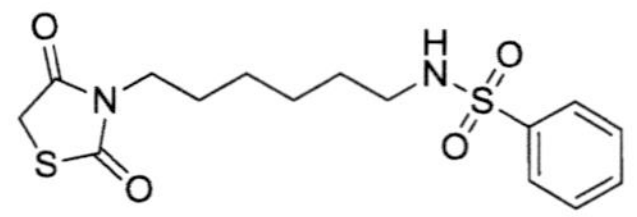

43

5. ROLE OF TZD AS MULTITARGETING AGENTS IN MULTIFACTORIAL DISEASES

As discussed earlier, TZD scaffold has been used successfully in several drug development projects where it was found to be useful for the generation of new lead compounds. Moreover, TZD derivatives have also been used in designing multitargeted agents. Drug repurposing or repositioning are other applications of polypharmacology where various recognized drugs for a given pathology are being rediscovered for new applications, which is either serendipitous or through exploration of the original mechanism of action. For instance, glitazones have been now clinically evaluated for their effectiveness as neuroprotective, antihypertensive, and anticancer agents.

5.1 Drug Repurposing for Glitazones

5.1.1 Glitazones as Neuroprotective Agents

As discussed earlier, pioglitazone (**4**) and rosiglitazone (**5**) are TZD PPARγ agonists, which are orally administered insulin-sensitizing agents used to treat type II diabetes [87]. Later, it was shown that pioglitazone and rosiglitazone exhibit neuroprotective properties through several possible mechanisms. For instance, it has been suggested that they have neuroprotective activity in cerebral ischemic stroke and traumatic brain injury [88–91]. In Parkinson's disease models, it has been shown that the TZDs are neuroprotective in various animal models [92–96]. In addition to their neuroprotective effects through PPAR-γ agonists, TZDs including troglitazone, ciglitazone, and rosiglitazone also exhibit MAO-B inhibitory activity, which is of significant interest in multitargeted drug designing [97,98]. The most potent of the group is pioglitazone, followed by rosiglitazone [98]. For MAO-B activity, crystallographic studies have shown that pioglitazone and rosiglitazone bind in the substrate cavity of MAO-B with the TZD moiety facing the flavin adenine dinucleotide, while the tail part spans into the entrance cavity, with Ile199, the "gate-keeper" residue, flipped into the out/open position. In comparison to the MAO-B binding of the glitazones, these compounds bind to a Y-shaped binding pocket in the ligand-binding domain of PPAR-γ to elicit PPAR-γ activity [99–101]. The tail of glitazones such as pioglitazone and rosiglitazone plays an important role in binding and possibly activation of PPAR-γ, where it stretches into the second arm of the Y-shaped binding pocket [99,101,102]. Taken together, this suggests that in a dual PPAR-γ/MAO-B activity paradigm the longer tail attached to the TZD moiety generally enhances the potential to uncover multiple pharmacological activities.

5.1.2 Glitazones as Anticancer Agents

As PPAR-γ receptors are expressed not only in normal but also in cancer cells [103–105], their ligands may alter tumor growth and progression. Numerous (mostly in vitro) reports showed that glitazones inhibit cancer cell growth, but some data are contradictory and indicate enhancement of carcinogenesis by PPAR-γ activation [106,107]. PPAR-γ ligands may inhibit carcinogenesis through cell cycle arrest and apoptosis induction [108–110], angiogenesis suppression [111], and antiinflammatory actions [112].

Pioglitazone inhibited glioma growth both in vitro and in vivo [113]; rosiglitazone reduced glioma growth and cell invasiveness in vitro [114]. Pioglitazone inhibited human gastric cancer cell proliferation in vitro [109], and rosiglitazone suppressed cell invasion and metastasis in human gastric cell lines [115]. Pancreatic cancer cell proliferation was inhibited by pioglitazone [116] and rosiglitazone [117] both in vitro and in vivo. Colon cancer cell growth was inhibited by pioglitazone and 15d-PGJ2 [118]; the same effect was reported for rosiglitazone [119] and troglitazone. However, their proof of clinical efficacy is not available and is under investigation. In breast cancer patients, rosiglitazone administration (stages 0–II, 2–6 week treatment) [120] and troglitazone administration (stage IV, 8-week median treatment) [23] showed no anticancer effect.

5.1.3 Glitazones as Antihypertensive Agents

PPAR-γ nuclear receptors are mainly expressed in adipose tissue. However, they are also expressed in renal glomerular tissue and in vascular walls, thus participating through various and complex mechanisms, contributing to glomerular and vascular sclerosis and to nephropathy development and progression. Studies carried out with glitazones have demonstrated their favorable effects on arterial blood pressure and on prevention and/or progression of diabetic nephropathy. The few clinical studies conducted in type 2 diabetic patients to assess these effects are also in favor of a beneficial effect of glitazones on blood pressure and nephropathy in these patients [121].

5.2 Dual Inhibitors of PI3K-α and MEK

The significance of PI3K/Akt and Raf/MEK/ERK pathways in the etiology of human cancers has led to the consideration of different members of these signaling pathways for drug discovery research. Engelman and coworkers demonstrated that the simultaneous attenuation of PI3K and MEK is highly effective, particularly in K-Ras G12D and H1047R murine lung cancers [122]. The dual activity (PI3K-α and MEK) of TZD derivatives was reported by Li and coworkers, and molecule **44** was reported as the most potent one [36]. Thereafter, Liu and coworkers also reported the PI3K-α and MEK inhibitory potential of TZD derivatives [36]. The modifications of molecule **44**, including replacement of amine moiety of ethylamine with carboxyl (**45**) and tertiary amino (**46**) groups, diminished biological activity suggesting the importance of primary amine; however, the presence of hydroxyl group (**47**) at the same position resulted in equipotent anticancer activity to **44** in U937 cancer cell line. These results suggested the presence of hydrogen bonding interactions with active site amino acid residues instead of ionic interactions. Overall, molecule **48** exhibited the highest efficacy in U937 cancer cell lines and in in vitro assay of PI3K-α and MEK1 enzyme inhibition.

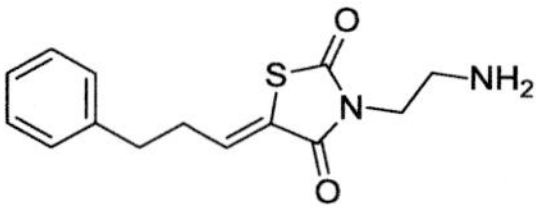

44 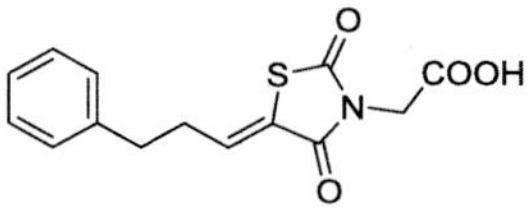   **45** 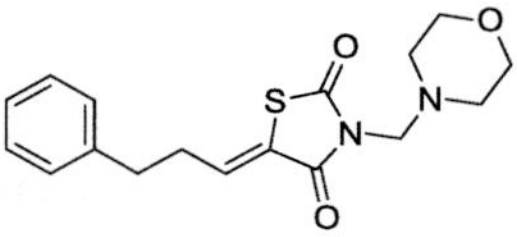   **46**

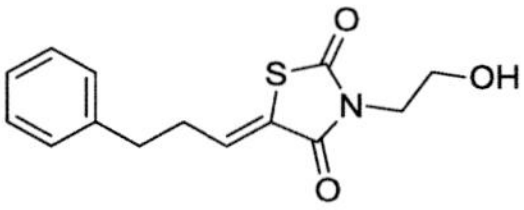

47 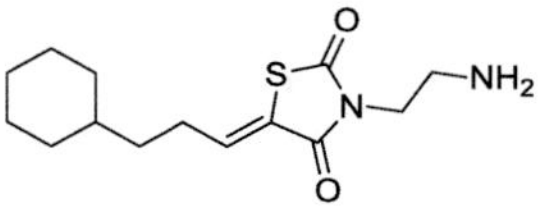   **48**

5.3 Dual Inhibitors of Bacterial Mur Ligases

Mur ligases catalyze the formation of an amide or peptide bond between the UDP substrate and the condensing amino acid. Initially, the terminal carboxyl group of the UDP substrate is activated by ATP phosphorylation, resulting in the formation of an acyl phosphate intermediate that is subsequently attacked by the amino group of the incoming amino acid or dipeptide. The tetrahedral high-energy intermediate formed collapses with elimination of inorganic phosphate and concomitant formation of the amide or peptide bond [123–125]. Moreover, based on biochemical studies of MurC and MurF, Mur ligases exhibit an ordered kinetic mechanism in which ATP binds first to the free enzyme, followed by the corresponding UDP substrate, and finally the condensing amino acid or dipeptide [126]. All the Mur ligases share the same three-domain topology, with the N-terminal and central domains binding UDP precursor and ATP, respectively, while the C-terminal domain binds the condensing amino acid or dipeptide residue [127,128].

Glutamic-acid-based selective MurD inhibitors containing a rhodanine moiety that act as MurD product mimics have been reported [52]. Since compounds bearing a rhodanine ring often exhibit antibacterial activity due to inhibition of bacterial enzymes [129], rhodanine-based inhibitors of Mur ligases were explored.

The most potent compound **49** inhibited MurD and MurF with IC_{50} values of 2 μm, and MurE with an IC_{50} value of 6 μm. However, the TZD derivative **50** was found to be a potent inhibitor of MurE and MurF with IC_{50} values of 3 μM against both enzymes. The observed inhibition of different Mur enzymes involved in the biosynthesis of bacterial peptidoglycan makes these compounds interesting leads in the search for multitarget antibacterial agents [130].

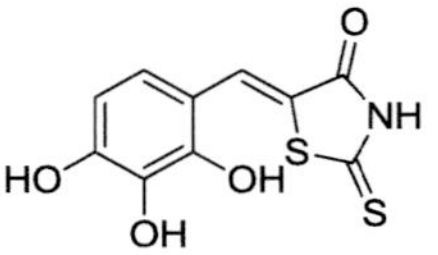

49 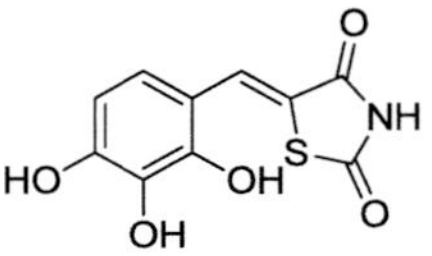   **50**

5.4 Multitargeted Inhibitors of COX and 5-LOX

Atherosclerosis is a complex and chronic inflammatory disease involving elastic and muscular arteries [131]. COX and 5-lipoxygenase (5-LOX), which play pivotal roles in atherogenesis, have been reported to be involved in plaque stability. One of these dual COX and 5-LOX inhibitors, BHB-TZD (5-(3,5-di-tert-butyl-4-hydroxybenzylidene)thiazolidin-2,4-dione), has been found to reduce the number of lesions in adjuvant induced polyarthritis, a chronic model of inflammation, with least gastric ulceration [132,133]. It has been shown that BHB-TZD reduced the formation of atherosclerotic lesions. Moreover, within these lesions, BHB-TZD increased the number of smooth muscle cells and the amount of collagen, while decreasing macrophage recruitment. These findings indicate that this dual COX and 5-LOX inhibitor not only decreased the size of atherosclerotic lesions, but also increased plaque stability, suggesting that dual inhibitors may be beneficial in the treatment of atherosclerosis [134].

51, BHB-TZD

5.5 PPAR-γ and GLUT-4 Modulator

Multitarget therapies, which at the same time control hyperglycemia and inhibit progression of cardiovascular complications, may be attractive options for the therapeutic treatment of diabetes. The fibrate drugs (e.g., clofibrate and fenofibrate), which are hypolipidemic agents, exert their effect by agonist action on PPAR-α. On the other hand, thiazolidine-2,4-diones (rosiglitazone and pioglitazone) function as insulin-sensitizing drugs, through the activation of PPAR-γ [135,136]. PPAR-α, -γ dual agonists are new class of drugs that have been developed to target both PPARs in order to produce antidiabetic and hypolipidemic effects [137]. Navarrete-Vázquez et al. reported preparation of {4-[({4-[(Z)-(2,4-dioxo-1,3-thiazolidin-5-ylidene)methyl]phenoxyacetyl) amino]phenoxy}acetic acid (**52**) and its potential ethyl ester prodrug (**53**), as well as the in vitro relative expression of PPAR-α, PPAR-γ, and GLUT-4. The designing of compound **1** was based on pharmacophoric pattern for PPAR-α, -γ dual agonists with the presence of two acid heads, two aromatic/hydrophobic rings, and a flexible linker. Compound **1** was demonstrated to be a specific PPAR-γ ligand as it exhibited a marked induction of the target gene GLUT-4 [138].

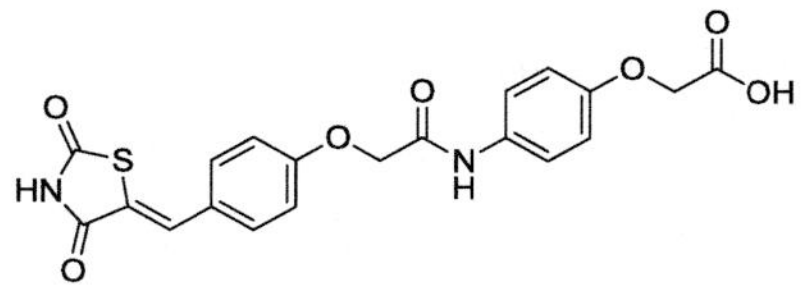

52

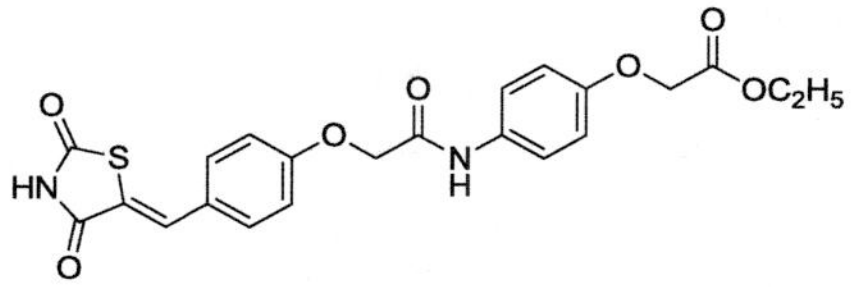

53

5.6 ALR2 and PARP-1 Inhibitors

Diabetic retinopathy is a common reason of vision loss characterized by retinal capillary cell loss, capillary basement membrane thickening, increased vascular permeability, and increased leukocyte adhesion to endothelial cells. Literature reports highlight that increased aldose reductase activity is responsible for enhancing oxidative stress, and up-regulates retinal vascular endothelial growth factor and activation of PARP in diabetic retinal cells, which may lead to cataract formation and diabetic retinopathy [139–142]. Based on these reports, Chadha et al. designed and evaluated novel dual inhibitors (**54**) for ALR2 and PARP-1 using a structure-based design approach [143]. To evaluate the inhibitory activities of the synthesized compounds, IC_{50} values against ALR2 and PARP-1 were calculated using colorimetric and ELISA-based enzymatic assay. With 4-fluoro and 2-chloro substitutions at R position, the corresponding compounds exhibited better ALR2 and PARP-1 inhibitory activities, with IC_{50} values of 1.34–5.03 μM. The substitution of bigger groups (4-ethyl, 3-trifluromethyl, 4-isopropyl, 2,4-dichloro and 4-tert-butyl) at R position of benzyl ring significantly lowered ALR2 inhibitory activities. On the other hand, PARP-1 inhibitory activities were significantly improved with these substitutions.

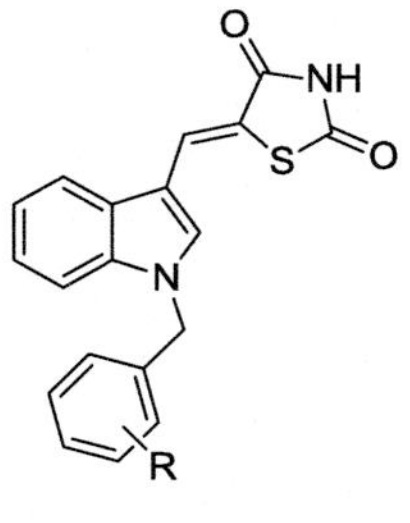

54

6. CONCLUSION

The chemical architecture of the TZD nucleus makes it a valuable scaffold for the targeting of various proteins (receptors and enzymes) involved in complex and multifactorial pathological conditions. Apart from the pharmacophoric features of TZD core, introduction of required structural fragments at free –NH and –CH$_2$

groups of the TZD core facilitates the designing of selective and potent ligands for different targets. The improved efficacy of multitargeted drugs produced by modulation of network/pathways of targets has led to increased popularity over single-targeted drugs. Moreover, there are efforts for drug repurposing (e.g., pioglitazone has exhibited neuroprotective, antihypertensive, anticancer properties) and TZD is under clinical evaluation for the same. There are other studies that define the role of TZD in multitargeting in multifarious diseases such as diabetic complications, cancer, chronic inflammatory diseases, arthritis, and microbial infection. Thus, successful multitargeting of various proteins with TZD derivatives confer their beneficial therapeutic potential in complex pathological conditions associated with these targets.

REFERENCES

[1] T. Fujita, Y. Sugiyama, S. Taketomi, T. Sohda, Y. Kawamatsu, H. Iwatsuka, Z. Suzuoki, Reduction of insulin resistance in obese and/or diabetic animals by 5-[4-(1-methylcyclohexylmethoxy)benzyl]-thiazolidine-2,4-dione (ADD-3878, U-63,287, ciglitazone), a new antidiabetic agent, Diabetes 32 (9) (1983) 804–810.

[2] M.M. Ip, P.W. Sylvester, L. Schenkel, Antitumor efficacy in rats of CGP 19984, a thiazolidinedione derivative that inhibits luteinizing hormone secretion, Cancer Research 46 (4 Pt 1) (1986) 1735–1740.

[3] L.F. da Rocha Junior, M.J. Rego, M.B. Cavalcanti, M.C. Pereira, M.G. Pitta, P.S. de Oliveira, S.M. Goncalves, A.L. Duarte, C. de Lima Mdo, R. Pitta Ida, Synthesis of a novel thiazolidinedione and evaluation of its modulatory effect on IFN- gamma, IL-6, IL-17A, and IL-22 production in PBMCs from rheumatoid arthritis patients, BioMed Research International 2013 (2013) 926060.

[4] A.M. Youssef, M.S. White, E.B. Villanueva, I.M. El-Ashmawy, A. Klegeris, Synthesis and biological evaluation of novel pyrazolyl-2,4-thiazolidinediones as anti-inflammatory and neuroprotective agents, Bioorganic and Medicinal Chemistry 18 (5) (2010) 2019–2028.

[5] U. Albrecht, D. Gordes, E. Schmidt, K. Thurow, M. Lalk, P. Langer, Synthesis and structure-activity relationships of 2-alkylidenethiazolidine-4,5-diones as antibiotic agents, Bioorganic and Medicinal Chemistry 13 (14) (2005) 4402–4407.

[6] N.D. Oakes, C.J. Kennedy, A.B. Jenkins, D.R. Laybutt, D.J. Chisholm, E.W. Kraegen, A new antidiabetic agent, BRL 49653, reduces lipid availability and improves insulin action and glucoregulation in the rat, Diabetes 43 (10) (1994) 1203–1210.

[7] T. Sohda, Y. Momose, K. Meguro, Y. Kawamatsu, Y. Sugiyama, H. Ikeda, Studies on antidiabetic agents. Synthesis and hypoglycemic activity of 5-[4-(pyridylalkoxy)benzyl]-2,4-, Arzneimittelforschung 40 (1) (1990) 37–42.

[8] J.W. Kim, J.R. Kim, S. Yi, K.H. Shin, H.S. Shin, S.H. Yoon, J.Y. Cho, D.H. Kim, S.G. Shin, I.J. Jang, K.S. Yu, Tolerability and pharmacokinetics of lobeglitazone (CKD-501), a peroxisome proliferator-activated receptor-gamma agonist: a single- and multiple-dose, double-blind, randomized control study in healthy male Korean subjects, Clinical Therapeutics 33 (11) (2011) 1819–1830.

[9] T.P. Ciaraldi, A. Gilmore, J.M. Olefsky, M. Goldberg, K.A. Heidenreich, In vitro studies on the action of CS-045, a new antidiabetic agent, Metabolism 39 (10) (1990) 1056–1062.

[10] N. Chadha, M.S. Bahia, M. Kaur, O. Silakari, Thiazolidine-2,4-dione derivatives: programmed chemical weapons for key protein targets of various pathological conditions, Bioorganic and Medicinal Chemistry 23 (13) (2015) 2953–2974.

[11] P. Visentini, Antituberculosis chemotherapeutic action of derivatives of 2-4-thiazolidinedione. I. Action in vitro of the 2-phenylhydrazone of 2,4-thiazolidinedione, Farmaco Scientifica 9 (5) (1954) 274–277.

[12] P.G. Marshall, D.K. Vallance, Anticonvulsant activity; derivatives of succinimide, glutarimide, thiazolidinedione and methanol, and some miscellaneous compounds, Journal of Pharmaceutics and Pharmacology 6 (10) (1954) 740–746.

[13] P.K. Das, G.B. Singh, P.K. Debnath, S.B. Acharya, S.N. Dube, A study of anticonvulsant activity of N-substituted derivatives of succinimides, thiazolidinediones and their structural congeners, Indian Journal of Medical Research 63 (2) (1975) 286–301.

[14] R.S. Dewey, E.F. Schoenewaldt, H. Joshua, W.J. Paleveda Jr., H. Schwam, H. Barkemeyer, B.H. Arison, D.F. Veber, R.G. Strachan, J. Milkowski, R.G. Denkewalter, R. Hirschmann, The synthesis of peptides in aqueous medium. VII. The preparation and use of 2,5-thiazolidinediones in peptide synthesis, Journal of Organic Chemistry 36 (1) (1971) 49–59.

[15] M. Salama Kh, E.V. Vladzimirskaia, Derivatives of 1,4-di-(thiazolidinedione-2′,4′-yl-3′)-butane, Farmatasavtychnyl Zhurmal (4) (1978) 26–29.

[16] A.A. Sirotenko, P.A. Ciavarri Jr., Some hydrazones of 5-phenyl-2,4-thiazolidinedione, Journal of Medicinal Chemistry 9 (4) (1966) 642–643.

[17] S.A. Tawab, A. Mustafa, M.M. Sallam, A contribution to the pharmacology of 2-mercapto-5-benzylidenethiazol-4-one, Archives Internationales de Pharmacodynamie et de Therapie 128 (1960) 11–13.

[18] E.V. Vladzimirskaia, V.A. Zdorenko, Synthesis of thiazolidinedione-2,4 derivatives with 2 heterorings with a bridge bond to a benzene ring, Farmatasavtychnyl Zhurmal (3) (1977) 37–40.

[19] V.I. Zapadniuk, Pharmacological studies on 3(-beta-aminoethyl)-2,4-thiazolidinedione hydrochloride, Farmakologiya i Toksikologiya 24 (1961) 665–670.

[20] T. Sohda, K. Mizuno, E. Imamiya, Y. Sugiyama, T. Fujita, Y. Kawamatsu, Studies on antidiabetic agents. II. Synthesis of 5-[4-(1-methylcyclohexylmethoxy)-benzyl]thiazolidine-2,4-dione (ADD-3878) and its derivatives, Chemical and Pharmaceutical Bulletin 30 (10) (1982) 3580–3600.

[21] R.R. Henry, Effects of troglitazone on insulin sensitivity, Diabetic Medicine 13 (9 Suppl. 6) (1996) S148–S150.

[22] H. Richard, US5002953 (A) - Novel Compounds. 1991, Hindley Richard M.

[23] H.J. Burstein, G.D. Demetri, E. Mueller, P. Sarraf, B.M. Spiegelman, E.P. Winer, Use of the peroxisome proliferator-activated receptor (PPAR) γ ligand troglitazone as treatment for refractory breast cancer: a phase II study, Breast Cancer Research and Treatment 79 (3) (2003) 391–397.

[24] S. Chorbadjiev, V. Enchev, B. Jordanov, Comparative study of the structure of rhodanine, isorhodanine, thiazolidine-2,4-dione, and thiorhodanine, Chemistry of Heterocyclic Compounds 38 (9) (2002) 1110–1120.

[25] S. Santhosh Kumar, B.R. Prashantha Kumar, Ashish Wadhwani, Patel Viral, R. Vadivelan, M.N. Satish Kumar, M.J. Nanjan, Novel glitazones: glucose uptake and cytotoxic activities, and structure–activity relationships, Medicinal Chemistry Research 21 (2012) 2689–2701.

[26] S. Kallenbeg, Stereochemische Untersuchungen der Diketo-thiazolidine, Berichte der Deutschen Chemischen Gesellschaft 56 (1) (1923) 316–331.

[27] O. Bozdag, G. Ayhan-Kilcigil, M. Tuncbilek, R. Ertan, Studies on the synthesis of some substituted flavonyl thiazolidinedione derivatives-I, Turkish Journal of Chemistry 23 (1999) 163–170.

[28] J. Taylor, The chloroacetates of S-alkylthiocarbamides, Journal of the Chemical Society Transactions 117 (1920) 4–11.

[29] W. Heintz, Beiträge zur Kenntniss der Glycolamidsäuren, European Journal of Organic Chemistry 136 (2) (1865) 213–223.

[30] H. Shinkai, S. Onogi, M. Tanaka, T. Shibata, M. Iwao, K. Wakitani, I. Uchida, Isoxazolidine-3,5-dione and noncyclic 1,3-dicarbonyl compounds as hypoglycemic agents, Journal of Medicinal Chemistry 41 (11) (1998) 1927–1933.

[31] Q. Li, A. Al-Ayoubi, T. Guo, H. Zheng, A. Sarkar, T. Nguyen, S.T. Eblen, S. Grant, G.E. Kellogg, S. Zhang, Structure–activity relationship (SAR) studies of 3-(2-amino-ethyl)-5-(4-ethoxy-benzylidene)-thiazolidine-2, 4-dione: development of potential substrate-specific ERK1/2 inhibitors, Bioorganic and Medicinal Chemistry Letters 19 (21) (2009) 6042–6046.

[32] R. Romagnoli, P.G. Baraldi, M.K. Salvador, M.E. Camacho, J. Balzarini, J. Bermejo, F. Estevez, Anticancer activity of novel hybrid molecules containing 5-benzylidene thiazolidine-2,4-dione, European Journal of Medicinal Chemistry 63 (2013) 544–557.

[33] B. Božić, J. Rogan, D. Poleti, M. Rančić, N. Trišović, B. Božić, G. Ušćumlić, Synthesis, characterization and biological activity of 2-(5-arylidene-2, 4-dioxotetrahydrothiazole-3-yl) propanoic acid derivatives, Arabian Journal of Chemistry 10 (2013) S2637–S2643.

[34] R. Mohan, A.K. Sharma, S. Gupta, C. Ramaa, Design, synthesis, and biological evaluation of novel 2, 4-thiazolidinedione derivatives as histone deacetylase inhibitors targeting liver cancer cell line, Medicinal Chemistry Research 21 (7) (2012) 1156–1165.

[35] R. Maccari, R. Ottanà, C. Curinga, M.G. Vigorita, D. Rakowitz, T. Steindl, T. Langer, Structure–activity relationships and molecular modelling of 5-arylidene-2, 4-thiazolidinediones active as aldose reductase inhibitors, Bioorganic and Medicinal Chemistry 13 (8) (2005) 2809–2823.

[36] Q. Li, J. Wu, H. Zheng, K. Liu, T.L. Guo, Y. Liu, S.T. Eblen, S. Grant, S. Zhang, Discovery of 3-(2-aminoethyl)-5-(3-phenyl-propylidene)-thiazolidine-2, 4-dione as a dual inhibitor of the Raf/MEK/ERK and the PI3K/Akt signaling pathways, Bioorganic and Medicinal Chemistry Letters 20 (15) (2010) 4526–4530.

[37] K. Liu, W. Rao, H. Parikh, Q. Li, T.L. Guo, S. Grant, G.E. Kellogg, S. Zhang, 3, 5-Disubstituted-thiazolidine-2, 4-dione analogs as anticancer agents: design, synthesis and biological characterization, European Journal of Medicinal Chemistry 47 (2012) 125–137.

[38] D. Havrylyuk, B. Zimenkovsky, O. Vasylenko, C.W. Day, D.F. Smee, P. Grellier, R. Lesyk, Synthesis and biological activity evaluation of 5-pyrazoline substituted 4-thiazolidinones, European Journal of Medicinal Chemistry 66 (2013) 228–237.

[39] A.M. Ali, G.E. Saber, N.M. Mahfouz, M.A. El-Gendy, A.A. Radwan, M.A. Hamid, Synthesis and three-dimensional qualitative structure selectivity relationship of 3,5-disubstituted-2,4-thiazolidinedione derivatives as COX2 inhibitors, Archives of Pharmacal Research 30 (10) (2007) 1186–1204.

[40] L. Nagarapu, B. Yadagiri, R. Bantu, C.G. Kumar, S. Pombala, J. Nanubolu, Studies on the synthetic and structural aspects of benzosuberones bearing 2, 4-thiazolidenone moiety as potential anti-cancer agents, European Journal of Medicinal Chemistry 71 (2014) 91–97.

[41] Z. Wang, Z. Liu, W. Lee, S.-N. Kim, G. Yoon, S.H. Cheon, Design, synthesis and docking study of 5-(substituted benzylidene) thiazolidine-2, 4-dione derivatives as inhibitors of protein tyrosine phosphatase 1B, Bioorganic and Medicinal Letters 24 (15) (2014) 3337–3340.

[42] F.W. Barros, T.G. Silva, M.G. da Rocha Pitta, D.P. Bezerra, L.V. Costa-Lotufo, M.O. de Moraes, C. Pessoa, M.A. de Moura, F.C. de Abreu, C. de Lima Mdo, S.L. Galdino, R. Pitta Ida, M.O. Goulart, Synthesis and cytotoxic activity of new acridine-thiazolidine derivatives, Bioorganic and Medicinal Chemistry 20 (11) (2012) 3533–3539.

[43] Y.M. Ha, Y.J. Park, J.-A. Kim, D. Park, J.Y. Park, H.J. Lee, J.Y. Lee, H.R. Moon, H.Y. Chung, Design and synthesis of 5-(substituted benzylidene) thiazolidine-2, 4-dione derivatives as novel tyrosinase inhibitors, European Journal of Medicinal Chemistry 49 (2012) 245–252.

[44] A.J. Russell, I.M. Westwood, M.H. Crawford, J. Robinson, A. Kawamura, C. Redfield, N. Laurieri, E.D. Lowe, S.G. Davies, E. Sim, Selective small molecule inhibitors of the potential breast cancer marker, human arylamine *N*-acetyltransferase 1, and its murine homologue, mouse arylamine *N*-acetyltransferase 2, Bioorganic and Medicinal Chemistry 17 (2) (2009) 905–918.

[45] G. Thirupathi, M. Venkatanarayana, P. Dubey, Y. Bharathi Kumari, Facile and green syntheses of substituted-5-arylidene-2, 4-thiazolidine diones using l-tyrosine as an eco-friendly catalyst in aqueous medium, Der Pharma Chemica 4 (5) (2012).

[46] L. Ma, C. Xie, Y. Ma, J. Liu, M. Xiang, X. Ye, H. Zheng, Z. Chen, Q. Xu, T. Chen, Synthesis and biological evaluation of novel 5-benzylidenethiazolidine-2, 4-dione derivatives for the treatment of inflammatory diseases, Journal of Medicinal Chemistry 54 (7) (2011) 2060–2068.

[47] U.R. Pratap, D.V. Jawale, R.A. Waghmare, D.L. Lingampalle, R.A. Mane, Synthesis of 5-arylidene-2, 4-thiazolidinediones by Knoevenagel condensation catalyzed by baker's yeast, Nouveau Journal de Chimie 35 (1) (2011) 49–51.

[48] Z. Xia, C. Knaak, J. Ma, Z.M. Beharry, C. McInnes, W. Wang, A.S. Kraft, C.D. Smith, Synthesis and evaluation of novel inhibitors of Pim-1 and Pim-2 protein kinases, Journal of Medicinal Chemistry 52 (1) (2009) 74–86.

[49] S.H. Kim, Y.M. Ha, K.M. Moon, Y.J. Choi, Y.J. Park, H.O. Jeong, K.W. Chung, H.J. Lee, P. Chun, H.R. Moon, Anti-melanogenic effect of (Z)-5-(2, 4-dihydroxybenzylidene) thiazolidine-2, 4-dione, a novel tyrosinase inhibitor, Archives of Pharmacal Research 36 (10) (2013) 1189–1197.

[50] M. Azizmohammadi, M. Khoobi, A. Ramazani, S. Emami, A. Zarrin, O. Firuzi, R. Miri, A. Shafiee, 2*H*-chromene derivatives bearing thiazolidine-2, 4-dione, rhodanine or hydantoin moieties as potential anticancer agents, European Journal of Medicinal Chemistry 59 (2013) 15–22.

[51] H. Chen, Y.-H. Fan, A. Natarajan, Y. Guo, J. Iyasere, F. Harbinski, L. Luus, W. Christ, H. Aktas, J.A. Halperin, Synthesis and biological evaluation of thiazolidine-2, 4-dione and 2, 4-thione derivatives as inhibitors of translation initiation, Bioorganic and Medicinal Chemistry Letters 14 (21) (2004) 5401–5405.

[52] T. Tomašić, N. Zidar, V. Rupnik, A. Kovač, D. Blanot, S. Gobec, D. Kikelj, L.P. Mašič, Synthesis and biological evaluation of new glutamic acid-based inhibitors of MurD ligase, Bioorganic and Medicinal Chemistry Letters 19 (1) (2009) 153–157.

[53] N. Zidar, T. Tomasic, R. Sink, V. Rupnik, A. Kovac, S. Turk, D. Patin, D. Blanot, C. Contreras Martel, A. Dessen, M. Muller Premru, A. Zega, S. Gobec, L. Peterlin Masic, D. Kikelj, Discovery of novel 5-benzylidenerhodanine and 5-benzylidenethiazolidine-2,4-dione inhibitors of MurD ligase, Journal of Medicinal Chemistry 53 (18) (2010) 6584–6594.

[54] N. Zidar, T. Tomasic, R. Sink, A. Kovac, D. Patin, D. Blanot, C. Contreras-Martel, A. Dessen, M.M. Premru, A. Zega, S. Gobec, L.P. Masic, D. Kikelj, New 5-benzylidenethiazolidin-4-one inhibitors of bacterial MurD ligase: design, synthesis, crystal structures, and biological evaluation, European Journal of Medicinal Chemistry 46 (11) (2011) 5512–5523.

[55] V. Pomel, J. Klicic, D. Covini, D.D. Church, J.P. Shaw, K. Roulin, F. Burgat-Charvillon, D. Valognes, M. Camps, C. Chabert, Furan-2-ylmethylene thiazolidinediones as novel, potent, and selective inhibitors of phosphoinositide 3-kinase γ, Journal of Medicinal Chemistry 49 (13) (2006) 3857–3871.

[56] L.A. Dakin, M.H. Block, H. Chen, E. Code, J.E. Dowling, X. Feng, A.D. Ferguson, I. Green, A.W. Hird, T. Howard, E.K. Keeton, M.L. Lamb, P.D. Lyne, H. Pollard, J. Read, A.J. Wu, T. Zhang, X. Zheng, Discovery of novel benzylidene-1,3-thiazolidine-2,4-diones as potent and selective inhibitors of the PIM-1, PIM-2, and PIM-3 protein kinases, Bioorganic and Medicinal Chemistry Letters 22 (14) (2012) 4599–4604.

[57] H. Yanagisawa, M. Takamura, E. Yamada, S. Fujita, T. Fujiwara, M. Yachi, A. Isobe, Y. Hagisawa, Novel oximes having 5-benzyl-2,4-thiazolidinedione as antihyperglycemic agents: synthesis and structure-activity relationship, Bioorganic and Medicinal Chemistry Letters 10 (4) (2000) 373–375.

[58] S. Nazreen, M.S. Alam, H. Hamid, M.S. Yar, A. Dhulap, P. Alam, M.A. Pasha, S. Bano, M.M. Alam, S. Haider, C. Kharbanda, Y. Ali, K.K. Pillai, Thiazolidine-2,4-diones derivatives as PPAR-gamma agonists: synthesis, molecular docking, in vitro and in vivo antidiabetic activity with hepatotoxicity risk evaluation and effect on PPAR-gamma gene expression, Bioorganic and Medicinal Chemistry Letters 24 (14) (2014) 3034–3042.

[59] S. Nazreen, M.S. Alam, H. Hamid, M.S. Yar, S. Shafi, A. Dhulap, P. Alam, M.A. Pasha, S. Bano, M.M. Alam, S. Haider, Y. Ali, C. Kharbanda, K.K. Pillai, Design, synthesis, in silico molecular docking and biological evaluation of novel oxadiazole based thiazolidine-2,4-diones bis-heterocycles as PPAR-gamma agonists, European Journal of Medicinal Chemistry 87 (2014) 175–185.

[60] J.H. Lee, C.K. Noh, C.S. Yim, Y.S. Jeong, S.H. Ahn, W. Lee, D.D. Kim, S.J. Chung, Kinetics of the absorption, distribution, metabolism, and excretion of lobeglitazone, a novel activator of peroxisome proliferator-activated receptor gamma in rats, Journal of Pharmaceutical Sciences 104 (9) (2015) 3049–3059.

[61] G.R. Madhavan, R. Chakrabarti, K. Anantha Reddy, B. Rajesh, V. Balraju, P. Bheema Rao, R. Rajagopalan, J. Iqbal, Dual PPAR-α and-γ activators derived from novel benzoxazinone containing thiazolidinediones having antidiabetic and hypolipidemic potential, Bioorganic and Medicinal Chemistry 14 (2) (2006) 584–591.

[62] A. Hashimoto, Y. Shi, K. Drake, J.T. Koh, Design and synthesis of complementing ligands for mutant thyroid hormone receptor TRβ (R320H): a tailor-made approach toward the treatment of resistance to thyroid hormone, Bioorganic and Medicinal Chemistry 13 (11) (2005) 3627–3639.

[63] C. Prabhakar, G. Madhusudhan, K. Sahadev, C.M. Reddy, M.R. Sarma, G.O. Reddy, R. Chakrabarti, C.S. Rao, T.D. Kumar, R. Rajagopalan, Synthesis and biological activity of novel thiazolidinediones, Bioorganic and Medicinal Chemistry Letters 8 (19) (1998) 2725–2730.

[64] T. Sohda, K. Mizuno, H. Tawada, Y. Sugiyama, T. Fujita, Y. Kawamatsu, Studies on antidiabetic agents. I. Synthesis of 5-[4-(2-methyl-2-phenylpropoxy)-benzyl]thiazolidine-2,4-dione (AL-321) and related compounds, Chemical and Pharmaceutical Bulletin 30 (10) (1982) 3563–3573.

[65] J.M. Lehmann, L.B. Moore, T.A. Smith-Oliver, W.O. Wilkison, T.M. Willson, S.A. Kliewer, An antidiabetic thiazolidinedione is a high affinity ligand for peroxisome proliferator-activated receptor gamma (PPAR gamma), Journal of Biological Chemistry 270 (22) (1995) 12953–12956.

[66] R. Maccari, P. Paoli, R. Ottanà, M. Jacomelli, R. Ciurleo, G. Manao, T. Steindl, T. Langer, M.G. Vigorita, G. Camici, 5-Arylidene-2, 4-thiazolidinediones as inhibitors of protein tyrosine phosphatases, Bioorganic and Medicinal Chemistry 15 (15) (2007) 5137–5149.

[67] J.-H. Lee, Y.-A. Woo, I.-C. Hwang, C.-Y. Kim, D.-D. Kim, C.-K. Shim, S.-J. Chung, Quantification of CKD-501, lobeglitazone, in rat plasma using a liquid-chromatography/tandem mass spectrometry method and its applications to pharmacokinetic studies, Journal of Pharmaceutical and Biomedical Analysis 50 (5) (2009) 872–877.

[68] A. Kong, A. Yamasaki, R. Ozaki, H. Saito, T. Asami, S. Ohwada, G. Ko, C. Wong, G. Leung, K. Lee, A randomized-controlled trial to investigate the effects of rivoglitazone, a novel PPAR gamma agonist on glucose–lipid control in type 2 diabetes, Diabetes, Obesity and Metabolism 13 (9) (2011) 806–813.

[69] K. Schimke, T. Davis, Drug evaluation: rivoglitazone, a new oral therapy for the treatment of type 2 diabetes, Current Opinion in Investigational Drugs (London, England: 2000) 8 (4) (2007) 338–344.

[70] V.B. Lohray, B.B. Lohray, R.B. Paraselli, R.M. Gurram, R. Ramanujam, R. Chakrabarti, S.K. Pakala, Heterocyclic Compounds, Process for Their Preparation and Pharmaceutical Compositions Containing Them and Their Use in the Treatment of Diabetes and Related Diseases, Google Patents, 2001.

[71] V. Lohray, B. Lohray, Drug Discovery and Development, Volume 2: Drug Development, Wiley-Interscience, 2007.

[72] D.S. Aschenbrenner, FDA lifts restrictions on rosiglitazone, The American Journal of Nursing 114 (4) (2014) 24.

[73] S.E. Nissen, Rosiglitazone: a case of regulatory hubris, British Medical Journal 347 (2013).

[74] J.Y. Lee, W.H. Park, M.K. Cho, H.J. Yun, B.H. Chung, Y.K. Pak, H.G. Hahn, S.H. Cheon, Design and synthesis of novel antidiabetic agents, Archives of Pharmacal Research 28 (2) (2005) 142–150.

[75] R. Jeon, Y.J. Kim, Y. Cheon, J.H. Ryu, Synthesis and biological activity of [[(heterocycloamino)alkoxy] benzyl]-2,4-thiazolidinediones as PPARgamma agonists, Archives of Pharmacal Research 29 (5) (2006) 394–399.

[76] X.F. Liu, C.J. Zheng, L.P. Sun, X.K. Liu, H.R. Piao, Synthesis of new chalcone derivatives bearing 2,4-thiazolidinedione and benzoic acid moieties as potential anti-bacterial agents, European Journal of Medicinal Chemistry 46 (8) (2011) 3469–3473.

[77] B.R. Bhattarai, B. Kafle, J.-S. Hwang, D. Khadka, S.-M. Lee, J.-S. Kang, S.W. Ham, I.-O. Han, H. Park, H. Cho, Thiazolidinedione derivatives as PTP1B inhibitors with antihyperglycemic and antiobesity effects, Bioorganic and Medicinal Chemistry Letters 19 (21) (2009) 6161–6165.

[78] B.R. Bhattarai, B. Kafle, J.-S. Hwang, S.W. Ham, K.-H. Lee, H. Park, I.-O. Han, H. Cho, Novel thiazolidinedione derivatives with anti-obesity effects: dual action as PTP1B inhibitors and PPAR-γ activators, Bioorganic and Medicinal Chemistry Letters 20 (22) (2010) 6758–6763.

[79] Z. Liu, Q. Chai, Y-y. Li, Q. Shen, L-p. Ma, L-n. Zhang, X. Wang, L. Sheng, J-y. Li, J. Li, Discovery of novel PTP1B inhibitors with antihyperglycemic activity, Acta Pharmacologica Sinica 31 (8) (2010) 1005–1012.

[80] G. Bruno, L. Costantino, C. Curinga, R. Maccari, F. Monforte, F. Nicolo, R. Ottana, M. Vigorita, Synthesis and aldose reductase inhibitory activity of 5-arylidene-2, 4-thiazolidinediones, Bioorganic and Medicinal Chemistry 10 (4) (2002) 1077–1084.

[81] O. Bozdağ-Dündar, E.J. Verspohl, N. Daş-Evcimen, R.M. Kaup, K. Bauer, M. Sarıkaya, B. Evranos, R. Ertan, Synthesis and biological activity of some new flavonyl-2, 4-thiazolidinediones, Bioorganic and Medicinal Chemistry 16 (14) (2008) 6747–6751.

[82] Z. Xia, C. Knaak, J. Ma, Z.M. Beharry, C. McInnes, W. Wang, A.S. Kraft, C.D. Smith, Synthesis and evaluation of novel inhibitors of Pim-1 and Pim-2 protein kinases, Journal of Medicinal Chemistry 52 (1) (2008) 74–86.

[83] Y.W. Lin, Z.M. Beharry, E.G. Hill, J.H. Song, W. Wang, Z. Xia, Z. Zhang, P.D. Aplan, J.C. Aster, C.D. Smith, A.S. Kraft, A small molecule inhibitor of Pim protein kinases blocks the growth of precursor T-cell lymphoblastic leukemia/lymphoma, Blood 115 (4) (2010) 824–833.

[84] S. Tyagi, S. Sharma, R.D. Budhiraja, Effect of phosphatidylinositol 3-kinase-gamma inhibitor CAY10505 in hypertension, and its associated vascular endothelium dysfunction in rats, Canadian Journal of Physiology and Pharmacology 90 (7) (2012) 881–885.

[85] T. Rückle, J. Shaw, D. Church, D. Covini, Preparation of 2-imino-4-(thio) oxo-5-Polycyclovinylazolines as PI3 Kinase Inhibitors, PCT Int. Appl. WO2005011686, 2005.

[86] T. Rueckle, A. Quattropani, V. Pomel, J. Dorbais, D. Covini, A. Bischoff, Pyridine Methylene Azolidinones and Use Thereof Phosphoinositide Inhibitors, Google Patents, 2010.

[87] P.S. Gillies, C.J. Dunn, Pioglitazone, Drugs 60 (2) (2000) 333–343 discussion 344–345.

[88] A.T. White, A.N. Murphy, Administration of thiazolidinediones for neuroprotection in ischemic stroke: a pre-clinical systematic review, Journal of Neurochemistry 115 (4) (2010) 845–853.

[89] J. Culman, Y. Zhao, P. Gohlke, T. Herdegen, PPAR-γ: therapeutic target for ischemic stroke, Trends in Pharmacological Sciences 28 (5) (2007) 244–249.

[90] Y. Luo, W. Yin, A.P. Signore, F. Zhang, Z. Hong, S. Wang, S.H. Graham, J. Chen, Neuroprotection against focal ischemic brain injury by the peroxisome proliferator-activated receptor-γ agonist rosiglitazone, Journal of Neurochemistry 97 (2) (2006) 435–448.

[91] K. Tureyen, R. Kapadia, K.K. Bowen, I. Satriotomo, J. Liang, D.L. Feinstein, R. Vemuganti, Peroxisome proliferator-activated receptor-γ agonists induce neuroprotection following transient focal ischemia in normotensive, normoglycemic as well as hypertensive and type-2 diabetic rodents, Journal of Neurochemistry 101 (1) (2007) 41–56.

[92] C. Laloux, M. Petrault, C. Lecointe, D. Devos, R. Bordet, Differential susceptibility to the PPAR-γ agonist pioglitazone in 1-methyl-4-phenyl-1, 2, 3, 6-tetrahydropyridine and 6-hydroxydopamine rodent models of Parkinson's disease, Pharmacological Research 65 (5) (2012) 514–522.

[93] T. Breidert, J. Callebert, M. Heneka, G. Landreth, J. Launay, E. Hirsch, Protective action of the peroxisome proliferator-activated receptor-γ agonist pioglitazone in a mouse model of Parkinson's disease, Journal of Neurochemistry 82 (3) (2002) 615–624.

[94] T. Dehmer, M.T. Heneka, M. Sastre, J. Dichgans, J.B. Schulz, Protection by pioglitazone in the MPTP model of Parkinson's disease correlates with IκBα induction and block of NFκB and iNOS activation, Journal of Neurochemistry 88 (2) (2004) 494–501.

[95] R.L. Hunter, N. Dragicevic, K. Seifert, D.Y. Choi, M. Liu, H.C. Kim, W.A. Cass, P.G. Sullivan, G. Bing, Inflammation induces mitochondrial dysfunction and dopaminergic neurodegeneration in the nigrostriatal system, Journal of Neurochemistry 100 (5) (2007) 1375–1386.

[96] P. Kumar, R.K. Kaundal, S. More, S.S. Sharma, Beneficial effects of pioglitazone on cognitive impairment in MPTP model of Parkinson's disease, Behavioural Brain Research 197 (2) (2009) 398–403.

[97] L. Quinn, B. Crook, M. Hows, M. Vidgeon-Hart, H. Chapman, N. Upton, A. Medhurst, D. Virley, The PPARγ agonist pioglitazone is effective in the MPTP mouse model of Parkinson's disease through inhibition of monoamine oxidase B, British Journal of Pharmacology 154 (1) (2008) 226–233.

[98] W.J. Geldenhuys, A.S. Darvesh, M.O. Funk, C.J. Van der Schyf, R.T. Carroll, Identification of novel monoamine oxidase B inhibitors by structure-based virtual screening, Bioorganic and Medicinal Chemistry Letters 20 (17) (2010) 5295–5298.

[99] R. Montanari, F. Saccoccia, E. Scotti, M. Crestani, C. Godio, F. Gilardi, F. Loiodice, G. Fracchiolla, A. Laghezza, P. Tortorella, Crystal structure of the peroxisome proliferator-activated receptor γ (PPARγ) ligand binding domain complexed with a novel partial agonist: a new region of the hydrophobic pocket could be exploited for drug design, Journal of Medicinal Chemistry 51 (24) (2008) 7768–7776.

[100] P. Cronet, J.F. Petersen, R. Folmer, N. Blomberg, K. Sjöblom, U. Karlsson, E.-L. Lindstedt, K. Bamberg, Structure of the PPARα and -γ ligand binding domain in complex with AZ 242; ligand selectivity and agonist activation in the PPAR family, Structure 9 (8) (2001) 699–706.

[101] R.T. Gampe, V.G. Montana, M.H. Lambert, A.B. Miller, R.K. Bledsoe, M.V. Milburn, S.A. Kliewer, T.M. Willson, H.E. Xu, Asymmetry in the PPARγ/RXRα crystal structure reveals the molecular basis of heterodimerization among nuclear receptors, Molecular cell 5 (3) (2000) 545–555.

[102] H.E. Xu, M.H. Lambert, V.G. Montana, K.D. Plunket, L.B. Moore, J.L. Collins, J.A. Oplinger, S.A. Kliewer, R.T. Gampe, D.D. McKee, Structural determinants of ligand binding selectivity between the peroxisome proliferator-activated receptors, Proceedings of the National Academy of Sciences 98 (24) (2001) 13919–13924.

[103] R.N. DuBois, R. Gupta, J. Brockman, B.S. Reddy, S.L. Krakow, M.A. Lazar, The nuclear eicosanoid receptor, PPARgamma, is aberrantly expressed in colonic cancers, Carcinogenesis 19 (1) (1998) 49–53.

[104] L. Fajas, M. Debril, J. Auwerx, Peroxisome proliferator-activated receptor-gamma: from adipogenesis to carcinogenesis, Journal of Molecular Endocrinology 27 (1) (2001) 1–9.

[105] E. Mueller, P. Sarraf, P. Tontonoz, R.M. Evans, K.J. Martin, M. Zhang, C. Fletcher, S. Singer, B.M. Spiegelman, Terminal differentiation of human breast cancer through PPARγ, Molecular cell 1 (3) (1998) 465–470.

[106] A.-M. Lefebvre, I. Chen, P. Desreumaux, J. Najib, J.-C. Fruchart, K. Geboes, M. Briggs, R. Heyman, J. Auwerx, Activation of the peroxisome proliferator-activated receptor γ promotes the development of colon tumors in C57BL/6J-APC Min/+ mice, Nature Medicine 4 (9) (1998).

[107] K. Yang, K.H. Fan, S.A. Lamprecht, W. Edelmann, L. Kopelovich, R. Kucherlapati, M. Lipkin, Peroxisome proliferator-activated receptor γ agonist troglitazone induces colon tumors in normal C57BL/6J mice and enhances colonic carcinogenesis in Apc1638 N/+ Mlh1+/− double mutant mice, International Journal of Cancer 116 (4) (2005) 495–499.

[108] E. Elstner, C. Müller, K. Koshizuka, E.A. Williamson, D. Park, H. Asou, P. Shintaku, J.W. Said, D. Heber, H.P. Koeffler, Ligands for peroxisome proliferator-activated receptorγ and retinoic acid receptor inhibit growth and induce apoptosis of human breast cancer cells in vitro and in BNX mice, Proceedings of the National Academy of Sciences 95 (15) (1998) 8806–8811.

[109] N. Takahashi, T. Okumura, W. Motomura, Y. Fujimoto, I. Kawabata, Y. Kohgo, Activation of PPARγ inhibits cell growth and induces apoptosis in human gastric cancer cells, FEBS Letters 455 (1–2) (1999) 135–139.

[110] T. Takashima, Y. Fujiwara, K. Higuchi, T. Arakawa, Y. Yano, T. Hasuma, S. Otani, PPAR-γ ligands inhibit growth of human esophageal adenocarcinoma cells through induction of apoptosis, cell cycle arrest and reduction of ornithine decarboxylase activity, International Journal of Oncology 19 (3) (2001) 465–471.

[111] B.C. Park, D. Thapa, J.S. Lee, S.-Y. Park, J.-A. Kim, Troglitazone inhibits vascular endothelial growth factor–induced angiogenic signaling via suppression of reactive oxygen species production and extracellular signal–regulated kinase phosphorylation in endothelial cells, Journal of Pharmacological Sciences 111 (1) (2009) 1–12.

[112] P. Mohanty, A. Aljada, H. Ghanim, D. Hofmeyer, D. Tripathy, T. Syed, W. Al-Haddad, S. Dhindsa, P. Dandona, Evidence for a potent antiinflammatory effect of rosiglitazone, Journal of Clinical Endocrinology and Metabolism 89 (6) (2004) 2728–2735.

[113] A. Papi, L. Tatenhorst, D. Terwel, M. Hermes, M.P. Kummer, M. Orlandi, M.T. Heneka, PPARγ and RXRγ ligands act synergistically as potent antineoplastic agents in vitro and in vivo glioma models, Journal of Neurochemistry 109 (6) (2009) 1779–1790.

[114] P. Wang, J. Yu, Q. Yin, W. Li, X. Ren, X. Hao, Rosiglitazone suppresses glioma cell growth and cell cycle by blocking the transforming growth factor-beta mediated pathway, Neurochemical Research 37 (10) (2012) 2076–2084.

[115] B. Bojková, P. Orendáš, P. Kubatka, M. Péč, M. Kassayová, T. Kisková, K. Kajo, Positive and negative effects of glitazones in carcinogenesis: experimental models vs. clinical practice, Pathology, Research and Practice 210 (8) (2014) 465–472.

[116] Y. Takeuchi, M. Takahashi, K. Sakano, M. Mutoh, N. Niho, M. Yamamoto, H. Sato, T. Sugimura, K. Wakabayashi, Suppression of N-nitrosobis (2-oxopropyl) amine-induced pancreatic carcinogenesis in hamsters by pioglitazone, a ligand of peroxisome proliferator-activated receptor γ, Carcinogenesis 28 (8) (2007) 1692–1696.

[117] Y.-W. Dong, X.-P. Wang, K. Wu, Suppression of pancreatic carcinoma growth by activating peroxisome proliferator-activated receptor γ involves angiogenesis inhibition, World Journal of Gastroenterology 15 (4) (2009) 441.

[118] D. Shen, C. Deng, M. Zhang, Peroxisome proliferator-activated receptor γ agonists inhibit the proliferation and invasion of human colon cancer cells, Postgraduate Medical Journal 83 (980) (2007) 414–419.

[119] K. Wu, B. He, Q. Zhou, Inhibitory effect of rosiglitazone on proliferation of human colon cancer lovo cells and relevant mechanism, Chinese Journal of Biology 24 (2011) 276–279.

[120] L.D. Yee, N. Williams, P. Wen, D.C. Young, J. Lester, M.V. Johnson, W.B. Farrar, M.J. Walker, S.P. Povoski, S. Suster, Pilot study of rosiglitazone therapy in women with breast cancer: effects of short-term therapy on tumor tissue and serum markers, Clinical Cancer Research 13 (1) (2007) 246–252.

[121] G. Deray, H. Izzedine, V. Launay-Vacher, C. Bagnis, Kidney and glitazones, in: Annales D'endocrinologie, 2005.

[122] J.A. Engelman, L. Chen, X. Tan, K. Crosby, A.R. Guimaraes, R. Upadhyay, M. Maira, K. McNamara, S.A. Perera, Y. Song, Effective use of PI3K and MEK inhibitors to treat mutant Kras G12D and PIK3CA H1047R murine lung cancers, Nature Medicine 14 (12) (2008) 1351–1356.

[123] J. Van Heijenoort, Recent advances in the formation of the bacterial peptidoglycan monomer unit, Natural Product Reports 18 (5) (2001) 503–519.

[124] J.A. Bertrand, G. Auger, L. Martin, E. Fanchon, D. Blanot, D. Le Beller, J. van Heijenoort, O. Dideberg, Determination of the MurD mechanism through crystallographic analysis of enzyme complexes, Journal of Molecular Biology 289 (3) (1999) 579–590.

[125] A. Bouhss, S. Dementin, J. van Heijenoort, C. Parquet, D. Blanot, MurC and MurD synthetases of peptidoglycan biosynthesis: borohydride trapping of acylphosphate intermediates, Methods in Enzymology 354 (2002) 189–196.

[126] J.J. Emanuele, H. Jin, J. Yanchunas, J.J. Villafranca, Evaluation of the kinetic mechanism of *Escherichia coli* uridine diphosphate-N-acetylmuramate: L-alanine ligase, Biochemistry 36 (23) (1997) 7264–7271.

[127] W. Vollmer, D. Blanot, M.A. De Pedro, Peptidoglycan structure and architecture, FEMS Microbiology Reviews 32 (2) (2008) 149–167.

[128] C.A. Smith, Structure, function and dynamics in the mur family of bacterial cell wall ligases, Journal of Molecular Biology 362 (4) (2006) 640–655.

[129] T. Tomasic, L.P. Masic, Rhodanine as a privileged scaffold in drug discovery, Current Medicinal Chemistry 16 (13) (2009) 1596–1629.

[130] T. Tomašić, N. Zidar, A. Kovač, S. Turk, M. Simčič, D. Blanot, M. Müller-Premru, M. Filipič, S.G. Grdadolnik, A. Zega, 5-Benzylidenethiazolidin-4-ones as multitarget inhibitors of bacterial mur ligases, ChemMedChem 5 (2) (2010) 286–295.

[131] P. Shah, Inflammation, inflammatory markers, and cardiovascular risk, Fundamental and clinical Cardiology 59 (2006) 121.

[132] P.C. Unangst, D.T. Connor, W.A. Cetenko, R.J. Sorenson, C.R. Kostlan, J.C. Sircar, C.D. Wright, D.J. Schrier, R.D. Dyer, Synthesis and biological evaluation of 5-[[3, 5-bis (1, 1-dimethylethyl)-4-hydroxyphenyl] methylene] oxazoles,-thiazoles, and-imidazoles: novel dual 5-lipoxygenase and cyclooxygenase inhibitors with antiinflammatory activity, Journal of Medicinal Chemistry 37 (2) (1994) 322–328.

[133] Y. Song, D.T. Connor, A.D. Sercel, R.J. Sorenson, R. Doubleday, P.C. Unangst, B.D. Roth, V.G. Beylin, R.B. Gilbertsen, K. Chan, Synthesis, structure– activity relationships, and in vivo evaluations of substituted di-tert-butylphenols as a novel class of potent, selective, and orally active cyclooxygenase-2 inhibitors. 2. 1, 3, 4-and 1, 2, 4-thiadiazole series 1, Journal of Medicinal Chemistry 42 (7) (1999) 1161–1169.

[134] J.-H. Choi, H.J. Jeon, J.-G. Park, S.K. Sonn, M.-R. Lee, M.-N. Lee, H.J. You, G.-Y. Kim, J.-H. Kim, M.H. Lee, Anti-atherogenic effect of BHB-TZD having inhibitory activities on cyclooxygenase and 5-lipoxygenase in hyperlipidemic mice, Atherosclerosis 212 (1) (2010) 146–152.

[135] D.K. Nevin, D.G. Lloyd, D. Fayne, Rational targeting of peroxisome proliferating activated receptor subtypes, Current Medicinal Chemistry 18 (36) (2011) 5598–5623.

[136] D.D. Sternbach, Modulators of peroxisome proliferator-activated receptors (PPARs), Annual Reports in Medicinal Chemistry 38 (2003) 71–80.

[137] P. Balakumar, M. Rose, S.S. Ganti, P. Krishan, M. Singh, PPAR dual agonists: are they opening Pandora's Box? Pharmacological Research 56 (2) (2007) 91–98.

[138] G. Navarrete-Vázquez, H. Torres-Gómez, S. Hidalgo-Figueroa, J.J. Ramírez-Espinosa, S. Estrada-Soto, J.L. Medina-Franco, I. León-Rivera, F.J. Alarcón-Aguilar, J.C. Almanza-Pérez, Synthesis, in vitro and in silico studies of a PPARγ and GLUT-4 modulator with hypoglycemic effect, Bioorganic and Medicinal Chemistry Letters 24 (18) (2014) 4575–4579.

[139] I.G. Obrosova, P. Pacher, C. Szabó, Z. Zsengeller, H. Hirooka, M.J. Stevens, M.A. Yorek, Aldose reductase inhibition counteracts oxidative-nitrosative stress and poly (ADP-ribose) polymerase activation in tissue sites for diabetes complications, Diabetes 54 (1) (2005) 234–242.

[140] I.G. Obrosova, A.G. Minchenko, R. Vasupuram, L. White, O.I. Abatan, A.K. Kumagai, R.N. Frank, M.J. Stevens, Aldose reductase inhibitor fidarestat prevents retinal oxidative stress and vascular endothelial growth factor overexpression in streptozotocin-diabetic rats, Diabetes 52 (3) (2003) 864–871.

[141] C. Harada, A. Okumura, K. Namekata, K. Nakamura, Y. Mitamura, H. Ohguro, T. Harada, Role of monocyte chemotactic protein-1 and nuclear factor kappa B in the pathogenesis of proliferative diabetic retinopathy, Diabetes Research and Clinical Practice 74 (3) (2006) 249–256.

[142] J. N Sangshetti, R. S Chouthe, N. S Sakle, I. Gonjari, D. B Shinde, Aldose reductase: a multi-disease target, Current Enzyme Inhibition 10 (1) (2014) 2–12.

[143] N. Chadha, O. Silakari, Identification of low micromolar dual inhibitors for aldose reductase (ALR2) and poly (ADP-ribose) polymerase (PARP-1) using structure based design approach, Bioorganic and Medicinal Chemistry Letters 27 (11) (2017) 2324–2330.

Chapter 6

Oxindole: A Nucleus Enriched With Multitargeting Potential Against Complex Disorders

Maninder Kaur

Punjabi University, Patiala, India

Chapter Outline

1. INTRODUCTION

Oxindole nucleus is an organic heterocyclic compound having a benzene ring fused with the pyrrole ring with a carbonyl group at 2-position. It can be obtained from synthetic or natural origin and displays a wide range of biological activities. Chemically named 1,3-dihydro-2H-indole-2-one, oxindole exists as two hydroxyl tautomers (I'' and I''') (Fig. 6.1). Being abundant in nature, it has been found in tissues and fluids of mammals as well as natural products produced by a range of plants, bacteria, and invertebrates. The oxindole in the form of alkaloids are extracted from the cat claw's plant *Uncaria tomentosa*, which is a woody, tropical vine indigenous to the Amazon rainforest and other tropical areas of South and Central America.

Several reports have been published in the literature covering the chemistry aspects [1,2] and its pharmacological properties [3]. A review compiling the chemistry, synthetic strategies, naturally occurring oxindole alkaloids, and detailed pharmacological profile including anticancer, anti-HIV, antidiabetic, antibacterial, antioxidant, kinase inhibitory, AChE inhibitory, antileishmanial,

Key Heterocycle Cores for Designing Multitargeting Molecules. https://doi.org/10.1016/B978-0-08-102083-8.00006-6

211

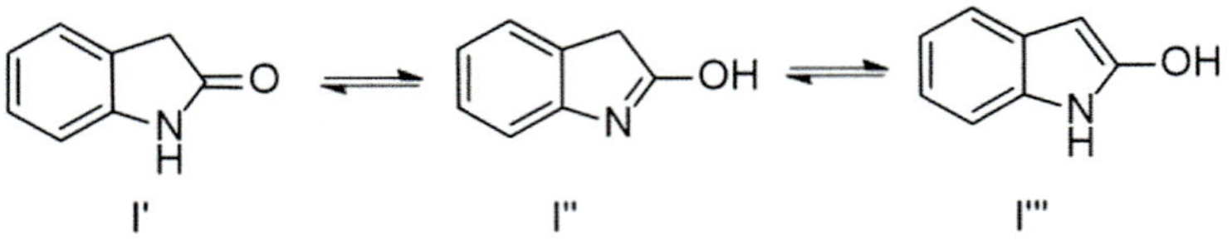

FIGURE 6.1   Tautomerism in oxindole.

β3 adrenergic receptor agonistic, phosphatase inhibitory, analgesic, spermicidal, vasopressin antagonists, progesterone antagonists, neuroprotection, NMDA blocker, and sleep-inducing activities of oxindole-based compounds emphasizing their structure activity relationship (SAR) has already been published [4]. This chapter focuses on its multitarget potential for the treatment of complex disorders.

2. BRIEF HISTORY

Oxindole has been used in the treatment of infection, cancer, gastric ulcers, arthritis, and other inflammatory processes as reported in traditional literature [5–7]. Its diverse pharmacological profile has encouraged industry and academia to develop novel synthetic oxindole derivatives with diverse biological activities. The development of synthetic oxindole derivatives has provided a marketed anticancer agent, sunitinib, employed in gastrointestinal stromal tumors and metastatic renal cell cancer [8]. Subsequent modification of the substituents around the oxindole nucleus lead to several oxindole-based kinase inhibitors that are in clinical trials including SU11248, SU5416, SU5614, SU6668, SU14813, SU4984, and others [9]. Additionally, indolidan and adibendan, oxindole derivatives, have also been used for the treatment of congestive heart failure due to their strong vasodilatory, ionodilatory, and positive ionotropic effects [10,11]. Also, 3-substituted and spiro-oxindole derivatives have been implicated in a wide spectrum of biological activities including antitumor [12], antioxidant [13], anti-Alzheimer's [14], kinase inhibitory activity [15], β3 adrenergic receptor agonist [16], antibacterial [17], neuroprotective [18], spermicidal [19], and analgesic activity [20].

3. CHEMISTRY

Chemically, oxindole nucleus heterocyclic compound has a benzene ring fused with the pyrrole ring, with a carbonyl group at 2-position (molecule **1**). As far as nomenclature of oxindole is considered, it is an indole derivative known as 2-indolinone. But instead of its more systematic name 2-indolinone, it is widely called by its common name, oxindole. The unsubstituted oxindole nucleus exists as off-white crystalline powder with melting point range of 124–126°C. The ^{1}H-NMR depicts a singlet for NH at δ 9.29 (ppm), which is due to the deshielding effect of the carbonyl group, and another singlet at δ 3.46 (ppm) for CH_2 protons of an indole ring. The mass spectrum of oxindole depicts a base peak at m/z 133 (100%).

Oxindole exists in tautomeric form, usually presented as the lactam (I′) of *o*-aminophenyl aceticacid. The other forms are lactim (I″), in which the H of N tautomerises, and the enol form (I‴), in which H of CH_2 group tautomerises (Fig. 6.1) [1].

1

4. SYNTHESIS

4.1 Possible Routes for Oxindole Nucleus Synthesis

In 1866, Bayer and Knop were the first to attempt synthesis of oxindole nucleus via reduction of isatin (1a) with sodium amalgam in an alkaline medium that gives dioxindole (1b), and it was further treated with tin and mineral acids to give oxindole (I) (Scheme 6.1-I). In another effort by Marschalk, isatin (1a) is reduced to dioxindole (1b) by sodium hydrosulfite followed by reduction to oxindole (I) by the action of sodium amalgam in a saturated solution of carbon dioxide (Scheme 6.1-II). On a similar pattern, Curtius and Thun synthesized oxindole by reduction of isatin to oxindole using hydrazine. Another method of synthesizing oxindole using 2-nitro-phenylacetic acid (2a) instead of isatin (1a) was established by Baeyer using tin and hydrochloric acid as reducing agents (Scheme 6.1-III). Oxindole (I) was also prepared by Suida by reduction of 2-acetaminomandelic acid (3a) by either hydroiodic acid and phosphorous or sodium amalgam (Scheme 6.1-IV). Heating β-acetylphenyl hydrazine (4a) with lime at 200–2200°C also gave oxindole (Scheme 6.1-V). Stolle gave another useful method for oxindole synthesis in which α-halogenated acid chloride (5b) is condensed with an aromatic amine (5a) and the resulting amide (5c) was cyclized to oxindole (I) by aluminum chloride (Scheme 6.1-VI). In 1952, oxindole was prepared by heating *o*-chloro-phenylacetic acid (6a) with concentrated ammonium hydroxide and copper powder in a sealed tube at 155–165°C (Scheme 6.1-VII) [1]. Further, Gassman and Bergen gave an unconventional method (i.e., Gassman synthesis) for synthesis of oxindole in which aniline (7a) is treated with tert-butyl hypochlorite, ethyl methyl thioacetate, and triethyl amine, resulting in an unstable amino ester (7b) that is subsequently treated with acid followed by reduction with Ni catalyst yielding oxindole (Scheme 6.1-VIII) [21].

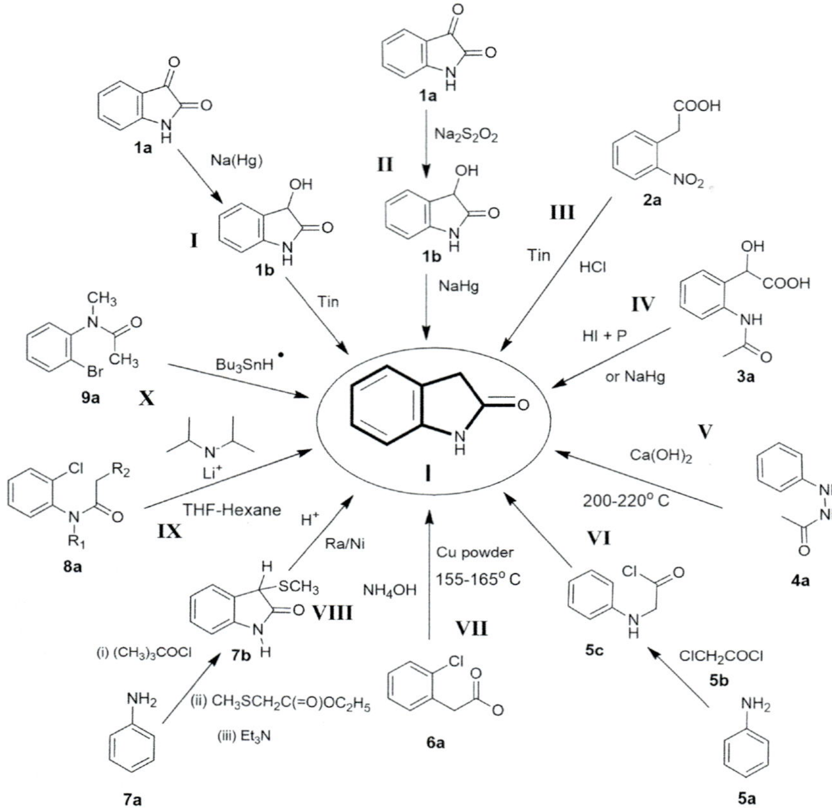

SCHEME 6.1

Following Gassman Synthesis, Pfizer Central Research Group synthesized oxindole with slight modification in the original procedure; that is, using oxalyl chloride to activate the sulfoxide to aid the formation of the key N–S bonded intermediate [22]. Besides these methods, another method for oxindole synthesis involves photoinduced cyclization of *N*-acyl-*o*-chloroanilines (8a) (Scheme 6.1-IX) [23]. Another synthetic procedure involves treatment of *o*-bromo-*N*-methylanilides (9a) with tributyl stannane at 160°C via tandem translocation and homolytic aromatic translocation (Scheme 6.1-X) [24]. Among all the synthetic methods, the reaction III in Scheme 6.1 using 2-nitrophenylacetic acid as starting material was found to be most suitable in terms of cost-effectiveness and % yield. The starting reactant (i.e., 2-nitrophenylacetic acid) is less costly, less time consuming, and easily available. It also won the race in terms of yield obtained (80%).

4.2 Synthesis of Various Oxindole Derivatives

A synthetic procedure for synthesis of 3-ethylidene-2-oxindole (10e) has been reported by Arumugam et al. Via the Heck reaction, in which the aniline

(10a) was alkylated using cyclohexane carboxaldehyde to give the amine (10b) that was subsequently acetylated with crotonyl chloride, tertiary amide (10c) thus formed was followed by cyclization, giving resin-bound compound 3-ethylidene-2-oxindole (10d). The obtained product was further cleaved from the resin by 25% trifluoroacetic acid (TFA) in dichloromethane (DCM) to give 3-ethylidene-2-oxindole (10e) (Scheme 6.2) [25]. Another reaction reported for the synthesis of oxindole was intramolecular amide arylation. In this reaction, N-benzyl-2-bromoacetanilide (11a) was treated with sodium tert-butoxide in the presence of Pd(dba)$_2$ and a chelating phosphine such as BINAP (2,2'-bis(diphenylphosphino)-1,1'-binaphthyl) and DPPF(1,1'-Bis(diphenylphosphino)ferrocene) that resulted in 1-benzyloxindole (11b) (Scheme 6.3) [26]. In 2003, a novel variant of the Friedel–Crafts procedure using palladium-catalyzed C–H functionalization has been developed. This reaction involved the conversion of α-chloroacetanilides (12a) to regioselective oxindoles (12b) in the presence of palladiumacetate, 2-(di-tert-butylphosphino) biphenyl, and triethylamine (Scheme 6.3) [27].

Another method for the synthesis of 3-substituted oxindole (13b) is via vinyl palladation of isocyanates (13a) (Scheme 6.3) [28]. Domino carbopalladation has been reported as another method for synthesis of unsymmetrically substituted 3-(diarylmethylenyl) oxindole (14b) fromanilides (14a) (Scheme 6.4) [29]. The strategy for the synthesis of enantioselective 3,3-disubstituted oxindoles (15) involves Pd catalyzed cynoamidation in the presence of phosphoramidite Pd(lba)$_2$ and N-N-dimethyl propyleneurea (DMPU) in decalin (Scheme 6.4) [30]. Besides the methods discussed earlier, the direct coupling of two C–H bonds can yield oxindoles (16b) from acetanilides (16a) in the presence of tBuONa base, oxidant CuCl$_2$, and Pd(OAc)$_2$ catalyst (Scheme 6.4) [31]. In another procedure, the 2-(alkynyl)arylisocyanates (17a) were cyclized to give 3-(amidoalkylidene)oxindoles (17b) in the presence of palladium(0)/diphosphine catalyst (Scheme 6.4) [32]. The Claisen rearrangement was another method given by a research group in which indole (18a) upon treatment with NCS (N-chlorosuccinimide) in the presence of 1,4-dimethylpiperazine gives chloroindolenine (18b), which is then treated with allyl alcohol under acidic conditions to give a product (18c) that subsequently undergoes [33] sigmatropic (Claisen) rearrangement to give 3-substituted oxindole (18d) (Scheme 6.5) [33].

Another novel enantioselective synthetic strategy has been explored by Hills et al. in which the substrate (19a) was treated with methyl chloroformate and catalyst PPY [4-(pyrrolidino) pyridine], which yields oxindole with a quaternary stereocenter (**19b**) (Scheme 6.5) [34]. Another method, the Meerwein–Eschenmoser–Claisen rearrangement, has been employed to include the transformation of 2-amino allylvinyl ethers (20a) to γ,δ-unsaturated amides (20b) (Scheme 6.5) [35]. Duguet et al. explored the Asymmetric Hetero-Claisen reaction in which phenyl nitrone (21a) is treated with asymmetric disubstituted ketenes (21b) resulting in 3-alkyl-3-aryloxindoles (21c) (Scheme 6.6) [36]. Enatiomerically enriched 3-substituted oxindole (22c) has been synthesized by employing the Morita-Baylis-Hillman reaction, where the isatin (22a) acts as

R₂
O
R₁
10c
Pd(OAc)₂, Ag₂CO₃,
PPh₃, DMF, 100-C, 16h,
R₂
O
R₁
10d
25% TFA,
DCM, 25-C, 16 h,
R₂
O
R₁
NH₂
10e
R₂CH=CHCOCl, DIPEA
DMAP, DMF, 60 °C, 16h
NH
R₁
10b
R₁CHO, DCM
NaBH(OAc)₃, 16 h
NH₂
10a
R₁= cyclohexyl
R₂= CH₃

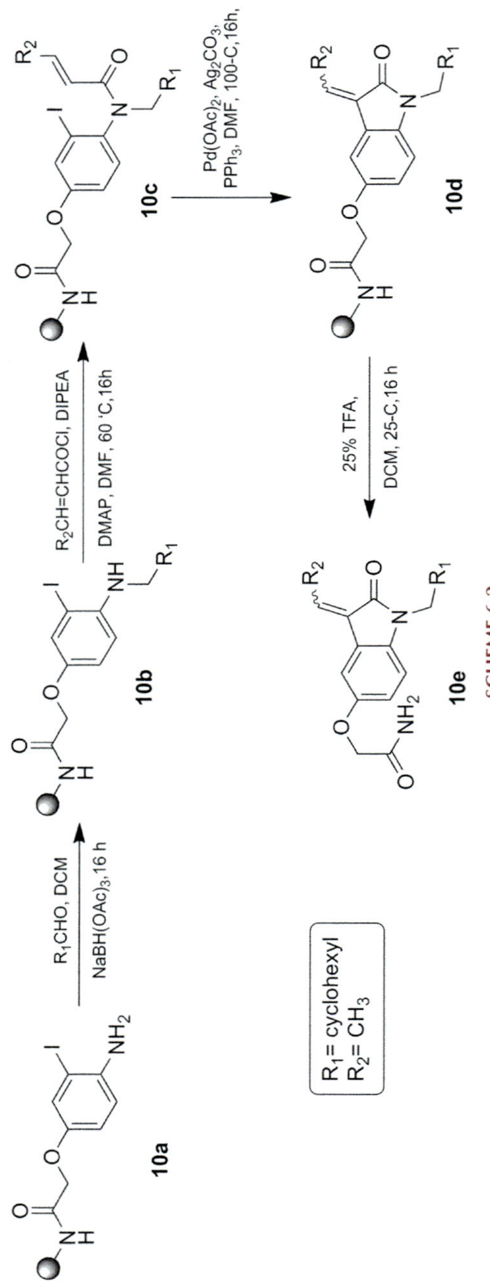

SCHEME 6.2

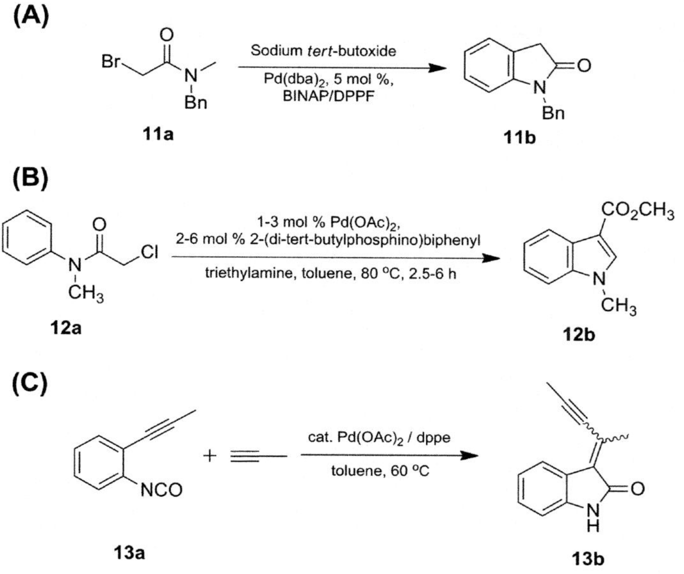

(A)

Sodium *tert*-butoxide

Pd(dba)$_2$, 5 mol %,
BINAP/DPPF

11a

11b

(B)

1-3 mol % Pd(OAc)$_2$,
2-6 mol % 2-(di-tert-butylphosphino)biphenyl

triethylamine, toluene, 80 °C, 2.5-6 h

CO$_2$CH$_3$

12a

12b

(C)

cat. Pd(OAc)$_2$ / dppe

toluene, 60 °C

NCO

13a

13b

SCHEME 6.3

electrophile while acrolein (22b) acts as nucleophile and reacts in the presence of β-isocupreidine (β-ICD) (Scheme 6.6) [37]. Aldol reaction has also been reported for the synthesis of oxindole derivatives (23d) using 3-diazooxindoles (23a) and anilines (23b) along with glyoxylates (23c) in the presence of rhodium complex and chiral phosphoric acid (Scheme 6.6) [38]. In another strategy, *N*-alkyl-*N*-arylacrylamide derivatives (24a) were cyclized to give 3-(iodomethyl)-3-substituted-indolin-2-ones (24b) in the presence of PhI(OAc)$_2$/I$_2$ (Scheme 6.7) [39]. Ju et al. gave another cyclization approach in which 2-bromoanilides (25a) were converted to 3,3-disubstituted oxindoles (25b) using high efficiency visible light irradiation and fac-Ir(ppy)$_3$ as the photoredox catalyst (Scheme 6.7) [40].

5. OXINDOLE AS A PRIVILEGED SUBSTRUCTURE

Oxindole nucleus has emerged as a key nucleus in several natural and synthetic compounds that elicit varied pharmacological responses. A number of research groups have designed, synthesized, and evaluated oxindole derivatives for numerous biological activities including anticancer, anti-HIV, antidiabetic, antibacterial, antioxidant, kinase inhibitory, AChE inhibitory, antileishmanial, β3 adrenergic receptor agonistic, phosphatase inhibitory, and analgesic (Fig. 6.2). Thus, oxindole is in agreement with Evan's idea of privileged substructure of "a single molecular framework able to provide ligands for diverse receptors."

Numerous natural and synthetic oxindole derivatives have been evaluated for their anticancer potential. Two endogenous molecules, 5-hydroxy-oxindole

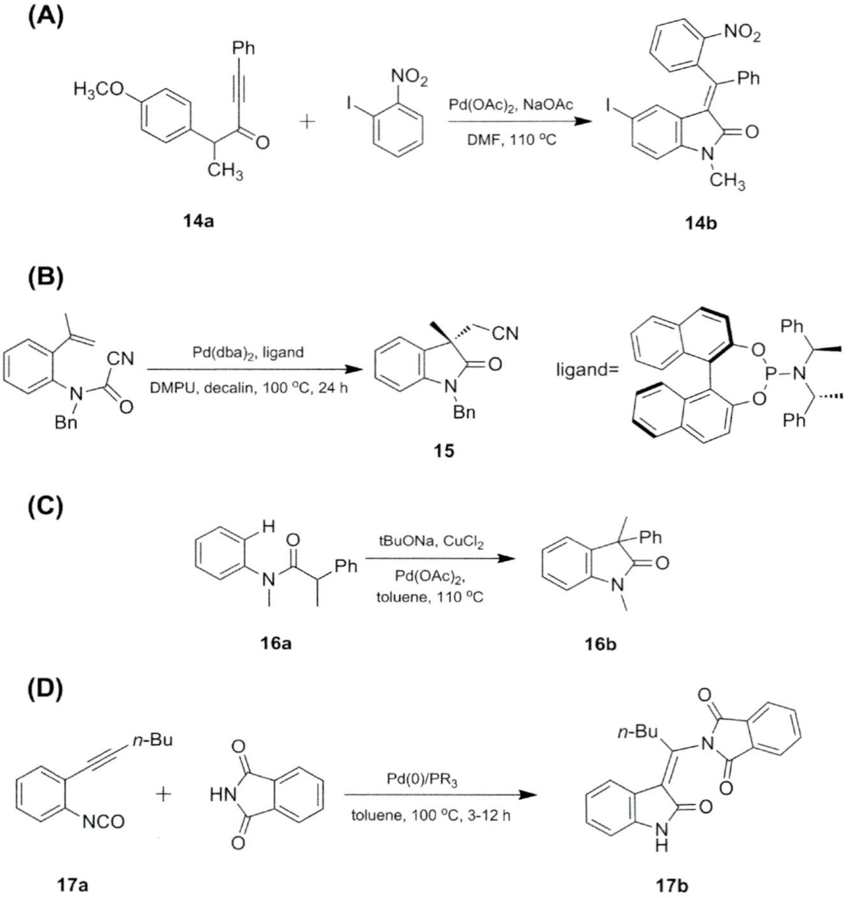

SCHEME 6.4

(**1**) and isatin (**2**), present in bodily fluids, were found to have antiproliferative activity exerted by inhibition of cell proliferation via interaction with extracellular signal-regulated kinases (ERKs) (35% inhibition at 100 μM) and promotion of apoptosis [41]. In 2001, Bramson and coworkers from GlaxoSmithKline Inc. identified two classes of oxindole-based compounds, 1H-indole-2,3-dione 3-phenylhydrazones and 3-(anilinomethylene)-1,3-dihydro-2H-indol-2-one as potent and selective CDK2 inhibitors [42]. Among the synthesized library of compounds, 3-(anilinomethylene)-1,3-dihydro-2Hindol-2-one analog (compound **3**) was found to be an extremely potent compound against CDK2 (IC$_{50}$ value of 0.54 nM) with excellent selectivity over CDK1 (IC$_{50}$ value = 12 nM), having additional ability to prevent hair loss caused by chemotherapy. In 2003, Dermatakis et al. synthesized novel and potent oxindole derivatives as CDK2 inhibitors, starting from compound **4** as lead molecule, with low nanomolar activity (IC$_{50}$ = 39 nM) against CDK2 enzymes [43]. The compound **5** obtained

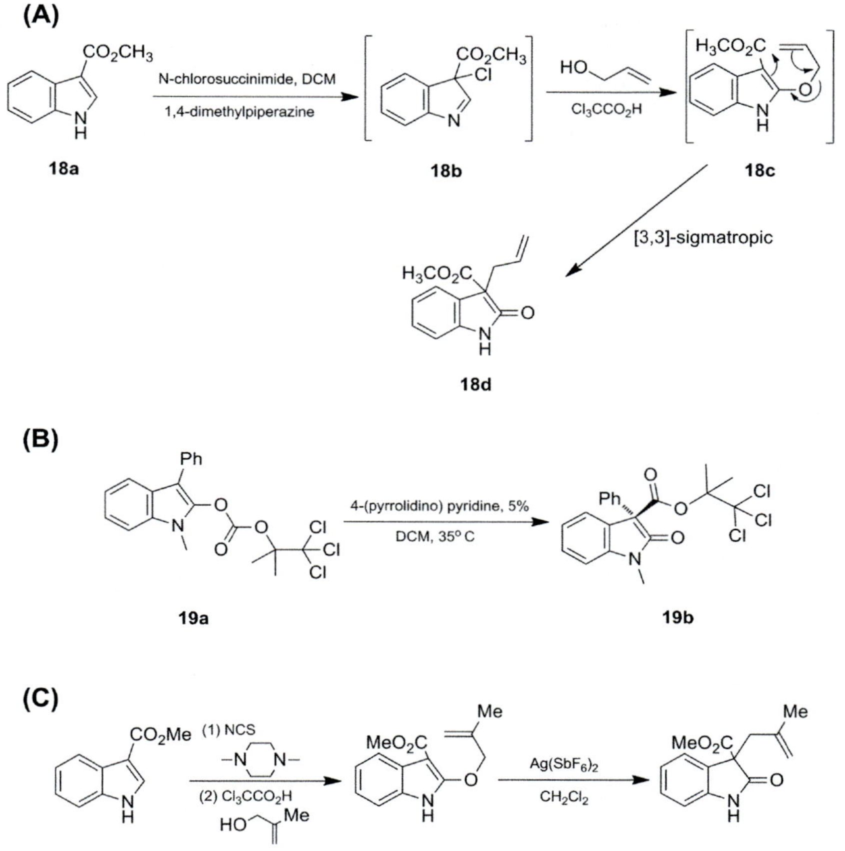

(A)

N-chlorosuccinimide, DCM

1,4-dimethylpiperazine

Cl_3CCO_2H

18a **18b** **18c**

[3,3]-sigmatropic

18d

(B)

4-(pyrrolidino) pyridine, 5%

DCM, 35° C

19a **19b**

(C)

(1) NCS

(2) Cl_3CCO_2H

$Ag(SbF_6)_2$

CH_2Cl_2

20a **20b**

SCHEME 6.5

with 3-aminopyrrolidine-1-yl substituted at C-4 position showed IC_{50} value of 3 nM against CDK2. Later in 2004, oxindole derivatives (**6, 7**) were further optimized by substituting with heteroatoms containing alkynyl moieties at C-4position of the oxindole nucleus that showed increased potency (IC_{50} value **6** = 3 nM, **7** = 2 nM) [44].

In 2004, Wood et al. identified selective oxindole- and aza-oxindole-based TrkA kinase inhibitors via focused screening [45]. The oxindole derivatives with substitution of phthalamide (**8**), 3-sulphonamide (**9**), and 4-triazole (**10**) groups at 3-position of the oxindole nucleus showed good TrkA inhibitory activity with an IC_{50} value of 0.007, 0.063, and 0.008 µM, respectively, along with selectivity over CDK and Raf1. The docking study revealed that the lactam portion of the oxindole ring acts as donor/acceptor motif that binds to the hinge region in the active site. These oxindole derivatives were found to be more potent and selective than published TrkA inhibitors like staurosporines and tyrphostins.

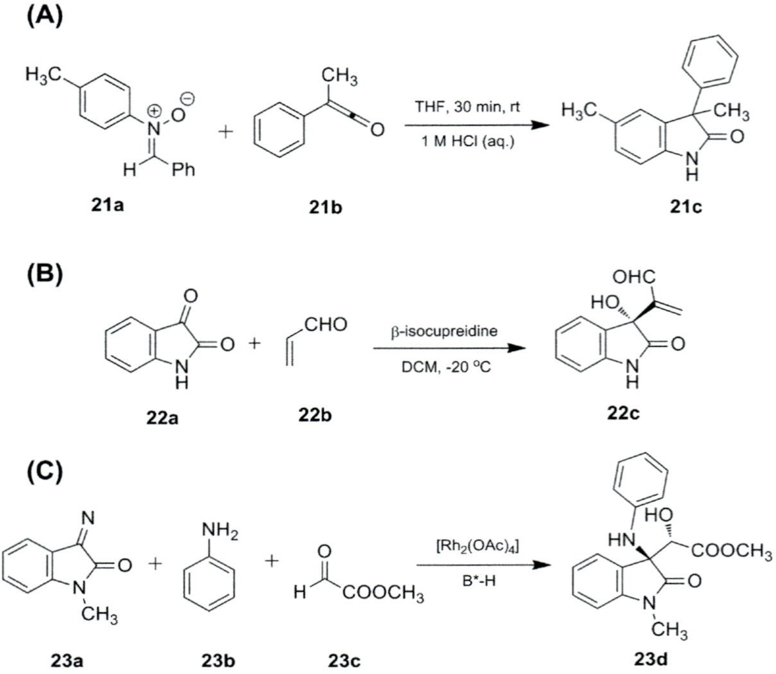

(A)

21a + **21b** →[THF, 30 min, rt; 1 M HCl (aq.)] **21c**

(B)

22a + **22b** →[β-isocupreidine; DCM, -20 °C] **22c**

(C)

23a + **23b** + **23c** →[[Rh₂(OAc)₄]; B*-H] **23d**

SCHEME 6.6

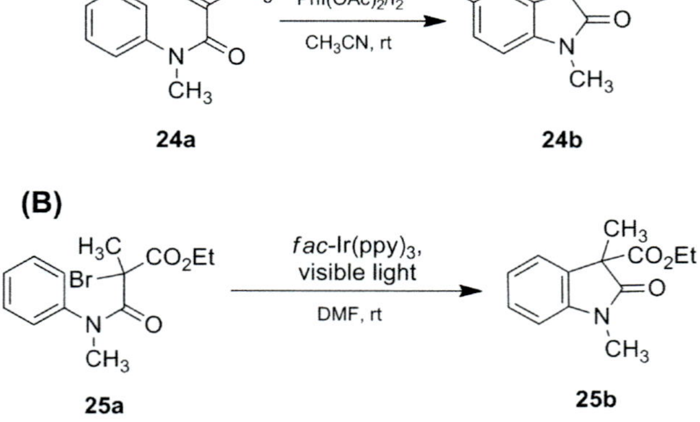

(A)

24a →[PhI(OAc)₂/I₂; CH₃CN, rt] **24b**

(B)

25a →[*fac*-Ir(ppy)₃, visible light; DMF, rt] **25b**

SCHEME 6.7

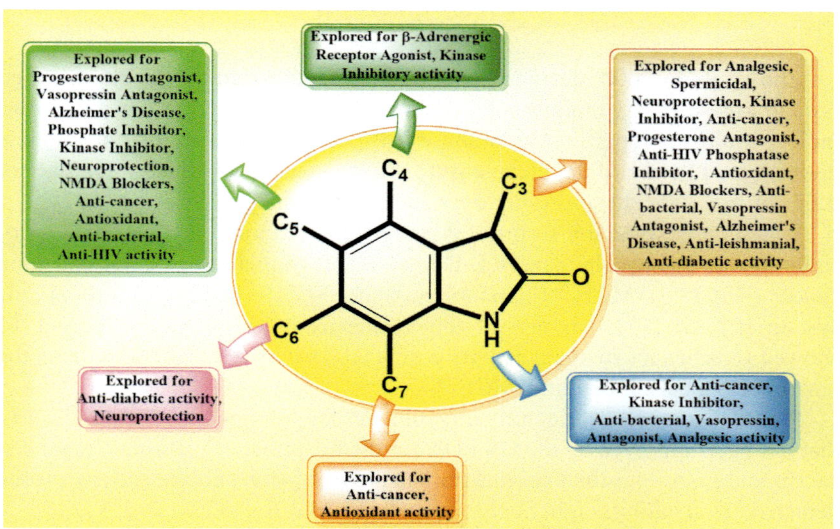

FIGURE 6.2   Pharmacological profile of oxindole.

Zhu et al., in 2006, designed potent and selective oxindole-pyridine-based protein kinase B/Akt inhibitors by replacing metabolically labile isoqunioline moiety in already reported Akt inhibitors [46]. Among the synthesized oxindole derivatives, **11** showed the highest Akt inhibitory activity with an IC_{50} value of 0.17 nM and 100-fold selectivity over other Akt isoenzymes.

Yong et al. explored a series of 3′-spirocyclic-oxindole derivatives as cytostatic agents against the cancer cell lines H460 (human non–small cell lung cancer [NSCLC]), MCF-7 (human breast) and SF268 (human central nervous system [CNS]) [12]. Among all the synthesized derivatives compound **12** showed significant cytostatic activity with a GI_{50} value of 2.6 μM activity on the human breast cancer cell line, MCF-7.

Mendel and coworkers identified a synthetic oxindole derivative, sunitinib (**13**), that has been approved by the US FDA for the treatment of metastatic renal cell cancer and gastrointestinal stromal tumors in 2006 [4]. Its kinase activity profile depicted that it inhibits eight kinases including VEGFR-1, VEGFR-2, VEGFR-3, PDGFRα, PDGFRβ, Kit, Flt-3, and CSF-1R. Sunitinib has been approved for tumors resistant to imatinib. Employing a random screening approach, Sun and colleagues identified another oxindole derivative, SU-5416 (**14**), which also showed inhibitory activity againstVEGFR-1, VEGFR-2, VEGFR-3, PDGFRα, PDGFRβ, Flt-3, and CSF-1R along with minimal Kit inhibitory activity [47,48]. It has been reported to be effective in cancer cell lines including breast, colon, glioblastoma, lung, melanoma, and prostate. Laird and coworkers established the kinase profile of SU-6668 (**15**), another oxindole-based anticancer agent. It showed a similar kinase inhibitory profile but was more potent than SU-5416. Likewise, Patel and coworkers carried out the kinase profiling of SU-10944 (**16**), which inhibited VEGFR-1 with an IC_{50} value of 96 nM but showed lower inhibitory activity against PDGFRβ, Kit, or FGF-R1 [8].

Prado and coworkers isolated a pentacyclic oxindole alkaloid mitraphylline from the *Uncaria tomentosa* inner bark that has been traditionally used as an antitumor agent [49]. Mitraphylline (**17**) showed growth-inhibitory and cytotoxic activity against glioma (GAMG) and neuroblastoma (SKN-BE) cancer cell lines with IC_{50} values of 12.3 and 20 μM, respectively.

In 2008, Silveria and coworkers discovered oxindole-Schiff base copper (II) complexes inspired from the ability of coordination compounds to intercalate between the bases of DNA and to act as oxidizing and reducing agents in various catalytic cycles in biological processes [50]. The research group analyzed their cytotoxic activity against tumor cells (neuroblastoma cells SH-SY5Y). It was suggested that these complexes cause apoptosis by triggering the migration of copper into the cells that lead to increased oxidative stress and apoptosis. Also they interact with membranes and organelles as delocalized lipophilic cations. Among all the complexes, complex $[Cu(isaepy)_2]^{2+}$ (**18**) showed maximum DNA nuclease activity.

Later in 2011, Hong et al. explored spirocyclicoxindole scaffold for antitumor activity against human lung cancer cell A549, human liver cell BEL7402, and human colon cancer cell HCT-8 [51]. Among the synthesized compounds, compound **19** with 2-(triflouromethoxy)benzylic substituents on the nitrogen of the piperidine ring showed nanomolar inhibitory activity against human lung cancer cells (IC_{50} = 50 nmol/L). In the same year Kamal and coworkers carried out the synthesis of oxindole-derived imidazo[1,5-a]pyrazines on the basis of their established anticancer potential and screened them against five human cancer cell lines obtained from nine different cancer types including leukemia, lung, colon, CNS, melanoma, ovarian, renal, prostate, and breast cancer [52]. Among the synthesized compounds, compound **20** showed significant cytotoxic activity with GI_{50} ranging from 1.54 to 13 μM. Compound **20** showed growth inhibitory activity at <2μM against colon, CNS, ovarian, and breast cancer.

Later in 2012, a de novo designing approach was explored to propose oxobenzimidazoles and oxindoles as novel and potent androgen receptor (AR) antagonists that can be used as therapeutics for the treatment of prostate cancer [53]. Initially they developed a series of oxobenzimidazole derivatives that showed lower solubility possibly due to high lipophillicity and the planarity of the core nucleus. In order to improve the pharmacokinetic profile, the planar structure of oxobenzimidazole was replaced with the less planar core nucleus dimethyl oxindole. The dimethyl oxindole derivative **21** (IC_{50} value 79.6 nM) showed improved solubility and good pharmacological and ADME properties, thus presenting it as novel core for AR full antagonists.

On the basis of reported anticancer potential of oxindole and tetrahydroisoquinoline (THI), hybrids of oxindole and THI have been discovered as antitumor agents against human NSCLC along with inhibitory activity against Akt [54]. The compounds **22**, **23**, and **24** (IC_{50} value 0.66, 2.75, 2.19 μM) showed two- to sixfold higher antitumor activity than the standard Akt inhibitor (perifosine) and 2.5- to 10-fold higher than sunitinib (FDA approved anticancer drugs). Thus, these derivatives can be explored as potential antitumor agents against NSCLC.

Ribeiro and coworkers targeted to inhibit p53-MDM2 interaction to design anticancer agents [55]. Compound **25**, obtained by substituting chloro at 6-position and *p*-methyl substituted phenyl group at 3-position of the oxazole ring was obtained as a lead with GI_{50} of 29.1 μM. On a similar pattern, Ivanenkov et al. designed dispirooxindole-based compounds as potent p53-MDM2 interaction inhibitors and evaluated them against hepatocellular carcinoma perpetual cell line (HepG2), human embryonic kidney 293 cells (Hek293), breast cancer cell line (MCF-7), human cervical cancer cells (SiHa), and human colon cancer cells (HCT116) using MTT-based assay [56]. Among the series, compound **26** displayed an IC_{50} value ranging from 4.88 to 10.46 μM against the respective cell lines. Another series of spiropyrazoline oxindoles was synthesized and evaluated for anticancer activity against breast cancer cell lines MCF-7 and MDA-MB-231 by Monteiro et al. [57]. Among the 19 synthesized derivatives, 6 compounds showed GI_{50} value < 12 μM (**31**, IC_{50} = 7.0 μM) with good therapeutic index as they were nontoxic against HEK293 normal cells. Owing to medicinal importance of oxindole and indole nucleus, both were hybridized to give 3-indolyl-3-hydroxy derivatives that were subjected to screening for four cancer cell lines, leukemia (U937, THP-1), lung (A549), and breast cancer (MCF-7) [58]. Among the synthesized derivatives, compounds **27** and **28** showed good cytotoxic activity against all four cell lines.

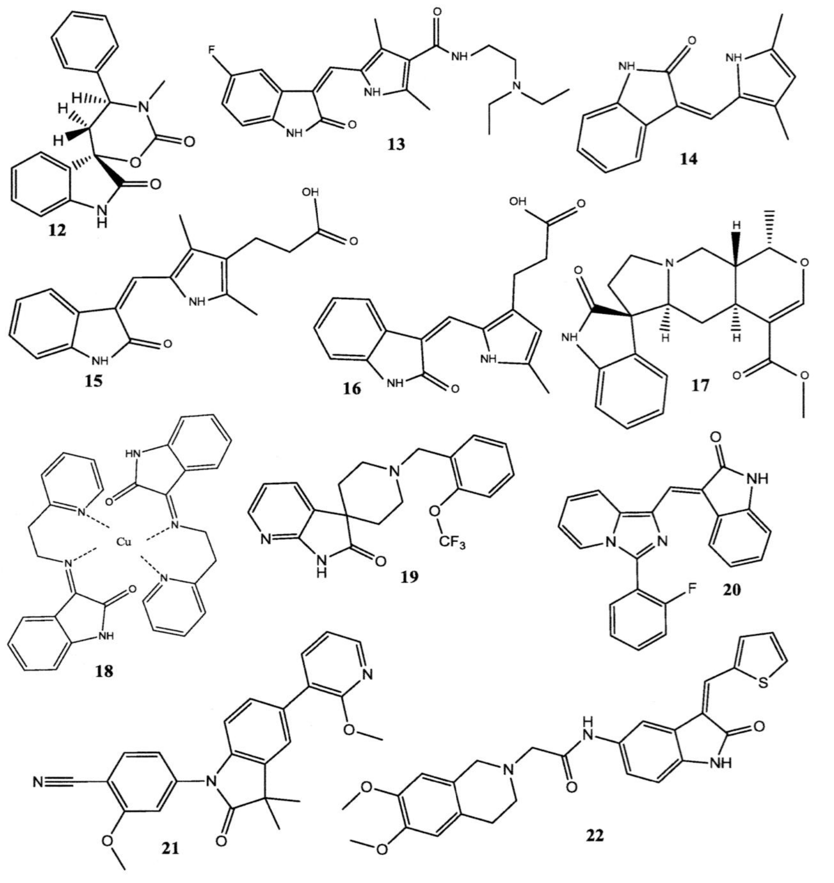

As the steroids have a well-established role in a wide range of biological processes and in human physiology, novel steroidal spiro-oxindoles have been synthesized and evaluated as potent antitumor agents against the human gastric cancer cell line (MGC-803), human breast cancer cell line (MCF7), human liver cancer cell line (SMMC-7721), and human esophageal cancer cell line (EC-109) [59]. Although all compounds showed good antiproliferative activity, some of the compounds possess potent antiproliferative activity compared to 5-FU (5-flourouracil). Compound **29** showed twofold potent cytotoxic activity than 5-FU against EC109 cell lines, compound **30** was more potent against

MGC-803 than 5-Fu, and compound **31** was more potent than 5-Fu against MCF-7 cell line. Interestingly, compound **32** was found to be one log order more potent than the standard drug 5-Fu against SMMC-7721.

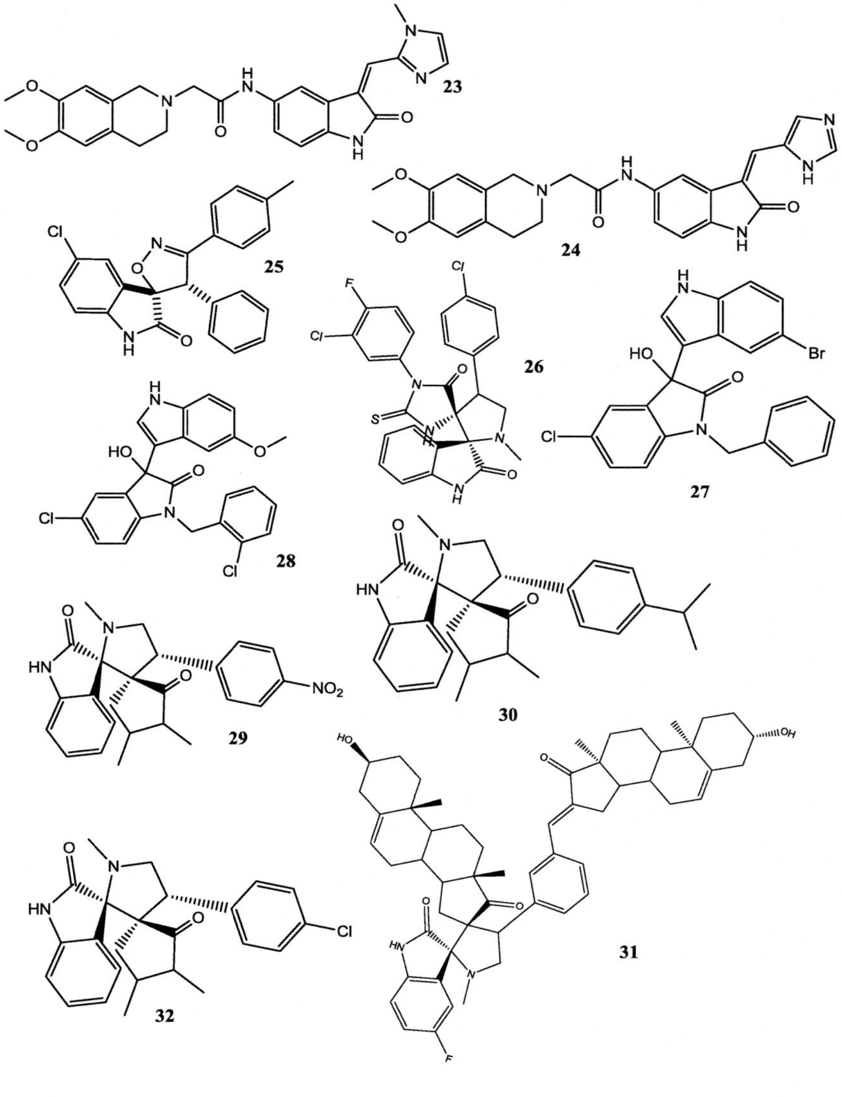

The same research group developed another steroidal pyran-oxindole hybrid as cytotoxic agents against human bladder cancer cell line (T24), human liver cancer cell line (SMMC-7721), human breast cancer cell line (MCF-7), and human gastric cancer cell line (MGC-803). Among the library of synthesized steroidal derivatives, **33** and **34** were found to be more potent than 5-Fu against T24 and MGC-803 cell lines with IC_{50} value of 4.43 and 8.45 μM, respectively.

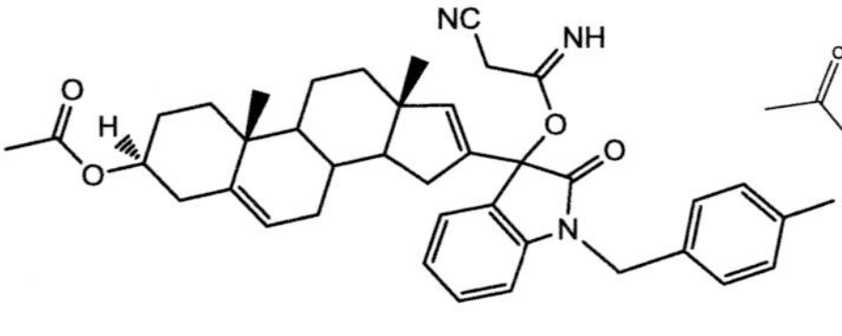

33

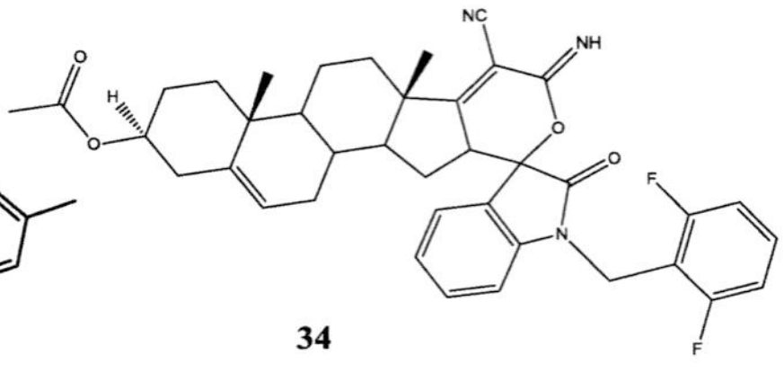

34

In 1992, a series of oxindole-1-acetic acids was investigated against AR enzyme, the key enzyme that causes diabetic complications [60]. The authors designed the molecules by incorporating two important pharmacophores, *N*-aryl glycine moiety and an adjacent carbonyl group, recommended by Kador et al. The synthesized oxindole derivatives showed potent in vitro AR inhibitory activity but poor in vivo efficacy, which may be due to a poor pharmacokinetic profile. In 2013, emerging pharmacological significance of oxindole nucleus encouraged Khan et al. to investigate oxindole derivatives their antiglycation activity [61]. Among the synthesized oxindole derivatives, two compounds, **35** (IC_{50} = 150.4 ± 2.50 μM) and **48** (IC_{50} = 194.5 ± 2.50 μM), showed potent antiglycation activity even better than the standard drug rutin (IC_{50} = 294.5 ± 1.50 μM). Thus, these can be considered promising lead molecules for antiglycation activity in the future.

Inspired by the revealed therapeutic potential of oxindole, another series of oxindole analogues have been discovered as potent α-glucosidase inhibitors [62]. The compounds were synthesized by condensation of 6-chlorooxindole with various aldehydes that resulted in potent α-glucosidase inhibitors. Among these inhibitors, seven inhibitors, **36–42** (IC_{50} = 2.71 ± 0.007– 37.93 ± 0.002 μM), showed inhibitory activity even greater than standard acarbose (IC_{50} = 38.25 ± 0.12 μM).

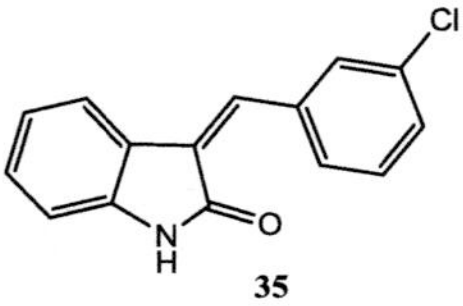

35

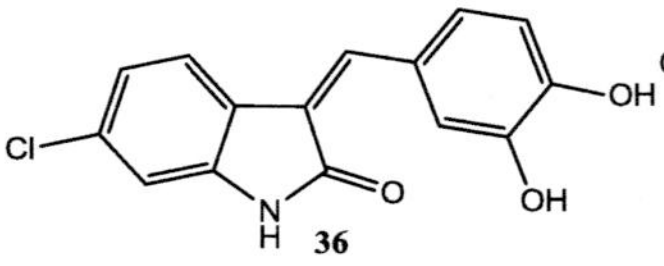

36

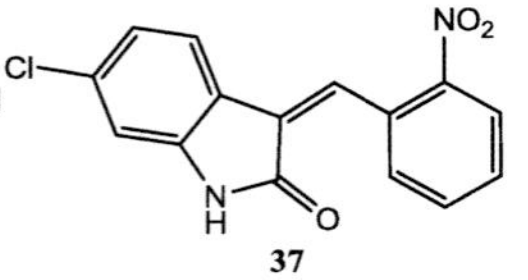

37

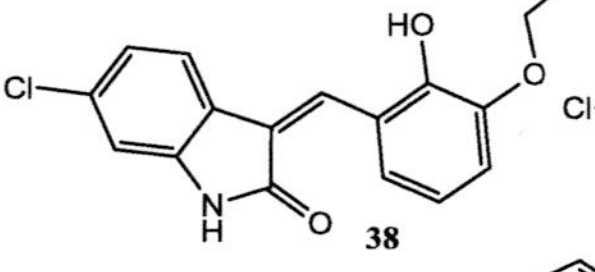

38

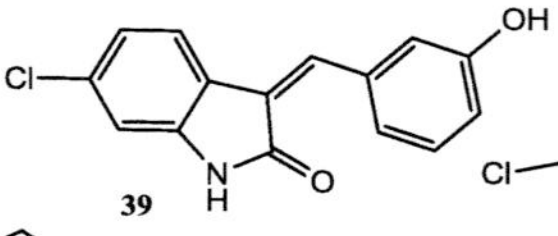

39

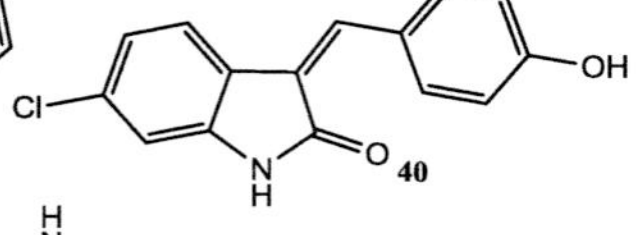

40

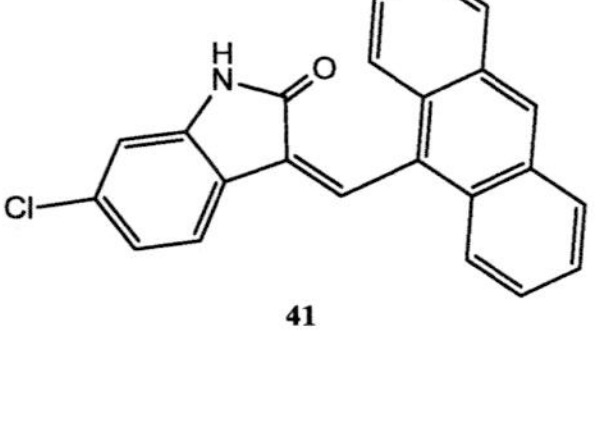

41

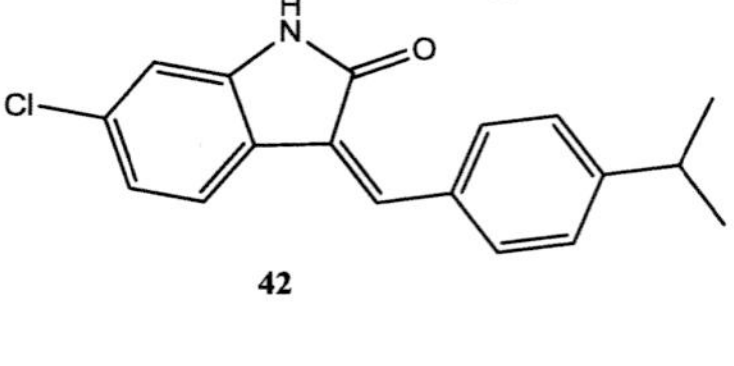

42

Jiang et al., in 2006, discovered oxindole as a novel nonnucleoside reverse transcriptase inhibitor (NNRTI) via screening using cell-based HIV reporter infection assay [63]. The lead molecule **43** ($IC_{50} = 0.066\,\mu M$) was further optimized to establish SAR and to get potent NNRTI. Br at 5-position of aromatic ring was found to be optimum for the inhibitory activity as the activity was reduced when the Br is replaced with a larger, smaller, electron-donating, or electron-withdrawing group (**44** $IC_{50} = 0.486\,\mu M$). Thus there is very limited space for the substitution at 5-position and only a small hydrophobic group can be accommodated. Also the substitution at any other position of aromatic ring led to decrease in activity (**45** $IC_{50} = 7.363\,\mu M$). Further the ester region was explored by substituting different esters like iso-propyl, tert-butyl, and benzyl groups. In case of both iso-propyl (**46**) and benzyl (**47**) ester the activity decreases, whereas tert-butyl (**48**) ester derivative showed slight improved activity.

Additionally the esters were replaced with amides and among amide derivatives **49** and **50** showed significantly improved activity. In order to further explore SAR, the substitutions were made at the cyclopropane ring. Among these synthesized derivatives, gem-dimethyl analog **51** showed a noticeable increase in activity (EC_{50} of 15 nM) and another analog **52** substituted with ethyl group showed EC_{50} of 30 nM. The research group further optimizes the lead molecule to improve the pharmacokinetic properties and metabolic stability by replacing esters since esters are metabolically unstable [64]. The ester group was replaced with aromatic rings including tetrazole, pyridine, and furan ring, and analog **53** with the pyridine ring showed EC_{50} of 8 nM along with an improved pharmacokinetic profile.

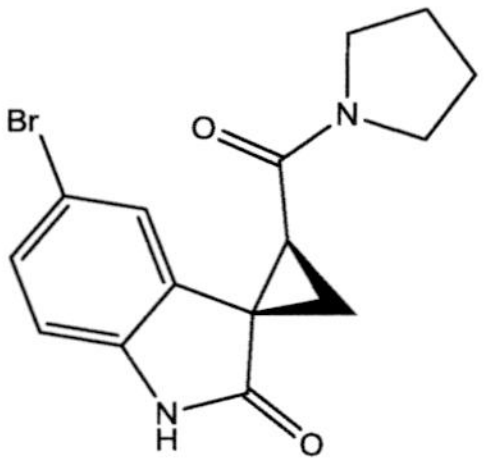

In 2010, Midoh et al. explored the antioxidant activities of oxindole-3-acetic acid derivatives from supersweet corn powder used in corn soups and snacks [65]. 7-(O-β-glucosyloxy)oxindole-3-acetic acid (GOA) **54** and its aglycone, 7-hydroxy-oxindole-3-acetic acid (HOA) **55** both contribute to antioxidant activity. It was found that HOA showed more DPPH radical scavenging activity compared to GOA. This might possibly be due to the presence of a hydroxyl group at 7-position instead of glucose. In 2013, Yasuda et al. synthesized derivatives of 5-hydroxyoxindole found in human serum and brain that possess MAO and ERK inhibitory activity [13]. All the synthesized derivatives were evaluated for lipid peroxidation inhibitory activity, DPPH radical scavenging activity, intracellular oxidative stress-suppressing effect, and cytotoxicity. The lead molecule **1**, endogenous oxidized indole (5-hydroxyoxindole), was substituted with lipophilic groups at C-3 position in order to increase the antioxidant effects and to retain the radical scavenging activity. The compound **56** showed stronger lipid peroxidation inhibitory activity than **1** whereas compound **57** showed lower inhibitory activity than **1**. This implies that substitution of a hydroxyl and phenacyl group was favorable, whereas hydroxyl and acetonyl group substitution resulted in decreased activity. In DPPH radical scavenging activity contrary effect was observed; that is, incorporation of hydroxyl or phenacyl group at C-3 position led to decreased radical scavenging activity. Additionally, intracellular oxidative stress-suppressing effect was assessed and found to be in complement with the lipid peroxidation inhibitory activity. Thus, 5-hydroxy oxindole can be substituted at C-3 position with a lipophillic group to increase its antioxidant activity.

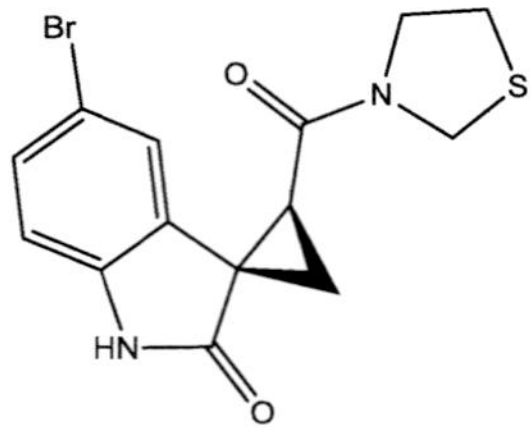

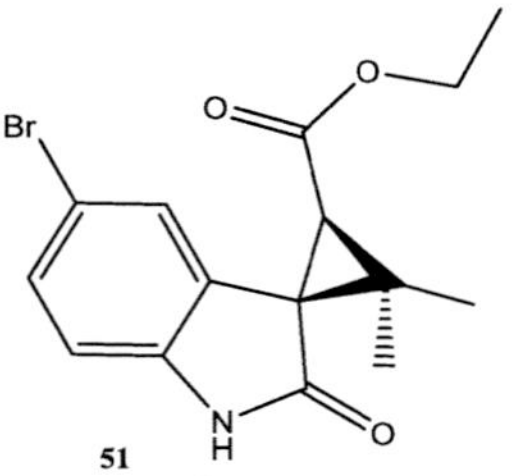

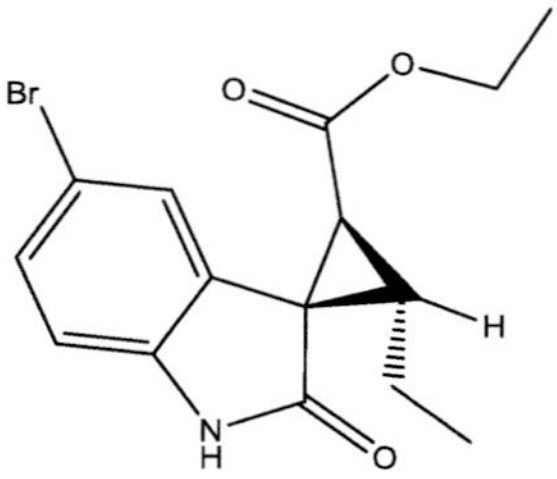

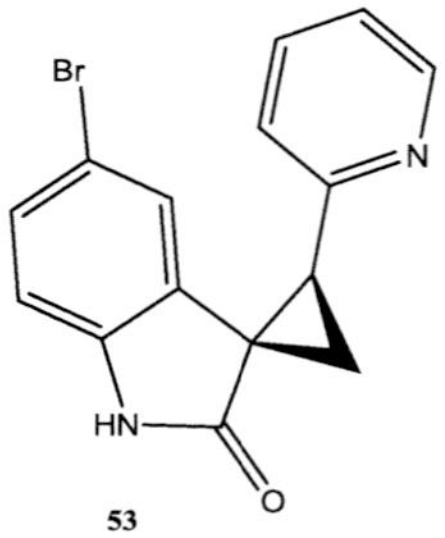

54   **55**   **56**   **57**

In 2002, Fensome et al. successfully explored 3,3-disubstituted-5-aryl oxindoles as progesterone antagonists [66]. The 3-position was substituted with alkyl, dialkyl, or spirocyclic groups and 5-position with substituted phenyl ring. The dimethyl derivatives (**58**) were found to be more potent as compared to mono-methyl derivatives (**59**) with an IC_{50} value of 30.6 nM and 102 nM, respectively, whereas diethyl derivatives were found to be less potent than the mono-ethyl analogues. Also, the spirocyclic ring substituted analogues showed good potency and the antagonism increases with an increase in ring size (**60**; 85 nM, 74; 32 nM). In addition to substitution at 3-position of the aryl ring, a small group can be added at 5-position of the ring. 3′cyano-5′flouro-substituted derivative (**61**) was found to be the most potent with an IC_{50} value of 13.2 nM.

58   **59**   **60**   **61**

In 2002, Bristol Myers Squibb recognized novel fluorooxindole **62** (BMS-204352, MaxiPost) as a potent maxi-K channel opener along with KCNQ4 potassium channel activator [67]. Further SAR was revealed by synthesizing and evaluating the analogues of compound **62** for K^+ channel opening potential. Unsubstituted phenyl analogs (**63, 64**), moving the triflouromethyl group from 6-position to 5-position (**65**) and also replacing with electron withdrawing groups (**66,67**) resulted in loss of activity. The BMS-204352 MaxiPost also showed excellent brain penetration with good solid state stability, which makes it a good candidate for further development of neuroprotective agents.

In 2010, Ali et al. explored a series of oxindoles as AChE inhibitors for the treatment of AD disease [61]. Among the synthesized oxindole derivatives, compound **68** showed potent inhibitory activity against acetyl cholinesterase enzyme with IC_{50} 0.10 μmol/L equipotent to donepezil. The same research group in 2012 synthesized a series of pyrrolothiazolyl oxindole and evaluated it for acetyl cholinesterase inhibitory activity [68]. Among the synthesized analogues, compound **69** was the most potent inhibitor against acetyl cholinesterase enzyme with IC_{50} 0.11 µmol/L. Later in 2013, Kia et al. discovered mono- and bis-spirooxindole-hexahydropyrrolizine analogues as novel and potent acetyl cholinesterase inhibitors [69]. The mono-spirooxindole hexahydropyrrolizine analogues were found to be more potent as compared to dipolarophiles and bisspiropyrrolizines. Among mono-spirooxindole-hexahydropyrrolizine, **70** showed highest AChE inhibitory activity with an IC_{50} value of 3.36 µM and five times selectivity over BuChE. Interestingly, Watanabe and his coworkers explored novel radioiodinated oxindole derivatives for imaging neurofibrillary tangles (NFTs), the diagnosis of Alzheimer's disease [70]. Among the series, 3-OI (**71**), a structural isomer of 2-OI, showed the highest binding affinity toward the tau aggregates and it also stained the NFTs in brain sections of AD patient.

A number of compounds with oxindole nucleus have been synthesized and evaluated against various kinases including cRaf1, Syk, TAK1, KDR, FGFR1, and AMPK. Lackey et al., in 2000, selected oxindole nucleus on the basis of simplistic synthetic routes and diversity of substitution pattern. The synthesized oxindole derivatives were evaluated against cRaf1 kinase [71]. The SAR revealed that donor/acceptor motif, double bond at 3-position, and

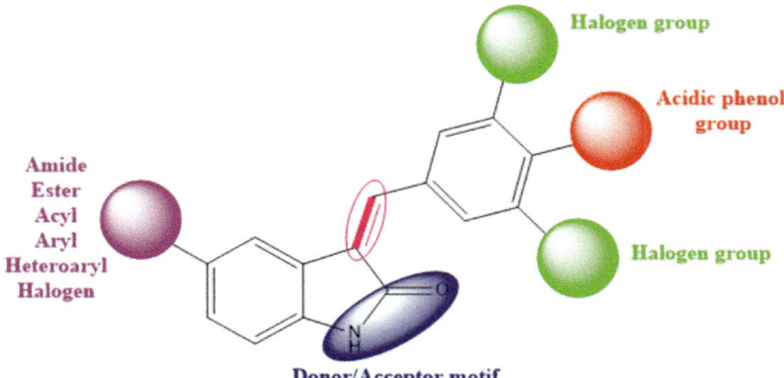

FIGURE 6.3   Structure–activity relationship for the oxindole-based cRaf1 kinase inhibitors.

acidic phenol flanked by two halogen groups are essential pharmacophoric features for cRaf1 inhibitory activity. The diverse range of groups can be substituted at 5-position (i.e., halogen, esters, amides, aryl, heteroaryl, and acyl groups) to give potent and selective cRaf1 inhibitors. Compound **72** showed the highest potency among all the synthesized compounds with an IC_{50} value of 0.009 µM (Fig. 6.3). Lai et al. identified the oxindole derivatives by exploring high throughput screening (HTS) and evaluated against Syk [15]. The oxindole derivative **73** with sulphonamide group substituted at 5-position of the oxindole nucleus and indole ring at 3-position showed the highest IC_{50} value, 5 nM.

Aventis Pharmaceuticals further identified oxindole hit exploring the homogeneous time resolved fluorescence assay as JAK3 inhibitors. Among the designed JAK3 inhibitors, compound **74** substituted with pyridyl group at 5-position showed 27 nM activity and showed efficacy comparable to dexamethasone in the ear odema model. Oxindole derivatives were also obtained as hits through against TAK1 by Lockman et al. [72]. Among the synthesized compounds, compound **75** showed the highest IC_{50} value of 8.9 nM, and the SAR depicted that hydroxyl group substituted at the para-position of the phenyl ring is important for TAK1 inhibitory activity owing to its hydrogen bond donating ability. A research group in 2012 identified oxindole-based hit (**76**) with moderate inhibitory activity against KDR, FGFR1, and PDGFRα that was involved in the promotion of angiogenesis required for the growth of tumors [73]. The hit (**76**) obtained through HTS was further modified by substituting ethyl piperidine at the 5-position, which leads to compound (**77**) with good potency in both biochemical and cellular processes. Further modification leads to compound **78** with desirable pharmacokinetic and toxicity profiles along with good efficacy in multiple xenograft tumor models. Later in 2013, oxindole nucleus emerged as potential therapeutics for diabetes and other metabolic disorders via activating AMP-activated protein kinase [65].

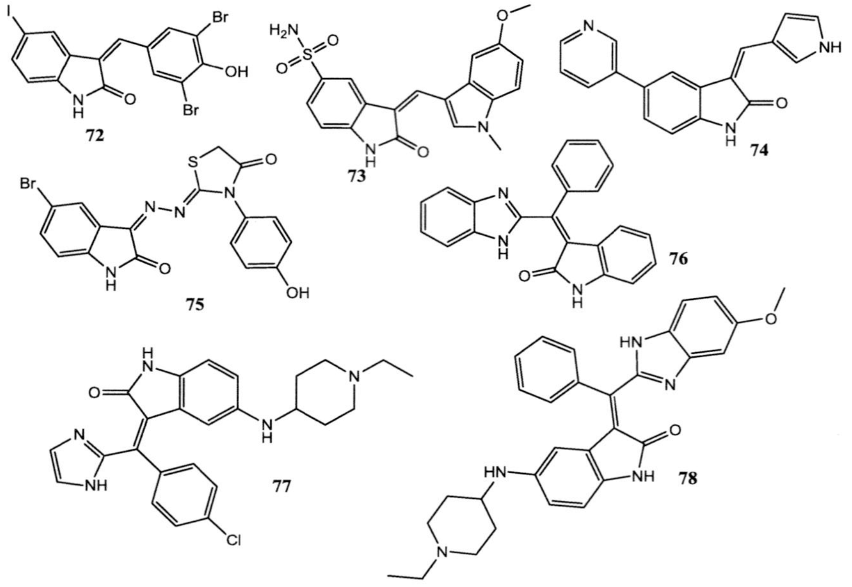

Oxindole derivatives are the most recent synthetic class among newer antibacterial agents against multiple resistant strains of bacteria. Rinde et al. reported a series of oxindole-based 3(Z)-{4-[4-(arylsulfonyl)piperazin-1-ylbenzylidene)-1,3-dihydro-2H-indol-2-one derivatives that were screened in vitro against different species of microorganisms like *Staphylococcus aureus, Streptococcus pyrogenes, Escherichia coli*, and *Pseudomonas aeruginosa*. Most of the compounds exhibited significant antibacterial activity. Among all, compound **79** showed significant antibacterial activity against all the tested strains of microbes with 25 μg/mL minimum inhibitory concentration (MIC) [74]. Karki et al. synthesized and evaluated a series of 1,3,5-trisubstituted-2-oxindole derivatives for preliminary in vitro antituberculosis activity against *Mycobacterium tuberculosis* H37Rv strain. The compound **80** was capable of exhibiting antitubercular activity against *M. tuberculosis* H37Rv at 25 μg/mL [75]. The spiro-oxindole system constitutes the basic nucleus of many pharmacological agents and natural alkaloids reported as antimicrobial agents. In 2013, Singh et al. synthesized and evaluated novel spiro-oxindole derivatives for their antimicrobial activities against two bacterial strains, *E. coli* and *S. aureus*, and were found to possess significant antimicrobial activity [76]. The synthesized spiro-oxindole **81** (MIC = 500 μg), showed excellent antibacterial activity against *E. coli* and *S. aureus* equivalent to standard drug streptomycin. The spiro framework with bromo-substituent and sulfur moiety was important for antibacterial activity [76]. Already reported antimicrobial potential of spiro-oxindoles,

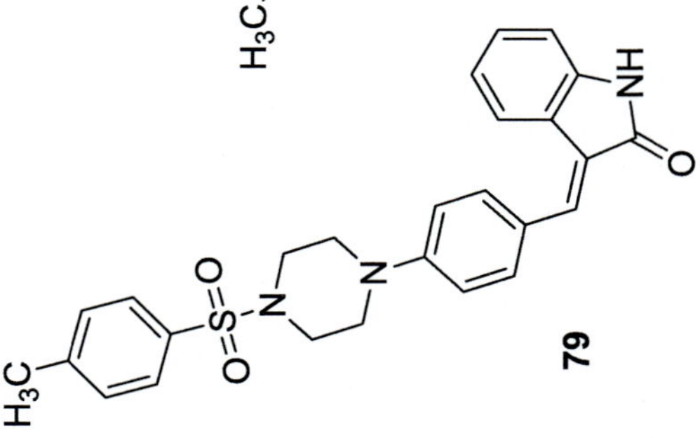

82

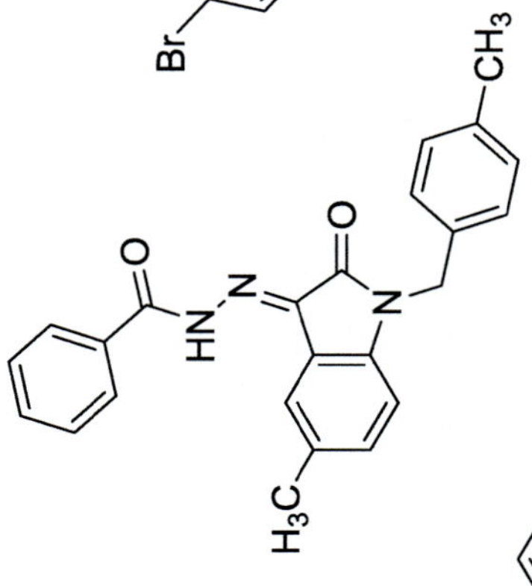

81

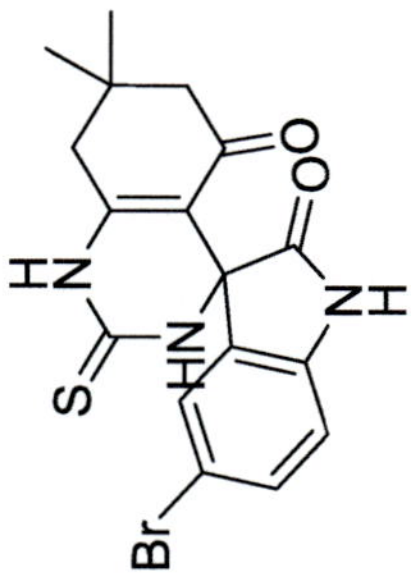

80

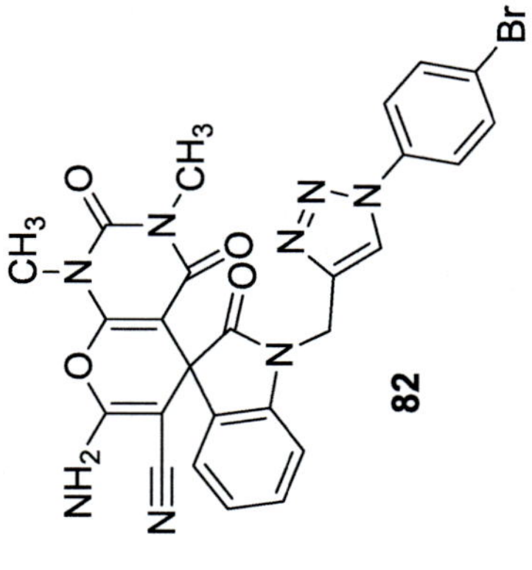

79

fused pyrans derivatives and 1,2,3-triazoles encouraged the authors to combine the spirocyclicoxindole, 2-amino-4H-pyranand 1,2,3-triazoles in a single matrix through a multicomponent reactions approach and studied their biological activity and mode of action against pathogenic microorganisms [77]. It revealed that all compounds showed good Gram-positive antibacterial activity against *S. aureus* (MIC value = 32–256 µg/mL) and *Bacillus subtilis* (MIC value = 16–128 µg/mL), though none of the compounds exhibit activity against Gram-negative (*E. coli* and *P. aeruginosa*) bacteria. Among all novel compounds, **82** was found to be most active against Gram-positive bacteria with an MIC value of 32 and 16 µg/mL against *S. aureus* and *B. subtilis*, respectively [17].

A novel series of oxindole-based compounds with antiproliferative potential against *Leishmania infantum* was reported by Scala et al. [78]. Among all the compounds, **83** and **84** showed good antileishmanial activity (16.2 µg/mL and 6.1 µg/mL) with quite high therapeutic index (4.4 and 10.2).

In 2007, Stevens et al. identified a compound by replacing benzimidazolone nucleus of a potent β3 adrenoreceptor agonist (**85**; EC_{50} 7.1 nM) with oxindole nucleus (**86**; EC_{50} 24.4 nM) that showed slightly decreased potency [16]. It was then optimized with different substitutions at 3-position of oxindole ring. The substitution of methyl group (**87**: EC_{50} 12 nM, **88**; EC_{50} 5.2 nM) at 3-position of oxindole lead to an increase in potency whereas bigger substituents like benzyl (**89**; EC_{50} 268.1 nM) at this position lead to decreased potency.

In 2008, Lawrence and coworkers identified oxindole **90** (NSC-117199) in the screening of NCI database that showed moderate Shp2 inhibitory activity with an IC_{50} value of 47 µM [79]. The docking analysis revealed that the introduction of a carboxyl group at both terminals to incorporate both polar and phosphate mimic groups leads to an inhibitor with 40-fold increase in inhibitory activity and desirable selectivity over Shp1 and PTP1B (**91**).

Oxindole nucleus has also been explored as a selective and potent NMDA blocker. Although ifenprodil is the first drug of a new class of NMDA blockers, it lacks selectivity and showed poor bioavailability [80]. Chenard et al. attempted to improve the selectivity by replacing the phenolic hydroxyl with chlorine, leading to compound **92** SL 82.0715, which showed little improvement

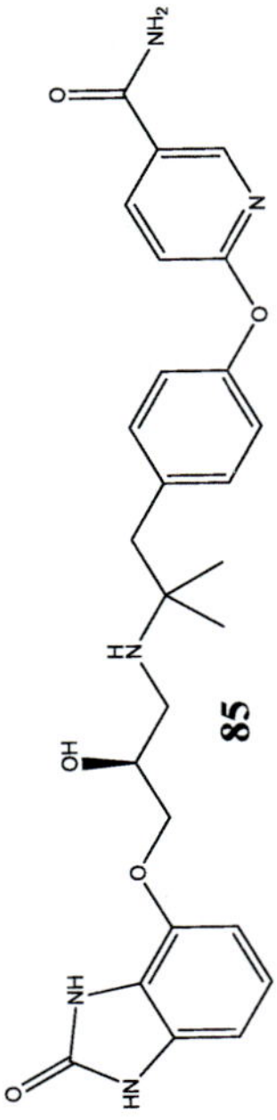

85

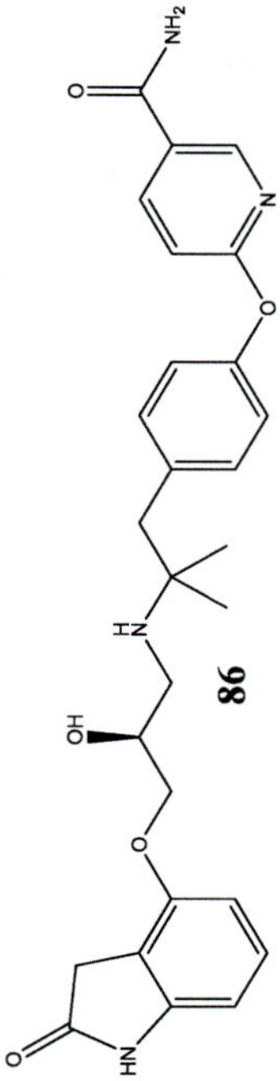

86

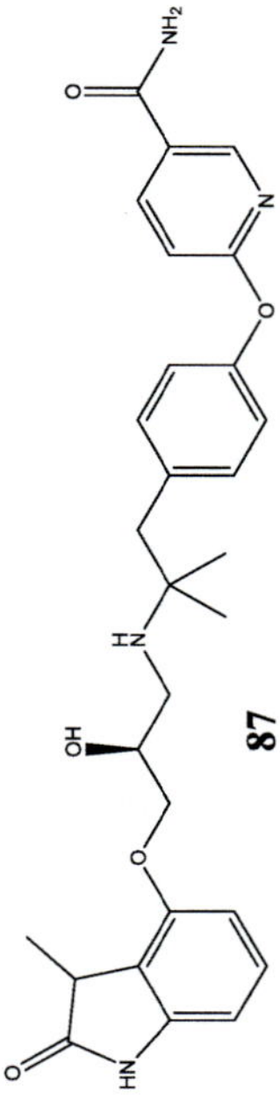

87

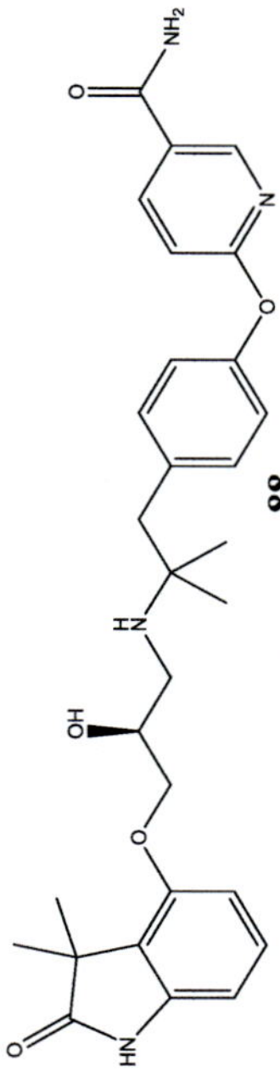

88

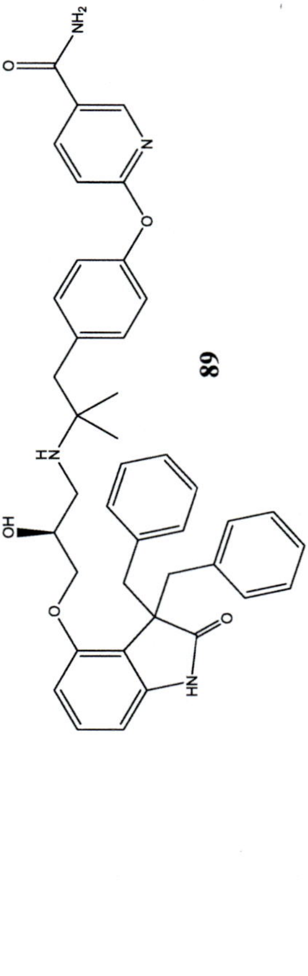

89

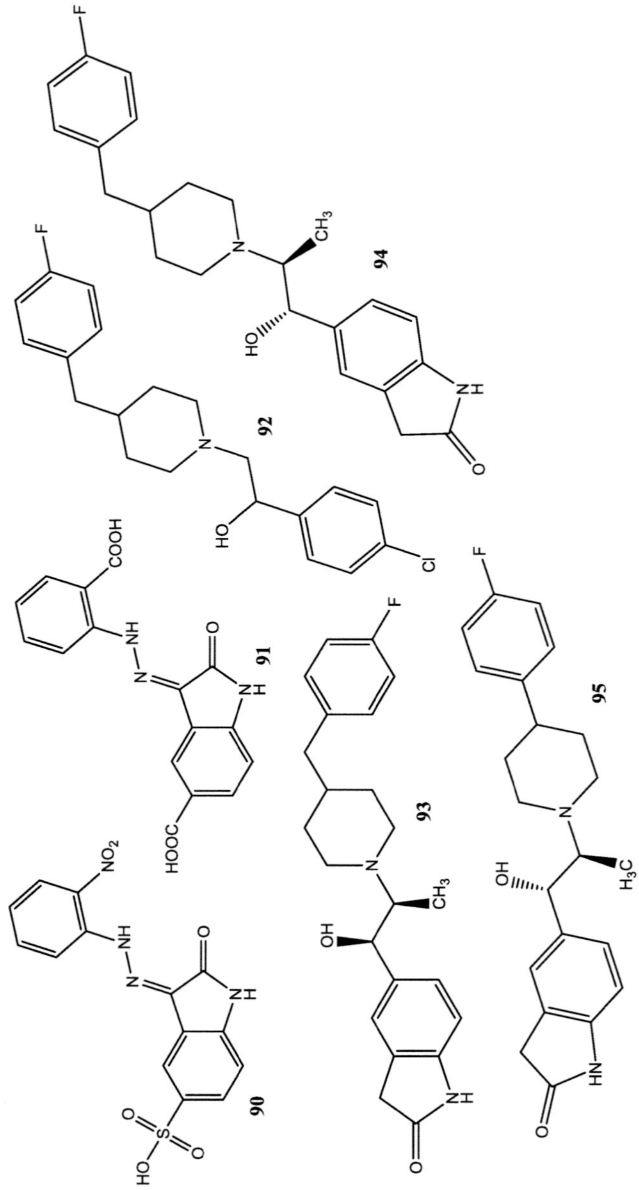

94
92
91
93
95
90

in selectivity but significant reduction in NMDA antagonist potency [81]. The phenolic hydroxyl group was further replaced with oxindole moiety, resulting in potent NMDA antagonists.

The derivative with oxindole in place of hydroxyl-substituted phenyl group retaining erythro stereochemistry (**93**) results in similar biological activity, whereas threo relative stereochemistry (**94**) yields improved NMDA antagonist along with improved selectivity. The antagonist activity was further increased by removing the methylene linker connecting the piperidine and phenyl ring, which also results in increased selectivity (**95**) [70].

Oxindole nucleus has also been explored for dual function of contraception and STD protection for women. Paira et al. synthesized 3,3-diheteroaromatic oxindole analogues and evaluated their in vitro potential as spermicidal agents [19]. Compound **96** displayed maximum efficacy with MEC 0.34 mg/mL, the maximum effective concentration, which is much more effective compared to standard spermicide N-9. It also showed a sufficiently large window between the spermicidal MEC and hemolytic index, indicating that it can be used as a spermicide without possibility of general membrane disruption.

SSR149415 oxindole derivative (**97**) has been identified as a potent, selective, orally active vasopressin V 1b receptor antagonist that showed antianxiety and antidepressant effects [82]. Despite its good efficacy, it was found to have a poor pharmacokinetic profile due to extensive first pass metabolism. To improve its pharmacokinetic profile, the proline *N,N*-dimethyl amide group was masked by oxazole moiety and northern phenyl ring with benzylic amine moiety. As a result of these modifications, compound **98** was obtained with equipotent antidepressant activity compared to SSR149415 along with improved metabolic stability, oral bioavailability, half-life, and brain penetration compared to SSR149415.

In 2011, Chowdhury et al. discovered XEN907, a spirooxindole (**99**; IC_{50} 0.003 μM), as a Nav1.7 channel blocker for the treatment of chronic pain [16]. Through high throughput screening, 3-hydroxyoxindole (**100**; IC_{50} 0.30 μM) analogue was identified as a promising candidate for further optimization. The *p*-chlorophenyl substituent at N-1 position in compound **100** was replaced by an n-pentyl group that lead to compound **101** (IC_{50} 0.03 μM), with a 10-fold increase in activity. Additionally, scaffold rigidification (i.e., introduction of spiroether at 3-position) leads to the identification of XEN907 (**99**), which showed further 10-fold increase in the activity compared to compound **101**.

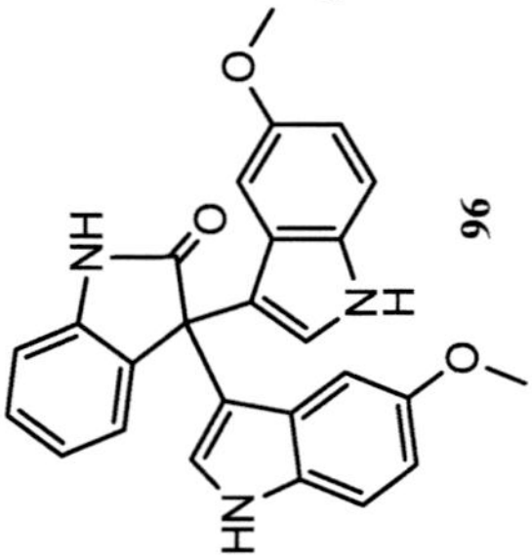

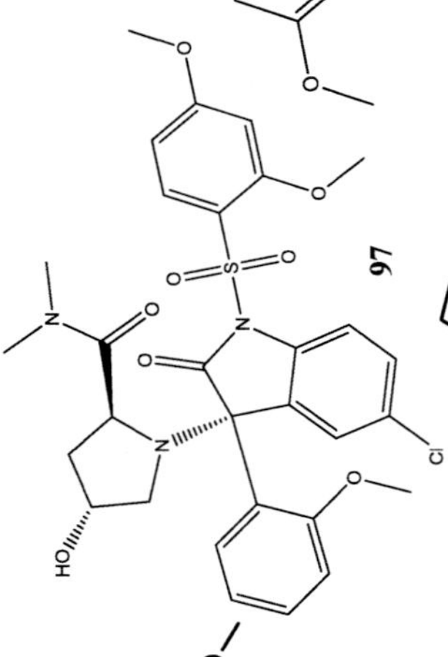

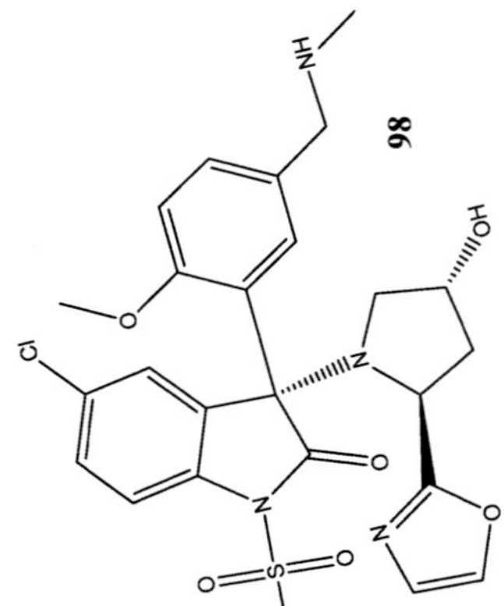

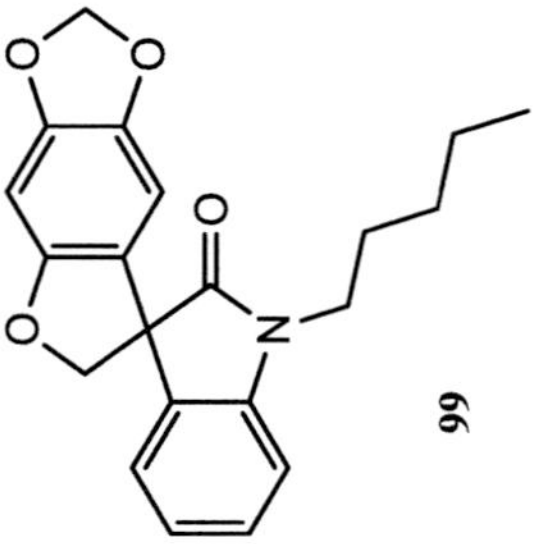

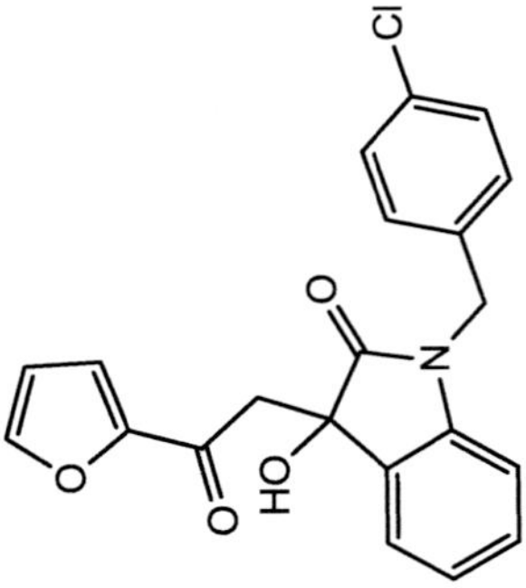

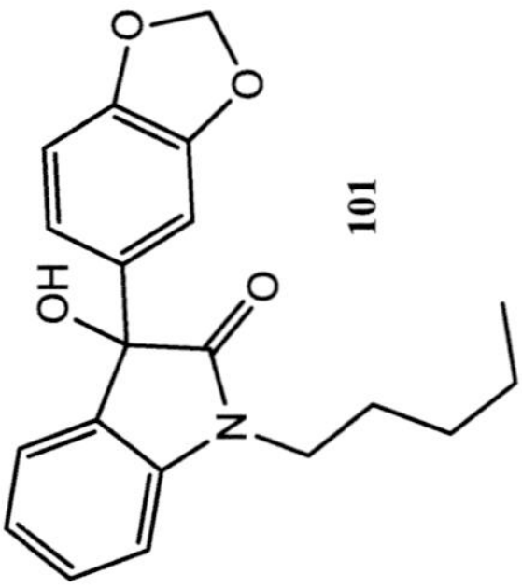

6. ROLE OF OXINDOLE AS MULTITARGETING AGENTS IN MULTIFACTORIAL DISEASES

In the light of the previous reports, oxindole nucleus has emerged as a successful lead that has been optimized to yield potential therapeutics possessing diverse pharmacological profiles. Apart from a single targeting agent with diverse biological activities, it has been implicated as a multitargeting agent as suggested by various literature reports. This section provides insight into the multitargeting potential of oxindole.

6.1 Multiple Kinase Inhibitors

Protein phosphorylation is the key step in almost all cellular processes. By transferring the phosphate group from adenosine triphosphate to substrate proteins, kinases express the activity, maintain overall function, and trigger the cellular signaling processes [71]. Anomalous expression of various kinases leads to disease condition. Thus, their implication in pathogenesis of a number of diseases and their highly druggable nature made them an important target class in drug development [71]. A number of orally active kinase inhibitors have been clinically approved and some kinase inhibitors are in later phases of clinical trials.

Sunitinib (**13**), an oxindole derivative, was approved by the FDA in January 2006 simultaneously for two different indications: stromal tumors and renal carcinoma. It is marketed as Sutent by Pfizer [83]. On a molecular level, it has been characterized as multiple tyrosine kinase inhibitor exhibiting potency against vascular endothelial growth factor receptor (VEGFR), platelet-derived growth factor receptor (PDGFR), KIT, FLT3, and colony stimulating factor receptor (CSF-1R). In June 2009, another oxindole analogue, Toceranib (**102**) (Palladia), developed by Pfizer, was approved by the FDA for canine cutaneous mast cell tumors (skin-based cancer) [72]. Earlier, it was approved as a dog-specific anticancer drug but later was approved for human therapies also. Palladia is a small molecule inhibitor displaying antiangiogenic and antitumor activity via inhibiting multiple tyrosine kinases including VEGFR, PDGFR, and KIT. Another oxindole derivative SU-5416 (**14**) also showed inhibitory activity against VEGFR-1, VEGFR-2, VEGFR-3, PDGFRα, PDGFRβ, Flt-3, and CSF-1R along with minimal Kit inhibitory activity [47,48]. It has been reported to be effective in cancer cell lines including breast, colon, glioblastoma, lung, melanoma, and prostate. Using a similar pattern, Laird and coworkers established the kinase profile of SU-6668 (**15**), another oxindole-based anticancer agent. It showed a similar kinase inhibitory profile (i.e., VEGFR-1, VEGFR-2, VEGFR-3, PDGFRα, PDGFRβ, Flt-3, and CSF-1R) but was more potent than SU-5416. Another oxindole-based multikinase inhibitor has been explored by Patel and coworkers (i.e., SU-10944 (**16**)), which inhibited VEGFR-1 with an IC_{50} value of 96 nM but showed lower inhibitory activity against PDGFRβ, Kit, or FGF-R1 (Fig. 6.4) [8].

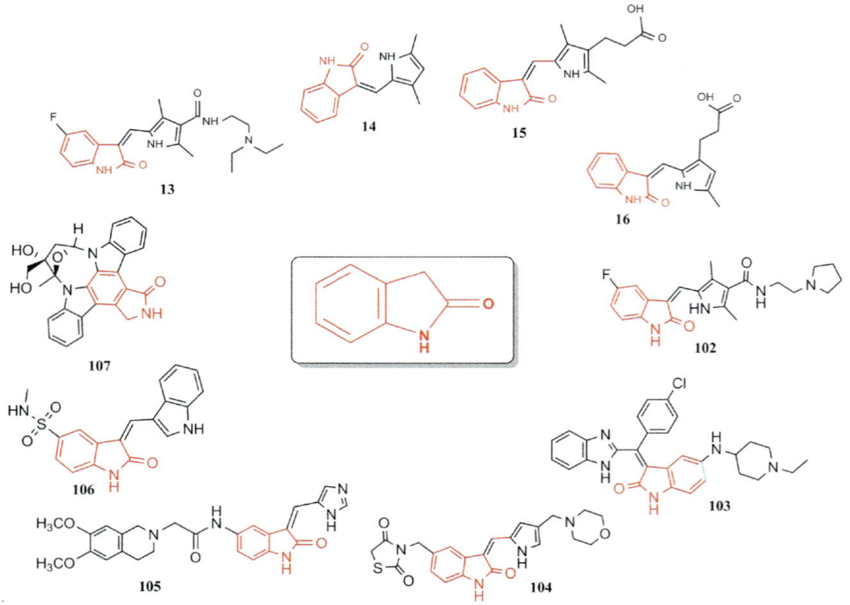

FIGURE 6.4 Various oxindole derivatives developed as multikinase inhibitors.

Singh et al. have identified oxindole nucleus as dual Cyclin-dependent kinase 1 (CDK1) and Cyclin-dependent kinase 2 (CDK2) inhibitors via a 3D-QSAR CoMFA study [73]. In another study, Kim et al. developed a potent, small molecule inhibitor of multiple angiogenic RTKs, optimizing a lead molecule with oxindole identified by screening an in-house library for inhibitors of KDR. Organized modification around the core led to the identification of compound **103** with improved biochemical activity along with favorable ADME/PK properties [84]. The compound displayed significant inhibitory activities against KDR, Flt-1, PDGFR, and FGFR1. Further it also demonstrated low nM inhibitory activity against several Src family tyrosine kinases (Src and Yes), and several cell cycle and mitogen-activated protein kinases (MAP4K3 and the Rsk family members).

S49076 (**104**) is another oxindole-based orally available multikinase inhibitor that has entered the first phase of clinical trials [85]. It has been identified as a potent inhibitor of MET, AXL/MER, and FGFR1/2/3. In cell models, S49076 inhibited the proliferation of MET- and FGFR2-dependent gastric cancer cells, blocked MET-driven migration of lung carcinoma cells, and inhibited colony formation of hepatocarcinoma cells expressing FGFR1/2 and AXL.

In 2015, Sestito et al. designed a series of 2-oxindole-based multitargeted inhibitors of PDK1/Akt signaling pathway for the treatment of glioblastoma multiforme [86]. PKD1/Akt pathway has been implicated in cell signaling, proliferation, and differentiation. Any abnormality in the PDK1/Akt signaling pathway may lead to many types of cancer. Thus it can be considered a significant target in the treatment of various cancers. Among the synthesized compounds, **105** was found to be a potential candidate in the treatment of chemoresistant glioblastoma as it inhibited PDK1 kinase as well as some downstream effectors such as CHK1, GS3Ka, and GS3Kb.

Another study revealed the dual Syk and JAK3 inhibitory potential of oxindole derivatives [87]. The dual inhibitors were designed by employing both structure-based and ligand-based drug designing. Among the designed and synthesized derivatives, compound **106** was found to be most potent against both the Syk (IC_{50} 0.36 μM) and JAK3 (IC_{50} 1.67 μM) kinases. Further, it has proved its potential in the treatment of complex autoimmune disorders like rheumatoid arthritis. It has shown better activity as compared to the standard drug, indomethacin.

CEP 701 (**106**) is another oxindole derivative that is a clinical trial candidate for numerous complex disorders like myelofibrosis, essential thrombocythemia, neuroblastoma, acute myeloid leukemia, psoriasis, and prostate cancer; it acts via inhibition of JAK2, FLT3, and TrKB kinases [88–93].

7. CONCLUSION

The bicyclic oxindole nucleus with one carbonyl (—C=O) and one amino (NH) group has emerged as valuable scaffold in medicinal chemistry and drug design. The lactam ring in the core scaffold imparts diverse pharmacological activities along with the substitution of different groups at N-1, C2, C3, and phenyl ring. C3 of the lactam ring and 5-position of the phenyl ring are highly explored positions to yield pharmacologically active oxindole derivatives. Apart from oxindole analogues, spirocyclic derivatives of oxindole have made an important contribution in therapeutic journey of oxindole derivatives with anticancer, anti-HIV, antibacterial, progesterone antagonistic, acetylcholinesterase inhibitory, antileishmanial, and analgesic activity. The proficient optimization of oxindole derivatives with multiple biological activities has led to potential multitargeted agents for the treatment of multifactorial diseases. The success of oxindole as a multitargeted agent has been evidenced by FDA approval of Sunitinib and Palladia. Both derivatives have exhibited multikinase inhibitory activity and have shown their potential in various multifactorial cancers. Despite synthetic and medicinal importance of oxindole nucleus, its multitargeted potential still needs to be explored for variety of pathological conditions as a clinical candidate. This chapter will definitely help researchers further explore the multitargeted potential of oxindole-based derivatives in the treatment of multifactorial complex diseases.

REFERENCES

[1] W.C. Sumpter, Chemistry Review 37 (1945) 443–479.

[2] G.M. Ziarani, P. Gholamzadeh, N. Lashgari, P. Hajiabbasi, Arkivoc 1 (2013) 470–535.

[3] S.R.S. Rudrangi, V.K. Bontha, V.R. Manda, S. Bethi, Asian Journal of Research in Chemistry 4 (2011) 335–338.

[4] M. Kaur, M. Singh, N. Chadha, O. Silakari, European Journal of Medicinal Chemistry 123 (2016) 858–894.

[5] A.A. Dreifuss, A.L. Bastos-Pereira, T.V. Avila, S. Soley Bda, A.J. Rivero, J.L. Aguilar, Acco, Journal of Ethnopharmacology 130 (2010) 127–133.

[6] D.G. Giménez, E.G. Prado, T.S. Rodríguez, A.F. Arche, R. De la Puerta, Planta Medica 76 (2010) 133–136.

[7] H. Guan, A.D. Laird, R.A. Blake, C. Tang, C. Liang, Bioorganic and Medicinal Chemistry Letters 14 (2004) 187–190.

[8] R. Roskoski, Biochemical and Biophysical Research Communications 356 (2007) 323–328.

[9] https://www.clinicaltrials.gov.

[10] R.F. Kauffman, D.W. Robertson, R.B. Franklin, G.E. Sandusky, F. Dies, J.L. McNay, J.S. Hayes, Cardiovascular Drug Reviews 8 (1990) 303–322.

[11] T. Bethke, D. Brunkhorst, H.V. der Leyen, W. Meyer, R. Nigbur, H. Scholz, Naunyn Schmiedebergs Arch, Pharmacology 337 (1988) 576–582.

[12] S.R. Yong, A.T. Ung, S.G. Pyne, B.W. Skelton, A.H. White, Tetrahedron Letters 63 (2007) 5579–5586.

[13] D. Yasuda, K. Takahashi, T. Ohe, S. Nakamura, T. Mashino, Bioorganic and Medicinal Chemistry 21 (2013) 7709–7714.

[14] H. Watanabe, M. Ono, H. Kimura, K. Matsumura, M. Yoshimura, Y. Okamoto, M. Ihara, R. Takahashi, H. Saji, Bioorganic and Medicinal Chemistry Letters 22 (2012) 5700–5703.

[15] J.Y. Lai, P.J. Cox, R. Patel, S. Sadiq, D.J. Aldous, S. Thurairatnam, K. Smith, D. Wheeler, S. Jagpal, S. Parveen, G. Fenton, T.K. Harrison, C. McCarthy, P. Bamborough, Bioorganic and Medicinal Chemistry Letters 13 (2003) 3111–3114.

[16] F.C. Stevens, W.E. Bloomquist, A.G. Borel, M.L. Cohen, C.A. Droste, M.L. Heiman, A. Kriauciunas, D.J. Sall, F.C. Tinsley, C.D. Jesudason, Bioorganic and Medicinal Chemistry Letters 17 (2007) 6270–6273.

[17] H. Singh, J. Sindhu, J.M. Khurana, C. Sharma, K.R. Aneja, European Journal of Medicinal Chemistry 77 (2014) 145–154.

[18] S. Ingrand, L. Barrier, C. Lafay-Chebassier, B. Fauconneau, G. Page, J. Hugon, FEBS Letters 581 (2007) 4473–4478.

[19] P. Paira, A. Hazra, S. Kumar, R. Paira, K.B. Sahu, S. Naskar, P. Saha, S. Mondal, A. Maity, S. Banerjee, N.B. Mondal, Bioorganic and Medicinal Chemistry Letters 19 (2009) 4786–4789.

[20] S. Chowdhury, M. Chafeev, S. Liu, J. Sun, V. Raina, R. Chui, W. Young, R. Kwan, J. Fu, J.A. Cadieux, Bioorganic and Medicinal Chemistry Letters 21 (2011) 3676–3681.

[21] P.G. Gassman, T.J. Van Bergen, Journal of the American Chemical Society 95 (1973) 2718–2719.

[22] S.W. Wright, L.D. McClure, D.L. Hageman, Tetrahedron Letters 37 (1996) 4631–4634.

[23] J.F. Wolfe, M.C. Sleevi, R.R. Goehring, Journal of the American Chemical Society 102 (1980) 3646–3647.

[24] L.J. Athelstan, M.D. John, Journal of the Chemical Society 9 (1995) 977–978.

[25] V. Arumugam, A. Routledge, C. Abell, S. Balasubramanian, Tetrahedron Letters 38 (1997) 6473–6476.

[26] K.H. Shaughnessy, B.C. Hamann, J.F. Hartwig, Journal of Organic Chemistry 63 (1998) 6546–6553.

[27] E.J. Hennessy, S.L. Buchwald, Journal of the American Chemical Society 125 (2003) 12084–12085.

[28] S. Kamijo, Y. Sasaki, C. Kanazawa, T. Schüßeler, Y. Yamamoto, Angewandte Chemie International Edition in English 44 (2005) 7718–7721.

[29] A. Pinto, L. Neuville, P. Retailleau, J. Zhu, Organic Letters 8 (2006) 4927–4930.

[30] Y. Yasui, H. Kamisaki, Y. Takemoto, Organic Letters 10 (2008) 3303–3306.

[31] Y.X. Jia, E.P. Kündig, Angewandte Chemie International Edition in English 48 (2009) 1636–1639.

[32] T. Miura, T. Toyoshima, Y. Takahashi, M. Murakami, Organic Letters 11 (2009) 2141–2143.

[33] K.I. Booker-Milburn, M. Fedouloff, S.J. Paknoham, J.B. Strachan, J.L. Melville, M.A. Voyle, Tetrahedron Letters 41 (2000) 4657–4661.

[34] I.D. Hills, G.C. Fu, Angewandte Chemie International Edition in English 42 (2003) 3921–3924.

[35] E.C. Linton, M.C. Kozlowski, Journal of the American Chemical Society 130 (2008) 16162–16163.

[36] N. Duguet, A.M. Slawin, A.D. Smith, Organic Letters 11 (2009) 3858–3861.

[37] Y.L. Liu, B.L. Wang, J.J. Cao, L. Chen, Y.X. Zhang, C. Wang, J. Zhou, Journal of the American Chemical Society 132 (2008) 15176–15178.

[38] L. Ren, X.L. Lian, L.Z. Gong, Chemistry 19 (2013) 3315–3318.

[39] H.L. Wei, T. Piou, J. Dufour, L. Neuville, J. Zhu, Organic Letters 13 (2011) 2244–2247.

[40] X. Ju, Y. Liang, P. Jia, W. Li, W. Yu, Organic and Biomolecular Chemistry 10 (2012) 498–501.

[41] A. Cane, M.C. Tournaire, D. Barritault, M. Crumeyrolle-Arias, Biochemical and Biophysical Research Communications 276 (2000) 379–384.

[42] H.N. Bramson, J. Corona, S.T. Davis, S.H. Dickerson, M. Edelstein, S.V. Frye, R.T. Gampe, P.A. Harris, A. Hassell, W.D. Holmes, R.N. Hunter, K.E. Lackey, B. Lovejoy, M.J. Luzzio, V. Montana, W.J. Rocque, D. Rusnak, L. Shewchuk, J.M. Veal, D.H. Walker, L.F. Kuyper, Journal of Medicinal Chemistry 44 (2001) 4339–4358.

[43] A. Dermatakis, K.C. Luk, W. DePinto, Bioorganic and Medicinal Chemistry 11 (2003) 1873–1881.

[44] K.C. Luk, M.E. Simcox, A. Schutt, K. Rowan, T. Thompson, Y. Chen, U. Kammlott, W. DePinto, P. Dunten, A. Dermatakis, Bioorganic and Medicinal Chemistry Letters 14 (2004) 913–917.

[45] E.R. Wood, L. Kuyper, K.G. Petrov, R.N. Hunter, P.A. Harris, K. Lackey, Bioorganic and Medicinal Chemistry Letters 14 (2004) 953–957.

[46] G.D. Zhu, V.B. Gandhi, J. Gong, Y. Luo, X. Liu, Y. Shi, R. Guan, S.R. Magnone, V. Klinghofer, E.F. Johnson, J. Bouska, A. Shoemaker, A. Oleksijew, K. Jarvis, C. Park, R.D. Jong, T. Oltersdorf, Q. Li, S.H. Rosenberg, V.L. Giranda, Bioorganic and Medicinal Chemistry Letters 16 (2006) 3424–3429.

[47] L. Sun, N. Tran, F. Tang, H. App, P. Hirth, G. McMahon, C. Tang, Journal of Medicinal Chemistry 41 (1998) 2588–2603.

[48] P.W. Manley, G. Bold, J. Bruggen, G. Fendrich, P. Furet, J. Mestan, C. Schnell, B. Stolz, T. Meyer, B. Meyhack, W. Stark, A. Strauss, J. Wood, Biochimica et Biophysica Acta 1697 (2004) 17–27.

[49] E. Garcia Prado, M.D. Garcia Gimenez, R. De la Puerta Vazquez, J.L. Espartero Sanchez, M.T. Saenz Rodriguez, Phytomedicine 14 (2007) 280–284.

[50] V.C. da Silveira, J.S. Luz, C.C. Oliveira, I. Graziani, M.R. Ciriolo, A.M. da Costa Ferreira, Journal of Inorganic Biochemistry 102 (2008) 1090–1103.

[51] H. Hong, L.J. Huang, D.W. Teng, Chinese Chemical Letters 22 (2011) 1009–1012.

[52] A. Kamal, G. Ramakrishna, P. Raju, A.S. Rao, A. Viswanath, V.L. Nayak, S. Ramakrishna, European Journal of Medicinal Chemistry 46 (2011) 2427–2435.

[53] C. Guo, M. Pairish, A. Linton, S. Kephart, M. Ornelas, A. Nagata, B. Burke, L. Dong, J. Engebretsen, A.N. Fanjul, Bioorganic and Medicinal Chemistry Letters 22 (2012) 2572–2578.

[54] G. Nesi, S. Sestito, V. Mey, S. Ricciardi, M. Falasca, R. Danesi, A. Lapucci, M.C. Breschi, S. Fogli, S. Rapposelli, ACS Medicinal Chemistry Letters 4 (2013) 1137–1141.

[55] C.J. Ribeiro, J.D. Amaral, C.M. Rodrigues, R. Moreira, M.M. Santos, Bioorganic and Medicinal Chemistry 22 (2014) 577–584.

[56] Y.A. Ivanenkov, S.V. Vasilevski, E.K. Beloglazkina, M.E. Kukushkin, A.E. Machulkin, M.S. Veselov, N.V. Chufarova, E.S. Chernyaginab, A.S. Vanzcool, N.V. Zyk, D.A. Skvortsov, A.A. Khutornenko, A.L. Rusanov, A.G. Tonevitsky, O.A. Dontsova, A.G. Majouga, Bioorganic and Medicinal Chemistry Letters 25 (2015) 404–409.

[57] A. Monteiro, L.M. Goncalves, M.M. Santos, European Journal of Medicinal Chemistry 79 (2014) 266–272.

[58] P. Sai Prathima, P. Rajesh, J. Venkateswara Rao, U. Sai Kailash, B. Sridhar, M. Mohan Rao, European Journal of Medicinal Chemistry 84 (2014) 155–159.

[59] B. Yu, X.J. Shi, P.P. Qi, D.Q. Yu, H.M. Liu, The Journal of Steroid Biochemistry and Molecular Biology 141 (2014) 121–134.

[60] H. Howard, R. Sarges, T. Siegel, T. Beyer, European Journal of Medicinal Chemistry 27 (1992) 779–789.

[61] M.A. Ali, R. Ismail, T.S. Choon, Y.K. Yoon, A.C. Wei, S. Pandian, R.S. Kumar, H. Osman, E. Manogaran, Bioorganic and Medicinal Chemistry Letters 2 (2010) 7064–7066.

[62] M. Khan, M. Yousaf, A. Wadood, M. Junaid, M. Ashraf, U. Alam, M. Ali, M. Arshad, Z. Hussain, K.M. Khan, Bioorganic and Medicinal Chemistry 22 (2014) 3441–3448.

[63] T. Jiang, K.L. Kuhen, K. Wolff, H. Yin, K. Bieza, J. Caldwell, B. Bursulaya, T.Y. Wu, Y. He, Bioorganic and Medicinal Chemistry Letters 16 (2006) 2105–2108.

[64] T. Jiang, K.L. Kuhen, K. Wolff, H. Yin, K. Bieza, J. Caldwell, B. Bursulaya, T. Tuntland, K. Zhang, D. Karanewsky, Y. He, Bioorganic and Medicinal Chemistry Letters 16 (2006) 2109–2112.

[65] L.F. Yu, Y.Y. Li, M.B. Su, M. Zhang, W. Zhang, L.N. Zhang, T. Pang, R.T. Zhang, B. Liu, J.Y. Li, ACS Medicinal Chemistry Letters 14 (2013) 475–480.

[66] A. Fensome, R. Bender, J. Cohen, M.A. Collins, V.A. Mackner, L.L. Miller, J.W. Ullrich, R. Winneker, J. Wrobel, P. Zhang, Z. Zhang, Y. Zhu, Bioorganic and Medicinal Chemistry Letters 12 (2002) 3487–3490.

[67] P. Hewawasam, V.K. Gribkoff, Y. Pendri, S.I. Dworetzky, N.A. Meanwell, E. Martinez, C.G. Boissard, Bioorganic and Medicinal Chemistry Letters 12 (2002) 1023–1026.

[68] M.A. Ali, R. Ismail, T.S. Choon, R.S. Kumar, H. Osman, N. Arumugam, A.I. Almansour, K. Elumalai, A. Singh, Bioorganic and Medicinal Chemistry Letters 22 (2012) 508–511.

[69] Y. Kia, H. Osman, R.S. Kumar, V. Murugaiyah, A. Basiri, S. Perumal, H.A. Wahab, C.S. Bing, Bioorganic and Medicinal Chemistry 21 (2013) 1696–1707.

[70] B. Chenard, T. Butler, I. Shalaby, M. Prochniak, B. Koe, C. Fox, Bioorganic and Medicinal Chemistry Letters 3 (1993) 91–94.

[71] P. Cohen, Nature Reviews Drug Discovery 1 (2002) 309–315.

[72] C.A. London, P.B. Malpas, S.L. Wood-Follis, J.F. Boucher, A.W. Rusk, M.P. Rosenberg, C.J. Henry, K.L. Mitchener, M.K. Klein, J.G. Hintermeister, Clinical Cancer Research 15 (2009) 3856–3865.

[73] S.S. Kumar, N. Dessalew, P.V. Bharatam, Medicinal Chemistry 3 (2007) 75–84.

[74] S. Rindhe, B. Karale, R. Gupta, M. Rode, Indian Journal of Pharmaceutical Sciences 73 (2011) 292–296.

[75] S.S. Karki, R. Hazare, S. Kumar, A. Saxena, A. Katiyar, Turkish Journal of Pharmaceutical Sciences 8 (2011) 169–178.

[76] S.B. Singh, K. Tiwari, P.K. Verma, M. Srivastava, K.P. Tiwari, J. Singh, Supramolecular Chemistry 25 (2013) 255–262.

[77] M.J. Genin, D.A. Allwine, D.J. Anderson, M.R. Barbachyn, D.E. Emmert, S.A. Garmon, D.R. Graber, K.C. Grega, J.B. Hester, D.K. Hutchinson, Journal of Medicinal Chemistry 43 (2000) 953–970.

[78] A. Scala, M. Cordaro, G. Grassi, A. Piperno, G. Barberi, A. Cascio, F. Risitano, Bioorganic and Medicinal Chemistry 22 (2014) 1063–1069.

[79] H.R. Lawrence, R. Pireddu, L. Chen, Y. Luo, S.S. Sung, A.M. Szymanski, M.R. Yip, W.C. Guida, S.M. Sebti, J. Wu, Journal of Medicinal Chemistry 51 (2008) 4948–4956.

[80] K. Williams, Current Drug Targets 2 (2001) 285–298.

[81] B. Chenard, I. Shalaby, B. Koe, R. Ronau, T. Butler, M. Prochniak, A. Schmidt, C. Fox, Journal of Medicinal Chemistry 34 (1991) 3085–3090.

[82] T. Oost, G. Backfisch, S. Bhowmik, M.M. van Gaalen, H. Geneste, W. Hornberger, W. Lubisch, A. Netz, L. Unger, W. Wernet, Bioorganic and Medicinal Chemistry Letters 21 (2011) 3828–3831.

[83] V.L. Goodman, E.P. Rock, R. Dagher, R.P. Ramchandani, S. Abraham, J.V. Gobburu, B.P. Booth, S.L. Verbois, D.E. Morse, C.Y. Liang, Clinical Cancer Research 13 (2007) 1367–1373.

[84] M.H. Kim, A.L. Tsuhako, E.W. Co, D.T. Aftab, F. Bentzien, J. Chen, W. Cheng, S. Engst, L. Goon, R.R. Klein, D.T. Le, Bioorganic and Medicinal Chemistry Letters 22 (2012) 4979–4985.

[85] M.F. Burbridge, C.J. Bossard, C. Saunier, I. Fejes, A. Bruno, S. Léonce, G. Ferry, G. Da, G. Violante, F. Bouzom, V. Cattan, A. Jacquet-Bescond, Molecular Cancer Therapeutics 12 (2013) 1749–1762.

[86] S. Sestito, G. Nesi, S. Daniele, A. Martelli, M. Digiacomo, A. Borghini, D. Pietra, V. Calderone, A. Lapucci, M. Falasca, P. Parrella, European Journal of Medicinal Chemistry 105 (2015) 274–288.

[87] M. Kaur, M. Singh, O. Silakari, Future Medicinal Chemistry 9 (2017) 1193–1211.

[88] https://clinicaltrials.gov/ct2/show/NCT00668421?term=LESTAURTINIB&rank=1.

[89] https://clinicaltrials.gov/ct2/show/NCT00079482?term=LESTAURTINIB&rank=4.

[90] https://clinicaltrials.gov/ct2/show/NCT00494585?term=LESTAURTINIB&rank=5.

[91] https://clinicaltrials.gov/ct2/show/NCT00084422?term=LESTAURTINIB&rank=6.

[92] https://clinicaltrials.gov/ct2/show/NCT00236119?term=LESTAURTINIB&rank=7.

[93] https://clinicaltrials.gov/ct2/show/NCT00081601?term=LESTAURTINIB&rank=12.

Chapter 7

Thiazine: A Versatile Heterocyclic Scaffold for Multifactorial Diseases

Shalki Choudhary, Om Silakari
Punjabi University, Patiala, India

Chapter Outline

1. INTRODUCTION

Compounds with heterocyclic nuclei remain the topic of discussion when it comes to therapy and have received more attention in the drug development process [1]. In the context of heterocyclic compounds, thiazine and its substituted derivatives have been the talking point due to their varied therapeutic profile [2].

Thiazines are organic compounds with molecular formula C_4H_5NS and molar mass 99.15 g/mol containing a heterocyclic ring with nitrogen and sulfur atoms.

Key Heterocycle Cores for Designing Multitargeting Molecules. https://doi.org/10.1016/B978-0-08-102083-8.00007-8

Thiazines may be 1,2-thiazine (**1**), 1,3-thiazine (**2**), and 1,4-thiazines (**3**) [3]. The reduced form of thiazine is thiomorpholine (**4**), which is the thio analog of morpholine with molecular formula C_4H_9NS [4]. Benzfused derivatives of thiazine, benzothiazine (**5**) with molecular formula C_8H_7NS and phenothiazine (**6**), where there is an additional benzene ring, with molecular formula $S(C_6H_4)_2NH$, are also of potential interest due to their dynamic therapeutic properties [5–8].

(1)   (2)   (3)   (4)

(5)   (6)

The uncondensed thiazine ring (1,4-thiazine) was first synthesized and investigated by Barkenbus in 1948. During the course of study, it was found that 1,4-thiazine theoretically may exist in two tautomeric forms [9]. However, the combined structure of 1,4-thiazine with a benzene ring i.e. phenothiazine, reported by Bernthsen in 1883, was found to be part of various thiazine dyes such as methylene blue and Lauth's violet [10].

During past few decades, diversely substituted thiazine and its derivatives embedded with a variety of functional groups came into light and a significant amount of research activity has been directed toward this class.

Thiazines were discovered with various properties such as agonist of cannabinoid receptor and anti-HIV activity [11]. Additionally, thiazines also improve vascularity and reduce excess water from the stomach and therefore used as anabolic agents [12]. Further, literature reports demonstrated their significance as antitubercular, analgesics, antiinflammatory, antimicrobial, antifungal, antiviral, antimalarial, antitumor and antidiabetic agents [13,14].

A number of reports have been published during past few years that explain the chemistry and pharmacological profile of thiazine and its derivatives, but have not yet exhaustively explored for their multitargeting potential. However, thiazine derivatives, phenothiazine and benzothiazine, have several reports that claim their multitargeting ability. Since there is no report in the literature that summarizes the multitargeting calibre of thiazine and its derivatives, it is desirable to outline this topic. This chapter is a brief overview of the chemistry, synthesis and biological profile of thiazine and its derivatives with special emphasis on their multitargeting ability.

1.1 Nomenclature

Thiazine mainly exists in three isomeric forms i.e. 1,2-thiazine (**1**), 1,3-thiazine (**2**) and 1,4-thiazine (**3**) due to varied arrangement of nitrogen and sulfur in the ring and thus have different numbering systems. Further, two tautomeric forms exist for 1,4-thiazine (2H-1,4-thiazine (**7**), 4H-1,4-thiazine (**8**)) and 1,3-thiazine (2H-1,3-thiazine (**9**), 6H-1,3-thiazine (**10**)) [15]. Other derivatives of thiazine, benzothiazine and phenothiazine, also have interesting nomenclature. Benzothiazine has two tautomers namely 2H-benzothiazine (**11**) and 4H-benzothiazine (**12**). On the other hand, nomenclature of phenothiazine suggests that it exist only as 10H-1,4-phenothiazine (**13**), which was initially named as thiodiphenylamine due to its synthesis from diphenylamine and sulfur. It was also given the names dibenzoparathiazine and 2,3,5,6-dibenzo-1,4-thiazine. However, in modern chemistry, its preferred name is phenothiazine [16].

(7)   (8)   (9)   (10)

(11)   (12)   (13)

Houston et al. has briefly discussed nomenclature and different problems involved in the numbering of phenothiazine ring system [17]. It is evident from literature that, the 10th position of phenothiazine ring is most explored and it is easy to synthesize 10-substituted derivatives of phenothiazine [18]. The numbering system of thiazines is depicted in Fig. 7.1.

1.2 History

The historical background of thiazines began in 1948, when Charles Barkenbus and Phillip S. Landis for the first time synthesized 1,4-thiazine by reducing thioglycolimide [9]. Chemistry of thiazines made its inception from the synthetic dye industry, when some aromatic amines had undergone various organic reactions [19].

On the other side, a benzfused derivative of thiazine, phenothiazine, was invented in the 19th century [20]. In 1883, Bernthsen synthesized and reported the first novel series of phenothiazine and thus is known as the father of phenothiazine [10].

Before the discovery of phenothiazine, Lauth (1876) carried out a reaction of *p*-phenylenediamine with sulfur, which on further treatment with ferric chloride

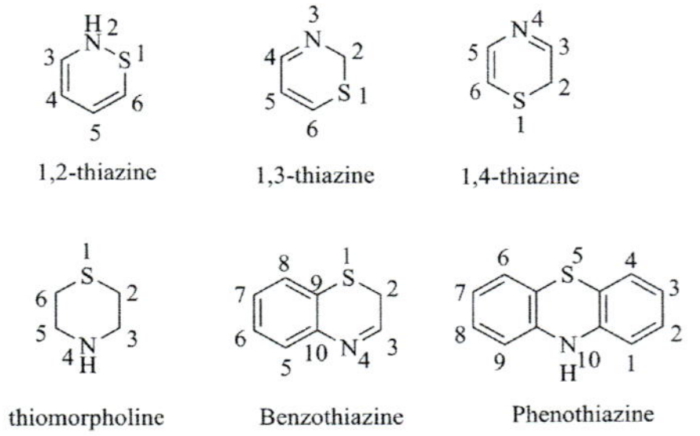

1,2-thiazine   1,3-thiazine   1,4-thiazine

thiomorpholine   Benzothiazine   Phenothiazine

FIGURE 7.1 Numbering system of thiazines.

produced a purple-colored dye called Lauth's violet [21]. Later, the same reaction was repeated by Caro using *p*-aminodimethylaniline as a reactant, which gave blue colored dye, which was later named methylene blue with phenothiazine nucleus in its structure [22,23]. This report suggest that some phenothiazine derivatives like Lauth's violet (**14**) and methylene blue (**15**) were commercially available and used as dyes even before the actual discovery of parent nucleus phenothiazine.

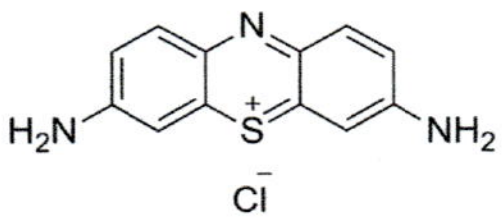

(14) 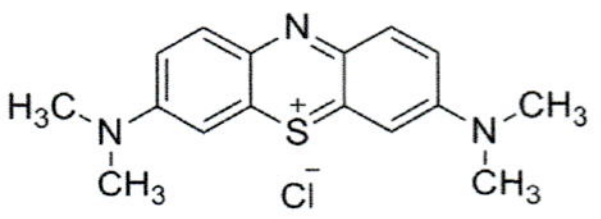   (15)

2. CHEMISTRY

Thiazine is a class of heterocyclic compounds with two hetero atoms; sulfur and nitrogen at different positions of six-membered ring, thus exist in various isomeric forms. Simple thiazines have a six membered-ring with sulfur and nitrogen at 1,2; 1,3 and 1,4 positions. Compounds having adjacent sulfur and nitrogen atoms, i.e. 1,2-thiazines, are extremely rare. 1,3-Thiazine is also not known itself, but cephalosporin C (**16**), a useful antibiotic, has 1,3-thiazine nucleus as an active core [24].

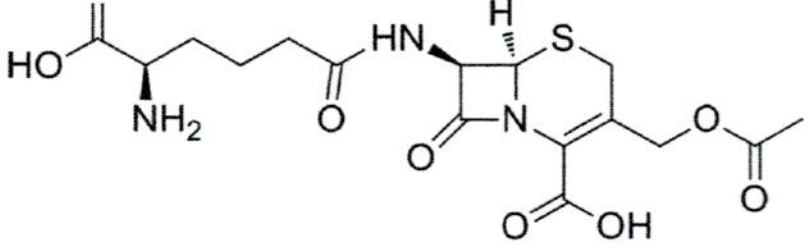

(16)

On the other hand, in phenothiazines and benzothiazines, the nitrogen and sulfur are present at 1,4 position of a six-membered ring with an additional benzene ring in the case of phenothiazine and two rings in benzothiazine. Some general physicochemical properties of different thiazines are summarized in Table 7.1.

TABLE 7.1 Physicochemical properties of thiazines

S. No	Properties	Thiazine	Thiomorpholine	Benzothiazine	Phenothiazine
1	Molecular formula	C_4H_5NS	C_4H_9NS	C_8H_7NS	$S(C_6H_4)_2NH$
2	Molar mass	99.15 g/mol	103.18 g/mol	149.21 g/mol	199.27 g/mol
3	Density	1.1 g/cm^3	1.01 g/cm^3	1.2 g/cm^3	1.22 g/cm^3
4	Appearance	Yellow crystals	Clear liquid	–	Light yellow crystalline solid
5	Melting point	183–186°C	51.2°C	60°C	185°C
6	Boiling point	235°C	170°C	247°C	326°C

Among all forms of thiazines, only 1,4-thiazine has been explored extensively for its chemistry. Since no sufficient chemical data is available on 1,2 and 1,3-thiazines, this section mainly focuses on the chemistry of 1,4-thiazine and its derivatives with little insights on 1,2 and 1,3-thiazine.

2.1 Chemistry of 1,4-Thiazines

1,4-Thiazine is a colorless liquid with a boiling point of 76.5–77°C, ammoniacal odor, and is completely miscible with water. These are weak base with K value 4×10^{-9} [9]. Many compounds of 1,4-thiazine are known and most of them are derivatives of phenothiazine. Reactivity of 1,4-thiazine is also an issue of special concern as it either reacts by losing the parent nucleus or retaining it. 1,4-Thiazine reacts in different ways including oxidative coupling, oxidative dimerization, nucleophilic substitution, acidic hydrolysis etc.

2.1.1 Tautomerism

As mentioned earlier, 1,4-thiazine principally exists in two tautomeric forms, 2H-thiazine (**7**) and 4H-1,4 thiazine (**8**) [25]. Literature reports suggest that, among all tautomeric forms of 1,4-thiazine, 2H form is most stable and frequently used in the synthesis. However, there is no spectral evidence in favor of this statement. Moreover, the stability of the 2H form of 1,4-thiazine was justified by conjugative stabilization depicted by **17** and **18** where **17** outweighs the structure **18** and sulfur atom does not act as an effective electron sink, which generally stabilizes the electron deficient transition state. The structure of 2H-1,4-thiazine was assigned on the basis of fact that it did not form sulphonamide when undergoing a Hinsberg reaction [9].

(17)   (18)

The chemistry of 4H tautomeric form of 1,4-thiazine is also interesting where substituted 1,4-thiaizne-1-oxide obtained from 4H-1,4-thiazines further exist in two tautomeric forms. On the other side, in the case of substituted 1,4-thiazine-1,1-dioxides, no tautomerism exists due to disturbance in conjugation, which further leads to loss of aromaticity. It was found in a study that two derivatives of 4H 1,4-thiazine are known where substitution at 2- and 3-positions of 1,4-thiazine with imino-, oxo-, thioxo- and methylene groups form tautomers of 4H-1,4-thiazine; however, there are no examples of compounds based upon the 1H-tautomer (Fig. 7.2) [26].

2.1.2 Spectral Analysis of 1,4-Thiazines

It is evident from reported IR spectral data that N-unsubstituted-1,4-thiazines (**19-23**) exist as the 2H-tautomers, and therefore show no absorptions in the

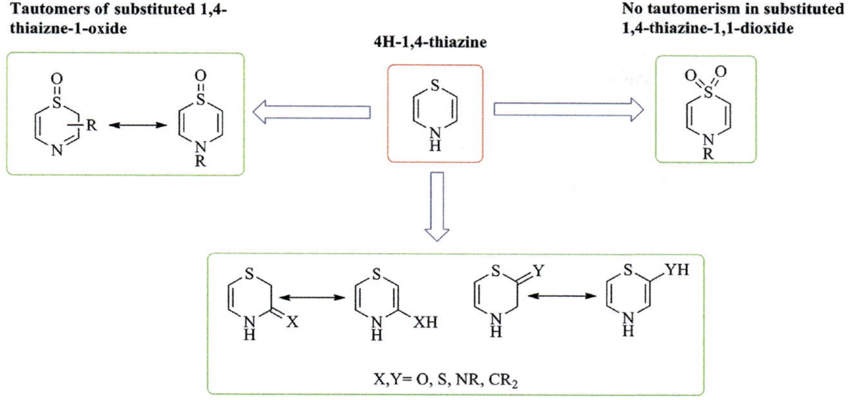

FIGURE 7.2   Different tautomers of substituted 1,4-thiazines.

region characteristic of N-H stretching. Further, stretching vibrations for C=N and C=C are observed in the 1550–1640 cm^{-1} region. Generally two bands of medium and strong intensity appear for these compounds, so compound **20** absorbs at 1573 and 1600 cm^{-1}. The C=O stretching of **21** and **22** can be shown in the 1715–1725 cm^{-1} region [25,27].

UV spectral analysis of 2H-1,4-thiazine can also be discussed by taking some examples. Parent compound 1,4-thiazine is a colorless liquid but most of its derivatives are yellow. It is expected that such compounds display $n + n*$ transitions for the S–C=C–N=C chromophore. The absorptions at 336 and 360 nm of the derivatives **19** and **21** and in the 370–385 nm region of 3,5-diaryl-2H-1,4-thiazines are possibly due to these transitions. Other absorptions at 251 and 300 nm in the case of compounds **19** and **21** and in the 261–274 nm region for 3,5-diaryl-2H-1,4-thiazines are also observed, which are of high intensity (ε 26,900–44,700) due to excitations of aryl groups [25,27].

The nuclear magnetic resonance (NMR) study of 1,4-thiazines suggests that the protons at position-2 of 2-unsubstituted 2H-1,4-thiazines resonate in the δ 3.36–3.40 region. The chemical shift of the 3-proton of a 3-unsubstituted 2H-1,4-thiazine has not been reported, and the only example of a 5-unsubstituted 2H-1,4-thiazine whose NMR spectrum is available is compound **21**; its 5-proton absorbs at δ 8.43. The 6-protons of 3,5-diaryl-2H-1,6-thiazines appear in the δ 6.30–6.53 region, and a slight deshielding of the 6-proton (to δ 6.13) is observed in the case of the bis(thiazine) **23**. 2,6-Unsubstituted 2H-1,4-thiazines display long-range coupling (J = 1.0–1.3 Hz) between the 2- and 6-protons [25].

Apart from spectral analysis, there are two reports displaying crystallographic studies and one report on the basic character of 1,4-thiazines. On the basis of X-ray diffraction data, the center of symmetry is observed in bis (thiazine) **(23)**, which establishes its mesoconfiguration. The crystal structure of the thiazinium chloride **25** has also been described [25].

Literature reports show that the parent compound **24** is the only 1,4-thiazine whose basic strength has been determined. The pKa value of **24** is 5.6, which is comparable to the pKa value of pyridine (pKa, 5.2) [9].

(19)   (20)   (21)

(22)   (23)   (24)

(25)

2.1.3 Dielectric Properties of 1,4-Thiazine

Well-known 1,4-thiazine-containing drugs like promethazine (**26**), chlorpromazine (**27**), trimeprazine (**28**), prochlorperazine (**29**), trifluoperazine (**30**), methotrimeprazine (**31**), and thioproperazine (**32**) exhibit similar dielectric behavior except the last two, which are subjected to severe supercoiling. Dielectric studies revealed polymeric behavior of these drugs in the supercooled state. Chlorpromazine shows a transition at about 32°C due to supercooled liquid microregions in solid cast slab of the material. There is no evidence for a transition in single crystals of chlorpromazine. Melting points, glass transition temperatures, polarizabilities, dipole moments, refractive indices, macroscopic and molecular dielectric relaxation times, and the activation energies associated with them are reported for the compounds [28].

Gutman and Netscheyl [29] revealed that chlorpromazine shows discontinuity in electrical properties at about 32°C due to crystallographic transition. Further, Kronick and Labes [30] investigated a single crystal of chlorpromazine and reported that there is no such discontinuity. They suggested that the effect observed previously was probably attributable to an impurity, as in the case of phthalocyanine. Freymann, when working on cast slabs and powdered samples of chlorpromazine, confirmed the dielectric results of Gutmann and Netschey, but claimed that solid-solid transition in chlorpromazine occurs at a lower temperature (19°C). Thus, the dielectric properties of thiazines and its derivatives were determined as a function of temperature on the basis of these reports.

Liquid samples of thiazine bases when run at low temperature (below 20°C) show abnormal dispersion at 1600 c/s. The permittivity is almost independent of frequency in the range 600–3000 c/s, between 20°C and the melting point, as well as below a critical temperature at which permittivity as well as dielectric loss start to rise steeply [28].

The thiazine bases in the supercooled state show the dielectric behavior like high polymers, such as polyvinylchloride and organic glasses (e.g., iso-butyl chloride), exhibiting well-defined glass transition temperatures [31,32].

(26)

(27)

(28)

(29)

(30)

(31)

(32)

3. SYNTHESIS

Different synthetic schemes have been developed and modified occasionally in order to synthesize different thiazines and their derivatives in high yield. Thiazines can be synthesized using synthetic schemes outlined in Fig. 7.3 (Schemes 1–7 for 1,3-thiazines) and Fig. 7.4 (Schemes 8–13 for 1,4-thiazines).

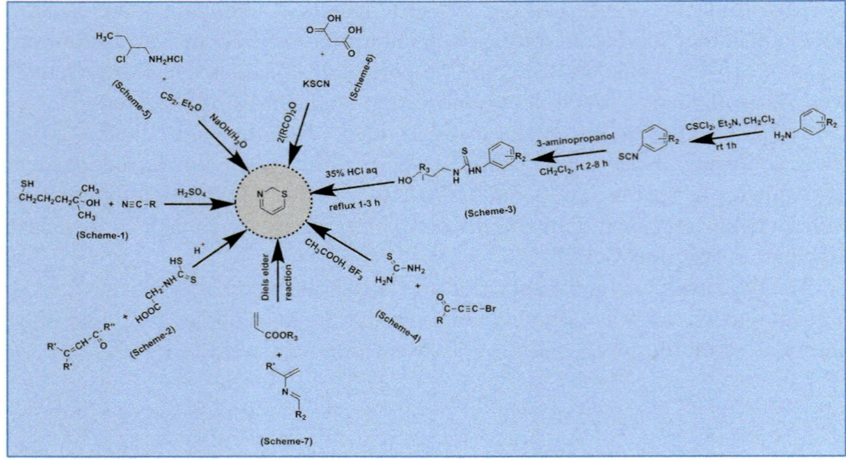

FIGURE 7.3 Synthetic schemes for 1,3-thiazines.

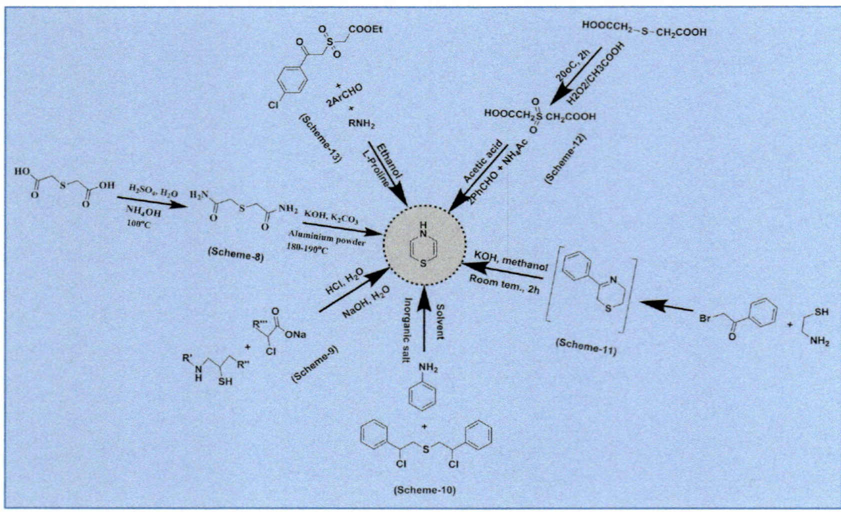

FIGURE 7.4 Synthetic schemes for 1,4-thiazines.

3.1 Synthesis of 1,3-Thiazines

In 1961, Meyers et al. developed a synthetic scheme to synthesize some dihydro-1,3-thiazines by the treatment of different tertiary alcohols with nitriles in concentrated sulfuric acid (Scheme 1) [33]. Further, a new scheme was designed by Shutalev et al. (1991) to afford 3,6-2H-1,3-thiazine-2-thiones by the reaction of α,β-unsaturated aldehydes or ketones with dithiocarbamic acid (Scheme 2) [34].

Dihydro-1,3-thiazine derivatives can also be prepared by the cyclization of phenylthioureas in 35% HCl (Scheme 3) [35,36]. In an attempt, Glotava et al. (2007) reported synthesis of 1,3-thiazine-6-thione hydro bromide by reacting 1-acyl-2-bromo-acetylene with two equivalent thiourea in the presence of BF_3 and glacial acetic acid at 40°C (Scheme 4) [37]. Trofimova et al. (2002) carried out the reaction of carbon disulfide with 2,3-dihalopropylamines in ether, which afforded 5-halo-3,4,5,6-tetrahydro-1,3-thiazine-2-thione (Scheme 5) [38]. Another scheme was developed to obtain 5-acetyl-4-hydroxy-3,6-dihydro-2*H*-1,3-thiazine-2,6-dione where malonic acid was reacted with potassium thiocyanate and acetic anhydride at room temperature with the aid of acetic acid (Scheme 6) [39]. Alternately, 1,4-Michael reaction has also been utilized to afford 1,3-thiazine derivatives using thiourea in alkaline solution [14]. Another addition-type reaction, Diels–Alder reaction, is also suggested for the synthesis of dihydro-1,3-thiazine using *N*-thiocarbonyl imine and acrylate (Scheme 7) [40].

3.2 Synthesis of 1,4-Thiazines

The first attempt to synthesize 1,4-thiazines was made by Charles Barkenbus in 1948, where mother nucleus 1,4-thiazine was synthesized by the reduction of thiodiglycolimide using a three-step reaction (Scheme 8) [9]. Another study suggested that direct addition of sodium salts of α-chlorocarboxylic acids under basic conditions on 1,2 aminopropanethiols affords respective tetrahydro-1,4-thiazine-3-ones as the major products (Scheme 9) [41]. Mihara et al. (2005) treated β,β′-dichloro sulfide with aqueous ammonia on silica gel to afford a complex mixture. On the other hand, it was also reported that the reaction of β,β′-dichloro sulfide with aniline produced 3,4,5-triaryltetrahydro-1,4-thiazine in good yield (Scheme 10) [42]. Literature also reported preparation of 3-phenylthiomorpholine via reaction of 2-aminoethanethiol with 2-bromo-1-phenylethanone in basic condition, followed by hydrogenation of the intermediate imine (Scheme 11) [43]. Later, *N*-hydroxymethylene thiazine derivatives were prepared by Sundari et al. using various *N*-methoxymethylene in dry acetone and dimethyl sulfoxide with anhydrous potassium carbonate as catalyst under reflux (Scheme 12) [44]. Another procedure for the synthesis of 1,4-thiazine derivatives reported in literature involves reaction of ethyl 2-[(2-oxo-2-arylethyl) sulfonyl] acetate, aromatic aldehyde and primary amine/ammonia (1:2:1) using 30 mol% L-proline as catalyst in ethanol, at room temperature (Scheme 13) [45].

4. THIAZINE AS A PRIVILEGED SUBSTRUCTURE

Thiazine can be considered as a key scaffold in various therapeutic agents, having the potential to strike different targets involved in various diseases like inflammation, CNS disorders, cardiovascular diseases, and various skin disorders. There are some well-known drugs available in the market with thiazine

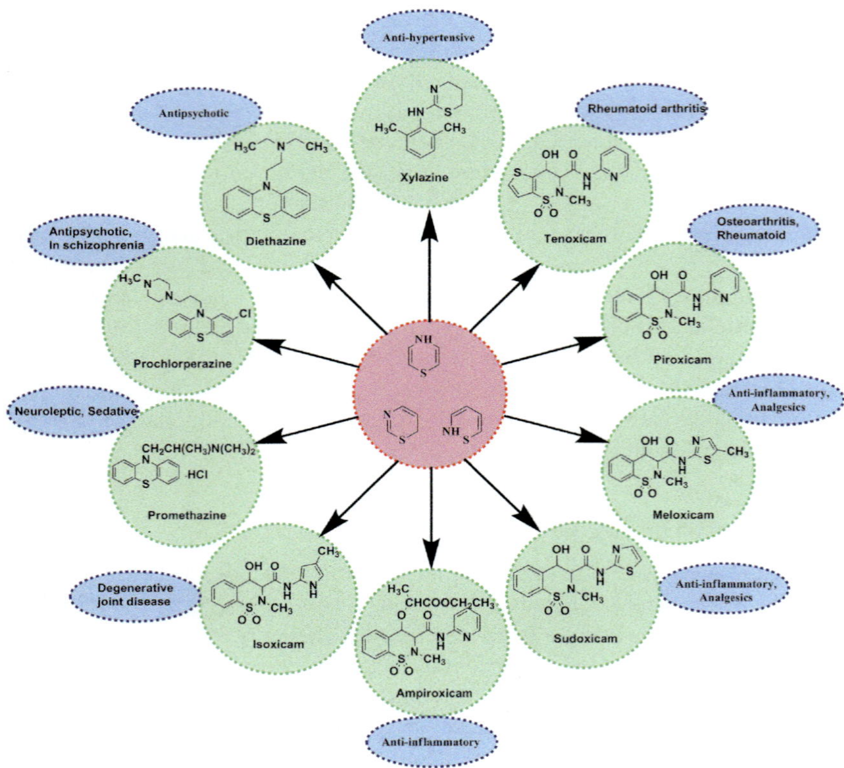

FIGURE 7.5 Thiazine: A resourceful scaffold.

as a key nucleus, displaying a promising therapeutic activity profile (Fig. 7.5). Diversely substituted thiazines have also been explored by the research community for various pharmacological activities. On the basis of their SAR data, thiazine has filtered as a lead molecule in various disease conditions such as microbial infections, fungal infections, viral diseases, CNS disorders, inflammatory diseases, cancer, tuberculosis (TB), malaria, and epilepsy.

4.1 Antimicrobial Agents

Conventional antimicrobial agents are losing their significance due to development of resistance in these drugs against various microbial strains as a result of mutation and other factors [46]. In the current scenario, the research community is attempting to find such new drug candidates in this category that can overcome most of the drawbacks of resistant drugs. Thiazine and its derivatives can be promising antimicrobial candidates that can fulfill this demand. Various derivatives of 1,2, 1,3, and 1,4-thiazines are reported in the literature with antimicrobial potential like antibiotic, antibacterial, antifungal, and antiviral agents.

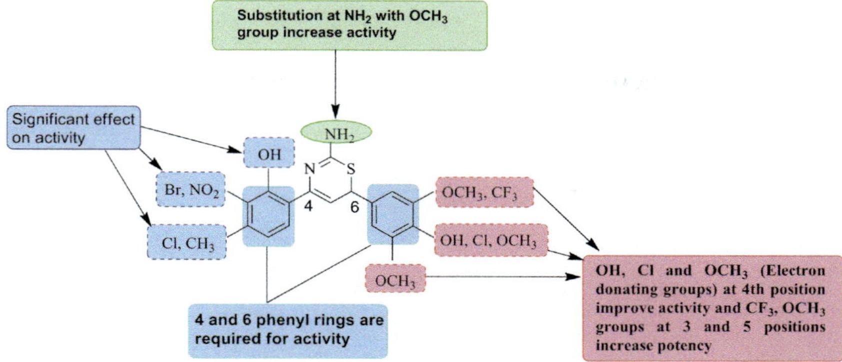

FIGURE 7.6 Structural requirements around 1,3-thiazines for antimicrobial activity.

1,3-thiazine derivative, cephalosporin, is an example of a well-known antibiotic. Antimicrobial agents with improved activity were obtained by exploring different positions of thiazine nucleus.

4.1.1 Antibacterial Agents

A number of 1,2/1,3/1,4-thiazine derivatives have been designed, synthesized and evaluated for antibacterial activities against various strains. 1,3 positional isomer was maximally explored for this purpose. Structural requirements around 1,3-thiazine for antibacterial activity are displayed in Fig. 7.6. SAR study of 1,3-thiazines revealed that a phenyl ring at 4- and 6-positions of 1,3-thiazine nucleus is required for antibacterial activity. Various substitutions at the phenyl ring further improve its activity (Fig. 7.6).

Malinka and colleagues reported a series of pyrido-1,2-thiazine derivatives and their antibacterial potential against strains of *M. fortuitum* and *S. aureus* using isoniazid as a reference drug. However, their study suggested that the majority of molecules were not efficient antibacterial agents. During in vivo evaluation, it was found that compounds **33** and **34** ($IC_{50}=>250\,\mu g/mL$) reduced the growth of microorganisms. It was also revealed that the modification of the α-dicarbonyl substructure of 3-acyl-4-hydroxypyrido-1,2-thiazine does not produce any potent antimicrobial effect [47].

(33)   (34)

Radhod et al. synthesized a few 2-imino-1,3-thiazine derivatives via phenyl thiourea and diphenyl thiourea; among all, compound **35** (MIC=>100 μg/mL) and compound **36** (MIC=>100 μg/mL) were found to have good antibacterial activity. SAR study showed that the presence of a phenyl ring at 4-position of the thiazine ring increased antimicrobial activity [2].

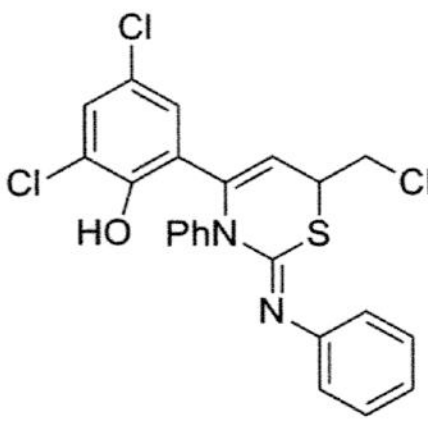

(35) 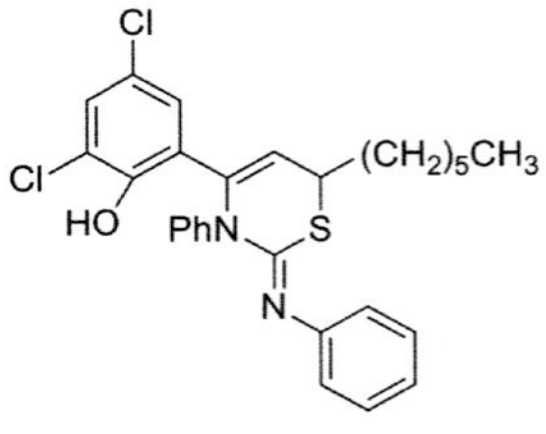　　　　　　　　　　(36)

In 2011, Sawant et al. synthesized 6-[4-substitutedphenyl]-4-phenyl-6H-1,3-thiazine-2-amines and N-[6-(4-substitutedphenyl)-4-phenyl-6H-1,3-thiazine-yl]acetamides to explore the thiazine nucleus according to the topliss approach. All designed compounds were evaluated against both gram-positive (*S. aureus*, *B. subtilus*) and gram-negative (*E. coli*, *P. aeruginosa*) strains to determine their antimicrobial activity. The study revealed that compound **37** (MIC=62.50 μg/mL) with hydroxyl group was a better antimicrobial than compound **38** with methoxy substituent. This observation clearly indicated that electron-donating substituent at 4th position of the phenyl ring and substitution at 6th position of 1,3-thiazine is required to improve antimicrobial activity [48].

A series of novel thiazine 4,6-di phenyl-6-H 1,3 thiazin-2 amine derivatives were prepared by treating a series of chalcones with thiourea. They were further evaluated against two gram-positive (*B. cereus*, *S. aureus*) and two gram-negative (*E. coli* and *P. vulgaris*) strains to determine their antibacterial activity using an agar plate diffusion method with penicillin and streptomycin as standard. The activity was maximum in compound **39** (MIC=25–100 μg/mL) having methoxy substitution at 4th position. Additionally, it was disclosed that chloro substitution at the phenyl ring enhanced the antibacterial potential of the derivatives [49].

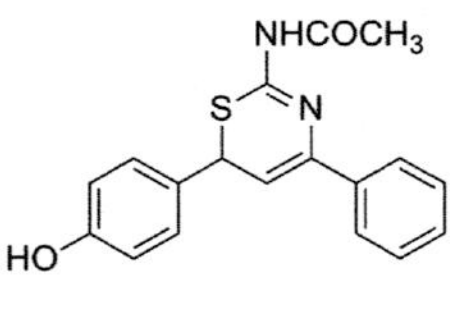

(37) 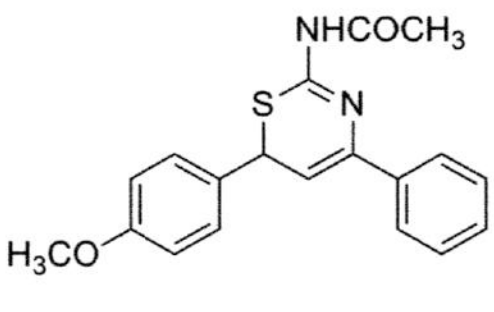　　　　　　　　　　(38) 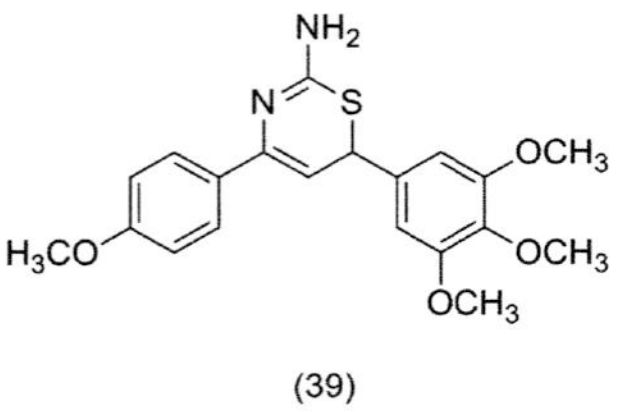　　　　　　　　　　(39)

Later, a series of novel thiazines was reported by Kalirajan et al. by the treatment of various chalcones with thiourea, where all of them were found to have potent antibacterial activity [50]. Similarly, synthesis of some 1,3-thiazine derivatives was reported; these derivatives were evaluated for their antibacterial activity against *E. coli*, *S. aureus*, and *P. aeruginosa*. The results were comparable to that of standard drugs vibromycin and ampicillin. Compounds in which the phenyl ring was substituted with hydroxyl **(40)**, dimethylamine **(41)**, and methoxy **(42)** were found to be most potent with significant zone of inhibition at concentration range 10–100 µg/mL [51].

(40)   (41)   (42)

Another report disclosed synthesis of some 1,3–thiazine derivatives that were tested to determine their antimicrobial activities using gram-positive (*S. aureus* and *B. subtilus*) and gram-negative (*E. coli* and *P. aeruginosa*) bacterial strains. In these compounds it was observed that the antimicrobial activity increases with an increase in the number of hetero atoms, and compound **43** (MIC=25 µg/mL) was found to be the most potent antimicrobial agent [52]. Another series of substituted phenyl thiazines was synthesized by Dipansu et al. and evaluated for their antibacterial activity against various bacterial strains (*E. coli*, *P. aeruginosa*, *K. pneumoniae*) by a paper disc diffusion technique using the standard drug ciprofloxacin. Among all the synthesized compounds, –CF3 substituted compound **44** (MIC=3.12 µg/mL) showed pronounced antibacterial activity [53]. Rathod and coworkers synthesized and screened newly designed 2-imino benzothiazines against different bacterial strains using ampicillin as the standard drug. Among all these compounds, *p*-methoxy phenyl substituted derivative **(45)** was found to possess maximum activity and zone of inhibition at concentration 100 µg/mL. However, none of the synthesized compounds showed better activity than ampicillin [54].

(43)   (44)   (45)

In another series, SAR study of 1,3-thiazine derivatives showed the bromo and nitro substitution at phenyl ring **(46)** produced more potent antimicrobial compounds against *E. coli*, *S. aureus*, *B. subtilis* and *Phaseolusargenosa* with significant zone of inhibition at 0.01 mol concentrations [55]. Some 1,3 thiazines have been recently synthesized and all compounds were screened against *E. coli*, *S. aureus*, *B. subtilis* and *P. aeruginosa*. Compound **47**, obtained from diphenyl thiourea, have shown more promising activity than dinitro diphenyl thiourea derived products [56].

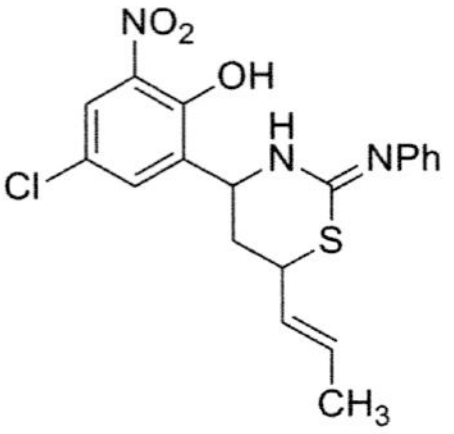

(46) 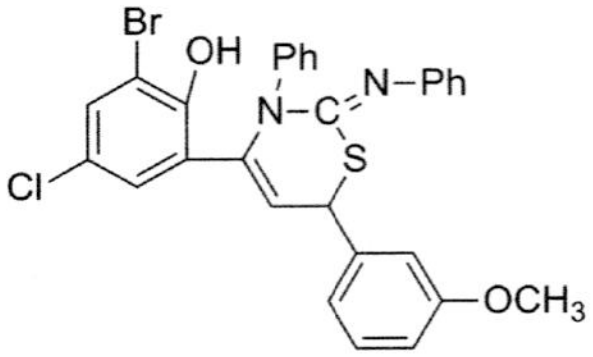   (47)

Later, Sundari and colleagues synthesized a series of 1,4-thiomorpholine-1, 1-dioxide derivatives and evaluated their antibacterial potential against various bacterial strains such as *E. coli*, *K. pneumoniae*, *S. aureus* and *B. subtilis* along with their antifungal potential against *A. niger* and *A. fumigatus*. Out of all the derivatives reported by them, compounds **48** and **49** were found to be the most potent (MIC = 25 µg/mL) antimicrobial molecules [57]. In another study, 7-fluoro-3,4-dihydro-2H-1,4-benzothiazine derivatives were synthesized and screened against gram-positive and gram-negative strains to determine their in vitro inhibitory potential. Compound **50** was found to possess promising activity against *S. aureus*, *B. subtilis* and *C. parapsilosis* [58].

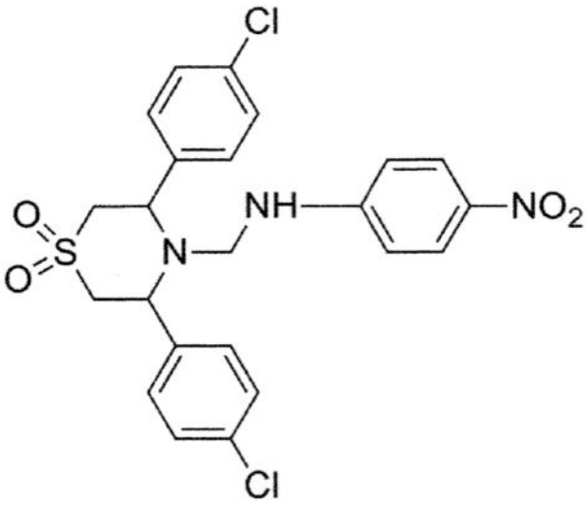

(48) 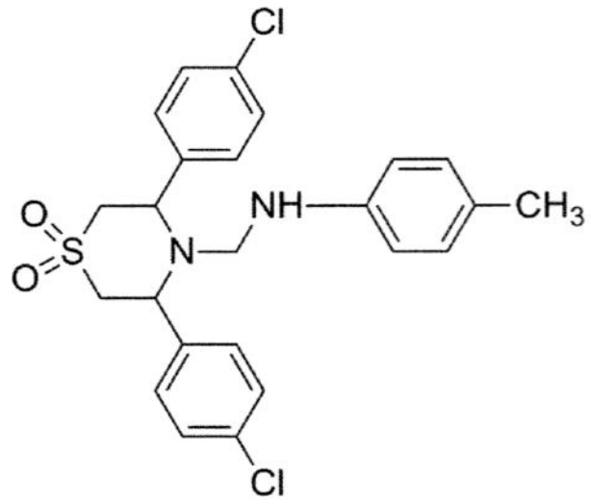   (49) 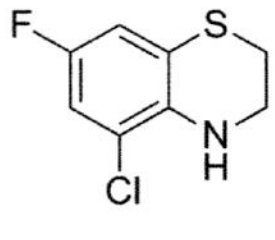   (50)

4.1.2 Antifungal Agents

There have been continuous efforts to design, synthesize and develop more potent and effective antifungal drugs. Current antifungal therapeutics mainly suffer from problems such as toxicity and safety [59–62]. Patients with impaired immunity are susceptible to severe fungal infections. There is a gradual increase in fungal disease and only small numbers of antifungal agents are available. Appearance of resistance to antifungal drugs further limits the effective treatment of fungal infections. Therefore, development of effective and less toxic antifungal drugs without the problem of resistance is a current objective and of interest in the medicinal chemistry community. No specific position of thiazine molecules has been explored for the development of antifungal agents. Antifungal activity of different thiazine derivatives has been studied at times but none of the derivatives has shown promising antifungal activity. A series of novel bis-thiazines in good yield was synthesized by Nagaraj et al. by the reaction of bis-chalcones with thiourea and evaluated for their antifungal potential using fluconazole as reference compound. Compound **51** (MIC = 25 µg/mL) was found to be highly active against *C. albicans* [63]. Later, Kadhim et al. reported novel thiazine derivatives, synthesized by reacting azachalcones with thiourea. The biological activities of these compounds against some fungal species like *C. albicans* and *A. flavus* were evaluated. This study revealed that m-NO$_2$ substituted compound **52** (MIC = 0.005 mol/mL) was the most potent; the remainder showed moderate antifungal activity [64].

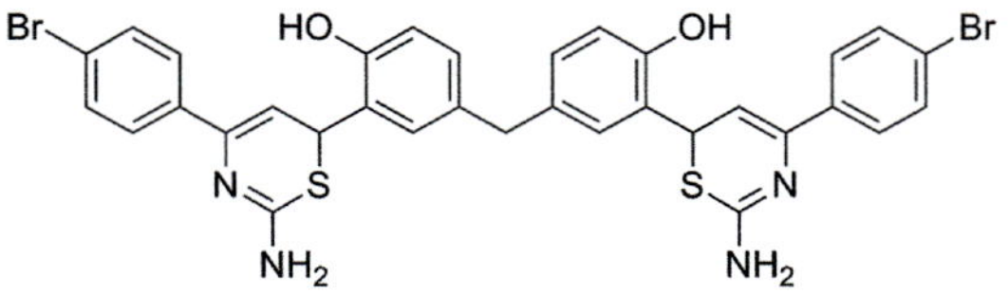

(51)

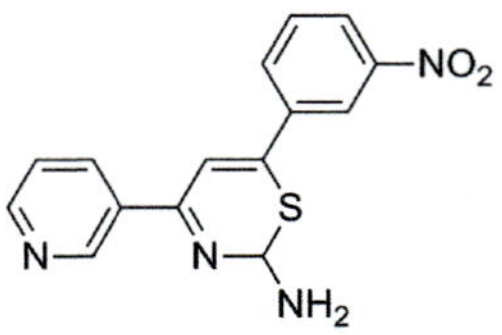

(52)

In 2014, Tony et al. synthesized some novel 1,3-thiazines from five different chalcones and thiobenzamide, analyzed them for antifungal activity by docking them on cytochrome p450 14 alpha sterol demethylase. Docking analysis in cytochrome p450 14 alpha demethylase (PDB ID: 1H5Z) showed that methyl substituted 1,3-thiazines (**53**) possess best antifungal activity [65].

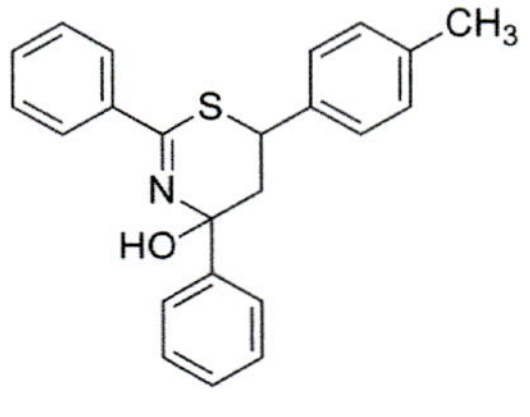

(53)

Sarmiento et al. synthesized some phenothiazine derivatives as antifungal agents and evaluated them against various strains of *Candida* and *Aspergillus*. α-chloro-*N*-acetyl phenothiazine (**54,** MIC = 2–32 µg/mL) was found to have very promising activity and was selected as lead compound in the search of antifungal agents [66].

(54)

Macchiarulo et al. explored benzothiazine nucleus by performing a docking study using a 3D crystal structure (CYP-51) of *C. albicans*. Considering the docking results, in vivo and in vitro antifungal activity were carried out against an experimental model of candidiasis. Although the docking study disclosed that all compounds show similar binding modes with the catalytic site of CA-CYP51, there were few differences in interactions and logP values, which was also reflected in the in vitro analysis of compounds. This study inferred that among all the synthesized compounds, S isomers are better than R isomers. Intramolecular and intermolecular interactions of *p*-chlorophenyl group and benzothiazine moiety with an aromatic ring of Tyr118 impose the binding of the S isomer over the R isomer (Fig. 7.7) [67].

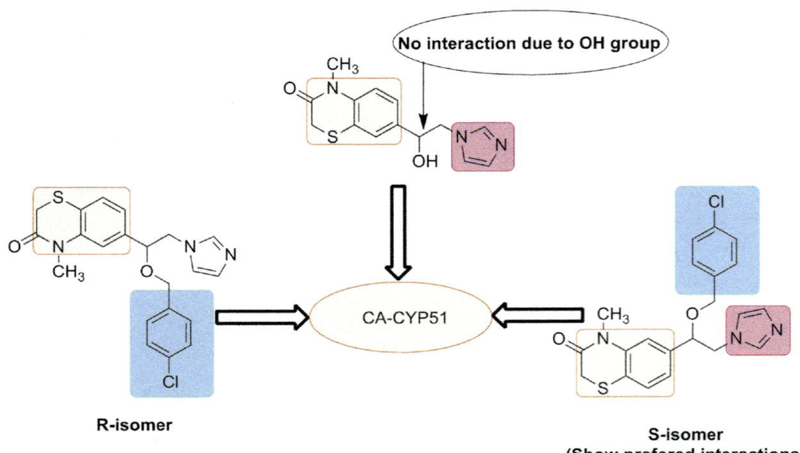

FIGURE 7.7 Docking interactions of R and S isomers of benzothiazine derivatives with CA-CYP51 (fungal protein).

4.1.3 Antiviral Agents

Antiviral drugs are most commonly designed with the purpose of combating various human infecting viruses such as HIV, herpes, hepatitis and influenza [68,69]. To counter severe conditions, antiviral drugs are administered in combination, usually three to four; this regimen is known as highly active anti-retroviral therapy [70]. McFarland et al. claimed in their patent that aryl vinyl derivatives of 5,6-dihydro-4H-1,3-thiazine **(55)** can inhibit viral RNA synthesis and exhibit antiviral activity at a concentration of 1–3 µg/mL, particularly in combating plant viral diseases. These compounds were found to be potent antiviral agents [71]. Later, Mizuhara et al. disclosed an approach to synthesize pyrimido-benzothiazin-6-imine **(56,** $EC_{50}=0.44\pm0.08$**)** and observed that the molecule was effective against different viral strains including HIV, hepatitis C virus and simian immunodeficiency virus. Substitution of 6-6 fused framework with 5-6-6 or 6-6-5 framework reduced the anti-HIV potential of the molecules. However, incorporation of a hydrophobic substituent at 9th or 10th position improved the potency [72].

A=Phenyl, 3-chlorophenyl
3-bromophenyl, 3-methylphenyl

(55)   (56)

Galal et al. synthesized carboxylic acid derivatives of thiazines and performed antiviral activity of the synthesized molecules against herpes simplex virus 1. Two compounds, **(57)** and **(58)** showed significant inhibitory potential and limited virus propagation by 94.7% and 91.3%. Compounds **57** (MIC=20 µg/mL) and **58** (MIC=50 µg/mL) showed higher potency than the standard drug acyclovir [73].

(57)   (58)

4.2 Analgesics and Antiinflammatory Agents

Thiazine and its derivatives have been reported for their various pharmacological actions; one includes reducing inflammatory symptoms in arthritic and postoperative conditions [74].

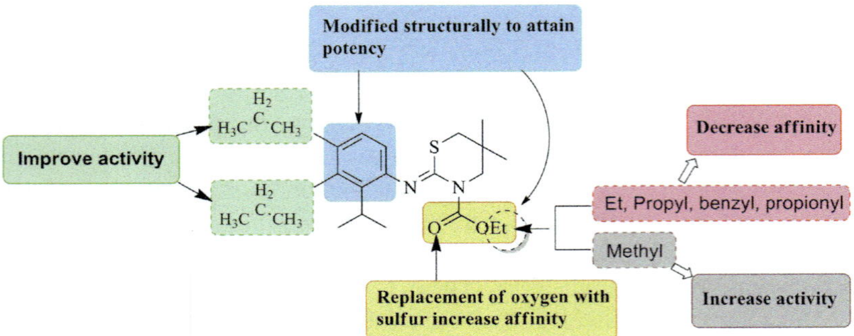

FIGURE 7.8 Structural requirements around 1,3-thiazine for cannabinoid receptor agonist activity.

Literature provided SARs of thiazine derivatives, establishing them as cannabinoid receptor agonists. The ethoxycarbonyl group and phenyl ring of thiazine were structurally modified in order to attain maximum potency. Methylthiocarbonyl group at position-3 of a thiazine nucleus increases affinity while replacement of a methyl group with bulky groups like ethyl, propyl, benzyl, and propionylthiocarbonyl group decreases the same. Isopropyl substitution at 2nd position of a phenyl ring is vital for activity. Substitution of 2nd position of a phenyl ring with electron withdrawing groups such as trifluorocarbon decreases activity. Structural requirements around the thiazine nucleus for cannabinoid receptor agonist activity are shown in Fig. 7.8.

This SAR analysis led to compound **59** having improved affinity for cannabinoid receptor 1 and 2, along with oral bioavailability [75]. A series of chalcones-based 1,3-thiazine derivatives were synthesized by Dabholkar et al. and evaluated for their antiinflammatory, analgesic and ulcerogenic activities. Compound **60** (30% inhibition at 15 mg/kg dose) showed noticeable antiinflammatory potential comparable to that of acetylsalicylic acid [11].

(59) (60)

Imidazolidin-4-one based thiazine derivatives and *N*-phenyl acetamide based thiazine derivatives were synthesized by Jupudi et al. and evaluated for their antiinflammatory activity. It was concluded that synthesized molecules **61** and **62** show dose-dependent activity in comparison to the standard drug diclofenac [76].

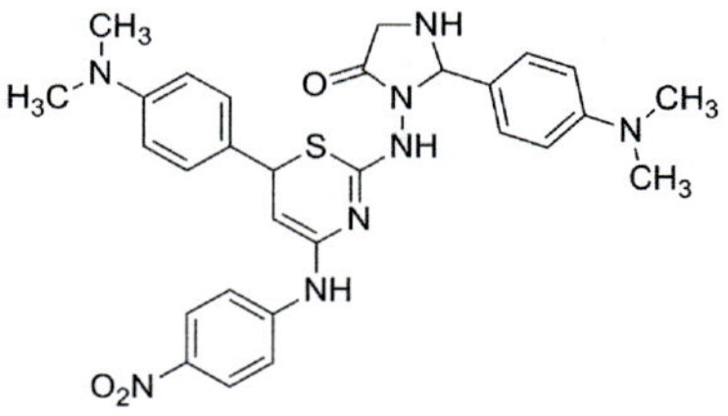

(61)

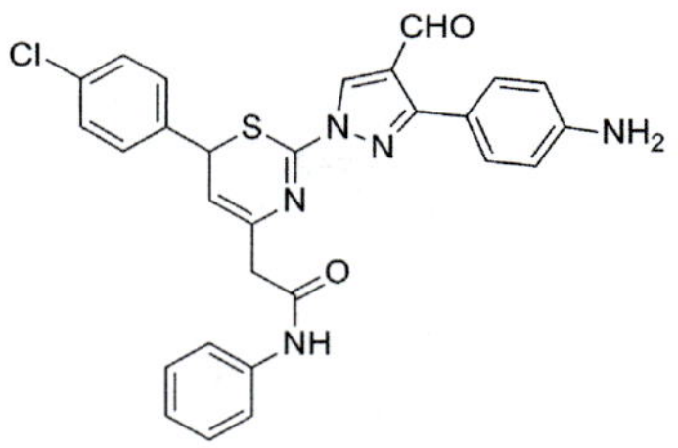

(62)

Rekka et al. synthesized thiomorpholine derivatives having 2,6-di-tert-butyl phenol moiety. Compound **63** was found to be the most active ($IC_{50} = 70\,\mu M$) COX-1 inhibitor, with significant antioxidant activity ($IC_{50} = 7.5\,\mu M$) and low toxicity [43]. In another effort, ascidiathiazones A (**64**) and B (**65**), thiazine-based quinolinequinone alkaloids, were isolated by Pearce et al. from ascidian *Aplidium* species and evaluated for antiinflammatory potential both in vitro and in vivo showing IC_{50} value 1.55 ± 0.32 and $0.44 \pm 0.09\,\mu M$. Compound **64** was also synthesized by them in the laboratory from 8-hydroxyquinoline-2-carboxylic acid [77].

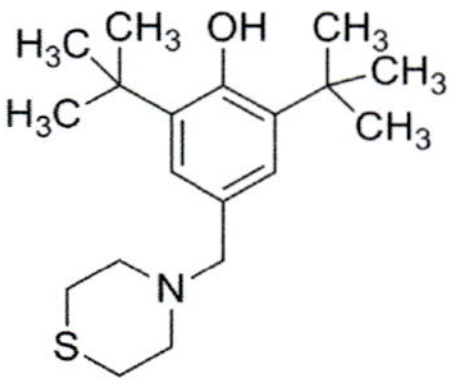

(63)

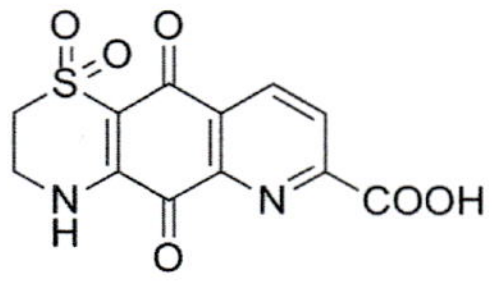

(64)

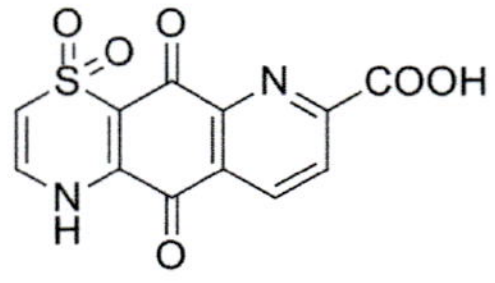

(65)

4.3 Anticancer Agents

Some compounds containing thiazine nucleus were found to possess significant activity against various type of tumors.

New imidazothiazine derivatives were synthesized and screened for anticancer activity against various human cancer cell lines. Among all screened compounds, **66** was found most potent and comparable to reference drugs like bicalutamide, fluorouracil, doxorubicin, cytarabin and gemcitabine on cancer cell lines such as cervical carcinoma cell lines (KB, $IC_{50} = 3.80 \times 10^{-1}\,\mu M$); ovarian carcinoma cell lines (SKOV-3, $IC_{50} = 1.03 \times 10^{-3}\,\mu M$); CNS cancer cell lines (SF-268, $IC_{50} = 2.97 \times 10^{-3}\,\mu M$); non–small cell lung cancer cell lines (NCI H460, $IC_{50} = 1.40 \times 10^{-3}\,\mu M$); colon adenocarcinoma cell lines (RKOP 27, $IC_{50} = 1.40 \times 10^{-4}\,\mu M$); leukemia cell lines (HL60, U937, K562, $IC_{50} = 4.24 \times 10^{-5}$, 4.83×10^{-5} and $1.18 \times 10^{-4}\,\mu M$ respectively); melanoma cell lines (G361, SK-MEL-28, $IC_{50} = 9.86 \times 10^{-5}$ and $3.45 \times 10^{-5}\,\mu M$); neuroblastoma cell lines (GOTO, NB-1, $IC_{50} = 1.25 \times 10^{-2}$ and $1.62 \times 10^{-3}\,\mu M$); cervical cancer cell lines

(HeLa, $IC_{50} = 5.93 \times 10^{-5}\,\mu M$); breast cancer cell lines (MCF-7, $IC_{50} = 1.21 \times 10^{-5}\,\mu M$); fibrosarcoma cell lines (HT1080, $IC_{50} = 2.18 \times 10^{-5}\,\mu M$ and liver cancer cell lines (HepG2, $IC_{50} = 5.15 \times 10^{-4}\,\mu M$) [78]. In another study, Wang et al. synthesized novel multithioether derivatives by combining thiazine with dibromides. Synthesized compounds were evaluated in vitro for their antitumor potential using MTT assay against A-549 and Bcap-37 cancer cells. Thiazine-derived bis-thioether compound **67** was found to be most potent with IC_{50} values 8.26 and 9.30 μg/mL against A-549 and Bcap-37 respectively [79]. 1,3-Thiazine-2,4-dione derivatives of thiazine are less frequently reported in the literature [80], except for dolitrone [81], which is an anaesthetic. In 2013, Ferreira et al. tried to explore this nucleus for anticancer activity. It was found that, introduction of phenyl and naphthyl group at 5th position of 1,3-thiazine-2,4-dione displayed selective antitumor activity [82].

Some carane- and apopinane-based thiazine derivatives were reported by Szakonyi et al. Carane-based compound **68** was found to have cell growth inhibitory activity on different cell lines at 10–30 μM concentration, which was found comparable to cisplatin. On the contrary, compounds with a pinane nucleus were found to be less potent. Moreover, introduction of a methyl group at 2nd position of the pinane ring improved the activity [83]. SAR detail of this nucleus for anticancer activity is depicted in Fig. 7.9.

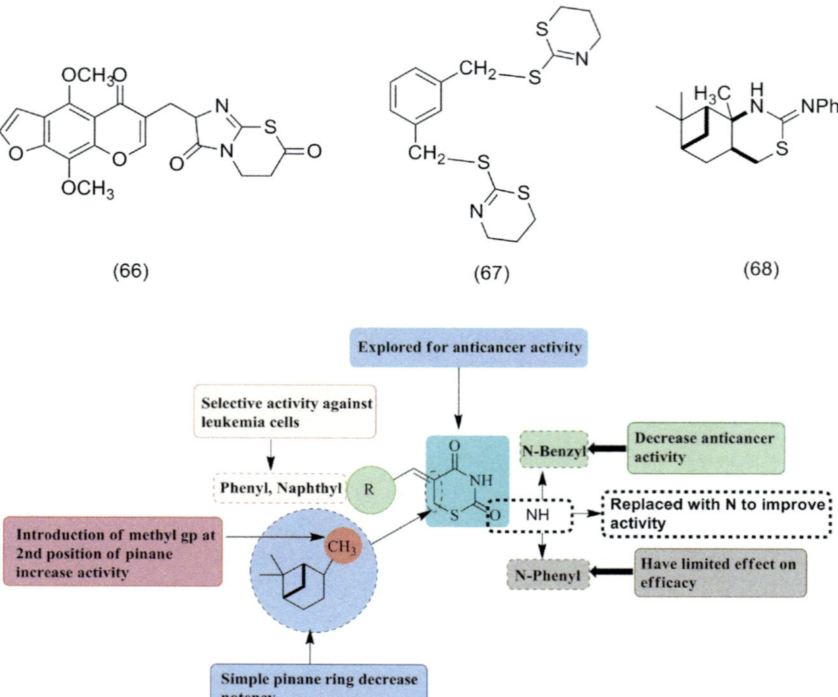

FIGURE 7.9 Structure–activity relationship profile of 1,3-thiazine for anticancer activity.

Zahra et al. synthesized tetrahydropyrido-thieno-thiazine by performing reductive lactamization by treatment with sodium dithionite. Compounds **69** (IC 76% = 10 μM) and **70** (IC 64% = 10 μM) were found to possess inhibitory activity against ovarian cancer cell lines [84]. Pluta et al. have synthesized some new azaphenothiazines, which were further tested for their anticancer activity. Compound **71** was found to be most active on cancer cell lines such as L1210 (GI_{50} = 2.28 μg/mL), SW948 (GI_{50} = 44.50 μg/mL), A-431 (GI_{50} = 2.84 μg/mL) and CX-1 (GI_{50} = 10.83 μg/mL) when screened against the standard drug cisplatin [85].

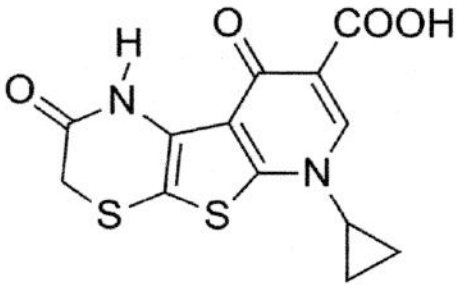

(69) 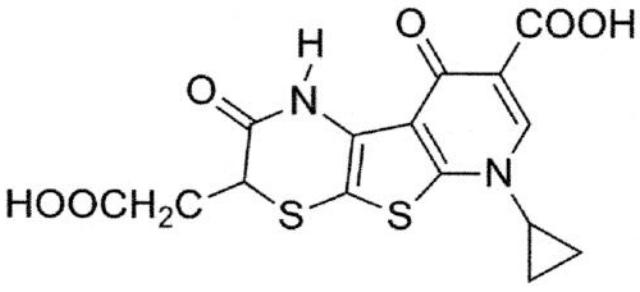   (70) 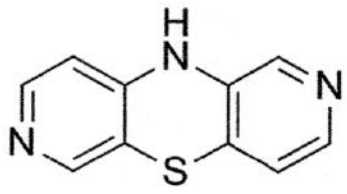   (71)

Novel azaphenothiazine derivatives were synthesized from 6H-9-fluoroquinobenzothiazine and evaluated for their anticancer potential. Results showed that compounds possess promising antiproliferative and cytotoxic effects. Two compounds of this series, **72** (IC70% = 10 μg/mL) and **73** (IC70% = 50 μg/mL), have shown strong antiproliferative activity. Compound **73** was found to be potent by suppressing growth of various tumor cell lines [86].

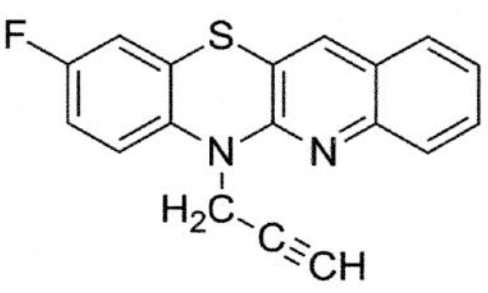

(72) 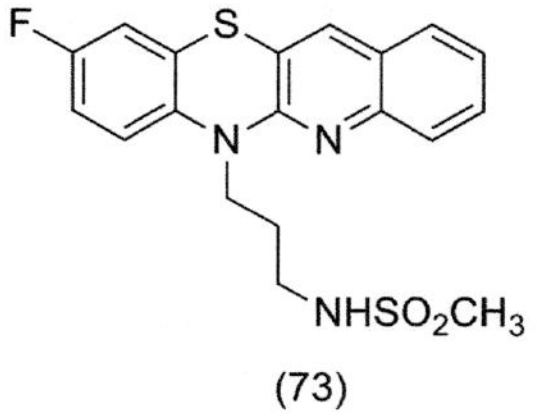   (73)

4.4 Antitubercular Agents

Occurrence of multidrug resistance in TB and failure of regular established therapies against resistant TB strains have underlined the need for the discovery of novel molecules for TB. New derivatives of 1,3-thiazines are reported as a class of antimycobacterial agents. Compound **74** is found to have best activity when evaluated in vitro against mycobacteria with low cytotoxicity [87].

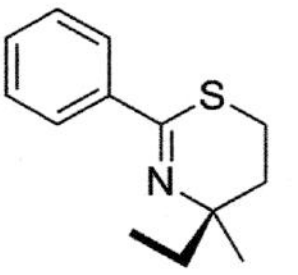

(74)

5,6-Dihydro thiazine derivatives are evaluated for their antimycobacterial potential using the alamar blue method. Antimycobacterial activity of compound **75** is found to be 97% at 6.25 µg/mL [88].

Novel derivatives of thiazine are also reported and screened for in vitro antimycobacterial activity. Compound **76** is reported to have potent antimycobacterial activity [89]. A series of 1,4-thiazine derivatives are reported by Perumal et al. and further evaluated for their antimycobacterial activity against *M. tuberculosis*. Compound **77** is reported as the most potent molecule with an MIC value of 0.68 mM, and is active against multidrug resistant (MDR) TB [45]. Some novel sulfadrug-substituted 1,4-thiazines and 1,3-thiazines are also reported. When screened for antimycobacterial activity, compound **78** was found to be most potent. Sulfa-substituted 1,4-thiazine derivatives are more effective (MIC 25 µg/mL) than 1,3-thiazine derivatives (MIC 50 µg/mL) [90].

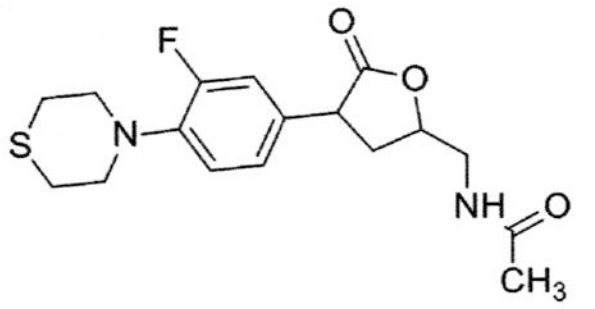

(75)

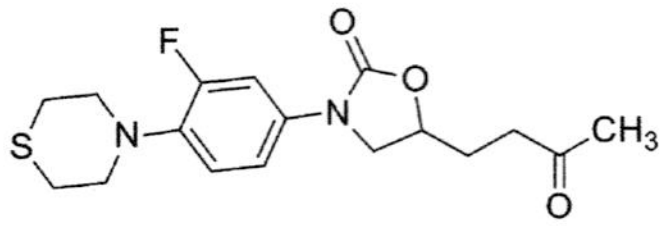

(76)

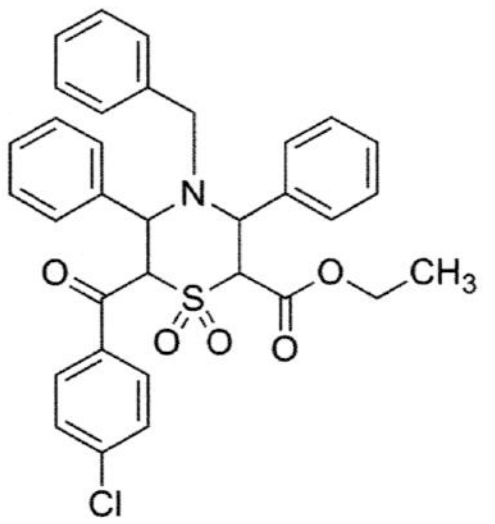

(77)

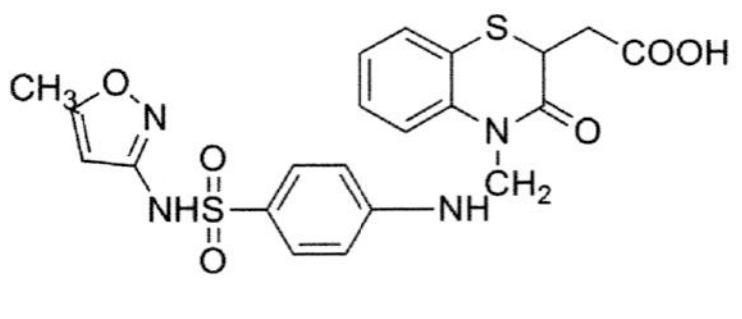

(78)

4.5 Antimalarial Agents

The World Health Organization has recommended the use of thiazine containing artimisinin-based combination therapy as first-line treatment for malaria. Haynes et al. reported that artemisinin (**79**) and its derivatives such as dihydroartemisinins, artesunate, artmisone and artmether are routinely used for the treatment of malaria in combination therapy with longer half life [91]. A series of new 3,5-Bis (m-nitro phenyl)-2,6-bis (*p*-chlorobenzoyl) tetrahydro 1,4-thiazine1, 1-dioxide is designed using docking analysis. It is observed that 3,5-Bis (m-nitro phenyl) and *p*-chlorobenzoyl groups in 1,4-thiazine are optimum for antimalarial activity as best Lib dock score and binding energy are obtained for compound **80** (IC$_{50}$ = 10 µg/mL) when docked with antimalarial proteins like Plasmodium falciparum dihydrofolate reductase-thymidylate synthase

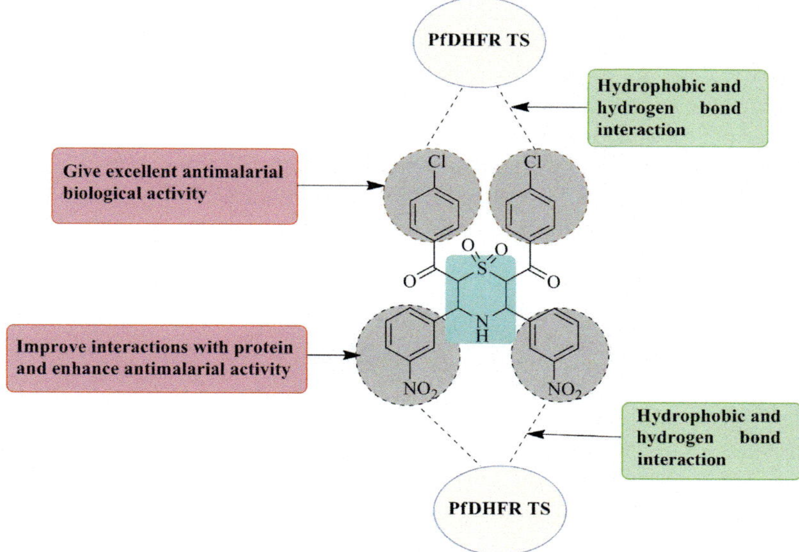

FIGURE 7.10 Structural requirements around thiazine nucleus for antimalarial activity.

(PfDHFR TS) [92]. Structural requirements for antimalarial activity around thiazine nucleus and their interactions with PfDHFR TS are shown in Fig. 7.10.

(79)

(80)

4.6 Antiepileptic Agents

There is limited exploration of a thiazine nucleus for its antiepileptic activity. In spite of that, a few reports display the potency of thiazine as an anticonvulsant agent.

In 1981, Soliman and coworkers for the first time synthesized 3-substituted-4-oxo-5,6-dihydro-1,3-thiazin-2-yl-(1-phthalazinyl)hydrazones (**81**, 1 mmol/kg) and evaluated them for anticonvulsant activity but the results were not promising [93]. 1,3-thiazine derivatives were further synthesized and screened for antiepileptic activity. Among screened compounds, **82** ($ED_{50} = 46$ mg/kg) displayed significant effects in minimum pentetrazole-induced seizures and was found most potent with less toxicity [94]. In 2014, Murthy et al. carried out the reaction of 4,4-dimethylcyclohexanone with different aromatic aldehydes to give 2,6-diarylidene-4,4-dimethylcyclohexanone, which upon reaction with thiourea offered some novel 1,3-benzothiazines. All synthesized compounds were evaluated for in vivo anticonvulsant activity by maximal electroshock. This study led to compounds **83** and **84** with reduced phase duration of convulsions at test dose 120 mg/kg and found to be most potent due to presence of the nitro, hydroxyl and fluoro groups, which are essential for anticonvulsant activity [95].

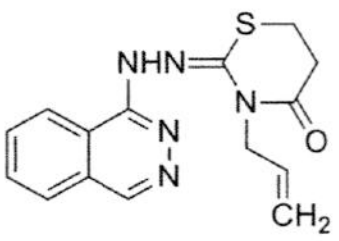

(81) 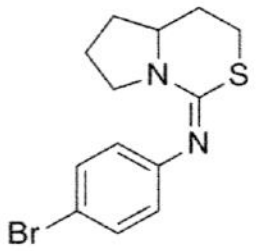 (82) 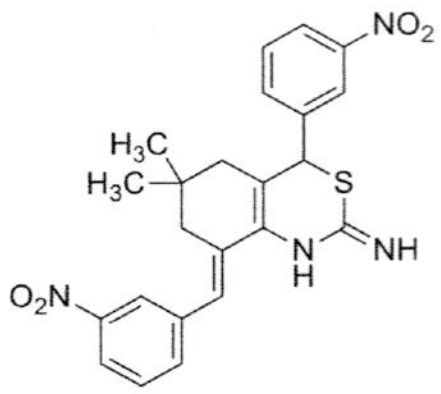 (83) 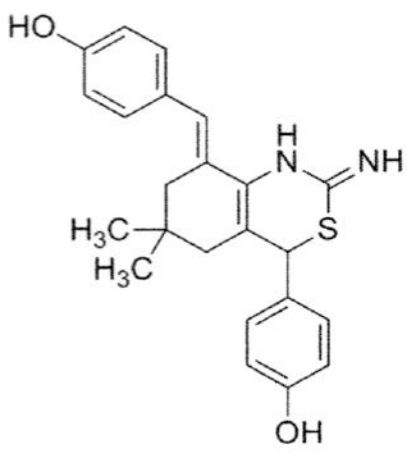 (84)

Two series of benzo-1,4-thiazine are reported by Zhang et al. and tested against seizures by the maximal electroshock (MES) method. Among all synthesized compounds, **85** ($ED_{50} = 34.08$ mg/kg) and **86** ($ED_{50} = 17.0$ mg/kg) are found to be the most useful and therapeutically active with minimum neurotoxicity when compared to carbamazepine [96]. 2,6-dicarbethoxy-3,5-diaryltetrahydro-1,4-thiazine-1,1-dioxides are also reported for their anticonvulsant activity using MES and scPTZ models. SAR study of these compounds revealed that anticonvulsant activity depends upon the position of substituent at the phenyl ring, and claimed that compound **87** ($ED_{50} = 10.2$ mg/kg) is the most effective in both models with lower neurotoxicity [97].

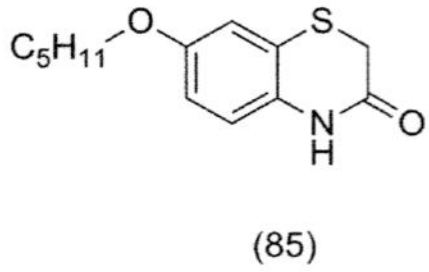

(85) 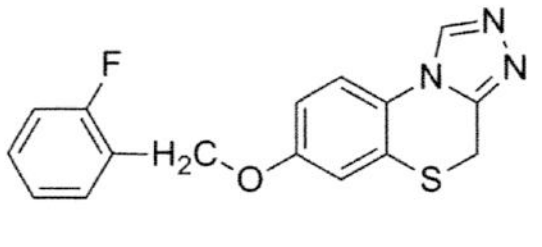 (86) 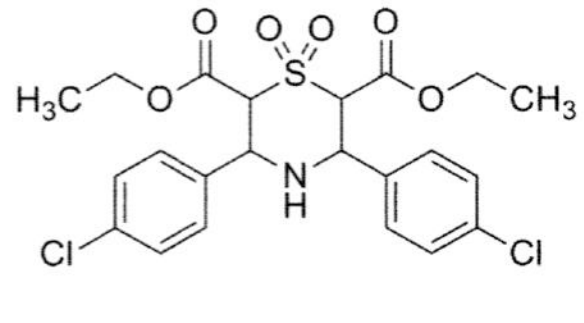 (87)

The fundamental structural elements required for anticonvulsant activity in thiazines are nitrogen containing heterocyclic system with a carbonyl group, an aryl substitution and electron donor atoms such as NH group in a specific arrangement. Introduction of a chloro group at para-position of phenyl rings increases potency, while the methoxy group at the same position decreases it. Moreover, chloro, nitro and methyl substitutions at meta- and ortho-positions decrease anticonvulsant activity. 3,5-diphenyl rings are essential for the activity and substituted phenyl rings act as lipophilic domain. A carbonyl group of ester

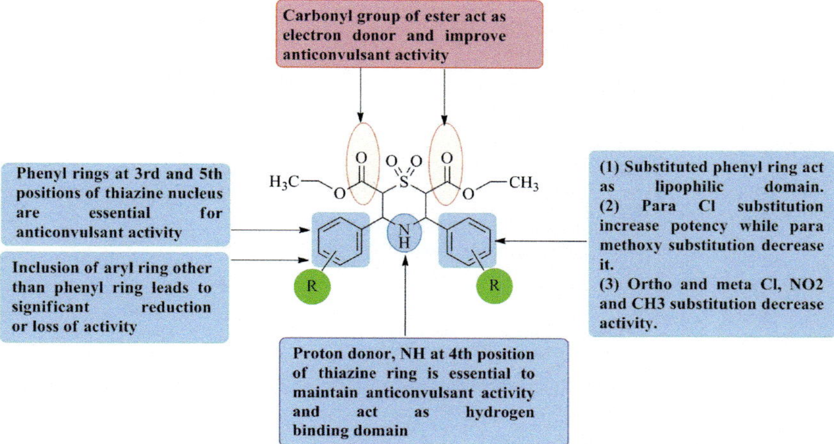

FIGURE 7.11 Fundamental structural requirements for antiepileptic activity of 1,4-thiazines.

and –NH group at 4th position of a thiazine ring act as electron donor groups and maintain anticonvulsant activity. aryl groups, other than a phenyl group, lead to significant reduction or permanent loss of activity (Fig. 7.11) [98].

4.7 Antipsychotic Agents

The antipsychotic activities of thiazine and phenothiazine are due to the electron-donating property of basic nitrogen of the thiazine ring, which donates electrons to the biological receptors through charge transfer mechanism, and replacement of the hydrogen attached to nitrogen with other substituents enhances pharmacological activities [99]. Some novel pentacyclic ring systems containing compounds 13H-5,14-dihydroquinoxalino[2,3-a]phenothiazine are synthesized and screened for neuropharmacological properties. All the synthesized compounds have been reported to show neurosedative activity. Neuroleptic effect of the compounds was also observed against ortwin-induced climbing behavior in mice. It was claimed that compound **88** (test dose 50–100 mg/kg) is the first angular pentacyclic phenothiazine to possess the neurosedative property [100].

(88)

In 2014, Blokhina et al. synthesized some novel spiro 1,3-thiazines as neuroprotectors. Mechanism of action of the compounds involves blockade of

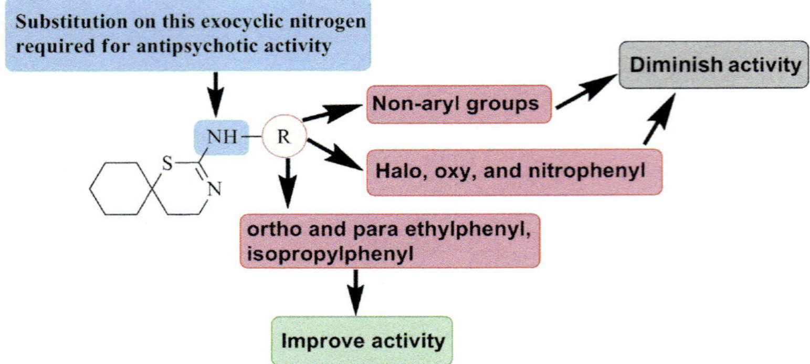

FIGURE 7.12 Essential structural components required for antipsychotic activity of 1,3-thiazines.

glutamate-induced calcium ion uptake into the brain cortex. SAR study disclosed that the chemical nature and structure of substituents on exocyclic nitrogen atoms were important factors for antipsychotic activity. Permeability of the compounds through the phospholipidic membrane was correlated with donor-acceptor properties of the substituents. It was also revealed from SAR study that, the alkyl group (ethyl and isopropyl) at ortho- and para-positions of the phenyl ring attached to exocyclic nitrogen increased inhibitory ability while a nonaryl group at the same exocyclic nitrogen decreased activity. Polar groups like halogen, nitrogen, and oxygen diminished potency (Fig. 7.12) [101].

5. THIAZINES AS MULTITARGETING AGENTS IN MULTIFACTORIAL DISEASES

Multifactorial disorders such as Alzheimer's disease, diabetic complications, cardiovascular diseases, inflammatory conditions, malaria, cancer, TB and so on involve highly complicated pathology [102]. These complex diseases remain the topic of special attention for researchers and require multitargeting molecules for their treatment. Thus, novel multitargeting therapeutics such as hybrid or conjugate molecules that can strike multiple pathways elicited in the etiology of these diseases is a current topic of interest.

It is evident from some literature reports that diversely substituted thiazines can play an effective role in multifactorial diseases.

1,3-thiazine is well known for its diverse pharmacological activities and can be utilized as one of the promoiety in a hybrid designing approach for multitargeting.

Wu et al. synthesized some novel 1,3-thiazine-pyrazine conjugates, carried out a molecular modeling study and screened them against human immunodeficiency virus (HIV-1), enterovirus (EV71 and coxsackie B3) and influenza viruses (hemagglutinin type 1 (H1) and neuraminidase type 1 (N1)) using enzyme-based assay to find out the interaction of these conjugates with key enzymes such as HIV-1 reverse transcriptase and neuraminidase, involved in viral infection progression. Among all conjugates, compound **89** was found to be the most potent

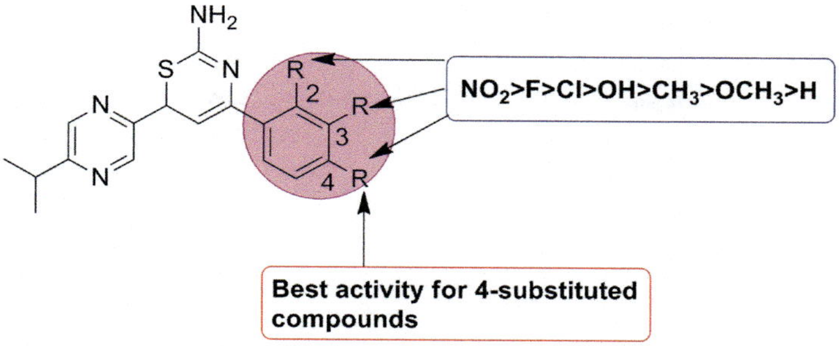

FIGURE 7.13 Structure–activity relationship profile of 1,3-thiazine for anti-HIV-1 activity.

against HIV-1 with $IC_{50}=3.26\pm0.2\,\mu M$. On the other hand, compound **90** inhibited both H1 and N1 virus with $IC_{50}=5.32\pm0.4\,\mu M$ along with N1 enzyme with $IC_{50}=11.24\pm1.1\,\mu M$ and proved to be a multitargeting molecule. SAR data obtained from molecular modeling study suggested that 4-phenyl substituted 1,3-thiazines are required for anti-HIV-1 activity. Further substitution at 2nd, 3rd and 4th positions of the phenyl ring improves antiviral activity (Fig. 7.13) [103].

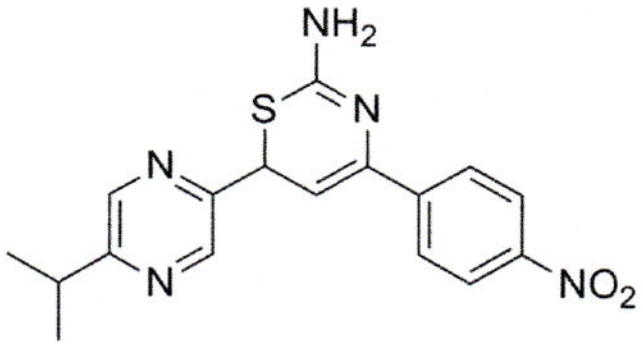

(89)

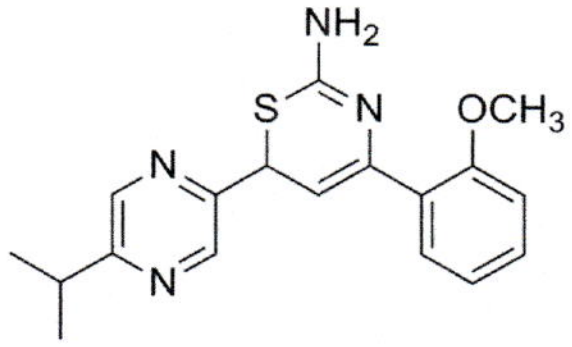

(90)

Another type of thiazine, thiomorpholine (reduced form of 1,4-thiazine) was conjugated with phenol-triazole using an ethyl spacer. Synthesized hybrid **91** was screened against various targets of Alzheimer's disease and found to have antioxidant character, Cu-binding affinity, Aβ peptide interaction and inhibition of Aβ1-42-induced neurotoxicity in human neuronal culture. It was unveiled from a 3D-NMR study that **91** interacts with Aβ-peptide but did not show notable interactions with other targets and hence failed as a multitargeting molecule [104].

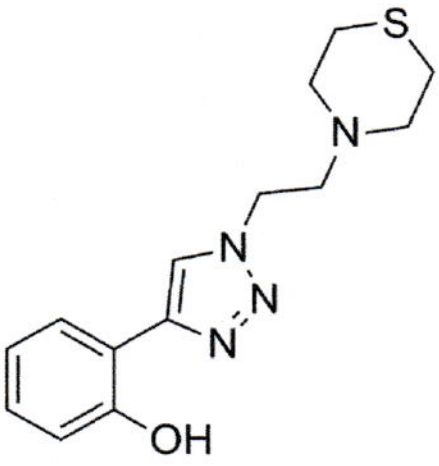

(91)

Benzfused thiazine moieties have shown better multitargeting potential than a single thiazine nucleus. Phenothiazine, which is 1,4-thiazine condensed with two benzene rings, has been explored against multifactorial diseases. Makhaeva et al. (2015) combined phenothiazine with γ-carboline using 1-oxo- and 2-hydroxypropylene as a linker. Synthesized hybrids were evaluated in silico and in vitro against acetylcholinesterase (AChE), butyrylcholinesterase (BChE), carboxylesterase (CE) and NMDA-receptors. Among all hybrids, 1-oxopropylene linker connected molecules **92** and **93** showed selective BChE inhibitory activity with $IC_{50} = 0.52 \pm 0.01$ and $0.58 \pm 0.06\,\mu M$ respectively and incurably inhibited AChE and CaE. It was also found that conjugates with 1-oxopropylene linker had greater affinity for both binding sites of NMDA-receptor. In silico study disclosed protein-ligand interactions and found that there is π-π stacking interaction of phenothiazine and indole group of γ-carboline with Trp82 [105].

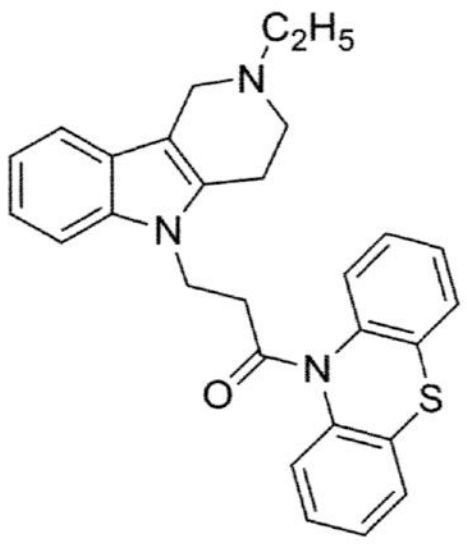

(92)

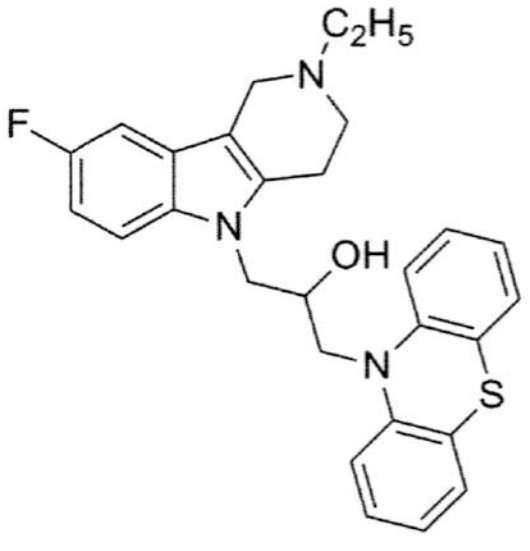

(93)

In another report, phenothiazine-based multitargeting neuroprotector hybrids were also constructed by conjugating bis (dimethylamino) phenothiazine with indole nucleus again using 1-oxopropylene as a spacer. Synthesized hybrids underwent screening as neuroprotective agents using radioligand binding assay on neuronal NMDA receptors. Compound **94** was found to be most active, which was further justified by in silico study. It was ascertained from computational study that the binding affinity of **94** with the ifenprodil binding site was comparable with that of γ-carboline conjugates and phenothiazine [106].

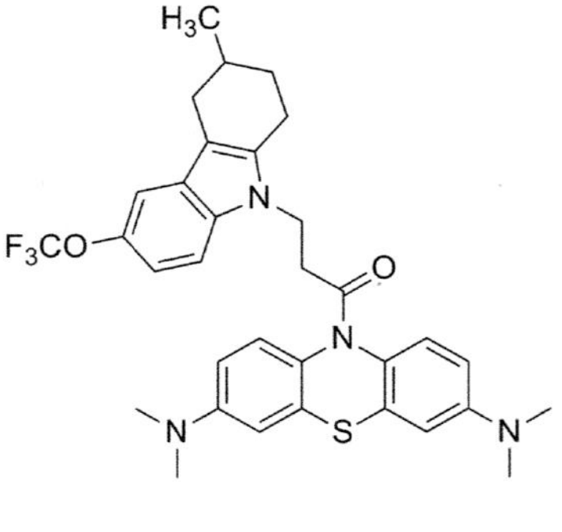

(94)

Phenothiazine can also avert tau filaments formation, one of the causes of Alzheimer's disease, therefore, this character of phenothiazine can be utilized in multitargeting approaches against neurodegenerative diseases. A well-known anti-Alzheimer's drug, Tacrine, was combined with phenothiazine with the aim to target the most important Alzheimer's targets, acetylcholinesterase and tau protein. Molecular modeling results filtered three hybrids for synthesis and biological evaluation against acetylcholinesterase and tau hyperphosphorylation. Among all hybrids, compound **95** was found to be the most potent molecule with $IC_{50} = 89\,nM$, and also showed interaction with beta amyloid tangles. It is suggested that phenothiazine-tacrine hybrids can be potential multitargeting ligands that simultaneously target acetylcholinesterase, beta amyloid and tau protein [107].

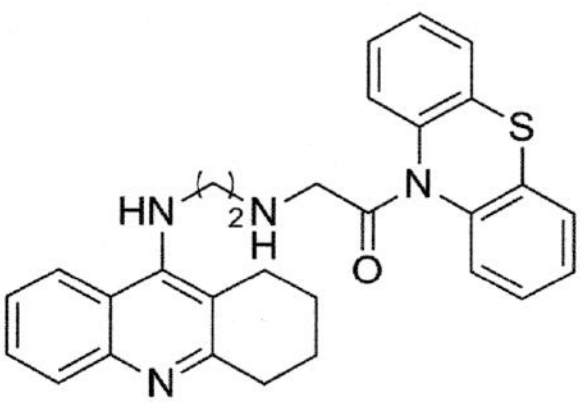

(95)

Phenothiazine-based novel molecules are also reported by Tin et al. (2015) aimed at targeting cholinergic, amyloid and oxidative stress pathways of Alzheimer's disease. Phenothiazine containing hybrid **96** was identified as the best multitargeting molecule with dual acetylcholinesterase and butyrylcholinestrase inhibitory activities with $IC_{50} = 5.9 \pm 0.6$ and $IC_{50} = 5.3 \pm 0.5\,\mu M$ respectively. SAR data suggest that N-10 unsubstituted phenothiazine exhibits multitargeting ability by demonstrating cholinesterase inhibition, beta-amyloid aggregation and antioxidant properties. It was found that N-10 acyl substitution with 4-methoxyphenyl group provided good cholinesterase inhibition. In contrast, N-10 nonacylated derivatives were able to target and inhibit both AChE and BuChE enzymes, prevent Aβ-aggregation and scavenge DPPH radicals [108].

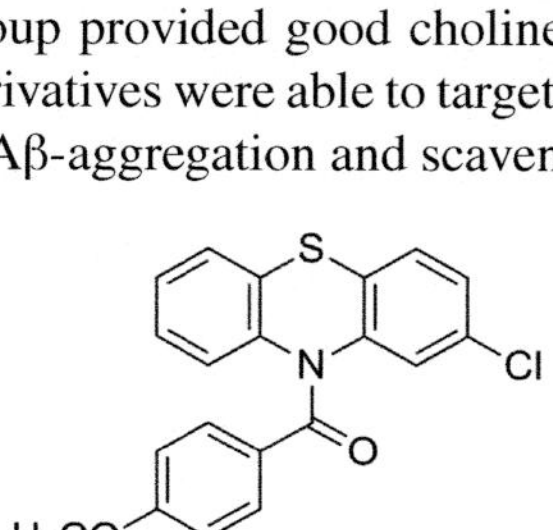

(96)

These reports suggest that phenothiazine is useful in developing hybrid molecules to target multiple pathological routes associated with Alzheimer's disease.

Phenothiazine-based novel hybrids may emerge as encouraging and safe multitargeting lead molecules for drug discovery against neurodegenerative disorders and other pathological conditions.

Phenothiazines can also inhibit multidrug efflux pump activity in *S. aureus*. The inhibitory effects of some phenothiazines are observed on strains of *S. aureus* having unique efflux-related MDR phenotypes. Thioridazine (**97**) is reported as most potent among phenothiazines against all strains of *S. aureus* with an MIC range of 25–100 μg/mL. These molecules have intrinsic antimicrobial activity and give synergistic interactions when combined with some common MDR efflux pump substrates. Thus, in conclusion, phenothiazines inhibit PMF-dependent efflux pumps through a multifactorial mechanism, which involves an interaction with the pump itself and reduction in the transmembrane potential [109].

(97)

6. CONCLUSION/PROSPECTIVE

Thiazine has been a noticeable entity and a very important part of heterocyclic compounds with interesting chemistry and a broad pharmacological profile. This nucleus has potential as an aspirant lead molecule to be designed as a novel therapeutic against challenging diseases like Alzheimer's disease, cancer, diabetes, TB and others. In spite of being extensively explored against various diseases, its multitargeting ability has yet to come into light to an considerable extent. Thiazine remains one of the key nuclei in most of the drugs acting on the CNS and antiinflammatory conditions. However, thiazines still need to achieve the status of multitargeting agents. Although, there are few literature reports in favor of the multitargeting potential of thiazine and its derivatives, this intriguing nucleus needs to be further explored in order to get more multifunctional molecules.

REFERENCES

[1] R.M. Acheson, Introduction to the Chemistry of Heterocyclic Compounds, 1967.
[2] S. Rathod, A. Charjan, P. Rajput, Synthesis and antibacterial activities of chloro-substituted-1,3-thiazines, Rasayan Journal of Chemistry 3 (2) (2010) 363–367.

[3] M. Asif, Chemical and pharmacological potential of various substituted thiazine derivatives, Journal of Pharmaceutical and Applied Chemistry 1 (2015) 49–64.

[4] A.C. White, M.M. Hann, Thiomorpholine Derivatives, Google Patents, 1984.

[5] N. Rajiv, J. Rajan, L.K. Pappachen, A review on synthesis of benzothiazine analogues, Research Journal of Pharmacy and Technology 10 (6) (2017) 1791–1797.

[6] S. Mor, et al., Synthesis and biological activities of 1,4-benzothiazine derivatives: an overview, Chemistry and Biology Interface 7 (1) (2017).

[7] G. Cimbura, Review of methods of analysis for phenothiazine drugs, Journal of Chromatographic Science 10 (5) (1972) 287–293.

[8] J. Blazek, A. Dymes, Z. Stejskal, Analysis of Phenothiazine Derivatives. 3. Review, 1975.

[9] C. Barkenbus, P.S. Landis, The preparation of 1,4-thiazine, Journal of the American Chemical Society 70 (2) (1948) 684–685.

[10] A. Bernthsen, Zur kenntniss des methylenblau und verwandter farbstoffe, European Journal of Inorganic Chemistry 16 (2) (1883) 2896–2904.

[11] V.V. Dabholkar, S.D. Parab, Synthesis of chalcones, 1,3-thiazines and 1,3-pyrimidines derivatives and their biological evaluation for anti-inflammatory, analgesic and ulcerogenic activity, Pharmaceutical Research 5 (2011) 127–143.

[12] S.K. Doifode, M. Wadekar, S. Rewatkar, Synthesis of 1-3 thiazines from aurone, Oriental Journal of Chemistry 27 (3) (2011) 1265.

[13] L.D.S. Yadav, V.K. Rai, B.S. Yadav, The first ionic liquid-promoted one-pot diastereoselective synthesis of 2,5-diamino-/2-amino-5-mercapto-1,3-thiazin-4-ones using masked amino/mercapto acids, Tetrahedron 65 (7) (2009) 1306–1315.

[14] L. Fu, et al., Synthesis and calming activity of 6H-2-amino-4-aryl-6-(4-β-D-allopyranosyloxyphenyl)-1,3-thiazine, Chemistry of Natural Compounds 46 (2) (2010) 169–172.

[15] J.E. Rice, Organic Chemistry Concepts and Applications for Medicinal Chemistry, Academic Press, 2014.

[16] S.P. Massie, The chemistry of phenothiazine, Chemical Reviews 54 (5) (1954) 797–833.

[17] D.F. Houston, E. Kester, F. DeEds, Phenothiazine derivatives: mono-oxygenated compounds, Journal of the American Chemical Society 71 (11) (1949) 3816–3818.

[18] H. Gilman, R.D. Nelson, J.F. Champaigne Jr., Some 10-substituted phenothiazines, Journal of the American Chemical Society 74 (16) (1952) 4205–4207.

[19] K. Hunger, Industrial Dyes: Chemistry, Properties, Applications, John Wiley & Sons, 2007.

[20] C. Okafor, The chemistry and applications of angular phenothiazine derivatives, Dyes and Pigments 7 (4) (1986) 249–287.

[21] H. Lauth, Uber eine neue Klasse von Farbstoffen, Ber 9 (1876) 1035.

[22] C. Reinhardt, A.S. Travis, Heinrich Caro and the Creation of Modern Chemical Industry, vol. 19, Springer Science & Business Media, 2013.

[23] O. Müller, Methylene Blue for Therapy of Malaria, Department of Medical Laboratory Sciences and Pathology, Jimma University, 1876.

[24] L.D.S. Yadav, A. Singh, Microwave activated solvent-free cascade reactions yielding highly functionalised 1,3-thiazines, Tetrahedron Letters 44 (30) (2003) 5637–5640.

[25] R.J. Stoodley, 1,4-thiazines and their dihydro derivatives, Advances in Heterocyclic Chemistry 24 (1979) 293–361.

[26] J. Teller, E. Schaumann, Houben–weyl Methods of Organic Chemistry, Hetarenes IV, Six-Membered and Larger Hetero-Rings with Maximum Unsaturation, Georg Thieme, Stuttgart, 1997, pp. 141–177.

[27] C.R. Johnson, C. Thanawalla, New derivatives of 1,4-thiazine, Journal of Heterocyclic Chemistry 6 (2) (1969) 247–249.

[28] F. Gutmann, H. Keyzer, Dielectric properties of thiazine derivatives, Electrochimica Acta 13 (4) (1968) 693–704.

[29] F. Gutmann, A. Netschey, Electrical properties of chlorpromazine, The Journal of Chemical Physics 36 (9) (1962) 2355–2360.

[30] P. Kronick, M. Labes, Conduction in hexamethylbenzene through the crystalline transition, The Journal of Chemical Physics 38 (3) (1963) 776–777.

[31] F. Bueche, Physical Properties of Polymers, 1962.

[32] A. Turkevich, C.P. Smyth, The dielectric behavior, supercooling, and vitrification of certain chlorobutanes and chloropentanes, Journal of the American Chemical Society 64 (4) (1942) 737–745.

[33] A.I. Meyers, An infrared examination of the C=N link in dihydro-1,3-oxazines and dihydro-1,3-thiazines1, Journal of Organic Chemistry 26 (1) (1961) 218–220.

[34] A. Shutalev, M. Pagaev, L. Ignatova, Addition of dithiocarbaminic acid to alpha, beta-unsaturated aldehydes and ketones-simple synthesis of 4-alkoxytetrahydro-1,3-thiazine-2-thiones, Zhurnal Organicheskoi Khimii 27 (6) (1991) 1274–1285.

[35] M. Ichinari, T. Sato, Y. Hayase, Formation and reaction of mesoionic imidazolo [2, 3-b] [1, 3] thiazines and [1, 2, 4, 6] thiatriazino [3, 2-b][1, 3] thiazin-4-one 2, 2-dioxides from 2-arylimino-1,3-thiazines, Heterocycles 27 (1) (1988) 227–235.

[36] H. Kai, et al., 2-Arylimino-5,6-dihydro-4H-1,3-thiazines as a new class of cannabinoid receptor agonists. Part 1: discovery of CB 2 receptor selective compounds, Bioorganic and Medicinal Chemistry Letters 17 (14) (2007) 4030–4034.

[37] T. Glotova, et al., Synthesis of substituted 2-amino-1,3-thiazine-6-thiones, Russian Chemical Bulletin 49 (11) (2000) 1917–1918.

[38] T. Trofimova, et al., Synthesis and study of NOS-inhibiting activity of 2-N-acylamino-5, 6-dihydro-4 H-1,3-thiazine, Moscow University Chemistry Bulletin 63 (5) (2008) 274–277.

[39] V.N. Yuskovets, B.A. Ivin, A new synthesis of 1,3-thiazines and their transformation into 1-substituted-6-alkyluracils by extrusion of carbonyl sulfide, Tetrahedron Letters 44 (28) (2003) 5279–5280.

[40] J.-P. Wan, et al., Microwave-assisted three-component reaction for rapid synthesis of some 5,6-dihydro-4 H-1,3-thiazine derivatives under solvent-free conditions, Synthetic Communications 40 (5) (2010) 709–716.

[41] A. Magerramov, et al., New synthesis of tetrahydro-1, 4-thiazine-3-ones from 1,2-aminopropanethiols and α-chlorocarboxylic acids, Synthetic Communications 29 (4) (1999) 721–728.

[42] M. Mihara, et al., Silica gel-promoted synthesis of 3,4,5-triaryltetrahydro-1,4-thiazine derivatives from β, β′-dichloro sulfides and aromatic amines, Tetrahedron Letters 46 (47) (2005) 8105–8108.

[43] G.N. Ziakas, et al., New analogues of butylated hydroxytoluene as anti-inflammatory and antioxidant agents, Bioorganic and Medicinal Chemistry 14 (16) (2006) 5616–5624.

[44] V. Sundari, G. Nagarajan, R. Valliappan, Synthesis and biological activities of some N-hydroxymethylene and N-methoxymethylene thiazines, Medicinal Chemistry Research 16 (7–9) (2007) 402–407.

[45] S. Indumathi, et al., 1-Proline-catalysed facile green protocol for the synthesis and antimycobacterial evaluation of [1,4]-thiazines, European Journal of Medicinal Chemistry 44 (12) (2009) 4978–4984.

[46] H.S. Gold, R.C. Moellering Jr., Antimicrobial-drug resistance, New England Journal of Medicine 335 (19) (1996) 1445–1453.

[47] W. Malinka, et al., Pyrido-1,2-thiazines and their in vitro antibacterial evaluation, Acta Poloniae Pharmaceutica 65 (1) (2007) 71–74.

[48] R.L. Sawant, L.P. Bhangale, J.B. Wadekar, Topliss modified approach for design and synthesis of 1,3 thiazines as antimicrobials, International Journal of Drug Design and Discovery 2 (4) (2011) 637–641.

[49] P.B.R. Kumar, Synthesis Characterization and Biological Studies of Some New Thiazolidinones and Thiazines, 2014.

[50] R. Kalirajan, et al., Synthesis and biological evaluation of some heterocyclic derivatives of chalcones, International Journal of ChemTech Research 1 (1) (2009) 27–34.

[51] M.J. Elarfi, H.A. Al-Difar, Synthesis of some heterocyclic compounds derived from chalcones, Scientific Reviews and Chemical Communications 2 (2) (2012).

[52] F.H.Z. Haider, Synthesis and antimicrobial screening of some 1,3-thiazines, Journal of Chemical and Pharmaceutical Research 4 (4) (2012) 2263–2267.

[53] G. Dipansu, B. Mander, Synthesis, characterization and biological evaluation of some novel 4,6-disubstituted-1,3-thiaizne derivatives for their antibacterial activity, International Journal of Health Pharmaceutical Sciences 1 (1) (2012) 27–33.

[54] A.K. Rathod, Synthesis and characterization of 4-aryl-8-arylidene thiazines derivatives using microwave irradiation, International Journal of Pharmaceutical and Chemical Science 3 (2) (2012) 111–1114.

[55] R.S. Ganorkar, R.P. Ganorkar, V. Parhate, Synthesis, characterisation and antibacterial activities of some new bromo/nitro 1,3-thiazenes, Culture 6 (1) (2013) 65–67.

[56] S. Begum, et al., Therapeutic Utility of 1,3-Thiazines-Mini Review, Saudi Journal of Medical and Pharmaceutical Sciences 2 (2016) 326–338.

[57] V. Sundari, et al., Synthesis, characterization and biological activities of 3, 5-diaryltetrahydro-N-[(phenylamino) methyl]-1,4-thiazine-1, 1-dioxide, Journal of Chemistry 6 (1) (2009) 177–182.

[58] D. Armenise, et al., 4H-1,4-benzothiazine, dihydro-1,4-benzothiazinones and 2-Amino-5-fluorobenzenethiol derivatives: design, synthesis and in vitro antimicrobial screening, Archiv der Pharmazie 345 (5) (2012) 407–416.

[59] P.C. Sharma, et al., Microwave-assisted efficient synthesis and anti-fungal evaluation of some N-phenyl-3-(substituted phenyl) propenamides, Indian Journal of Pharmaceutical Education and Research 41 (2) (2007) 140–145.

[60] P. Sharma, S. Sharma, B. Suresh, CoMFA (3D-QSAR) studies on certain 3-aryl-4-[α-(1H-imidazol-1-yl) aryl methyl] pyrroles as potent anti-Candida agents, Indian Journal of Pharmaceutical Sciences 70 (2008) 154–158.

[61] P.C. Sharma, A. Jain, S. Jain, Fluoroquinolone antibacterials: a review on chemistry, microbiology and therapeutic prospects, Acta Poloniae Pharmaceutica 66 (6) (2009) 587–604.

[62] P.C. Sharma, et al., Ciprofloxacin: review on developments in synthetic, analytical, and medicinal aspects, Journal of Enzyme Inhibition and Medicinal Chemistry 25 (4) (2010) 577–589.

[63] A. Nagaraj, C.S. Reddy, Synthesis and biological study of novel bis-chalcones, bis-thiazines and bis-pyrimidines, Journal of the Iranian Chemical Society 5 (2) (2008) 262–267.

[64] M. Kadhim, Synthesis and chemical characterization of some novel azachalcones compounds and evaluation of their biological activity, Journal of University of Anbar for Pure Science 4 (3) (2010) 40–43.

[65] G. Tony, et al., Molecular docking studies: 1,3-thiazine and 1,3-oxazine derivatives, Journal of Pharmacy Research 8 (2) (2014) 136–138.

[66] G.P. Sarmiento, et al., Synthesis and antifungal activity of some substituted phenothiazines and related compounds, European Journal of Medicinal Chemistry 46 (1) (2011) 101–105.

[67] A. Macchiarulo, et al., 1,4-Benzothiazine and 1,4-benzoxazine imidazole derivatives with antifungal activity: a docking study, Bioorganic and Medicinal Chemistry 10 (11) (2002) 3415–3423.

[68] M.L. Barreca, et al., Discovery of 2,3-diaryl-1,3-thiazolidin-4-ones as potent anti-HIV-1 agents, Bioorganic and Medicinal Chemistry Letters 11 (13) (2001) 1793–1796.

[69] R. Kharb, M. Shahar Yar, P. Chander Sharma, Recent advances and future perspectives of triazole analogs as promising antiviral agents, Mini Reviews in Medicinal Chemistry 11 (1) (2011) 84–96.

[70] A. Rao, et al., Synthesis and anti-HIV activity of 2,3-diaryl-1,3-thiazolidin-4-ones, Il Farmaco 58 (2) (2003) 115–120.

[71] J.W. Mcfarland, V.A. Ray, Thiazoline and 5 6-Dihydro-4h-1 3-Thiazine Antiviral Agents, Google Patents, 1971.

[72] T. Mizuhara, et al., Concise synthesis and anti-HIV activity of pyrimido [1, 2-c][1, 3] benzothiazin-6-imines and related tricyclic heterocycles, Organic and Biomolecular Chemistry 10 (33) (2012) 6792–6802.

[73] S.A. Galal, et al., Novel benzimidazo [2, 1-c][1,4] thiazinone derivatives with potent activity against HSV-1, Archiv der Pharmazie 344 (4) (2011) 255–263.

[74] J.G. Lombardino, E.H. Wiseman, Piroxicam and other anti-inflammatory oxicams, Medicinal Research Reviews 2 (2) (1982) 127–152.

[75] H. Kai, et al., 2-Arylimino-5,6-dihydro-4H-1,3-thiazines as a new class of cannabinoid receptor agonists. Part 3: synthesis and activity of isosteric analogs, Bioorganic and Medicinal Chemistry Letters 18 (24) (2008) 6444–6447.

[76] S. Jupudi, et al., Screening of in vitro antiinflammatory activity of some newly synthesized 1,3-thiazine derivatives, International Journal of Research in Pharmacy and Chemistry 3 (2013) 2231–2781.

[77] A.N. Pearce, et al., Anti-inflammatory thiazine alkaloids isolated from the New Zealand ascidian Aplidium sp.: inhibitors of the neutrophil respiratory burst in a model of gouty arthritis, Journal of Natural Products 70 (6) (2007) 936–940.

[78] A.A. Magd-El-Din, H. Yosef, M. Abdalla, Synthesis of potent antitumor oxo quinazoline, pyrazole and thiazine derivatives, Australian Journal of Basic and Applied Sciences 6 (3) (2012) 675–685.

[79] W. Wang, et al., Synthesis and antitumor activity of the thiazoline and thiazine multithioether, International Journal of Organic Chemistry 2 (02) (2012) 117.

[80] E. Campaigne, P. Nargund, 3-Alkyl-1,3-thiazane derivatives and precursors as antiradiation agents, Journal of Medicinal Chemistry 7 (2) (1964) 132–135.

[81] A.B. Dobkin, G.M. Wyant, The physiological effects of intravenous anaesthesia on man, Canadian Anaesthetists' Society Journal 4 (3) (1957) 295–337.

[82] M. Ferreira, et al., Synthesis of 1,3-thiazine-2,4-diones with potential anticancer activity, European Journal of Medicinal Chemistry 70 (2013) 411–418.

[83] Z. Szakonyi, et al., Stereoselective synthesis and cytoselective toxicity of monoterpene-fused 2-imino-1,3-thiazines, Molecules 19 (10) (2014) 15918–15937.

[84] M.H. Al-Huniti, et al., Facile synthesis of some novel pyrido [3′, 2′: 4, 5] thieno [2, 3-b][1,4] thiazine-8-carboxylic acids, Molecules 12 (3) (2007) 497–503.

[85] K. Pluta, et al., Anticancer activity of newly synthesized azaphenothiazines from NCI's anticancer screening bank, Pharmacological Reports 62 (2) (2010) 319–332.

[86] M. Jeleń, et al., 6-Substituted 9-fluoroquino [3, 2-b] benzo [1,4] thiazines display strong antiproliferative and antitumor properties, European Journal of Medicinal Chemistry 89 (2015) 411–420.

[87] S. Choi, et al., De novo design and in vivo activity of conformationally restrained antimicrobial arylamide foldamers, Proceedings of the National Academy of Sciences 106 (17) (2009) 6968–6973.

[88] M. Koketsu, et al., Synthesis of 1,3-thiazine derivatives and their evaluation as potential antimycobacterial agents, European Journal of Pharmaceutical Sciences 15 (3) (2002) 307–310.

[89] A. Flemming, New strategy to target pseudomonas, Nature Reviews Drug Discovery 9 (4) (2010).

[90] T. Sindhu, et al., Comparative anti-tubercular activity of sulfa drug substituted 1,4-thiazines and 1,3-thiazines, International Journal for Pharmaceutical Research Scholars 3 (2014) 24–30.

[91] R.K. Haynes, From artemisinin to new artemisinin antimalarials: biosynthesis, extraction, old and new derivatives, stereochemistry and medicinal chemistry requirements, Current Topics in Medicinal Chemistry 6 (5) (2006) 509–537.

[92] P. Bhavani, K. Meena, Synthesis, Characterization, Antimicrobial Screening, and Molecular Docking Studies of Some Novel 2, 6-Bis (4-Chlorobenzoyl)-3, 5-Bi's (Substituted Aryl) Tetrahydro-1,4-Thiazine 1, 1-Dioxide Derivatives, 2015.

[93] R. Soliman, et al., Formation of thiazoles, thiazines, and thiadiazines from 1-phthalazine thiosemicarbazides as potential anticonvulsants, Journal of Pharmaceutical Sciences 70 (1) (1981) 94–96.

[94] T.S. Jagodzinski, et al., Synthesis and Biological Activity of Certain Novel DerivativeS of L#-Pyrrolo [1, 2-c][1, 3] Thiazine, 2003.

[95] S. Murthy, Synthesis, characterization and evaluation of anticonvulsant activity of some new 4-aryl-8-arylidene-5,6-dihydro-2-imino-6,6-dimethyl-4H, 7H-[3, 1] benzothiazine derivatives, Journal of Global Trends in Pharmaceutical Sciences 5 (4) (2014) 2199–2203.

[96] L.-Q. Zhang, et al., Synthesis and anticonvulsant activity of some 7-alkoxy-2H-1,4-benzothiazin-3 (4H)-ones and 7-alkoxy-4H-[1, 2, 4] triazolo [4, 3-d] benzo [b][1,4] thiazines, Chemical and Pharmaceutical Bulletin 58 (3) (2010) 326–331.

[97] N. Edayadulla, P. Ramesh, Synthesis of 2,6-dicarbethoxy-3,5-diaryltetrahydro-1,4-thiazine-1, 1-dioxide derivatives as potent anticonvulsant agents, European Journal of Medicinal Chemistry 106 (2015) 44–49.

[98] B. Malawska, Application of pharmacophore models for the design and synthesis of new anticonvulsant drugs, Mini Reviews in Medicinal Chemistry 3 (4) (2003) 341–348.

[99] E. Mofarrah, et al., The Hindered Internal Rotations in Isomerism Forms of a Particular Phosphorane Involving 2-Chloro-phenothi Azine: Dynamic 1 H NMR Study, 2015.

[100] E. Odin, P. Onoja, J. Saleh, Synthesis, characterization and neuropharmacological activity of novel angular pentacyclic phenothiazine, International Journal of the Physical Sciences 8 (26) (2013) 1374–1381.

[101] S.V. Blokhina, et al., Synthesis, biological activity, distribution and membrane permeability of novel spiro-thiazines as potent neuroprotectors, European Journal of Medicinal Chemistry 77 (2014) 8–17.

[102] Y. Bansal, O. Silakari, Multifunctional compounds: smart molecules for multifactorial diseases, European Journal of Medicinal Chemistry 76 (2014) 31–42.

[103] H.M. Wu, et al., Synthesis of pyrazine-1,3-thiazine hybrid analogues as antiviral agent against HIV-1, influenza a (H1N1), enterovirus 71 (EV71), and coxsackievirus B3 (CVB3), Chemical Biology and Drug Design 88 (3) (2016) 411–421.

[104] M.R. Jones, et al., Multi-target-directed phenol-triazole ligands as therapeutic agents for Alzheimer's disease, Chemical Science 8 (8) (2017) 5636–5643.

[105] G.F. Makhaeva, et al., Conjugates of γ-carbolines and phenothiazine as new selective inhibitors of butyrylcholinesterase and blockers of NMDA receptors for Alzheimer disease, Scientific Reports 5 (2015) 13164.

[106] S. Bachurin, et al., Molecular construction of multitarget neuroprotectors 1. Synthesis and biological activity of conjugates of substituted indoles and bis (dimethylamino) phenothiazine, Russian Chemical Bulletin 64 (6) (2015) 1354–1361.

[107] A-l. Hui, et al., Design and synthesis of tacrine-phenothiazine hybrids as multitarget drugs for Alzheimer's disease, Medicinal Chemistry Research 23 (7):) (2014) 3546–3557.

[108] G. Tin, et al., Tricyclic phenothiazine and phenoselenazine derivatives as potential multitargeting agents to treat Alzheimer's disease, MedChemComm 6 (11) (2015) 1930–1941.

[109] G.W. Kaatz, et al., Phenothiazines and thioxanthenes inhibit multidrug efflux pump activity in *Staphylococcus aureus*, Antimicrobial Agents and Chemotherapy 47 (2) (2003) 719–726.

Chapter 8

Indoles: As Multitarget Directed Ligands in Medicinal Chemistry

Navriti Chadha, Om Silakari

Punjabi University, Patiala, India

Chapter Outline

1. INTRODUCTION

With identification dating back to the 1860s by Baeyer and coworkers while studying the structure of indigo [1], indole gained immense popularity as a pharmacophore in numerous pharmacological conditions. Indole is a fusion of a six-membered benzene and five-membered pyrrole ring. Its interesting molecular architecture attracts the eyes of organic and medicinal chemists to design derivatives of medicinal interest. Electrophilicity of the nucleus has been well described in literature that leads to synthesis of various indole derivatives via nucleophilic addition and cycloaddition [2]. This property of indole improves the portfolio for synthetic manipulation. Moreover, indole is a prominent phytoconstituent across various plant species and is produced by a variety of

Key Heterocycle Cores for Designing Multitargeting Molecules. https://doi.org/10.1016/B978-0-08-102083-8.00008-X

bacteria. The natural occurrence of this nucleus can be owed to its presence in the essential amino acid tryptophan [3]. The indole-derived phytoconstituents and bacterial metabolites are a result of biosynthesis via coupling of tryptophan with other amino acids. For this reason, it is a constituent of flower perfumes, pharmacologically active indole alkaloids, and some animal hormones such as serotonin and melatonin. Some naturally occurring indole alkaloids have gained FDA approval, including vincristine, vinblastine, vinorelbine, and vindesine for antitumor activity; ajmaline for antiarrhythmic activity; and physostigmine for glaucoma and Alzheimer's disease (AD). Taking inspiration from these natural compounds several synthetic drugs have reached the patient's bedside such as indomethacin (NSAID), ondansetron (chemotherapy-induced nausea and vomiting), fluvastatin (hypercholesterolemia), and zafirlukast (leukotriene receptor antagonist) (Fig. 8.1). The success of these compounds indicates the importance

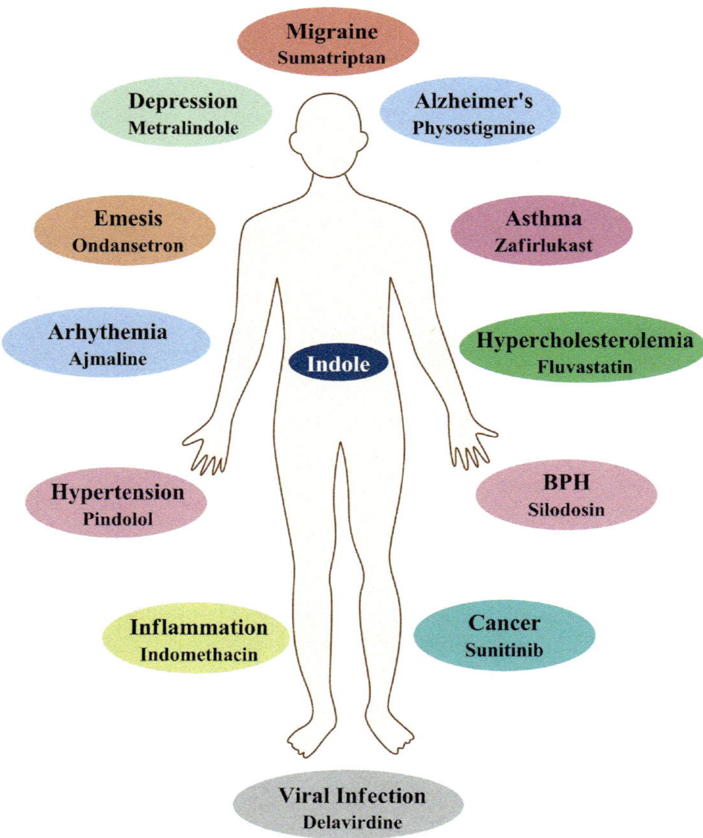

FIGURE 8.1 Multitargeted activity of indole nucleus.

of the ring system in multidisciplinary fields including pharmaceutical and agrochemical industries.

Tryptophan   Serotonin   Melatonin

The detailed description of the various pharmacological activities displayed by the indole nucleus have already been discussed in our previous publication [4]. Our effort, in this chapter, is an exhaustive compilation of the multitarget directed ability of indole moiety present in literature. This compiled information may be beneficial for medicinal chemists working in this area to design derivatives with good pharmacological activity.

2. CHEMISTRY

The indole nucleus is a planar bicyclic molecule containing 10π electrons (8π electrons from double bonds and 2π from a lone pair of electrons from nitrogen), thus it is aromatic according to Huckel's rule. It acts as a feeble base and protonates only in the presence of strong acids. The 3-position of the nucleus has the highest electron density and is the most reactive position for electrophilic substitution while the slightly acidic nature of the NH makes it susceptible to N-substitution reactions under basic conditions [5].

The unsubstituted indole (C_8H_7N) is a colorless crystalline solid with an unpleasant odor. The compound has a melting point range of 52–54°C (126–129°F) and boiling point of 254°C. The molar mass of the compound is 117.15 g/mol and a density of 1.17 g/cm^3 [6]. The mass spectrum of indole has been reported, which shows a molecular ion peak at *m/z* 117 (base peak). Two strong peaks appear at m/z 90 (relative abundance 40%) and 89 (24%) due to the loss of HCN and H_2CN, respectively [7]. The ^{1}H- NMR spectra of indole show the presence of seven peaks due to seven protons. The proton attached to the nitrogen of the ring appears maximally downfield as a singlet at 7.81 ppm. The aromatic protons showed a multiplet in the range of 7.64–6.52 ppm (7.64, 7.27, 7.18, 7.11, 7.04, and 6.52).

3. SYNTHESIS

The synthesis of indole has been carried out using various starting materials across the literature reports of organic chemistry. The literature reports revealed that different strategies have been explored and widely used for indole construction that include Bartoli indole synthesis [8], Bischler indole synthesis [9], Fischer indole

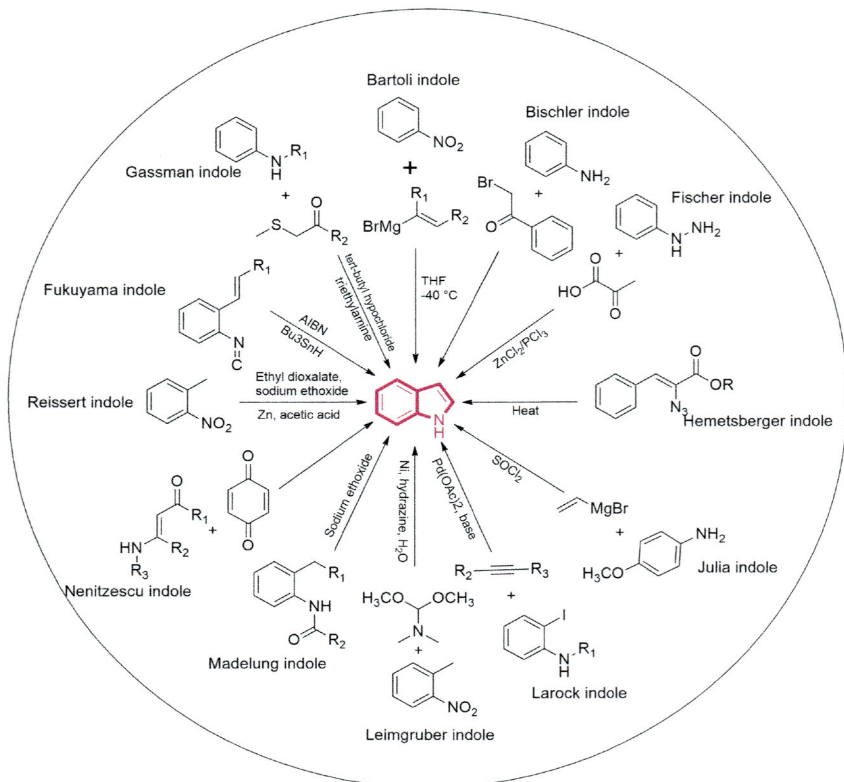

FIGURE 8.2 Synthetic routes for indole synthesis.

synthesis [10], Hemetsberger indole synthesis [11], Julia indole synthesis [12], Larock indole synthesis [13], Leimgruber indole synthesis [14,15], Madelung indole synthesis [16], Nenitzescu indole synthesis [17], Reissert indole synthesis [18], Fukuyama indole synthesis [19], Gassman indole synthesis [20], and Sundberg indole synthesis [21]. The details of these reactions are summarized in Fig. 8.2.

4. INDOLE AS A PRIVILEGED SUBSTRUCTURE

In this section, the various activities associated with the indole derivatives, including antitumor, antimicrobial, antiviral, antiinflammatory, antidepressant, anticholinergic, antimigraine, antiemetic, antihypertensive, and others are described.

4.1 Antitumor Activity

Cancer is one of the major causes of mortality across the globe affecting billions of people worldwide [22]. Various anticancer agents are reported that act via varying mechanisms. A number of molecules are approved by the FDA and even more are undergoing clinical evaluation containing indole nucleus.

From the category of indole alkaloids, a vast number of compounds have been isolated and evaluated for their cytotoxic potential. Among those vincristine, vinblastine, vindesine, and vinorelbine, isolated from *Catharanthus roseus*, achieved success and were approved for the treatment of various cancerous conditions including leukemia, lymphoma, melanoma, breast cancer, non–small cell lung cancer (NSCLC), and so on. These drugs are currently listed in the World Health Organization's (WHO) list of essential medicines [23]. They show their antitumor effect via inhibition of the polymerization of tubules in cancer cells.

First isolated in 1961, vincristine (Oncovin, marketed by Eli Lilly) got FDA approval in 1963 as a part of a cancer chemotherapy regimen for the treatment of Hodgkin's, non-Hodgkin's lymphoma, acute lymphoblastic leukemia, nephroblastoma, large B-cell lymphoma, retinoblastoma, rhabdomyosarcoma, follicular lymphoma, and others [24]. Vinblastine (Velban, marketed by Eli Lilly) was approved in 1963 for the chemotherapy regimen of Hodgkin's lymphoma [24]. Vinorelbine (Navelbine, marketed by Pierre Fabre Group) is a semisynthetic vinca alkaloid approved in 1994 for the treatment for NSCLC, breast cancer, and rhabdomyosarcoma. However, these drugs are associated with various side effects including chemotherapy-induced peripheral neuropathy, nausea, vomiting, hair loss, gastrointestinal problems, and depression [25].

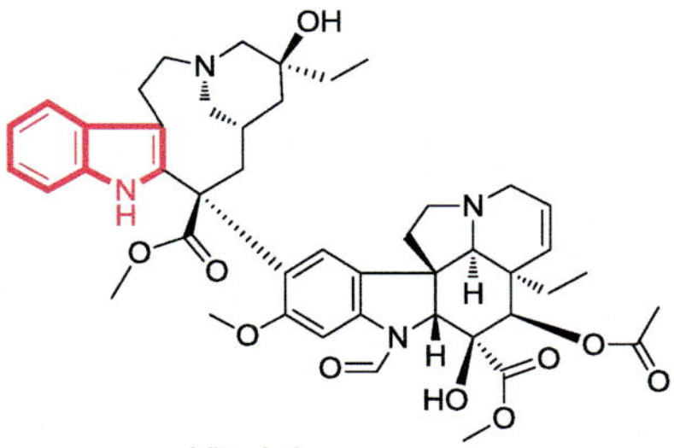

Vincristine

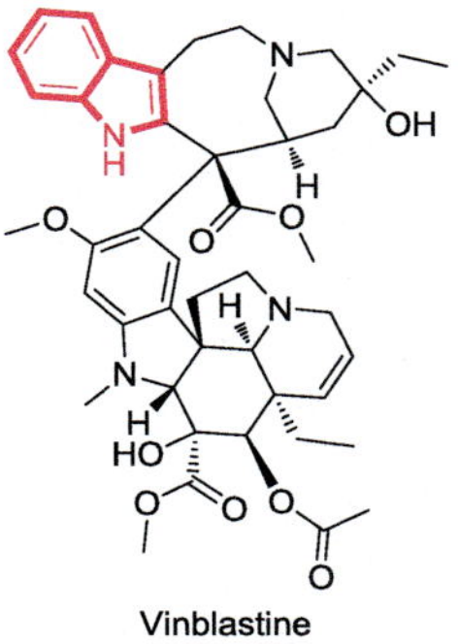

Vinblastine

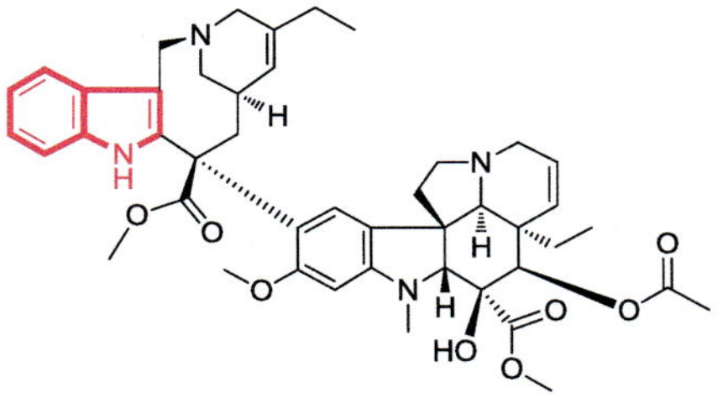

Vinorelbine

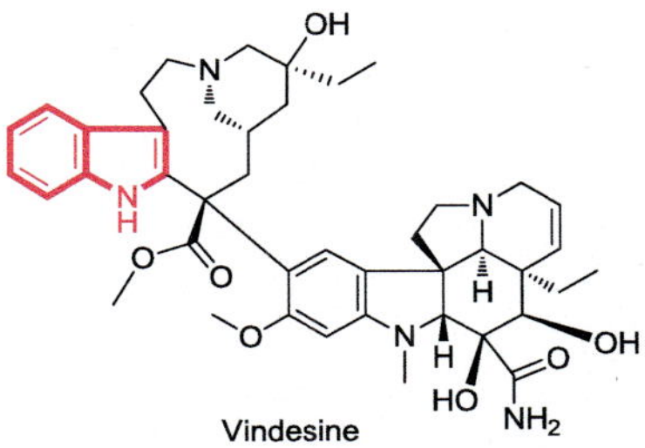

Vindesine

Numerous synthetic derivatives have been synthesized and evaluated for antitumor activity. Sunitinib and osimertinib are two marketed drugs with indole nucleus implicated for the treatment of renal cell carcinoma, gastrointestinal stromal tumor, and NSCLC, respectively. Sunitinib (Sutent) is a multitargeted receptor tyrosine kinase (RTK) inhibitor of platelet-derived growth factor receptor (PDGFR) and vascular endothelial growth factor receptor (VEGFR) along with RET (rearranged during transfection). In 2008, the drug received approval for the treatment of renal cell carcinoma and gastrointestinal stromal tumor [26]. The crystal structure of this molecule available with VEGFR (PDB: 4AGD) shows that nitrogen of the indole nucleus forms a hydrogen bond with carbonyl of Glu917 while the carbonyl group on the ring interacts with NH of Cys919 of the pocket. The oxindole ring system substituted with fluoro group forms hydrophobic interaction with Leu1035 residue of the pocket [27]. Osimertinib (Tagrisso) is a third-generation inhibitor of mutated (T790M) epidermal growth factor receptor. In 2015, AstraZeneca received license for marketing osimertinib for the treatment of T790M mutation positive NSCLC [24,28,29]. Panobinostat (Farydak) received FDA approval for the treatment of multiple myeloma [30]. The drug has been developed by Novartis to act as a pan-selective histone deacetylase (HDAC) inhibitor. It is under Phase III clinical trial for the treatment of Hodgkin's lymphoma, cutaneous T-cell lymphoma, and under Phase II clinical trial against myelodysplastic syndrome, breast cancer, and prostate cancer [31,32]. The drug is also being evaluated for its antiretroviral potential and is in Phase I/II trial for the same [33]. Another oral drug by AstraZeneca Alectinib (Alecensa) is ALK inhibitor implicated for crizotinib-resistant NSCLC that was approved by the FDA in 2015 [34].

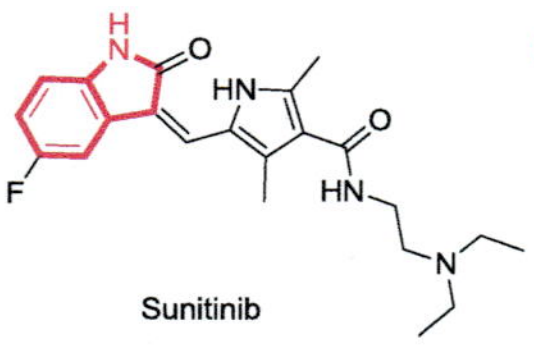

Sunitinib

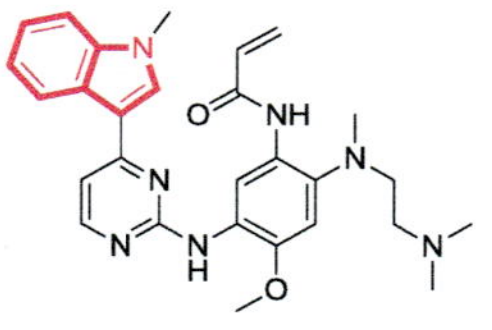

Osimertinib

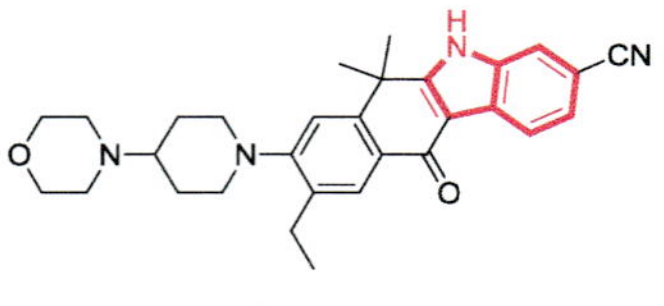

Alectinib

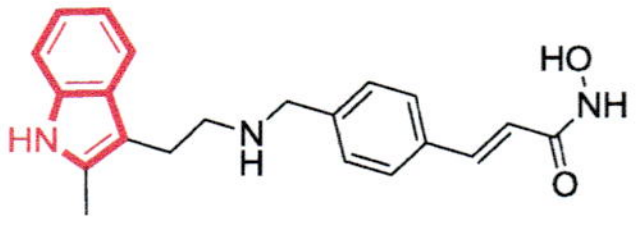

Panobinostat

Some compounds are under evaluation against a plethora of cancerous conditions. These include AstraZeneca's ATR/mTOR inhibitor AZ-20, which showed activity against the HT29 tumor cancer cell line (IC_{50} value of 50 nM) [35]. LAQ-824 (Dacinostat) is another HDAC inhibitor (32 nM) developed by Novartis whose clinical effectiveness is being evaluated for the treatment of NSCLC [36]. PCI-34051 is another investigational inhibitor of HDAC8 implicated for the

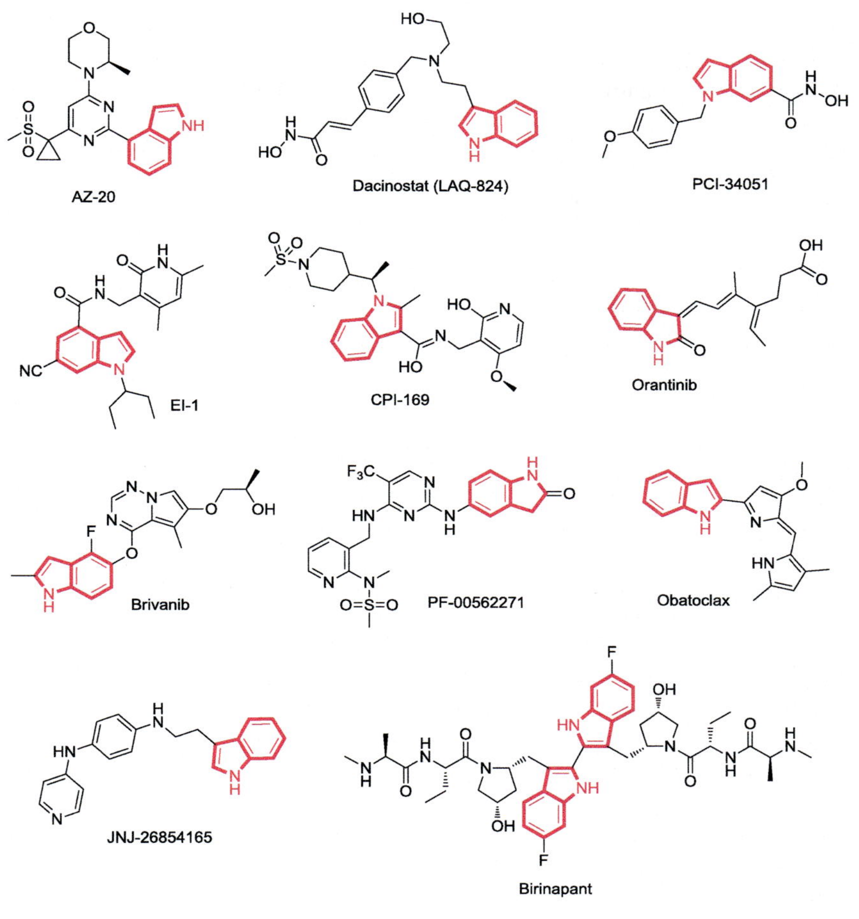

AZ-20

Dacinostat (LAQ-824)

PCI-34051

EI-1

CPI-169

Orantinib

Brivanib

PF-00562271

Obatoclax

JNJ-26854165

Birinapant

treatment of T-cell lymphoma or leukemia [37]. EI1 (developed by Novartis) is a selective EZH2 (Enhancer of zeste homolog 2) inhibitor effective against wild and mutated (Y641F) forms of the enzyme [38]. CPI-169 (developed by Constellation Pharmaceuticals) is another inhibitor with the ability to block EZH2 enzyme [39]. The EZH2 inhibitors are implicated in the treatment of various types of lymphomas. Orantinib (developed by Taiho Pharmaceuticals) is a potent and bioavailable RTK inhibitor capable of blocking VEGFR2, PDGFR, and fibroblast growth factor receptor, thus inhibiting tumor growth and angiogenesis. However, the drug failed in Phase III clinical trials for the treatment of hepatocellular carcinoma, and it is now being considered for advanced solid tumors [40]. Motesanib is an experimental drug candidate showing inhibitory activity on VEGFR, PDGFR, and stem cell factor receptor. In clinical evaluation, the drug failed to show promising results in Phase II evaluation for effectiveness in advanced NSCLC and metastatic breast cancer [41]. However, it showed positive results in Phase II clinical trial against thyroid cancer [42]. The crystal structure of VEGFR2 kinase domain with motesanib (PDB ID: 3EFL) depicts that NH of the amide group of the molecule forms a hydrogen bond with Glu885 and carbonyl group with Asp1046; nitrogen of the terminal pyridine ring with Cys919 and pyridine ring linked with amide forms π-cation interaction with Lys868 amino acid residue. PF-00562271 is a potent focal adhesion kinase (FAK; IC_{50} value = 1.5 nM) and proline rich tyrosine kinase (PYK2; IC_{50} value = 13 nM) inhibitor implicated for the treatment of hepatocellular carcinoma [43]. Obatoclax (developed by Gemin X) is an investigational drug with BCl-2 inhibitory activity. The drug has shown effectiveness in various cancerous conditions including leukemia, lymphoma, small-cell lung cancer, non-Hodgkin's lymphoma, multiple myeloma, myelofibrosis, and mastocystosis [44–47]. It is currently undergoing clinical evaluation for various cancerous states. JNJ-26854165 (Serdemetan), a product of Johnson and Johnson Pharmaceutical and Research Development Pvt. Ltd., has been clinically evaluated for the treatment of advanced stage or refractory solid tumors. It shows its pharmacological activity via inhibition of HDM2 ubiquitin ligase and activation of p53 [48]. Birinapant (TL32711, Teralogic Pharmaceuticals) is SMAC mimetic antagonist having inhibitory effects on apoptosis. It is a clinical candidate for the treatment of hematological malignancies, solid tumors, and acute myelogenous leukemia [49,50].

GSK2606414 and GSK2656157 are products of GlaxoSmithKline that act as protein kinase R like ER kinase inhibitors that are implicated in cell proliferation and differentiation [51,52]. Enzastaurin is a protein kinase C beta (PKCβ) inhibitor with antineoplastic activity. The drug is undergoing clinical evaluation for the treatment of lymphoma, breast cancer, prostate cancer, NSCLC, leukemia, colorectal cancer, ovarian cancer, renal cell carcinoma, and pancreatic cancer [53,54]. Sotrastaurin (developed by Novartis) is a pan-PKC inhibitor that is being clinically evaluated for effectiveness in the treatment of lymphocytic leukemia, lymphoma, melanoma, and kidney transplantation [55,56]. Another PKC inhibitor, Go 6976, is a PKCα and β inhibitor (2.3 and 6.2 nM, respectively) that has been reported to inhibit invasion of urinary bladder cancer cells [57]. Rucaparib (developed by Agouron Pharmaceuticals) is an investigational candidate for advanced solid tumor, breast and ovarian cancer with BRCA1 and BRCA2 mutation that shows its effect via inhibition of poly (ADP-ribose) polymerase (PARP-1) [58]. The catalytic domain of PARP-1 in complex with rucaparib shows (PDB ID: 4RV6) that amide linkage in the molecule form hydrogen bonds with Gly863 and Ser904 amino acid residues while indole nucleus forms π-π interaction withTyr907. TAK-901 (developed by Millennium Pharmaceuticals, Inc.) is an investigational multitargeted Aurora A/B kinase inhibitor implicated for hematologic malignancies or lymphoma. In addition, the drug has also shown to inhibit JAK2, c-src, and Abl kinases [59]. Tivantinib (ARQ-197) is a c-MET inhibitor implicated against solid tumor, NSCLC, colorectal cancer, prostate cancer, hepatocellular carcinoma, and others [60]. AZD-3463, a preclinical candidate designed by AstraZeneca, is an orally bioavailable ALK inhibitor showing effectiveness in crizotinib-resistant NSCLC cell line [61]. YH-239-EE is a highly potent p53-MDM2 antagonist with antimyeloid leukemia activity [62]. THZ1 is a selective and irreversible CDK7 inhibitor that has activity reported in leukemia. The compound has been shown to bind covalently to the Cys312 residue of the active site of CDK7 protein [63].

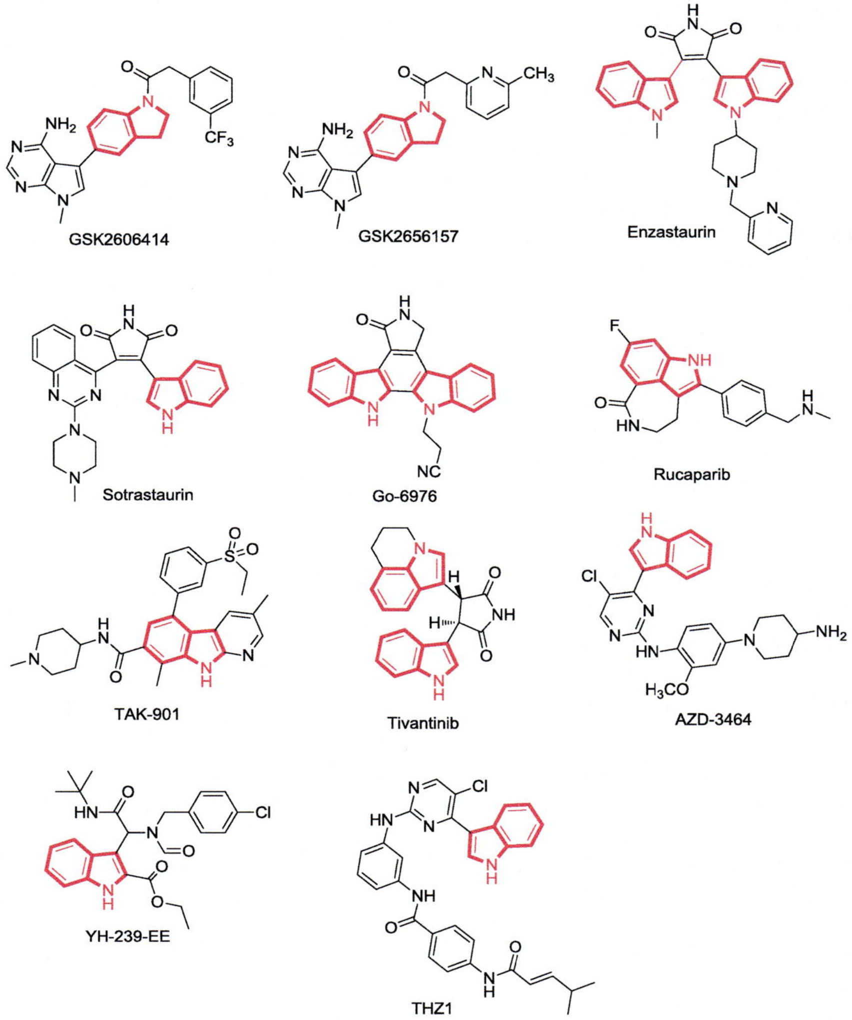

GSK2606414

GSK2656157

Enzastaurin

Sotrastaurin

Go-6976

Rucaparib

TAK-901

Tivantinib

AZD-3464

YH-239-EE

THZ1

The chemical structures of the marketed drugs when observed critically show that substitution of an indole ring at 3-position yielded highly potent molecules. Sunitinib and osimertinib show the substitution of nitrogen-containing ring systems with diethylamino substitution attached via different linkers. On the other hand, Alectinib contains a tetracyclic ring system with a fused indole ring. This ring system further has a nitrogen-containing ring (piperdine) followed by morpholine moiety. Panobinostat also contains the indole substituted at 3-position with aromatic benzyl substituent attached via an ethyl amino linker.

Other molecules in clinical trials are indole derivatives substituted at 3-position including AZ-20, dacinostat, CPI-169, orantinib, JNJ-26854165, birnapant, enzastaurin, sotrastaurin, tivantinib, AZD-3464, YH-239-EE, and THZ1. Various types of substitutions have been tried at this position to yield potent derivatives. For the other derivatives, 2- (obatoclax, rucaparib) and 5- (EI-1, brivanib, PF-00562271, GSK2606414, GSK2656157, TAK-901) positions of indole have been explored. In addition, substitution at nitrogen of indole as well as cyclic derivatives has yielded potent molecules. These substitutions have also yielded potent anticancer agents. Considering the substitution pattern, potency, and mechanism of action of these indole derivatives, medicinal chemists may design new derivatives with better potency and lesser toxicological implications.

4.2 Antiviral Activity

Viral diseases are one of the widespread infections around the globe that include common cold, influenza, chicken pox, herpes, gastroenteritis, human immunodeficiency virus (HIV), hepatitis, Ebola virus, among others. Antiviral therapy plays a key role in controlling the outbreak of the viral infections. To date, a number of indole containing molecules have been reported in the literature.

Arbidol (Umifenovir), manufactured by Russian pharmaceutical company Pharmstandard, is a broad-spectrum antiviral agent that has been demonstrated to possess activity against a number of enveloped and nonenveloped viruses by inhibiting the fusion of viral capsid with the host cell membrane. The drug has been shown to possess potency against influenza A, B, and C viruses; respiratory syncytical virus; hepatitis B and C viruses; human rhinovirus type 14; coxsackie B3 virus; and adenovirus type 7 [64]. The drug in complex with H3N2 influenza virus hemagglutinin (PDB ID: 5T6N) shows that a carbonyl group of ethyl ester group forms a hydrogen bonding interaction with Lys307 amino acid residue [65].

Delavirdine (Rescriptor), an antiretroviral agent marketed by ViiV healthcare, was approved by the FDA in 1997 [66]. It acts as a nonnucleoside reverse transcriptase inhibitor (NNRTI) and is used for the treatment of HIV. It is an inhibitor of chytochrome P450 enzyme CYP3A4 and interacts with many medications. The drug has been found to be hepatotoxic [67,68]. The cocrystal structure of delavirdine complexed with HIV-1 reverse transcriptase (PDB ID: 1KLM) shows that the carbonyl group and NH of the indole ring forms a hydrogen bonding interaction with amino acid Lys103 of the protein [69].

Atevirdine (U-87201E) is a bis-heteroarylpiperazine with in vitro activity against HIV-1. It is a Phase I clinical candidate that acts as NNRTI against HIV-1 as well as zidovudine-resistant HIV-1 [70].

GSK2248761 (Fosdevirine) is another NNRTI developed by GlaxoSmith-Kline that is under Phase II clinical evaluation. The compound has been reported to have subnanomolar activity against wild type as well as NNRTI-resistant mutant HIV [71]. The evaluation of clinical efficacy of the compound has been put on hold due to reports of seizures in treatment-experienced patients [72].

Golotimod (SCV-07) is an orally bioavailable synthetic peptide containing the amino acids D-glutamine and L-tryptophan connected by a gamma-glutamyl linkage with potential immune stimulating, antimicrobial, and antineoplastic activities. SciClone Pharmaceuticals Inc. carried out Phase II clinical trials of the drug for the treatment of hepatitis C (HCV). The clinical data demonstrated SCV-07 to be safe and well-tolerated at both administered doses. Results showed that SCV-07 did not meet the study's primary efficacy endpoint of a 2-log reduction in viral load from baseline level. A Phase II study is still on-going with SCV-07 in attenuating oral mucositis in subjects with head and neck cancer; however, no further study is listed for HCV [73].

Panobinostat (LBH589) is an experimental drug developed by Novartis as a nonselective histone deacetylase inhibitor (HDAC inhibitor) for treatment of multiple myeloma (Phase III) and acute myeloid leukemia (Phase II). It is currently being used in a Phase I/II clinical trial that aims at curing AIDS in patients on highly active antiretroviral therapy (HAART). In this technique, panobinostat is used to drive the HIV DNA out of the patient's DNA, with the expectation that the patient's immune system in combination with HAART will destroy it. This is the first proof of a viral "kick" leading to consistent plasma release of viral particles [30].

BILB-1941 is a NS5B inhibitor that demonstrated antiviral activity in patients chronically infected with genotype 1 HCV. It belongs to the category of allosteric nonnucleoside inhibitor of HCV NS5B polymerase that inhibits replication in replicon systems. It also displayed improved absorption, distribution, metabolism, and excretion profiles, and showed the most optimal balance between antiviral potency and a consistent cross-species pharmacokinetic profile [74].

BMS-79132, a cyclopropyl-fused indolobenzazepine HCV NS5B RNA-dependent polymerase inhibitor, developed by Bristol-Myers Squibb, was found to perform distinguishing antiviral, safety, and pharmacokinetic properties that resulted in its selection for clinical evaluation [75]. The X-ray crystallographic structure of RNA directed RNA polymerase in complex with BMS-79132 shows that sulphamoyl and carbonyl group of carboxamide attached to 5-position of the indole ring forms hydrogen bonding interaction with Arg503 amino acid residue present in the binding site [75].

MK-8742, a second generation tetracyclic indole-based NS5A inhibitor, is currently under Phase II clinical evaluation for the treatment of HCV infection. In combination with MK-5172, an NS3/4A protease inhibitor, this drug exhibited improvements in the genetic barrier while maintaining potency, yielding amazing results in terms of efficacy (90%–100%), tolerability, and safety [76].

(7R)-14-Cyclohexyl-7-{[2-(dimethylamino)ethyl](methyl)amino}-7,8-dihydro-6H-indolo[1,2-e][1,5]benzoxazocine-11-carboxylic Acid (MK-3281), a Phase II clinical candidate discovered by Merck Research Laboratories, is a potent and orally bioavailable second-generation tetracyclic allosteric finger-loop inhibitor of the HCV NS5B polymerase with indolo-benzoxazocine scaffold [77].

Sulfonylindolecarboxamide (L-737126), reported by Merck Laboratories, is a NNRTI active against NNRTI-resistant mutant carrying in vitro activity against wild and mutant HIV-1 in the low nanomolar range and cytotoxicity for MT-4 cells [78].

Enfuvirtide (T-20; Fuzeon), the peptide anti-HIV drug targeting gp41N-terminal heptad repeat, was approved by the FDA in 2003 as the first HIV fusion/entry inhibitor for treatment of HIV/AIDS patients who fail to respond to the current antiretroviral drugs. However, because T20 lacks the pocket-binding domain, it exhibits low anti-HIV-1 activity and short half-life [79].

TMC647055, a nonzwitterionic 17-membered macrocyclic indole, is a potent and selective inhibitor of the HCV NS5B polymerase. This compound was identified to possess nanomolar potency ($EC_{50} = 77$ nM) in HCV replicon cells, limited toxicity and off-target side effects, and encouraging preclinical pharmacokinetic profiles characterized by high liver distribution. It is currently being evaluated in Phase II clinical trials in combination with simeprevir [80].

Chemically, indole derivatives demonstrating antiviral activity are substituted at 2-, 3-, 5-, and 6-positions of the nucleus. Arbidol, delavirdine, and atevirdine are the derivatives with substitutions at 2-position of indole. GSK2248761, golotimod, and panobinostat are antiviral molecules that possess electronegative substitutions at 2-position. BIBL1941 is a derivative possessing α, β-unsubstituted carboxylic acid attached via a linker at 5-position of the indole ring. BMS-791325, MK-8742, MK-3281, and TMC-647055 are indole derivatives cyclized at 1- and 2-positions of the ring. In addition, these derivatives are also substituted at 5-position.

4.3 Antiinflammatory Activity

Inflammation is a complex biological response of body tissues to the harmful stimuli such as pathogens, irritants, and damaged cells. Antiinflammatory agents find their use in a wide range of pathological conditions including rheumatoid arthritis, osteoarthritis, migraine, gout, spondylitis, and so on.

Only a few natural antiinflammatory agents containing indole scaffold have been reported. Cycloexpansamine A and B, and penicillinolide A isolated from marine cultures of the *Penicillium* species are among the few. These are spiroindolinone alkaloids having antiinflammatory properties [81].

Indole-based indomethacin (Indocid) is a nonsteroidal antiinflammatory drug (NSAID) that acts as a nonselective inhibitor of cyclooxygenase (COX-1 and COX-2), which in turn blocks the production of prostaglandins [82]. It was approved by the FDA in 1965 for the treatment of fever, pain, and swelling, and has been implicated in a number of clinical indications including ankylosing spondylitis, gout, migraine, osteoarthritis, rheumatoid arthritis, Paget's disease of bone, juvenile arthritis, and so on [83]. The drug has been shown to have adverse effects such as peptic ulcers, ranitidine, dyspepsia, heartburn, diarrhea, and hyperkalemia, which occur due to nonselective inhibition of COX.

Tenidap (developed by Pfizer), a COX/5-LOX inhibitor, is a cytokine modulating antiinflammatory drug candidate with antirheumatoid activity. The drug was observed to inhibit interleukin 1 synthesis [84]. However, it was rejected by the FDA in 1996 due to reported liver and kidney toxicity [85].

Acemetacin (Emflex, manufactured by Merck KGaA) is a glycolic acid ester prodrug of indomethacin that acts as an NSAID. It has the advantage of reduction of gastric damage over indomethacin and has an implication in the treatment of osteoarthritis and rheumatoid arthritis [86].

Etodolac (Lodine, manufactured by Almirall Limited), an FDA-approved (1991) NSAID acting as selective COX-2 inhibitor, results in a decrease of prostaglandin levels in the body with better gastrointestinal tolerability. The drug is implicated in the treatment of osteoarthritis and rheumatoid arthritis [87].

The marketed indole-based antiinflammatory molecules possess substitutions at 1-, 2-, and 3-positions. At 3-position of indomethacin carboxylic acid substitution is present, which is esterified with glycolic acid in acemetacin, while in tenidap this position is substituted with a carboxamide group. The 3-positions of indomethacin and acemetacin are substituted with a parachlorobenzoyl group, while the thiophen-2-carbonyl group is substituted in a tenidap molecule. Etodolac is another marketed molecule in which an indole ring is cyclized with tetrahydropyran at 2- and 3-position of indole. Acetic acid group is substituted at the 1-position of the cyclized tetrahydropyran. Thus, presence of acidic moiety can be considered to be important in the antiinflammatory activity of the molecule.

4.4 Antidepressant Activity

Antidepressants are the class of drugs used in the treatment of major depressive disorder characterized by pervasive and persistent low mood. Depression is a disorder that is rising at an alarming rate, especially in young people. Various indole-containing antidepressant drugs that have been reported in the literature are discussed in the following.

Some of the synthetic antidepressants containing an indole nucleus have been synthesized in the 1990s, and act as reversible inhibitors of monoamine oxidase A (MAO-A). MAO-A is an enzyme that catalyzes the oxidative deamination of serotonin, dopamine, and norepinephrine. Thus, inhibitors of this enzyme prevent the breakdown of monoamine neurotransmitters and increase their availability. Three drugs, namely, metralindole (Inkazan), pirlindole (Pyrazidol), and terindole were synthesized in Russia as MAO-A inhibitors [88,89]. These drugs belong to the class of tetracyclic antidepressants.

Metralindole · Pirlindole · Terindole

Metralindole, pirlindole, and terindole are tertracyclic derivatives of indole. Metralindole is a triaza derivative with methoxy substitution at 5-position, while pirlindole and terindole are diaza derivatives with methyl and cyclohexyl group substituted at 5-position of indole, respectively.

4.5 Anticholinergic Activity

Cholinergic drugs are a class of drugs that modulate the functioning of the neurotransmitter acetylcholine. These drugs find their application in diseases such as glaucoma, AD, delayed gastric emptying, asthma, chronic bronchitis, and so on.

Physostigmine is an important reversible acetylcholinestrase inhibitor derived from *Physostigma venenosum* (calabar bean). It indirectly stimulates nicotinic and muscarinic acetylcholine receptor and is used in the treatment of glaucoma, AD, and delayed gastric emptying. It has been shown to improve long-term memory, and is used in the treatment of orthostatic hypotension, myasthenia gravis, and cholinergic disorders. It is an antidote for *Datura stramonium* and *Atropa belladonna* poisoning [90].

Physostigmine

4.6 Antimigraine Activity

Migraine is a severe neurological disorder characterized by throbbing pain in the head. According to a survey by WHO, between 50% and 75% of adults around the world suffer from migraine and 14% of adults report severe headaches. The triptans and NSAIDs are the most widely used treatment of migraine disorders. Triptan is a class of tryptamine-based drugs most widely used for the treatment of migraine that acts as an agonist on serotonin 5-HT_{1B} and 5-HT_{1D} receptors. These molecules have an indole ring as their basic scaffold. Some of the FDA-approved triptans include sumatriptan, rizatriptan, naratriptan, eletriptan, almotriptan, frovatriptan, avitriptan, and zolmitriptan.

Sumatriptan (Treximet, Imitrex, Imigran), developed by GlaxoSmithKline received FDA approval for the treatment of migraine in 1992 [91]. Naratriptan (Amerge) is another drug of the same class marketed by GlaxoSmithKline after getting approved in February, 1998 [92]. Rizatriptan (Maxalt), a product of Merck, also received approval in 1998 as a second-generation triptan [93]. Almotriptan (Axert) received marketing approval in 2001 for the management of heavy migraine attacks (Janssen Pharms) [94]. Pfizer in 2002 got approval for eletriptan (Relpax), which is used as an abortive medication in migraine attacks [95]. Frovatriptan (Frova) has been licensed to Endo Pharms (2001) for the treatment of migraine and short-term menstrual migraine [96]. Zolmitriptan (Zomig), marketed by AstraZeneca, was approved in 2001 for the treatment of acute migraine attacks [97]. Another triptan class candidate, Avitriptan (BMS-108, 048), never received FDA approval and is used as an investigational drug [98].

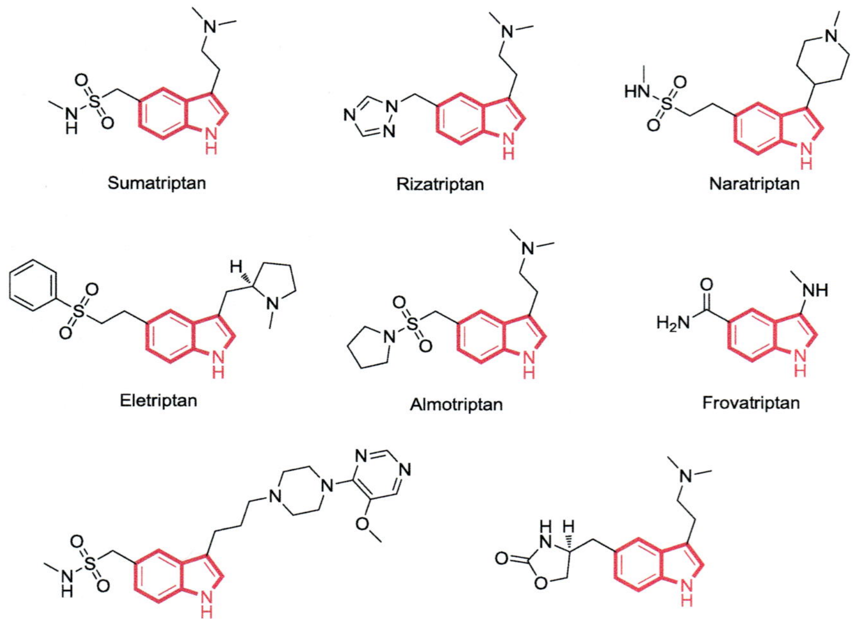

Sumatriptan

Rizatriptan

Naratriptan

Eletriptan

Almotriptan

Frovatriptan

Avitriptan

Zolmitriptan

Chemically, the triptan class of drugs contains an indole ring as their basic nucleus. These drugs show their action by acting as an agonist on a 5-HT$_{1B/1D}$ receptor, which is a G-protein coupled receptor with a ligand gated ion channel. These drugs have some common features that are required for their action on the receptor. The nitrogen (NH) of indole nucleus acts as a hydrogen bond donor to the Ser352 residue in the serotonin receptor pocket. In addition, the ring provides lipophilicity to the drug to form hydrophobic as well as π-π interaction with Phe residues of the pocket. The presence of protonated nitrogen is also sorted for ionic interaction with acidic residue Asp118 of the pocket. Other groups are required to form hydrogen bonds within the receptor site [99].

Methysergide (UML-491) was a prescription drug for prophylaxis of cluster headaches/migraine headaches, but no longer recommended due to retroperitoneal/retropulmonary fibrosis [100]. It was approved by the FDA in 1962, and acts as a 5-HT$_{2B}$ receptor inhibitor. It is one of the most effective medications for the prevention of migraine, but not for the treatment of an acute attack. Nicergoline (Sermion), is an ergot derivative used for the treatment of migraine of vascular origin, senile dementia, cerebral thrombosis, atherosclerosis, Raynaud's disease, bradycardia, and interstitial nephritis. Chemically, methysergide and nicergoline are cyclic derivatives of indole cyclized at 3- and 4-positions of the ring.

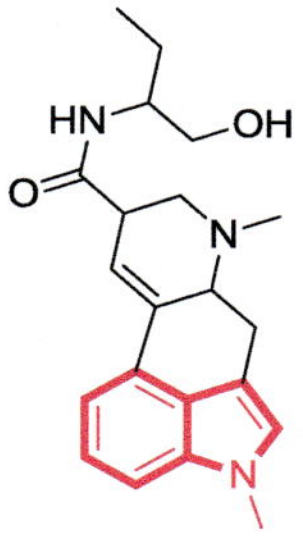

Methysergide

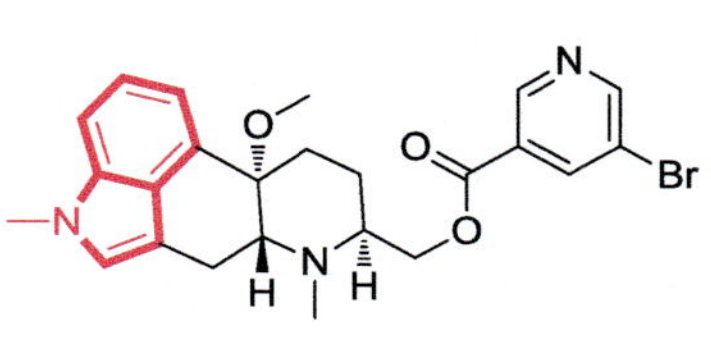

Nicergoline

4.7 Antiemetic Activity

Antiemetics are a class of drugs effective in the treatment of nausea and vomiting. These drugs are particularly used for the treatment of motion sickness, cancer chemotherapy, and drug-induced nausea and vomiting. Various serotonin 5-HT$_3$ antagonists have been approved and prescribed as antiemetics, especially for the treatment of cancer chemotherapy-induced as well as postoperative nausea and vomiting. These antagonists are generally categorized as -setrons and antagonize the activity of 5-HT$_3$ receptors present at the terminals of the vagus nerve. These are sometimes also prescribed in irritable bowel syndrome. The marketed drugs of this category including ondansetron, alosetron, ramosetron, dolasetron, and tropisetron have indole as a basic nucleus in their structure. The structure–activity relationship developed for this class of drugs highlighted the importance of an aromatic center (to form hydrophobic interactions), a basic amine moiety (form hydrogen bonds with the receptor), and a carbonyl linker (provide proper distance between the two moieties) [101].

Ondansetron (Zofran), a carbazole containing prototypic antiemetic generally used in cancer chemotherapy, radiation therapy, or surgery-induced nausea and vomiting, and is also indicated in morning sickness and hyperemesis gravidarus of pregnancy and gastroenteritis. It is a product of GlaxoSmithKline approved by the FDA in 1991. The side effects associated with the drug include diarrhea, headache, QT prolongation, and allergic reaction [102]. Dolasetron

(Anzemet, approved in 1997), is an indole containing drug used in the treatment of chemotherapy-induced, postoperative, and postradiation nausea and vomiting, and acute gastroenteritis [103]. It is a prodrug with hydrodolasetron being the active metabolite formed by the action of carbonyl reductase enzyme. Tropisetron (Navoban), marketed by Novartis, is another antiemetic that is also used as an analgesic in fibromyalgia. The common side effects associated with the drug include headache, constipation, and dizziness [104]. Ramosetron (Nasea), is implicated as an antiemetic as well as for irritable bowel syndrome symptoms [105]. Alosetron (Lotronex, GlaxoSmithKline), a 5-HT3 antagonist implicated in irritable bowel syndrome, was approved by the FDA in 2000 [106].

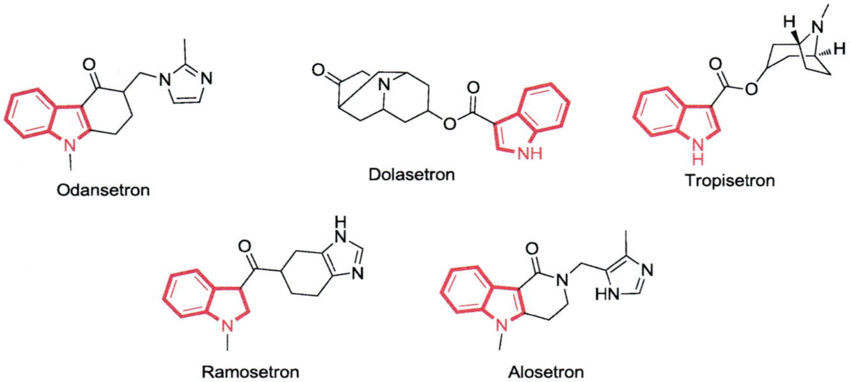

Odansetron

Dolasetron

Tropisetron

Ramosetron

Alosetron

Chemically, setrons are indole derivatives possessing heterocyclic ring substitution at 3-position of the ring linked with a carbonyl group. In ondansetron and alosetron, 2- and 3-positions of indole nucleus are cyclized with 6-membered lactone. The 1-position is also substituted with a methyl group in ondansetron, ramosetron, and alosetron. The indole ring substituted with a basic ring system linked with a carbonyl group at 3-position seems to be an important pharmacophore for 5-HT$_3$ antagonist activity.

4.8 Antihypertensive Activity

These are a class of drugs that are used for the treatment of hypertension (high blood pressure). Various indole based antihypertensive agents have been reported in literature. The reported drugs have varying mechanisms of action: α/β blocker, ACE inhibitor, thiazide-like diuretic, AT$_1$ antagonist, and so on. Perindopril (Coversyl, Coversum, or Aceon) and Trandopril (Mavik) are two marketed angiotensin converting enzyme (ACE) inhibitors having octahydro-indole nucleus in their structures. They act by inhibiting an angiotensin converting enzyme, a key component of a renin-angiotensin-aldosterone system [107,108]. The side effects associated with these drugs include hypotension, dry cough, headache, dizziness, fatigue, nausea, and renal impairment. Trandolapril is a prodrug with trandolaprilat (deesterified form of ethyl ester) being the active metabolite. Trandolaprilat is around eight times more active and has more half-life than the parent drug. The crystal structure of trandolaprilat in complex with angiotensin converting enzyme (PDB ID: 2X93) shows that a carboxylic acid group attached to 2- position of the octahydro-indole ring of the ligand forms hydrogen bonding interaction with Gln265, Tyr504, and Lys495 amino acid residues; another side chain carboxylic acid group forms hydrogen bonds with Glu368, Tyr507, and coordination complex with Zn^{2+} metal ion; and the quaternarized NH^{2+} interacts with His337 and Ala338 amino acid residues of the ACE enzyme. A similar type of interaction pattern was observed for the prodrug perindoprilat in complex with ACE (PDB ID: 2X94).

Indapamide (marketed by Servier) is a dihydro-indole based thiazide-like diuretic used in the treatment of hypertension as well as management of heart failure [109]. The common side effects include fatigue, orthostatic hypotension, allergies, and hypokalemia. Carvedilol (Coreg) is a well-established beta blocker with implication in the treatment of congestive heart failure and hypertension. First approved for use in 1995, the drug has a carbazole-based architecture [110]. Pindolol (Visken, Novartis) is another selective beta blocker that received FDA approval in 1982 for the treatment of hypertension [111]. It is also implicated in angina pectoris, arrhythmia, acute stress, and depression.

Chemically, perindopril and trandolapril are octahydro-indole derivatives substituted with carboxylic acid at 2-position and a hydrophobic group and ethyl carboxylic ester attached at 1-position with a linker group. Indapamide is also substituted at 1-position with a 4-chloro-3-methylsulphonylphenyl group attached via carboxamide linker. Carvedoilol and pindolol are beta blockers that have substituted ether groups at 4-position of the indole ring.

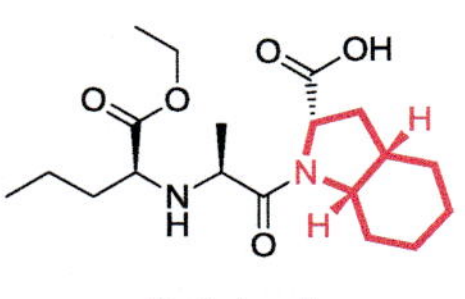

Perindopril

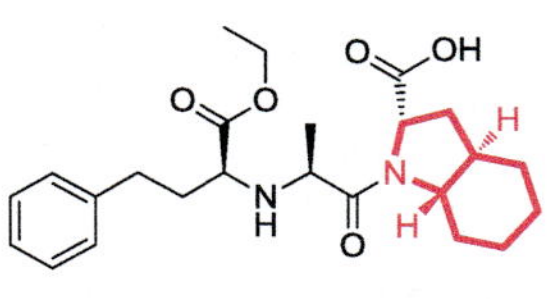

Trandolapril

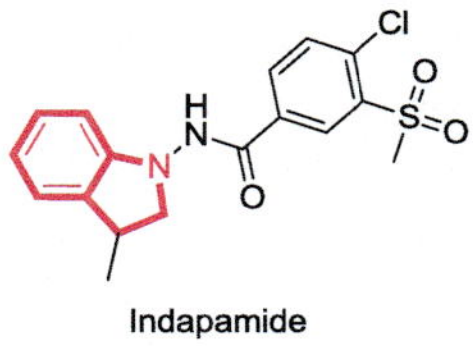

Indapamide

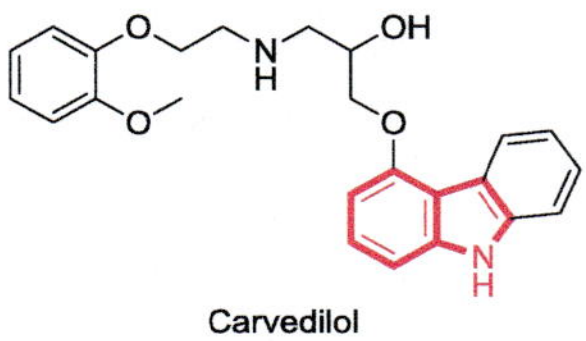

Carvedilol

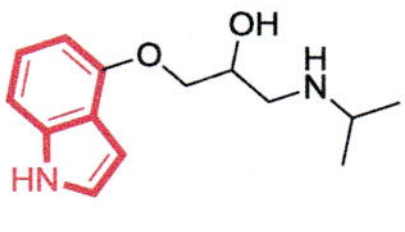

Pindolol

Some indole-based natural phytoconstituents have also been reported to possess antihypertensive activities. Reserpine (Raudixin) is an indole alkaloid with antihypertensive and antipsychotic properties. The antihypertensive action of reserpine is a result of depleted catecholamines from peripheral sympathetic nerve endings [112]. Hirsutine is an indole alkaloid isolated from *Uncaria rhynchophylla* which has shown to relieve headache and dizziness due to hypertension. It has also reported to possess sedative and antiarrhythmic pharmacological activities [113].

Reserpine

Hirsutine

4.9 Miscellaneous

Various drugs with indole in their architecture have been implicated in various other disorders. Antiarrhythmic drugs ajmaline and vinopocetine with indole in their structures are marketed drugs. Ajmaline (Gilurytmal), an indole alkaloid isolated from the roots of *Rauwolfia serpentina* as well as *C. roseus*, is a Class Ia antiarrhythmic agent. The drug has shown to lengthen the refractory period of the heart by blocking sodium ion channels and also interfering with human-ether-a-go-go-related (hERG) gene potassium ion channels. Ajmaline is implicated in the treatment of Wolff–Parkinson–White syndrome, which is characterized by arrhythmias with ventricles contracting prematurely resulting in tachycardia and shortened refractory period [114]. Vinopocetine (Cavinton), a semisynthetic derivative of vinca alkaloid vincamine is marketed for vasodilation and nooptropic for memory impairment and cerebral metabolism. It is also used as an antiinflammatory agent in the treatment of Parkinson's disease and AD. It acts by selective inhibition of voltage-sensitive sodium channels, resulting in decreased extracellular calcium ions in striatal nerve endings [115].

Silodosin (Rapaflo), α1-adrenoceptor antagonist, is indicated in the symptomatic treatment of benign prostatic hyperplasia [116]. It received FDA approval in 2008. Bazedoxifene (Viviant, Pfizer), third-generation selective estrogen receptor modulator, is used in the prevention of postmenopausal osteoporosis. FDA approved the combination of bazedoxifene and premarin (conjugated estrogens) for menopausal osteoporosis and treatment of moderate to severe hot flushes. The drug is under trial for use in dyspareunia (painful sexual intercourse), breast cancer, and pancreatic cancer [117]. Fluvastatin (Lescol) belongs to the statin class of drugs for hypercholestrolomia and prevention of cardiovascular disease. The drug shows its effect by blocking the HMG-CoA reductase enzyme that catalyzes an important step in cholesterol synthesis. It is also known to exhibit antiviral activity against hepatitus C virus (HCV) [118]. Icatibant (Firazyr) is a peptidomimetic orphan drug consisting of 10 amino acids, which is a selective and specific antagonist of bradykinin B2 receptors, received FDA approval in 2011. It is used in the treatment of acute attacks of hereditary angioedema in adults with C1-estrase inhibitor deficiency [119]. Zafirlukast (Accolate) is an oral leukotriene receptor antagonist for the maintenance treatment of asthma, which acts by blocking the activity of 5-lipoxygenase. It blocks the action of cysteinyl leukotrienes on the CysLT1 receptors, thus reducing the constriction of airways, build-up of mucus in the lungs, and inflammation of the breathing passages [120].

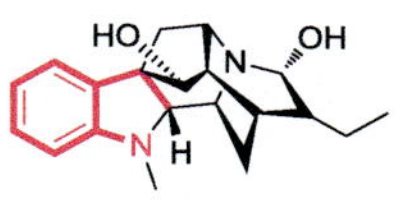

Ajmaline

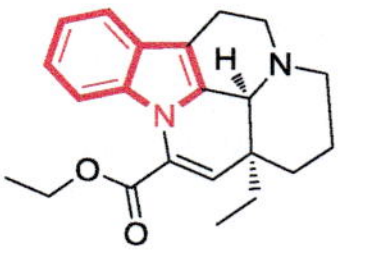

Vinopocetine

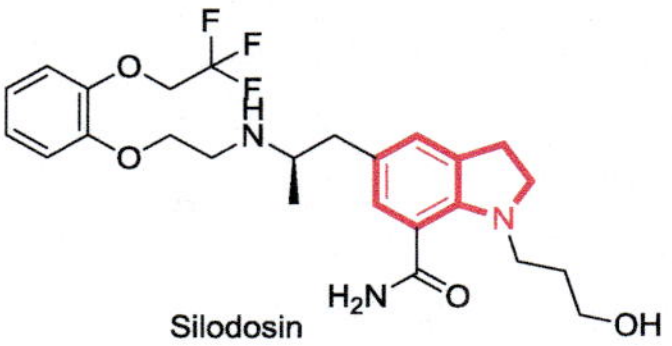

Silodosin

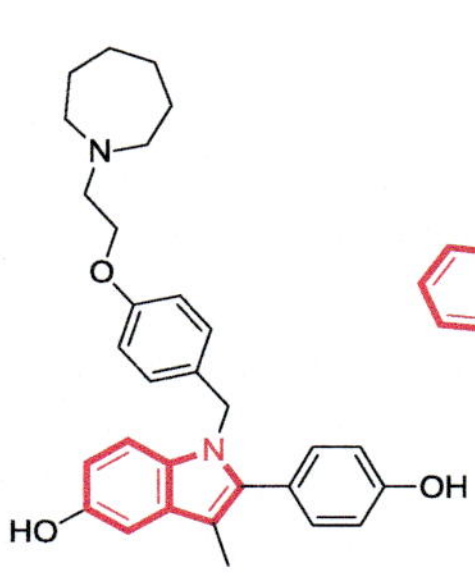

Bazedoxifene

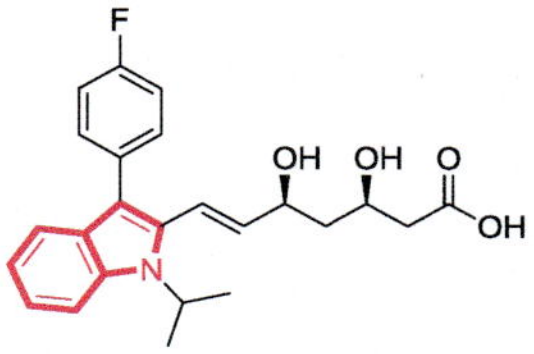

Fluvastatin

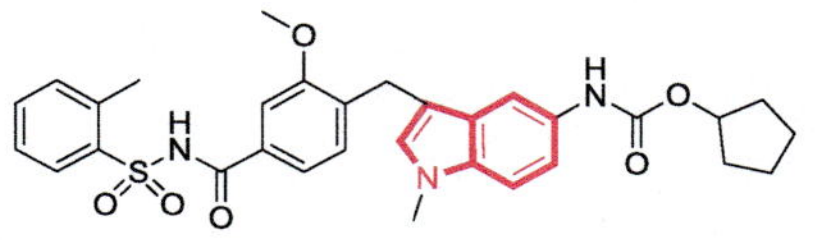

Zafirlukast

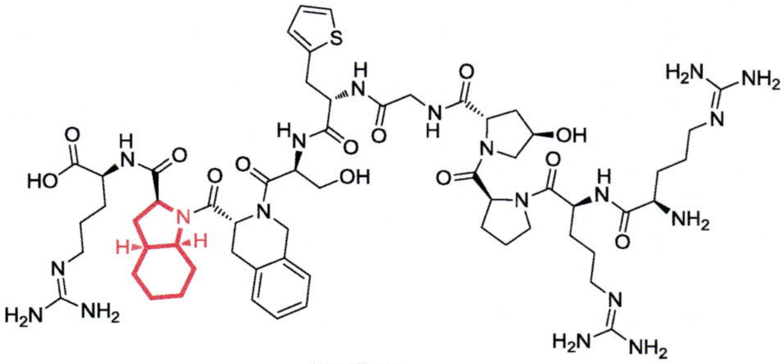

Icantibant

5. ROLE OF INDOLE AS A MULTITARGETING AGENT IN MULTIFACTORIAL DISEASES

Many diseases are highly variable and heterogeneous, involving multiple organ systems, tissues, and potential targets. These multifactorial diseases such as atherosclerosis, AD, metabolic syndrome, asthma, rheumatoid arthritis, osteoarthritis, diabetic complications, malaria, tuberculosis, cancer, neurotrauma, various CNS disorders, and multiple sclerosis have extremely complex etiopathology and involve two or more pathophysiological indications. Such diseases are mitigated or treated by a multitude of drugs belonging to different therapeutic classes. However, the current paradigm has been shifted toward development of a single entity with multiple pharmacological activities. One important example containing indole nuclei is sunitinib (Sutent), which is a multitargeted

RTK inhibitor of PDGFR and VEGFR along with RET. In 2008, the drug received approval for the treatment of renal cell carcinoma and gastrointestinal stromal tumor [26]. TAK-901 (developed by Millennium Pharmaceuticals, Inc.) is an investigational multitargeted Aurora A/B kinase inhibitor implicated for hematologic malignancies or lymphoma. In addition, the drug has also shown to inhibit JAK2, c-src, Abl kinases [59]. Various other indole derivatives that act as multitarget-directed molecules for complex disease conditions are discussed in the following section.

5.1 Multitargeted Agents Against Alzheimer's Disease

AD is an incurable neurodegenerative disorder characterized by deterioration of memory and cognitive functions in elderly people. With research advances in molecular biology and technology, multiple credible hypotheses about the progress of AD have been proposed; multitarget drugs have emerged as an innovative therapeutic approach for AD. Cholinergic hypothesis is a classic hypotheses of AD. Acetylcholinesterase (AChE) accounts for approximately 80% ACh hydrolysis in healthy brain [121]. The crystallographic structure of human AChE indicates that the active pocket of AChE is composed of two separate ligand binding sites, a peripheral cationic site (PAS) at the entrance and a catalytic active site (CAS) at the bottom. Inhibitors binding to these sites can interdict combination with AChE [122].

Zhu et al. synthesized and evaluated indole-2,3-dione derivatives as dual inhibitors of AChE and phosphodiesterase 5A (PDE5A) for the treatment of AD [123]. Molecule **1** appeared as the most active compound from the series and showed IC_{50} values of 79.43 nM and 50.00 μM against AChE and PDE5A enzymes, respectively, and additionally showed low cell toxicity to A549 cells. Docking studies showed that carbonyl of indole-2,3-dione plays a key role in the interaction with the active site of the enzymes. Campagna et al. synthesized and evaluated a series of isatin-3-arylhydrazones inhibitors of Aβ1-40 aggregation. The most active compounds carry methoxy- or hydroxy-substituents in the indolinone 5,6-positions and lipophilic groups such as iPr and Cl at 4′- and 3′-positions, respectively, of the phenylhydrazone moiety, suggesting a significant role of hydrogen bonding and hydrophobic interactions in the binding of arylhydrozono-1H-2-indolinone derivatives [124]. Pisani et al. developed a Mannich base of 5-methoxyisatin-3-(4-isopropylphenyl)hydrazone as its prodrug (**2**). It displayed a high solubility along with a good hydrolytic stability in acidic medium. In addition, it allowed a fast and efficient release in human serum of the parent hydrazone compound. The neuroprotective effect was also well correlated with the radical scavenging properties. The compound showed AChE and BuChE inhibition of 10.4 μM and 5.43 μM (IC_{50} value) and Aβ aggregation inhibition of 1.30 μM [125].

Tacrine was the first drug approved by the FDA for the treatment of AD. However, it has a drawback of poor selectivity due to simultaneous BuChE and hepatic toxicity due to oxidative stress. Rodriguez Franco et al. developed novel tacrine and melatonin hybrids having selective human AChE inhibitory activity and antioxidant properties. The most potent inhibitor of series **3** had around 40,000-fold more potency than tacrine and in oxygen radical absorbance capacity assay, this compound was 2.5-fold higher than trolox, a vitamin E analogue [126]. Benek et al. developed indolotacrine derivatives with AChE and MAO inhibitory activity. For this purpose, the 2-aminoindole-3-carbonitrile scaffold of the indole series was fused with the structure of the potent AChEI tacrine or 7-methoxytacrine (7-MEOTA). Moreover, the resulting indolotacrines also resemble β-carboline alkaloids (e.g., harmine), which are known MAO inhibitors. The most potent compound from the series, compound **4**, showed IC_{50} values of 0.49, 53.90, and 1.5 μM against MAO-A, MAO-B, and AChE enzymes. The overall inhibitory activities and potential to cross the blood–brain barrier make indolotacrine **4** a promising lead compound [127]. Bolea et al. synthesized a new family of multitarget molecules able to interact with AChE and butyrylcholinesterase (BuChE), as well as with MAO-A and -B [128]. These novel compounds were designed using a conjunctive approach that combines the benzylpiperidine moiety of the AChE inhibitor donepezil and the indolyl propargylamino moiety of the MAO inhibitor *N*-[(5-benzyloxy-1-methyl-1H-indol-2-yl)methyl]-*N*-methylprop-2-yn-1-amine (**5**), connected through an oligomethylene linker. The most promising hybrid (**5**) is a potent inhibitor of both MAO-A ($IC_{50}=5.2\pm1.1$ nM) and MAO-B ($IC_{50}=43\pm8.0$ nM) and is a moderately potent inhibitor of AChE ($IC_{50}=0.35\pm0.01$ μM) and BuChE ($IC_{50}=0.46\pm0.06$ μM). Thus, compound **5** emerged as a potential lead with disease modifying properties in AD.

Dominguez et al. obtained some indole derivatives from structure-based screening that were further synthesized and binding assays performed for BACE-1, AChE, and BuChE indicate their improved ligand efficiency and multitarget behavior across amyloid cascade and cholinergic pathways. The multitarget pharmacophore constituted the presence of hydroxyethylamine as an Asp dyad anchor for interaction with BACE-1 and one or two aromatic moieties to interact through π-stacking interactions with clusters of aromatic residues present both in the CAS and PAS of AChE. Thus, compound **6** was found to be the most potent inhibitor of AChE and BACE-1 with IC_{50} values of 9.1 and 2.5 μM, respectively [129].

López-Iglesias et al. reported Melatonin–*N,N*-dibenzyl (*N*-methyl)amine hybrids as potent neurogenic agents that show a balanced multifunctional profile covering neurogenic, antioxidant, cholinergic, and neuroprotective properties at low micromolar concentrations. The melatonin framework demonstrates antioxidant and neuroprotective features and could also interact with the AChE-PAS because of its aromatic character, while the protonable *N,N*-dibenzyl(*N*-methyl) amine, which is present in the well-known AChE inhibitor AP2238, interacts with the AChE-CAS. Compound **7** was found to be the most potent among the series for early neurogenesis, being twice as potent as melatonin with IC_{50} values of 4.7 μM and 2.9 μM in AChE inhibitory assay and oxygen radical absorbance capacity by fluorescence (ORAC-FL) assay, respectively [130]. Further, Luo et al. reported new multifunctional melatonin-derived benzylpyridinium bromides with potent cholinergic, antioxidant, and neuroprotective properties. The melatonin-based hybrids were designed considering the structure of benzylpyridinium salt, which may represent a privileged scaffold for the development of AChE inhibitors. In vitro studies showed that most of these compounds exhibited potent inhibitory activity toward h-AChE and h-BuChE, and good antioxidant capacity in the ORAC assay. In particular, compound **8** was the most attractive derivative, showing the highest potency to inhibit ChEs (AChE: $IC_{50}=0.11\,\mu M$; BuChE: $IC_{50}=1.1\,\mu M$) and good antioxidant ability (ORAC (trolox)$=3.41$) [131].

Lajarín-Cuesta et al. describe the synthesis of gramine derivatives and their pharmacological evaluation as multipotent drugs for the treatment of AD. They presented multitargeting voltage-gated Ca^{2+} channels studied for neurodegenerative diseases and Ser/Thr phosphatases with a key role in protein τ dephosphorylation would lead to discovery of new drugs for neurodegenerative diseases. The most active derivative, 1-benzyl-5-methyl-3-(piperidin-1-ylmethyl)-1H-indole (compound **9**, ITH12657), acts on a multitarget-based mechanism of action, aimed at neuronal Ca^{2+} control and Ser/Thr phosphatase activity, providing a promising strategy for the treatment of neurodegenerative diseases [132].

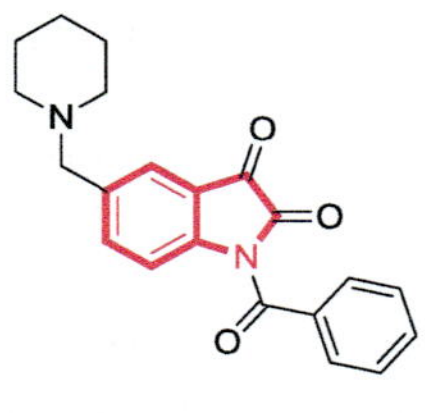

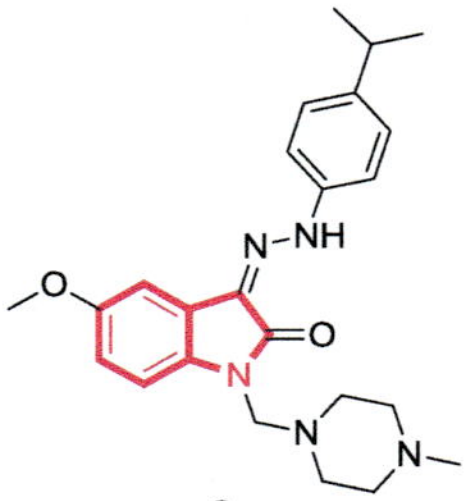

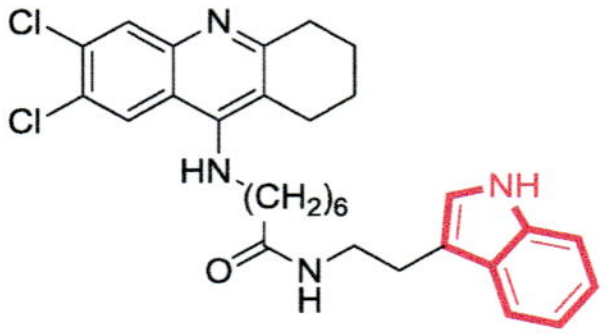

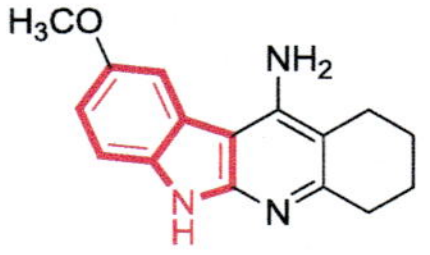

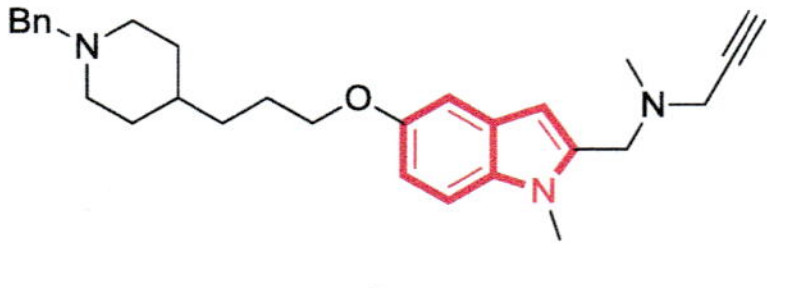

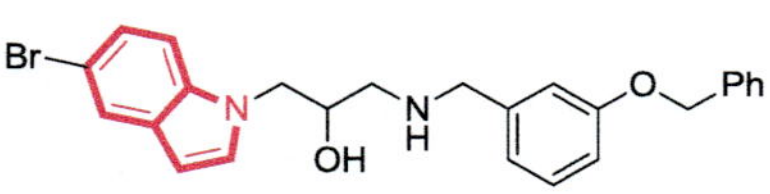

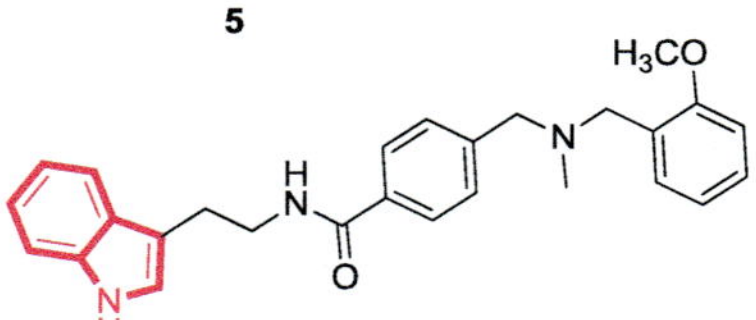

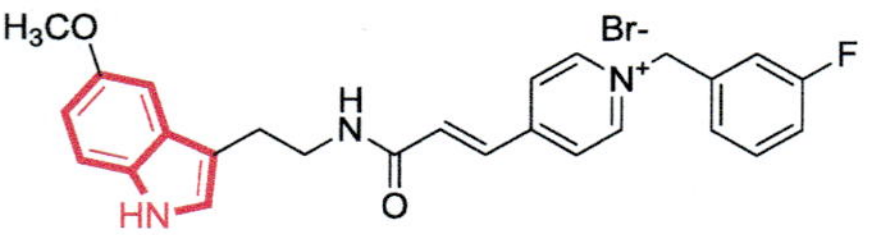

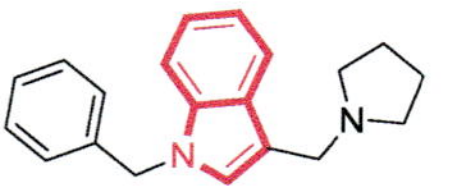

5.2 Multitargeted Agents for Cancer

Cancer evolves through a multistage process that requires the progressive accumulation of genetic lesions, activating oncogenes, and inactivating tumor suppressor genes [133]. Multitargeting is gaining popularity as it has ability to improve therapeutic efficacy, prevent drug resistance, and reduce therapeutic target-related adverse effects. Sunitinib (SU11248, marketed as Sutent by Pfizer) is an oral small-molecule inhibitor of multiple tyrosine kinases with antitumor and antiangiogenetic effects, which is approved for the treatment of gastrointestinal stromal tumor after disease progression or intolerance of imatinib therapy and for advanced renal cell carcinoma. The activity of sunitinib is based on the inhibition of VEGFRs 1–3, PDGFRα and PDGFRβ, steam cell factor receptor, FMS-related tyrosine kinase-3 (FLT3), and colony-stimulating factor-1 receptor [134–136]. Targeting protein kinases has a greater specificity and fewer side effects than the traditional cytotoxic therapy [137]. TAK-901 is another investigational multitargeted Aurora A/B kinase inhibitor implicated for hematologic malignancies or lymphoma. In addition, the drug has also shown to inhibit JAK2, c-src, and Abl kinases [59].

Other studies also indicate development of indole-based multitargeted agents for the treatment of gliomas and glioblastomas (GBMs). Daniele et al. reported structure–activity relationship of 2-phenylindolylglyoxylyldipeptide derivatives as murine double minute (MDM2)/translocator protein (TSPO) dual inhibitors for the treatment of gliomas considering indole as a privileged structure. As a result, compound **10** was obtained as the most potent compound from the series with reactivated the p53 functionality; inhibited viability of two human GBM cells; impaired proliferation of glioma cancer stem cells (CSCs), more resistant to chemotherapeutics, and responsible for GBM recurrence; and sensitized GBM cells and CSCs to the activity of temozolomide and directed its effects preferentially toward tumor cells with respect to healthy ones. Thus, compound **10** may represent a promising cytotoxic agent that is worthy of being further developed for a therapeutic approach against GBM, where the downstream p53 signaling is intact and TSPO is overexpressed [138]. In another study, Daniele et al. reported compound SA16 for its ability to inhibit the Phosphoinositide-dependent kinase-1 (PDK1) and Aurora kinase A (AurA) pathways at once, thus proving to be a useful tool for the simultaneous inhibition of the two kinases. SA16 was first identified as a PDK1 inhibitor designed by combining two pharmacophoric moieties known to bind the ATP binding site and the DFG-out pocket of PDK1 through a phenylglycine linker. The compound was further investigated for inhibition against 56 kinases, where it was found to be active against AurA, suggesting it was a prototype of a dual PDK1/AurA inhibitor. SA16 possessed a significant inhibition potency against both PDK1 ($IC_{50}=416\,nM$) and AurA ($IC_{50}=35\,nM$), suggesting that this compound could be studied as a prototype of a dual PDK1/AurA inhibitor. Moreover, it also blocked GBM cell proliferation, reduced tumor invasiveness, and triggered cellular apoptosis [139].

Huang et al. demonstrated naftopidil-related derivatives containing indole groups as alpha1-adrenoceptors antagonists and treatment of benign prostatic hyperplasia and a preventive medication against human prostate cancer. The modification of naftopidil was carried out by replacing the naphthyl ring with indole-3-carbinol to enhance anticancer property of the compound and 2-hydroxypropane moiety with amide structure to improve $\alpha_{1D/1A}$ binding affinity. Compound **11** demonstrated a potent and selective $\alpha_{1D/1A}$ antagonistic activity (47.9- and 19.1-fold for α_{1D} and α_{1A} compared with α_{1B}) and a potent antiproliferative activity in PC-3 cells ($IC_{50} = 15.70\,mM$) [140].

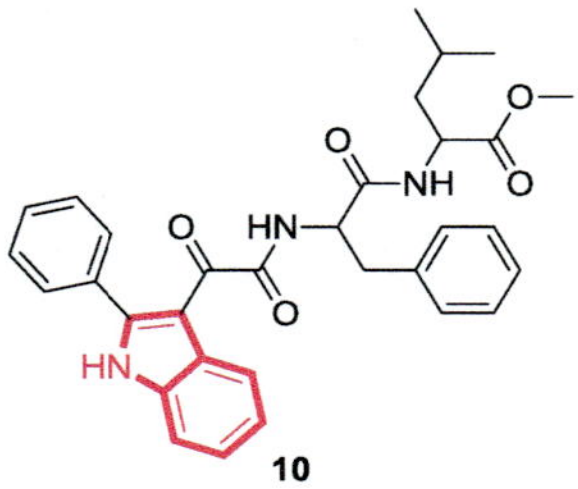

10

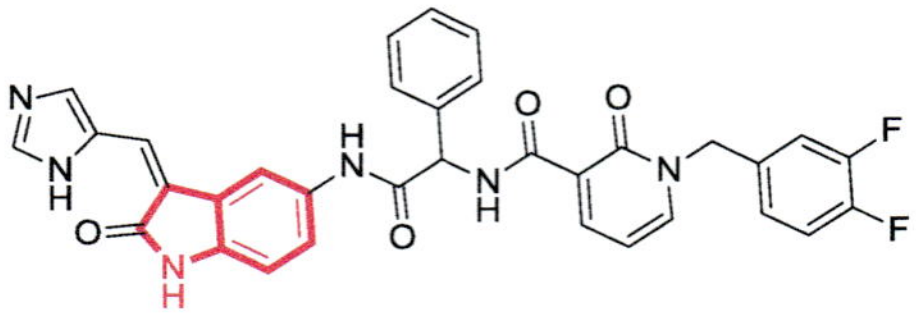

SA16

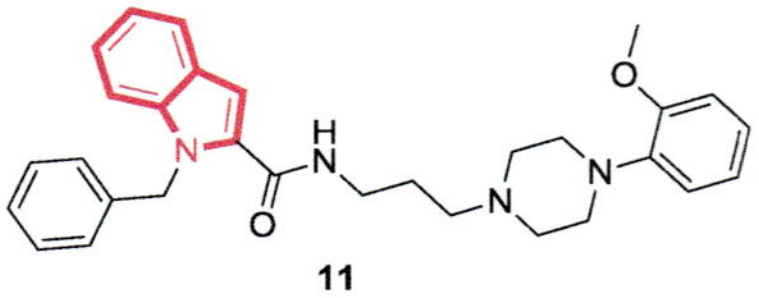

11

5.3 Multitargeted Agents for Diabetic Complications

Diabetic complication is a complex metabolic disorder developed primarily due to prolonged hyperglycemia in the body. It is characterized by the development of microvascular and macrovascular pathology. Pathophysiology of diabetic complications has been reported to occur via varying mechanisms, including polyol pathway, PKC pathway, AGEs pathway, and hexosamine pathway [141]. Stefek et al. reported that novel carboxymethylated pyridoindoles, characterized by antioxidant activity combined with the ability to inhibit aldose reductase, represent an example of a multitarget approach to the treatment of diabetic complications, severe diabetes-related health disorders of a multifunctional nature. One of the novel carboxymethylated pyridoindoles, (2-benzyl-2,3,4,5-tetrahydro-1H-pyrido[4,3-b]indol-8-yl)-acetic acid (compound **12**), was found to inhibit aldose reductase with the IC_{50} value $18.2 \pm 1.2\,\mu M$. Aldose reductase inhibitors contain acidic protons as a pharmacophoric feature. However, its ionization at physiological pH causes poor bioavailability. Thus, tertiary nitrogen was introduced in the structure to form double-charged zwitterionic species. This zwitterionic nature increases bioavailability and adds antioxidant activity to the compound, making it multifunctional for the treatment of diabetic complications [142].

Chadha et al. identified hybrids to thiazolidine-2,4-dione and indole as dual inhibitors of aldose reductase (ALR2) and poly (ADP-ribose) polymerase (PARP-1) using a structure-based design approach. A molecular docking and molecular dynamics simulation-based study led to identification of ALR2 and PARP-1 inhibitor **13**, which exhibited highest activity against PARP-1 enzyme ($IC_{50} = 1.34\,\mu M$) and low micromolar ALR2 inhibitory activity ($IC_{50} = 4.72\,\mu M$). The insight provided from the study concluded that close monitoring of the N-benzyl group on indolylated thiazolidine-2,4-dione is required to provide dual inhibitors of therapeutic significance [143].

12

13

5.4 Multitargeted Agents for Neuropathic Pain

Neuropathic pain is a complex phenomenon characterized by burning pain coupled with hyperalgesia and allodynia that involves several mechanisms in both the peripheral and central nervous systems. At present, first-line treatment options for this pathology are represented by antidepressants, anticonvulsants, and local anesthetics (gabapentin, lidocaine, tramadol, nortriptyline, doxepine) [144–147]. Both norepinephrine and nitric oxide are involved in the progression of pain. Since it is reported that bulkier groups show better potency and selectivity for the nNOS inhibitors, Mladenova et al. designed and synthesized 3,5-disubstituted indole derivatives having 6-membered rings as dual action selective neuronal nitric oxide synthase (nNOS) inhibitors with NERI activity. The highest active compound of the series, cis-(+)-**14**, exhibited dual nNOS and NET inhibition (IC_{50} of 0.56 and $1.0\,\mu M$, respectively) and excellent selectivity (88-fold and 12-fold) over eNOS and iNOS, respectively. Additionally, the compound was effective at reversing both allodynia and thermal hyperalgesia in a standard spinal nerve ligation rat neuropathic pain model [148].

14

5.5 Multitargeted Agents for Inflammatory Diseases

The arachidonic acid pathway is the root cause of several inflammatory diseases including asthma, atherosclerosis, irritable bowel syndrome, rheumatoid arthritis, and cancer. This cascade is catalyzed by the enzymes lipoxygenases and prostaglandins, respectively [149,150]. Singh et al. reported indole-based peptidomimetics as highly effective 5-LOX and COX-2 inhibitors. The IC_{50} value of compound **15** for COX-2 and 5-LOX was 2 nM and 6.3 nM, respectively, and significant reversal in the dextran-induced swelling and capsaicin-induced lickings in the hind paw of the mice. The general observations support the fact that stereochemistry of the molecule plays a critical role during its interaction with the cellular target. Moreover, the compounds with a free carboxyl group at the end of C-3 substituent were more potent than their corresponding esters [151].

Wei et al. screened out dual-target inhibitors against both the human leukotriene A4 hydrolase (LTA4H-h) and the human nonpancreatic secretory phospholipase A2 (hnps-PLA2) via a common pharmacophore mapping technique. In the arachidonic acid cascade, phospholipase A2 (PLA2) cleaves membrane phospholipids to release arachidonic acid, which is the precursor to a large family of proinflammatory eicosanoids, including prostaglandins and leukotrienes [152]. Inhibition of PLA2 depletes the sources of arachidonic acid and controls the generation of downstream metabolites that are not beneficial to antiinflammatory processes. On the other hand, leukotriene B4 (5(S),12(R)-dihydroxy-6,14-cis-8,10-transeicosatetraenoic acid, LTB4a), one of the leukotrienes, plays a significant role in the amplification of many inflammatory disease states, and it would be useful to inhibit its biosynthesis. LTB4 is generated by the hydrolysis of the epoxide LTA4 (5(S)-trans-5,6-oxido-7, 9-trans-11,14-cis-eicosatertraenoic acid) with catalysis by the human leukotriene A4 hydrolase (LTA4H-h). Therefore, simultaneous inhibition of PLA2 and LTA4H-h is a particularly attractive therapeutic approach [153]. Wei et al. developed a common pharmacophore for these two proteins containing two hydrophobic pharmacophores, and a pharmacophore that coordinated with a metal were common to both proteins. Compounds whose binding conformations in the proteins accommodate the common pharmacophores were expected to inhibit both proteins simultaneously. A synthetic indole derivative **16** was obtained, matching the common pharmacophores and inhibiting LTA4H-h and hnps-PLA2 with IC_{50} values of 12.5 and 52.5 µM, respectively [154].

15

16

6. CONCLUSION

A large number of drug molecules possessing indole nucleus, whether from natural origin or synthesized in the laboratory, have been reported for the treatment of various disease conditions. Many of these molecules have been approved by the FDA and are being currently utilized in drug therapies. However, despite extensive research, the full potential of indole based molecules is yet to be disclosed. Thus, this scaffold may be utilized as a versatile building block in drug discovery due to the wide spectrum of biological activities possible via varying structural modifications that govern the major interactions with the receptor relevant to develop selective and potent drug candidates with specific pharmacological activity. The nucleus may be further used in scaffold reevolution/refining such as scaffold hopping, multitarget-directed ligand designing using computational tools, focused libraries, and diversity-oriented synthesis.

REFERENCES

[1] A. Baeyer, Ueber die Reduction aromatischer Verbindungen mittelst Zinkstaub, Justus Liebigs Annalen der Chemie 140 (3) (1866) 295–296.

[2] M. Bandini, Electrophilicity: the "dark-side" of indole chemistry, Organic and Biomolecular Chemistry 11 (32) (2013) 5206–5212.

[3] F.G. Hopkins, S.W. Cole, A contribution to the chemistry of proteids: Part I. A preliminary study of a hitherto undescribed product of tryptic digestion, The Journal of Physiology 27 (4–5) (1901) 418.

[4] N. Chadha, O. Silakari, Indoles as therapeutics of interest in medicinal chemistry: Bird's eye view, European Journal of Medicinal Chemistry 134 (2017) 159–184.

[5] J. Barluenga, C. Valdés, Five-membered heterocycles: indole and related systems, Modern Heterocyclic Chemistry (2011) 377–531.

[6] W.J. Houlihan, The Chemistry of Heterocyclic Compounds, Indoles, vol. 25, John Wiley & Sons, 2009.

[7] J.C. Powers, Mass spectrometry of simple indoles, The Journal of Organic Chemistry 33 (5) (1968) 2044–2050.

[8] G. Bartoli, et al., The reaction of vinyl Grignard reagents with 2-substituted nitroarenes: a new approach to the synthesis of 7-substituted indoles, Tetrahedron Letters 30 (16) (1989) 2129–2132.

[9] A. Bischler, Ueber die entstehung einiger substituirter indole, Berichte der Deutschen Chemischen Gesellschaft 25 (2) (1892) 2860–2879.

[10] E. Fischer, F. Jourdan, Ueber die hydrazine der brenztraubensäure, Berichte der Deutschen Chemischen Gesellschaft 16 (2) (1883) 2241–2245.

[11] H. Hemetsberger, D. Knittel, Synthese und thermolyse von α-azidoacrylestern, Monatshefte für Chemie/Chemical Monthly 103 (1) (1972) 194–204.

[12] J.-B. Baudin, S.A. Julia, Synthesis of indoles from N-aryl-1-alkenylsulphinamides, Tetrahedron Letters 27 (7) (1986) 837–840.

[13] R.C. Larock, E.K. Yum, Synthesis of indoles via palladium-catalyzed heteroannulation of internal alkynes, Journal of the American Chemical Society 113 (17) (1991) 6689–6690.

[14] A. Batcho, W. Leimgruber, 1973. US Patent 3,732,245, May 8, 1973, in Chem. Abstr.

[15] A.D. Batcho, W. Leimgruber, Indoles from 2-methylnitrobenzenes by condensation with for-mamide acetals followed by reduction: 4-benzyloxyindole, Organic Syntheses (1985) 214.

[16] W. Madelung, Über eine neue darstellungsweise für substituierte indole. i, Berichte der Deutschen Chemischen Gesellschaft 45 (1) (1912) 1128–1134.

[17] C. Nenitzescu, Derivatives of 2-methyl-5-hydroxyindole, Bulletin De La Societe Chimique De Romania ll (1929) 37–43.

[18] A. Reissert, Einwirkung von oxalester und natriumäthylat auf nitrotoluole. synthese nitrirter phenylbrenztraubensäuren, Berichte der Deutschen Chemischen Gesellschaft 30 (1) (1897) 1030–1053.

[19] H. Tokuyama, et al., Radical cyclization of 2-alkenylthioanilides: a novel synthesis of 2, 3-disubstituted indoles, Journal of the American Chemical Society 121 (15) (1999) 3791–3792.

[20] P.G. Gassman, T. Van Bergen, G. Gruetzmacher, Use of halogen-sulfide complexes in the synthesis of indoles, oxindoles, and alkylated aromatic amines, Journal of the American Chemical Society 95 (19) (1973) 6508–6509.

[21] R. Sundberg, Deoxygenation of nitro groups by trivalent phosphorus. Indoles from o-nitrostyrenes, The Journal of Organic Chemistry 30 (11) (1965) 3604–3610.

[22] P.K. Singh, H. Singh, O. Silakari, Kinases inhibitors in lung cancer: from benchside to bedside, Biochimica et Biophysica Acta (BBA) - Reviews on Cancer 1866 (1) (2016) 128–140.

[23] Y.K. Syrkin, M. Dyatkina, The Chemical Bond and the Structure of Molecules, Goskhimizdat, Moscow-Leningrad, 1946, p. 219.

[24] P. Naik, et al., Angiotensin II receptor type 1 (AT 1) selective nonpeptidic antagonists—a perspective, Bioorganic and Medicinal Chemistry 18 (24) (2010) 8418–8456.

[25] A. Coates, et al., On the receiving end—patient perception of the side-effects of cancer che-motherapy, European Journal of Cancer and Clinical Oncology 19 (2) (1983) 203–208.

[26] FDA, U. Available from: https://www.accessdata.fda.gov/scripts/cder/drugsatfda/index.cfm?fuseaction=Search.Overview&DrugName=SUTENT.

[27] M. McTigue, et al., Molecular conformations, interactions, and properties associated with drug efficiency and clinical performance among VEGFR TK inhibitors, Proceedings of the National Academy of Sciences 109 (45) (2012) 18281–18289.

[28] D.B. Nale, B.M. Bhanage, N-substituted formamides as C1-Sources for the synthesis of benzimidazole and benzothiazole derivatives by using zinc catalysts, Synlett 26 (20) (2015) 2835–2842.

[29] S.L. Greig, Osimertinib: first global approval, Drugs 76 (2) (2016) 263–273.

[30] P. Atadja, Development of the pan-DAC inhibitor panobinostat (LBH589): successes and challenges, Cancer Letters 280 (2) (2009) 233–241.

[31] A. Younes, et al., Panobinostat in patients with relapsed/refractory Hodgkin's lymphoma after autologous stem-cell transplantation: results of a phase II study, Journal of Clinical Oncology 30 (18) (2012) 2197–2203.

[32] H.M. Prince, M.J. Bishton, R.W. Johnstone, Panobinostat (LBH589): a potent pan-deacetylase inhibitor with promising activity against hematologic and solid tumors, Future Oncology 5 (5) (2009) 601–612.

[33] T.A. Rasmussen, et al., Panobinostat, a histone deacetylase inhibitor, for latent-virus reac-tivation in HIV-infected patients on suppressive antiretroviral therapy: a phase 1/2, single group, clinical trial, The Lancet HIV 1 (1) (2014) e13–e21.

[34] K. McKeage, Alectinib: a review of its use in advanced ALK-rearranged non-small cell lung cancer, Drugs 75 (1) (2015) 75–82.

[35] K.M. Foote, et al., Discovery of 4-{4-[(3R)-3-methylmorpholin-4-yl]-6-[1-(methylsulfonyl) cyclopropyl] pyrimidin-2-yl}-1H-indole (AZ20): a potent and selective inhibitor of ATR protein kinase with monotherapy in vivo antitumor activity, Journal of Medicinal Chemistry 56 (5) (2013) 2125–2138.

[36] L. Ellis, et al., The histone deacetylase inhibitors LAQ824 and LBH589 do not require death receptor signaling or a functional apoptosome to mediate tumor cell death or therapeutic efficacy, Blood 114 (2) (2009) 380–393.

[37] S. Balasubramanian, et al., A novel histone deacetylase 8 (HDAC8)-specific inhibitor PCI-34051 induces apoptosis in T-cell lymphomas, Leukemia 22 (5) (2008) 1026–1034.

[38] W. Qi, et al., Selective inhibition of Ezh2 by a small molecule inhibitor blocks tumor cells proliferation, Proceedings of the National Academy of Sciences 109 (52) (2012) 21360–21365.

[39] V. Balasubramanian, et al., CPI-169, a novel and potent EZH2 inhibitor, synergizes with CHOP in vivo and achieves complete regression in lymphoma xenograft models, Cancer Research 74 (19 Suppl.) (2014) 1697.

[40] J.-W. Park, et al., G06: a randomized, double-blind, placebo-controlled phase III trial of TSU-68 (orantinib) combined with transcatheter arterial chemoembolization in patients with unresectable hepatocellular carcinoma, Journal of Hepatology 62 (2015) S189–S190.

[41] G.V. Scagliotti, et al., International, randomized, placebo-controlled, double-blind phase III study of motesanib plus carboplatin/paclitaxel in patients with advanced nonsquamous non–small-cell lung cancer: MONET1, Journal of Clinical Oncology 30 (23) (2012) 2829–2836.

[42] M.J. Schlumberger, et al., Phase II study of safety and efficacy of motesanib in patients with progressive or symptomatic, advanced or metastatic medullary thyroid cancer, Journal of Clinical Oncology 27 (23) (2009) 3794–3801.

[43] J.R. Infante, et al., Safety, pharmacokinetic, and pharmacodynamic phase I dose-escalation trial of PF-00562271, an inhibitor of focal adhesion kinase, in advanced solid tumors, Journal of Clinical Oncology (2012). https://doi.org/10.1200/JCO.2011.38.9346.

[44] S.A. Parikh, et al., Phase II study of obatoclax mesylate (GX15-070), a small-molecule BCL-2 family antagonist, for patients with myelofibrosis, Clinical Lymphoma, Myeloma and Leukemia 10 (4) (2010) 285–289.

[45] S.M. O'Brien, et al., Phase I study of obatoclax mesylate (GX15-070), a small molecule pan–Bcl-2 family antagonist, in patients with advanced chronic lymphocytic leukemia, Blood 113 (2) (2009) 299–305.

[46] A.D. Schimmer, et al., A phase I study of the pan bcl-2 family inhibitor obatoclax mesylate in patients with advanced hematologic malignancies, Clinical Cancer Research 14 (24) (2008) 8295–8301.

[47] Y. Oki, et al., Experience with obatoclax mesylate (GX15-070), a small molecule pan–Bcl-2 family antagonist in patients with relapsed or refractory classical Hodgkin lymphoma, Blood 119 (9) (2012) 2171–2172.

[48] J. Tabernero, et al., Phase I pharmacokinetic (PK) and pharmacodynamic (PD) study of HDM-2 antagonist JNJ-26854165 in patients with advanced refractory solid tumors, in: ASCO Annual Meeting Proceedings, 2009.

[49] S.M. Condon, et al., Birinapant, a smac-mimetic with improved tolerability for the treatment of solid tumors and hematological malignancies, Journal of Medicinal Chemistry 57 (9) (2014) 3666–3677.

[50] C. Krepler, et al., The novel SMAC mimetic birinapant exhibits potent activity against human melanoma cells, Clinical Cancer Research 19 (7) (2013) 1784–1794.

[51] J.M. Axten, et al., Discovery of 7-methyl-5-(1-{[3-(trifluoromethyl) phenyl] acetyl}-2, 3-dihydro-1 H-indol-5-yl)-7 H-pyrrolo [2, 3-d] pyrimidin-4-amine (GSK2606414), a potent and selective first-in-class inhibitor of protein kinase R (PKR)-like endoplasmic reticulum kinase (PERK), Journal of Medicinal Chemistry 55 (16) (2012) 7193–7207.

[52] J.M. Axten, et al., Discovery of GSK2656157: an optimized PERK inhibitor selected for preclinical development, ACS Medicinal Chemistry Letters 4 (10) (2013) 964–968.

[53] M.J. Robertson, et al., Phase II study of enzastaurin, a protein kinase C beta inhibitor, in patients with relapsed or refractory diffuse large B-cell lymphoma, Journal of Clinical Oncology 25 (13) (2007) 1741–1746.

[54] M. Crump, et al., Randomized, double-blind, phase III trial of enzastaurin versus placebo in patients achieving remission after first-line therapy for high-risk diffuse large B-Cell lymphoma, Journal of Clinical Oncology 34 (21) (2016) 2484–2492.

[55] T.L. Naylor, et al., Protein kinase C inhibitor sotrastaurin selectively inhibits the growth of CD79 mutant diffuse large B-cell lymphomas, Cancer Research 71 (7) (2011) 2643–2653.

[56] S. Friman, et al., Sotrastaurin, a novel small molecule inhibiting protein-kinase C: randomized phase II Study in renal transplant recipients, American Journal of Transplantation 11 (7) (2011) 1444–1455.

[57] J. Koivunen, et al., Protein kinase C α/β inhibitor Go6976 promotes formation of cell junctions and inhibits invasion of urinary bladder carcinoma cells, Cancer Research 64 (16) (2004) 5693–5701.

[58] M. Ihnen, et al., Therapeutic potential of the poly (ADP-ribose) polymerase inhibitor rucaparib for the treatment of sporadic human ovarian cancer, Molecular Cancer Therapeutics 12 (6) (2013) 1002–1015.

[59] P. Farrell, et al., Biological characterization of TAK-901, an investigational, novel, multitargeted aurora B kinase inhibitor, Molecular Cancer Therapeutics 12 (4) (2013) 460–470.

[60] R. Katayama, et al., Cytotoxic activity of tivantinib (ARQ 197) is not due solely to c-MET inhibition, Cancer Research 73 (10) (2013) 3087–3096.

[61] E.H. Yang, et al., New pyrimidine derivatives possessing ALK inhibitory activities, Bulletin of the Korean Chemical Society 34 (10) (2013) 3129–3132.

[62] Y. Huang, et al., Discovery of highly potent p53-MDM2 antagonists and structural basis for anti-acute myeloid leukemia activities, ACS Chemical Biology 9 (3) (2014) 802–811.

[63] N. Kwiatkowski, et al., Targeting transcription regulation in cancer with a covalent CDK7 inhibitor, Nature 511 (7511) (2014) 616–620.

[64] Y. Boriskin, et al., Arbidol: a broad-spectrum antiviral compound that blocks viral fusion, Current Medicinal Chemistry 15 (10) (2008) 997–1005.

[65] R.U. Kadam, I.A. Wilson, Structural basis of influenza virus fusion inhibition by the antiviral drug Arbidol, Proceedings of the National Academy of Sciences (2016) 201617020.

[66] K. Markham, H. Geiger, The Flavonoids: Advances in Research since 1986, in: J.B. Harborne (Ed.), Chapman & Hall, London, 1994, p. 467.

[67] J.R. Palmer, et al., Diaromatic Substituted Compounds as Anti-HIV-1 Agents, Google Patents, 1996.

[68] D.L. Romero, et al., Bis (heteroaryl) piperazine (BHAP) reverse transcriptase inhibitors: structure-activity relationships of novel substituted indole analogs and the identification of 1-[(5-methanesulfonamido-1H-indol-2-yl) carbonyl]-4-[3-[(1-methylethyl) amino] pyridinyl] piperazinemonomethanesulfonate (U-90152S), a second-generation clinical candidate, Journal of Medicinal Chemistry 36 (10) (1993) 1505–1508.

[69] R.M. Esnouf, et al., Unique features in the structure of the complex between HIV-1 reverse transcriptase and the bis (heteroaryl) piperazine (BHAP) U-90152 explain resistance mutations for this nonnucleoside inhibitor, Proceedings of the National Academy of Sciences 94 (8) (1997) 3984–3989.

[70] G.D. Morse, et al., Concentration-targeted phase I trials of atevirdine mesylate in patients with HIV infection: dosage requirements and pharmacokinetic studies, Antiviral Research 45 (1) (2000) 47–58.

[71] S. Castellino, et al., Central nervous system disposition and metabolism of fosdevirine (GSK2248761), a non-nucleoside reverse transcriptase inhibitor: an LC-MS and matrix-assisted laser desorption/ionization imaging MS investigation into central nervous system toxicity, Chemical Research in Toxicology 26 (2) (2012) 241–251.

[72] C. Zala, et al., Safety and Efficacy of GSK2248761, a next-generation nonnucleoside reverse transcriptase inhibitor, in treatment-naive HIV-1-infected subjects, Antimicrobial Agents and Chemotherapy 56 (5) (2012) 2570–2575.

[73] R.J. Aspinall, P.J. Pockros, SCV-07 (SciClone pharmaceuticals/verta), Current Opinion in Investigational Drugs (London, England: 2000) 7 (2) (2006) 180–185.

[74] P.L. Beaulieu, et al., Discovery of the first thumb pocket 1 NS5B polymerase inhibitor (BILB 1941) with demonstrated antiviral activity in patients chronically infected with genotype 1 hepatitis C virus (HCV), Journal of Medicinal Chemistry 55 (17) (2012) 7650–7666.

[75] R.G. Gentles, et al., Discovery and preclinical characterization of the cyclopropylindolo-benzazepine BMS-791325, a potent allosteric inhibitor of the hepatitis C virus NS5B polymerase, Journal of Medicinal Chemistry 57 (5) (2014) 1855–1879.

[76] C.A. Coburn, et al., Discovery of MK-8742: an HCV NS5A inhibitor with broad genotype activity, ChemMedChem 8 (12) (2013) 1930–1940.

[77] F. Narjes, et al., Discovery of (7 R)-14-cyclohexyl-7-{[2-(dimethylamino) ethyl](methyl) amino}-7, 8-dihydro-6 H-indolo [1, 2-e][1, 5] benzoxazocine-11-carboxylic acid (MK-3281), a potent and orally bioavailable finger-loop inhibitor of the hepatitis C virus NS5B polymerase, Journal of Medicinal Chemistry 54 (1) (2010) 289–301.

[78] R. Silvestri, et al., Simple, short peptide derivatives of a sulfonylindolecarboxamide (L-737,126) active in vitro against HIV-1 wild type and variants carrying non-nucleoside reverse transcriptase inhibitor resistance mutations, Journal of Medicinal Chemistry 47 (15) (2004) 3892–3896.

[79] J.M. Kilby, et al., The safety, plasma pharmacokinetics, and antiviral activity of subcutaneous enfuvirtide (T-20), a peptide inhibitor of gp41-mediated virus fusion, in HIV-infected adults, AIDS Research and Human Retroviruses 18 (10) (2002) 685–693.

[80] B. Devogelaere, et al., TMC647055, a potent nonnucleoside hepatitis C virus NS5B polymerase inhibitor with cross-genotypic coverage, Antimicrobial Agents and Chemotherapy 56 (9) (2012) 4676–4684.

[81] C. Lee, et al., Cycloexpansamines A and B: spiroindolinone alkaloids from a marine isolate of Penicillium sp. (SF-5292), The Journal of Antibiotics 68 (2015) 715–718.

[82] S. Ferreira, S.t. Moncada, J. Vane, Indomethacin and aspirin abolish prostaglandin release from the spleen, Nature 231 (25) (1971) 237–239.

[83] F.D. Hart, P. Boardman, Indomethacin: a new non-steroid anti-inflammatory agent, British Medical Journal 2 (5363) (1963) 965.

[84] I.G. Otterness, et al., Inhibition of interleukin 1 synthesis by tenidap: a new drug for arthritis, Cytokine 3 (4) (1991) 277–283.

[85] L.S. Simon, Actions and toxicity of nonsteroidal anti-inflammatory drugs, Current Opinion in Rheumatology 8 (3) (1996) 169–175.

[86] H. Jacobi, H. Dell, On the pharmacodynamics of acemetacin (author's transl), Arzneimittel-Forschung 30 (8A) (1979) 1348–1362.

[87] Y.-F. Chen, et al., Cyclooxygenase-2 Selective Non-steroidal Anti-inflammatory Drugs (Etodolac, Meloxicam, Celecoxib, Rofecoxib, Etoricoxib, Valdecoxib and Lumiracoxib) for Osteoarthritis and Rheumatoid Arthritis: A Systematic Review and Economic Evaluation, 2008.

[88] A. Medvedev, et al., The Influence of the Antidepressant Pirlindole and its Dehydro-derivative on the Activity of Monoamine Oxidase a and GABAA Receptor Binding, Springer, 1998.

[89] N. Andreeva, et al., The comparative influence of pyrazidol, inkazan and other antidepressant monoamine oxidase inhibitors on the pressor effect of tyramine, Farmakologiia i toksikologiia 54 (2) (1990) 38–40.

[90] B.H. Rumack, Anticholinergic poisoning: treatment with physostigmine, Pediatrics 52 (3) (1973) 449–451.

[91] S.S.I.S. Group, Treatment of migraine attacks with sumatriptan, New England Journal of Medicine 325 (5) (1991) 316–321.

[92] H. Havanka, et al., Efficacy of naratriptan tablets in the acute treatment of migraine: a dose-ranging study, Clinical Therapeutics 22 (8) (2000) 970–980.

[93] W.H. Visser, et al., Rizatriptan vs sumatriptan in the acute treatment of migraine: a placebo-controlled, dose-ranging study, Archives of Neurology 53 (11) (1996) 1132–1137.

[94] C. Dahlöf, et al., Dose finding, placebo-controlled study of oral almotriptan in the acute treatment of migraine, Neurology 57 (10) (2001) 1811–1817.

[95] J. Brandes, et al., Eletriptan in the early treatment of acute migraine: influence of pain intensity and time of dosing, Cephalalgia 25 (9) (2005) 735–742.

[96] S.D. Silberstein, et al., A randomized trial of frovatriptan for the intermittent prevention of menstrual migraine, Neurology 63 (2) (2004) 261–269.

[97] S. Linder, A. Dowson, Zolmitriptan provides effective migraine relief in adolescents, International Journal of Clinical Practice 54 (7) (2000) 466–469.

[98] P.R. Brodfuehrer, et al., An efficient Fischer indole synthesis of avitriptan, a potent 5-HT1D receptor agonist, The Journal of Organic Chemistry 62 (26) (1997) 9192–9202.

[99] D. Bremner, N. Ringan, G. Wishart, Modeling of the agonist binding site of serotonin human 5-HT 1A, 5-HT 1Dα and 5-HT 1Dβ receptors, European Journal of Medicinal Chemistry 32 (1) (1997) 59–69.

[100] J.R. Graham, Cardiac and pulmonary fibrosis during methysergide therapy for headache, Transactions of the American Clinical and Climatological Association 78 (1967) 79.

[101] M.F. Hibert, et al., Conformation-activity relationship study of 5-HT3 receptor antagonists and a definition of a model for this receptor site, Journal of Medicinal Chemistry 33 (6) (1990) 1594–1600.

[102] P. Scuderi, et al., Treatment of postoperative nausea and vomiting after outpatient surgery with the 5-HT3 antagonist ondansetron, Anesthesiology 78 (1) (1993) 15–20.

[103] J.A. Balfour, K.L. Goa, Dolasetron, Drugs 54 (2) (1997) 273–298.

[104] J.E. Macor, et al., The 5-HT 3 antagonist tropisetron (ICS 205-930) is a potent and selective α7 nicotinic receptor partial agonist, Bioorganic and Medicinal Chemistry Letters 11 (3) (2001) 319–321.

[105] X. Rabasseda, Ramosetron, a 5-HT3 receptor mutagonist for the control of nausea and vomiting, Drugs Today (Barc) 38 (2) (2002) 75–89.

[106] T. Lembo, et al., Alosetron controls bowel urgency and provides global symptom improvement in women with diarrhea-predominant irritable bowel syndrome, American Journal of Gastroenterology 96 (9) (2001) 2662–2670.

[107] P.C. Group, Randomised trial of a perindopril-based blood-pressure-lowering regimen among 6105 individuals with previous stroke or transient ischaemic attack, The Lancet 358 (9287) (2001) 1033–1041.

[108] L. Køber, et al., A clinical trial of the angiotensin-converting–enzyme inhibitor trandolapril in patients with left ventricular dysfunction after myocardial infarction, New England Journal of Medicine 333 (25) (1995) 1670–1676.

[109] G.M. London, et al., Mechanism (s) of selective systolic blood pressure reduction after a low-dose combination of perindopril/indapamide in hypertensive subjects: comparison with atenolol, Journal of the American College of Cardiology 43 (1) (2004) 92–99.

[110] M. Packer, et al., Effect of carvedilol on survival in severe chronic heart failure, New England Journal of Medicine 344 (22) (2001) 1651–1658.

[111] J.-H. Atterhög, H. Duner, B. Pernow, Experience with pindolol, a betareceptor blocker, in the treatment of hypertension, The American Journal of Medicine 60 (6) (1976) 872–876.

[112] S.D. Shamon, M.I. Perez, Blood pressure lowering efficacy of reserpine for primary hypertension, Cochrane Database of Systematic Reviews (2009) 4.

[113] K. Zhu, et al., The novel analogue of hirsutine as an anti-hypertension and vasodilatary agent both in vitro and in vivo, PLoS One 10 (4) (2015) e0119477.

[114] H.J. Wellens, D. Durrer, Effect of procaine amide, quinidine, and ajmaline in the Wolff-Parkinson-White syndrome, Circulation 50 (1) (1974) 114–120.

[115] C. Pereira, P. Agostino, C.R. Oliveira, Vinpocetine attenuates the metabolic dysfunction induced by amyloid b-peptides in PC12 cells, Free Radical Research 33 (5) (2000) 497–506.

[116] K. Kawabe, M. Yoshida, Y. Homma, Silodosin, a new α1A-adrenoceptor-selective antagonist for treating benign prostatic hyperplasia: results of a phase III randomized, placebo-controlled, double-blind study in Japanese men, BJU International 98 (5) (2006) 1019–1024.

[117] B.S. Komm, et al., Bazedoxifene acetate: a selective estrogen receptor modulator with improved selectivity, Endocrinology 146 (9) (2005) 3999–4008.

[118] P. Jones, et al., Comparative dose efficacy study of atorvastatin versus simvastatin, pravastatin, lovastatin, and fluvastatin in patients with hypercholesterolemia (the CURVES study), The American Journal of Cardiology 81 (5) (1998) 582–587.

[119] M. Cicardi, et al., Icatibant, a new bradykinin-receptor antagonist, in hereditary angioedema, New England Journal of Medicine 363 (6) (2010) 532–541.

[120] J. Christian Virchow Jr., et al., Zafirlukast improves asthma control in patients receiving high-dose inhaled corticosteroids, American Journal of Respiratory and Critical Care Medicine 162 (2) (2000) 578–585.

[121] N.H. Greig, et al., A new therapeutic target in Alzheimer's disease treatment: attention to butyrylcholinesterase, Current Medical Research and Opinion 17 (3) (2001) 159–165.

[122] Y. Bourne, et al., Structural insights into ligand interactions at the acetylcholinesterase peripheral anionic site, The EMBO Journal 22 (1) (2003) 1–12.

[123] Y. Zhu, et al., Design, synthesis and biological evaluation of dual acetylcholinesterase and phosphodiesterase 5A inhibitors in treatment for Alzheimer's disease, Bioorganic and Medicinal Chemistry Letters 27 (17) (2017) 4180–4184.

[124] F. Campagna, et al., Synthesis and biophysical evaluation of arylhydrazono-1H-2-indolinones as β-amyloid aggregation inhibitors, European Journal of Medicinal Chemistry, 46 (1) (2011) 275–284.

[125] L. Pisani, et al., Mannich base approach to 5-methoxyisatin 3-(4-isopropylphenyl) hydrazone: a water-soluble prodrug for a multitarget inhibition of cholinesterases, beta-amyloid fibrillization and oligomer-induced cytotoxicity, European Journal of Pharmaceutical Sciences 109 (2017) 381–388.

[126] M.I. Rodríguez-Franco, et al., Novel tacrine– melatonin hybrids as dual-acting drugs for Alzheimer disease, with improved acetylcholinesterase inhibitory and antioxidant properties, Journal of Medicinal Chemistry 49 (2) (2006) 459–462.

[127] O. Benek, et al., Design, synthesis and in vitro evaluation of indolotacrine analogues as multitarget-directed ligands for the treatment of Alzheimer's disease, ChemMedChem 11 (12) (2016) 1264–1269.

[128] I. Bolea, et al., Synthesis, biological evaluation, and molecular modeling of donepezil and N-[(5-(Benzyloxy)-1-methyl-1 H-indol-2-yl) methyl]-N-methylprop-2-yn-1-amine hybrids as new multipotent cholinesterase/monoamine oxidase inhibitors for the treatment of Alzheimer's disease, Journal of Medicinal Chemistry 54 (24) (2011) 8251–8270.

[129] J.L. Domínguez, et al., Computer-aided structure-based design of multitarget leads for Alzheimer's disease, Journal of Chemical Information and Modeling 55 (1) (2014) 135–148.

[130] B. López-Iglesias, et al., New melatonin–n, n-dibenzyl (n-methyl) amine hybrids: potent neurogenic agents with antioxidant, cholinergic, and neuroprotective properties as innovative drugs for Alzheimer's disease, Journal of Medicinal Chemistry 57 (9) (2014) 3773–3785.

[131] X.-T. Luo, et al., New multifunctional melatonin-derived benzylpyridinium bromides with potent cholinergic, antioxidant, and neuroprotective properties as innovative drugs for Alzheimer's disease, European Journal of Medicinal Chemistry 103 (2015) 302–311.

[132] R. Lajarín-Cuesta, et al., Gramine derivatives targeting Ca^{2+} channels and Ser/Thr phosphatases: a new dual strategy for the treatment of neurodegenerative diseases, Journal of Medicinal Chemistry 59 (13) (2016) 6265–6280.

[133] D. Hanahan, R.A. Weinberg, The hallmarks of cancer, Cell 100 (1) (2000) 57–70.

[134] H. Izzedine, et al., Sunitinib malate, Cancer Chemotherapy and Pharmacology 60 (3) (2007) 357–364.

[135] M. Atkins, C.A. Jones, P. Kirkpatrick, Sunitinib maleate, Nature Reviews Drug Discovery 5 (4) (2006) 279–281.

[136] G.S. Papaetis, K.N. Syrigos, Sunitinib, BioDrugs 23 (6) (2009) 377–389.

[137] R. Roskoski, Sunitinib: a VEGF and PDGF receptor protein kinase and angiogenesis inhibitor, Biochemical and Biophysical Research Communications 356 (2) (2007) 323–328.

[138] S. Daniele, et al., Lead optimization of 2-phenylindolylglyoxylyldipeptide murine double minute (MDM) 2/translocator protein (TSPO) dual inhibitors for the treatment of gliomas, Journal of Medicinal Chemistry 59 (10) (2016) 4526–4538.

[139] S. Daniele, et al., Dual inhibition of PDK1 and aurora kinase A: an effective strategy to induce differentiation and apoptosis of human glioblastoma multiforme stem cells, ACS Chemical Neuroscience 8 (1) (2016) 100–114.

[140] J. Huang, et al., Novel naftopidil-related derivatives and their biological effects as alpha 1-adrenoceptors antagonists and antiproliferative agents, European Journal of Medicinal Chemistry 96 (2015) 83–91.

[141] M. Brownlee, The pathobiology of diabetic complications, Diabetes 54 (6) (2005) 1615–1625.

[142] M. Stefek, et al., (2-Benzyl-2, 3, 4, 5-tetrahydro-1H-pyrido [4, 3-b] indol-8-yl)-acetic acid: an aldose reductase inhibitor and antioxidant of zwitterionic nature, Bioorganic and Medicinal Chemistry 19 (23) (2011) 7181–7185.

[143] N. Chadha, O. Silakari, Identification of low micromolar dual inhibitors for aldose reductase (ALR2) and poly (ADP-ribose) polymerase (PARP-1) using structure based design approach, Bioorganic and Medicinal Chemistry Letters 27 (11) (2017) 2324–2330.

[144] H. Knotkova, M. Pappagallo, Adjuvant analgesics, Anesthesiology Clinics 25 (4) (2007) 775–786.

[145] V.K. Kong, M.G. Irwin, Adjuvant analgesics in neuropathic pain, European Journal of Anaesthesiology 26 (2) (2009) 96–100.

[146] M.-M. Backonja, Local anesthetics as adjuvant analgesics, Journal of Pain and Symptom Management 9 (8) (1994) 491–499.

[147] F. Coluzzi, C. Mattia, Mechanism-based treatment in chronic neuropathic pain: the role of antidepressants, Current Pharmaceutical Design 11 (23) (2005) 2945–2960.

[148] G. Mladenova, et al., First-in-class, dual-action, 3, 5-disubstituted indole derivatives having human nitric oxide synthase (nNOS) and norepinephrine reuptake inhibitory (NERI) activity for the treatment of neuropathic pain, Journal of Medicinal Chemistry 55 (7) (2012) 3488–3501.

[149] B. Samuelsson, An elucidation of the arachidonic acid cascade, Drugs 33 (1) (1987) 2–9.

[150] P. Davies, et al., The role of arachidonic acid oxygenation products in pain and inflammation, Annual Review of Immunology 2 (1) (1984) 335–357.

[151] P. Singh, et al., Indole based peptidomimetics as anti-inflammatory and anti-hyperalgesic agents: dual inhibition of 5-LOX and COX-2 enzymes, European Journal of Medicinal Chemistry 97 (2015) 104–123.

[152] R.C. Reid, Inhibitors of secretory phospholipase A2 group IIA, Current Medicinal Chemistry 12 (25) (2005) 3011–3026.

[153] T.D. Penning, Inhibitors of Leukotriene A4 (LTA4) hydrolase as potential anti inflammatory agents, Current Pharmaceutical Design 7 (3) (2001) 163–179.

[154] D. Wei, et al., Discovery of multitarget inhibitors by combining molecular docking with common pharmacophore matching, Journal of Medicinal Chemistry 51 (24) (2008) 7882–7888.

Triazoles: Multidimensional 5-Membered Nucleus for Designing Multitargeting Agents

Aanchal Kashyap, Om Silakari
Punjabi University, Patiala, India

Chapter Outline

1. INTRODUCTION

Triazole is an important class of heterocyclic compounds exhibiting a wide range of pharmacological activities. It is also known as pyrrodiazoles, and is a five-membered, diunsaturated ring system containing three nitrogen atoms in a heterocyclic core and occurs in two possible isomeric forms, 1,2,3 triazoles and 1,2,4 triazoles (Fig. 9.1).

Triazoles are white to pale yellow crystals soluble in water and alcohol with a melting point of 120°C. A wide array of drugs containing triazole as a core heterocyclic structural component proves its pharmacological significance such as anticonvulsants [1], antimalarial [2], antimicrobial [3], antitumor, antiviral [4], antiproliferative [5], anticancer [6], antioxidants, analgesics [7], antifungal [8], antiplasmodial [9], antibacterial [10], immunostimulants [11], and antidiabetic [12]. These diverse attributes of the triazole nucleus have driven the interest of researchers to develop novel triazole derivatives with promising biological activities [13].

Key Heterocycle Cores for Designing Multitargeting Molecules. https://doi.org/10.1016/B978-0-08-102083-8.00009-1

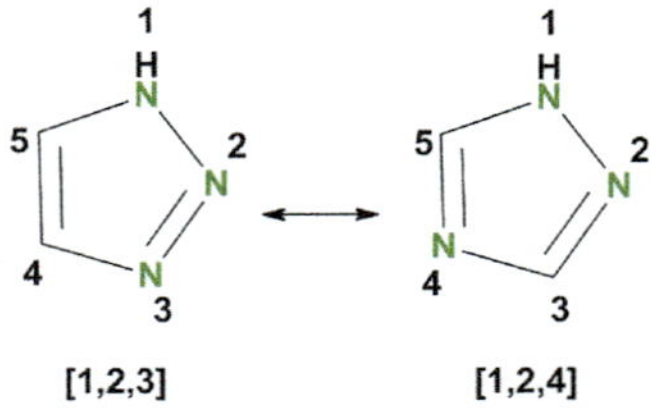

[1,2,3]   [1,2,4]

FIGURE 9.1

2. SYNTHETIC STRATEGIES OF 1,2,4-TRIAZOLE

1,2,4-Triazole usually exists in solid form, readily soluble in polar solvents and slightly soluble in nonpolar solvents, and exhibit structural isomerism [14]. Various methods for the synthesis of the 1,2,4-triazole nucleus have been reported in literature (Scheme 9.1). Scheme 9.1, I is a simple and efficient method involving microwave irradiation of hydrazines and formamides in the absence of catalyst for the synthesis of substituted 1,2,4-triazoles [15]. Another procedure, shown in Scheme 9.1, II, involves the facile synthesis of copper catalyzed N–C bond formation and oxidative N–N coupling [16].

Huang et al., has also reported the synthesis of substituted 1,2,4-triazole by using O_2 to reduce amidines and trialkylamines in the presence of K_3PO_4 as base (Scheme 9.1, III). The method was inexpensive and possessed high regioselectivity [17]. Another method involves microwave-induced one-pot synthesis to synthesize highly regioselective diverse triazoles by reacting carboxylic acids, primary amidines, and monosubstituted hydrazines using peptide coupling agents, base, and dimethylformamide (DMF) as solvent (Scheme 9.1, IV) [18]. Bechara et al. reported activation of secondary amides via triflic anhydride followed by addition of hydrazides to carry out microwave-induced cyclodehydration for the synthesis of substituted 1,2,4 triazole (Scheme 9.1, V) [19]. Nakka et al. reported another method involving oxidative cyclization of amidrazones in the presence of ceric ammonium nitrate using polyethylene glycol as recyclable reaction medium (Scheme 9.1, VI) [20]. Gogoi et al. accomplished the synthesis of substituted triazole using arylidenearylthiosemicarbazides via desulfurization in the presence of Cu(II) as catalyst (Scheme 9.1, VII) [21]. Another method involves the nitrilimine cycloaddition of oximes with hydrazonoyl hydrochlorides in the presence of triethylamine as a base to afford the synthesis of substituted triazoles (Scheme 9.1, VIII) [22]. Triazoles are also produced by allowing the salt of 1,3,4-oxadiazolium hexafluorophosphate to react with cyanamide in propan-2-ol using triethylamine as base in good yields (Scheme 9.1, IX) [23]. Cyanoimidation of aldehydes using N-bromosuccinimide as an oxidant and cyanamide as a nitrogen source was another method reported to synthesize 1,2,4-triazoles. The substituted N-cyanobenzimidates are further cyclized to form 1,2,4-triazoles by refluxing with phenyl hydrazine in the presence of methanol as solvent (Scheme 9.1, X) [24].

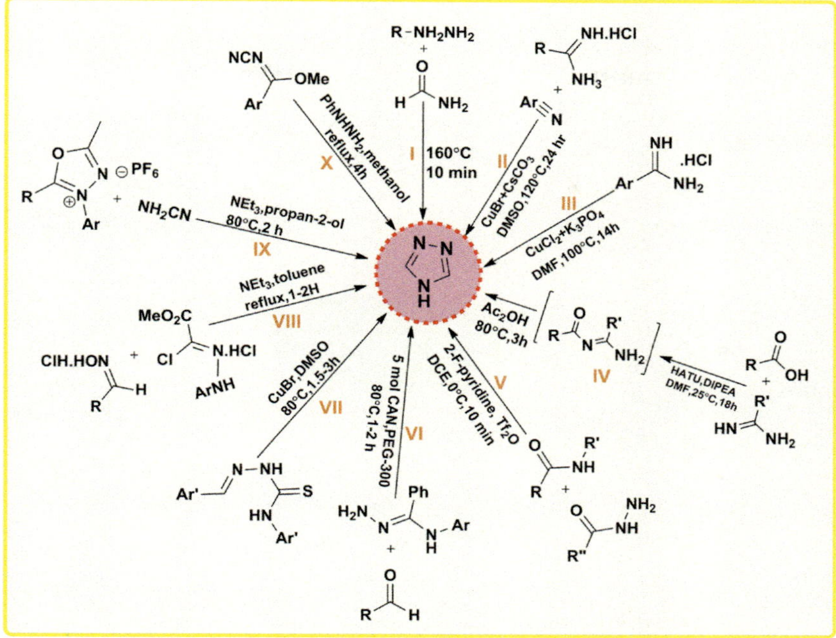

SCHEME 9.1   Synthetic strategies for the synthesis of 1,2,4-triazoles.

3. MEDICINAL ATTRIBUTES OF 1,2,4-TRIAZOLES

The biological investigation of 1,2,4-triazoles has revealed its significance in displaying various pharmacological attributes such as antibacterial, antimicrobial, anticancer, antitubercular, analgesic, antifungal, antiprotozoal, and anticonvulsant activity. Various drugs feature the triazole moiety highlighting its importance, and recent developments by various researchers have portrayed the significance of a triazole nucleus, which can be broadly divided into the categories described in the following sections.

3.1 Anticancer Activity

Various researchers have explored the 1,2,4-triazole nucleus for the antiproliferative and kinase inhibitory-based antitumor activity (Fig. 9.2). Various nucleoside-based triazole derivatives like gemcitabine (**1**) and cladribine (**2**), which mimic natural nucleosides, are in clinical practice as anticancer agents [25].

A series of various acetylated and benzoylated derivatives of 3,4-disubstituted-1,2,4-triazoline-5-thiones are synthesized, of which 1,4-dihydro-3-(3-acetyloxy-2-naphthyl)-4-ethyl-5*H*-1,2,4-triazoline-5-thione (**3**) is tested in vitro for its anticancer activity against 52 human tumor cell lines of leukemia, non–small cell lung cancer, colon, central nervous system, melanoma, ovarian, renal,

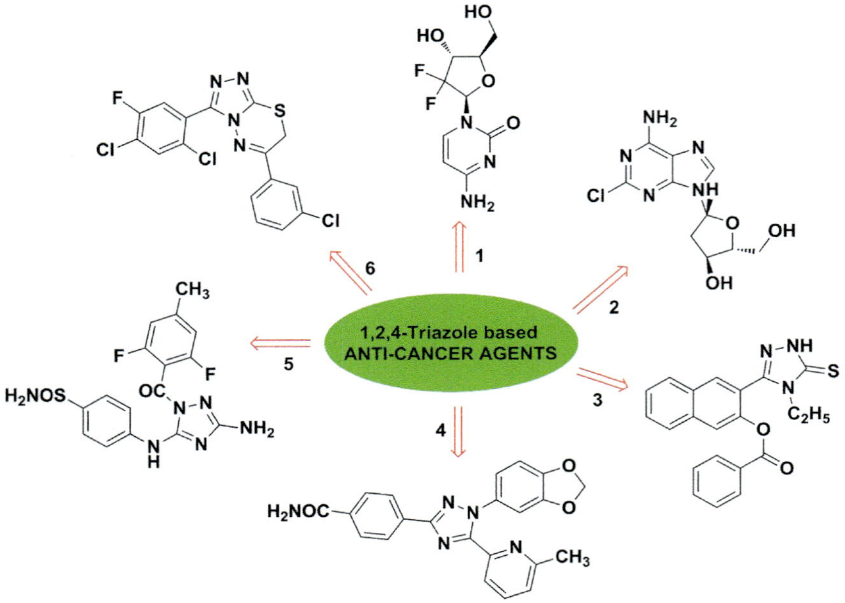

FIGURE 9.2 Various 1,2,4-triazole-based anticancer compounds.

prostrate, and breast cancer [26]. Series of novel 1-(2-pyridinyl)-[1,2,4]triazoles and 5-(2-pyridinyl)-[1,2,4]triazoles are synthesized and evaluated for activin-like kinase 5 (ALK5) inhibitory activity. The most active compound (**4**) displayed the inhibitory activity at a concentration of 5 μM and weak MAP kinase inhibition at 10 μM. The binding mode of the most active compound (**4**) is also studied by docking into the active site of ALK5 enzyme and it is observed that the inhibitor occupies the space for adenosine triphosphate (ATP) by interacting with amino acids present in the ATP binding pocket, (Leu278 and Ser280), which is poorly conserved in other kinases. This suggests that inhibitors are highly specific for the ALK5 enzyme [27].

Another series of 1-acyl-1*H*-[1,2,4]triazole-3,5-diamine analogues as ATP competitive, cyclin-dependent kinase (CDK) inhibitors are reported, which also inhibit in vitro cellular proliferation in various human cancer cells including melanoma, colon, prostate, ovarian, and breast cancer cells. CDKs are required for the regulation of eukaryotic cell division and its inhibition can result in controlling unregulated cell proliferation by inhibiting entry of cells into mitosis. Compound **5** shows significant in vivo efficacy at a concentration of 0.45 μM against A375 human melanoma tumor xenograft model in nude mice. The structure–activity relationship (SAR) study reveals that difluoro-substituted benzoyl derivatives displayed sevenfold more potent activity as compared to unsubstituted benzoyl derivative. Benzoyl moiety substituted with fluoro and methyl groups at 3-position results in analogues with better activity. Trisubstituted benzoyl derivatives abolish the cellular antiproliferative activity. Substitution of methyl and fluoro groups on the thiophene ring retains the potency for CDK1 inhibition and antiproliferative activity. Alkyl-substituted nitrogen on a sulfonamide group significantly reduced the CDK1 and cellular antiproliferative activity. Replacement of a carbonyl linker of an acyl group by a methylene or sulfonyl group greatly reduces CDK1 potency. The compounds also display modest inhibitory activity against CDK2 and CDK4 [28].

Fused derivatives of 1,2,4-triazole and 2,4-dichloro-5-fluorophenyl with substituted phenacyl bromides are also investigated for anticancer activity. Among all the tested compounds, analogues substituted with chlorine atom at various positions are found active, demonstrating that chlorine is essential for activity. Compound **6,** bearing a chlorine atom on the 4-position of the phenyl ring showed promising antiproliferative activity with the GI_{50} value in the range of 1.06–25.4 µM [29].

In another study, mannich bases of 1,2,4-triazole-3-thione containing adamantane moiety are also studied for antitumor activity along with molecular modeling studies. The synthesized compounds are evaluated against four cancer cell lines and displayed good cytotoxic activity against K562 and HL-60 cell lines and profound selectivity with low cytotoxicity against normal fibroblasts MRC-5 compared to cancer cells. Compounds favor ortho substitution on phenyl rings whereas meta substitution results in analogues with poor activity against K562 cells. However, electron donating and electron withdrawing substituent did not influence the activity and showed comparable effectiveness against HL-60 cell lines [30]. A novel series of triazole–spirodienone conjugates synthesized via copper salt-catalyzed oxidative amination are also screened and evaluated for their anticancer activities against MDA-MB-231, HeLa, A549, and MCF-7 cell lines [31].

3.2 Antifungal Activity

1,2,4-Triazoles are the promising drug candidates for the treatment of fungal infections. Triazoles are involved in inhibiting the synthesis of ergosterol, which forms a major component of fungal cytoplasmic membrane. Various marketed antifungal therapeutics contain triazole moiety such as fluconazole, voriconazole, ravuconazole, and terconazole (Fig. 9.3).

1,2,4-Triazole derivatives with a 4-(4-substitutedphenyl) piperazine side chain are evaluated by performing in vitro assay against eight human pathogenic fungi and determined minimum inhibitory concentration (MIC) for antifungal activity. All the compounds exhibit potent antifungal activity against the strain *Candida albicans* when compared to standard fluconazole. The SAR studies of the compounds revealed that increasing the length of the alkyl side chain resulted in low antifungal activity, suggesting the compounds with short or moderate length displayed better activity. Substitution of phenyl ring bearing a fluorine moiety results in moderate activity. Introduction of an aliphatic

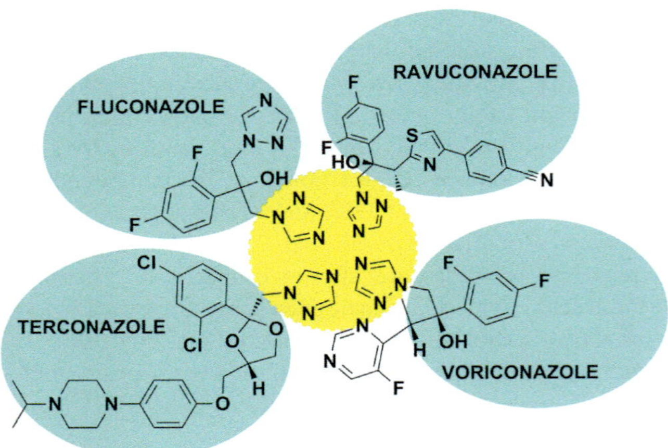

FIGURE 9.3 Marketed antifungal drugs containing triazole nucleus.

side chain shows better activity as aliphatic groups form hydrophobic interactions with the CYP51 enzyme that catalyzes the demethylation of lanosterol into ergosterol (Fig. 9.4). The docking of the most active compound into the active site of the CYP51 enzyme shows that fluoro-substituted phenyl group is positioned in the hydrophobic binding cleft composed of Phe126, Ile304, Met306, Gly307, and Gly308 amino acid residues. The piperazinyl side chain occupies a substrate access channel site and forms the hydrophobic along with van der Waals interactions with surrounding hydrophobic residues. The phenyl group attached to the piperazinyl of the side chain is involved in forming π-π stacking with Phe380 [32].

Quinoline derivatives containing 1,2,4-triazole moiety are also reported as antifungal agents against *Aspergillus flavus*, *Aspergillus fumigatus*, *Penicillium marneffei*, and *Trichophyton mentagrophytes*. Compounds with amine at 4-position of quinoline ring attributed good pharmacological activity. Substitution of –SH, –CH₂CH₂OCH₃ and –Ph groups at 4-position of triazole attached to the quinoline group also resulted in profound antifungal activity (Fig. 9.4) [33].

Stana et al. performed anticandida activity and docking studies of different thiazolyl-triazole Schiff bases, of which the compound shown in Fig. 9.4 displays most potent activity at a concentration of 15.2 μg/mL. Docking studies show that the *m*-nitro group interacts with the Arg98 from the access groove in the active site of the enzyme. Also the thiazole nitrogen forms the hydrogen bonds with the Met509 or Ser382, whereas three nitrogen atoms of triazole interact with the residues Phe506 and Ser508 [34].

Benzimidazole-1,2,4-triazole hybrids have been designed and tested against various *Candida* species. The difference in the structural variation of the compounds and antifungal activity helped in establishing the SAR. The analogues

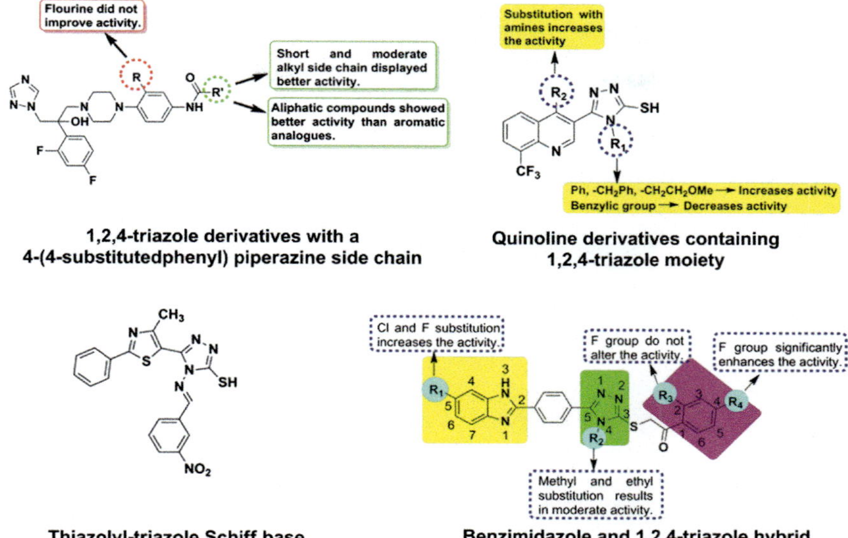

FIGURE 9.4 Various 1,2,4-triazole derivatives as antifungal agents.

bearing fluoro substituent at the C-4 position of the phenyl shows higher anti-candidal activity. Substitution of C-5 position of benzimidazole with fluoro or chloro substituent greatly increases the activity. Substitution of N-4 nitrogen of triazole with methyl or ethyl group does not alter the activity (Fig. 9.4) [35].

3.3 Antitubercular Activity

Triazole and its derivatives have attracted considerable attention as antitubercular agents. A large amount of effort has been invested in the past decade to develop triazole-based compounds as antitubercular agents, active on different clinically approved therapeutic targets showing excellent therapeutic potency.

Various 4-isopropylthiazole-4-phenyl-1,2,4-triazole derivatives are synthesized and screened for antimycobacterial activity against Mtb $H_{37}Rv$ strain using the broth dilution assay method. Compounds **9** and **10** displayed potent antitubercular activity as compared to standard drugs. It was observed that 3,4-dimethoxy and 4-methoxy substitution showed good antitubercular activity. This may be due to the presence of free electron pairs on the oxygen that attributes to increased electron density in the aromatic ring system (Fig. 9.5) [36].

A new class of fused pyrazolo [3′,4′:4,5] thiazolo [3,2-b] [1,2,4]-triazole, isoxazolo [3′,4′:4,5] thiazolo [3,2-b] [1,2,4]-triazole moieties are reported as antitubercular agents. The synthesized compounds displayed variable inhibitory activity and the compounds substituted with electron withdrawing group at C-4 position of the phenyl ring attached to the pyrazolo ring are found to be most active against $H_{37}Rv$ strain (Fig. 9.6) [37].

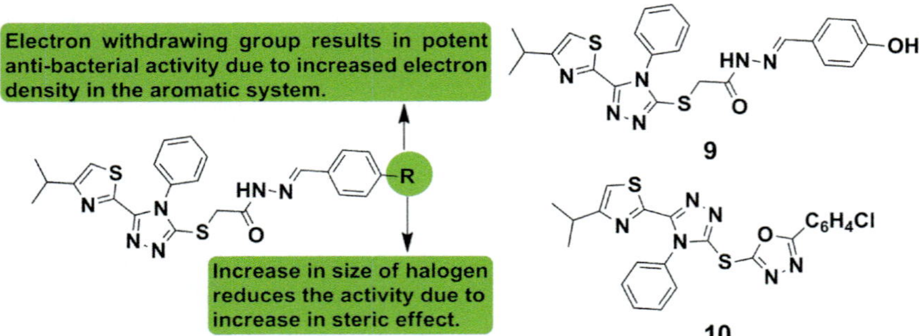

FIGURE 9.5 Novel 4-isopropylthiazole-4-phenyl-1,2,4-triazole series as antitubercular activity.

In another study series of N-alkyl/aryl-N'-[4-(4-alkyl/aryl-2,4-dihydro-3H-1,2,4-triazole-3-thione-5-yl)phenyl]thioureas derivatives are characterized for antimycobacterial activity against Mtb H37Rv and *Mycobacterium fortuitum*. Four compounds displayed the same MIC as that of tobramycin [38]. SAR studies revealed that substitution with a bulkier group at R_1 and R_2 positions resulted in loss of activity probably due to steric hindrance (Fig. 9.6).

A new series of diphenylamine fused with 1,2,4-triazole is also designed and along with its 3D quantitative SAR analysis are reported to be antitubercular agents. Series of triazolothiazolidinones, mannich bases, and triazoloquinazolinones derivatives of diphenylamine are synthesized and evaluated for their antimycobacterial activity. The SAR study shows that mannich bases bearing a morpholine moiety results in improved activity that piperidine substituted compounds. Substitution of phenyl ring at C-4 position decreases the activity. Triazolothiazolidinone derivatives show weaker activity than the mannich bases, may be due to loss of imine linkage that accounts for antitubercular activity. Triazoloquinazolinone derivatives display potent activity at concentration of 12.5 µM (Fig. 9.7) [39].

Kandemirli et al. synthesized and evaluated a novel series of 5-(4-aminophenyl)-4-alkyl/aryl-2,4-dihydro-3H-1,2,4-triazole-3-thiones **(16)** and their thiourea derivatives as antimycobacterial agents. Various compounds substituted with methyl, ethyl, and allyl groups showed significant antitubercular activity (Fig. 9.8) [40].

New 3-aryl-5-(alkyl-thio)-1*H*-1,2,4–triazoles are reported and are screened for antimycobacterial potency in both active and dormant stages of bacteria. The synthesized compounds displayed potent antitubercular activity in the range of $IC_{50} = 0.03–5.88$ µg/mL for dormant stage and 0.03–6.96 µg/mL for active stage. The biological data studies show that substitution of the nitro group on the aromatic ring substantially enhances the activity. Also the addition of electron withdrawing groups on the phenyl ring decreases the activity except for the CF_3 group present at meta- or para-position. Substitution of C-5 position of triazole with heteroatomic ring results in potent activity. S-alkylated compounds

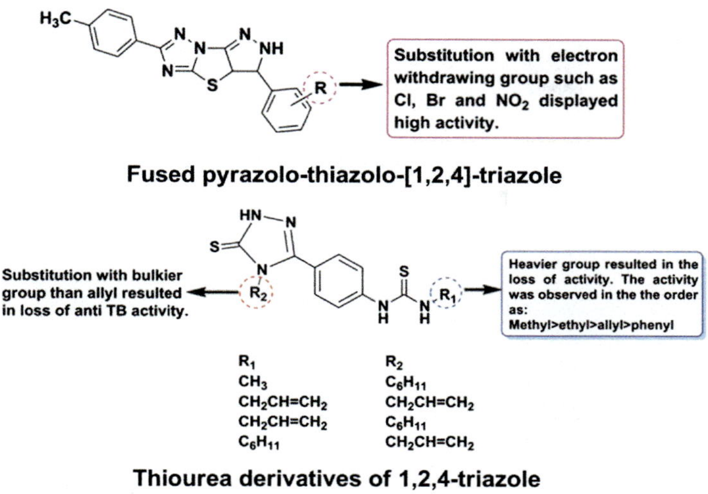

Fused pyrazolo-thiazolo-[1,2,4]-triazole

Thiourea derivatives of 1,2,4-triazole

FIGURE 9.6   1,2,4-Derivatives as antitubercular agents.

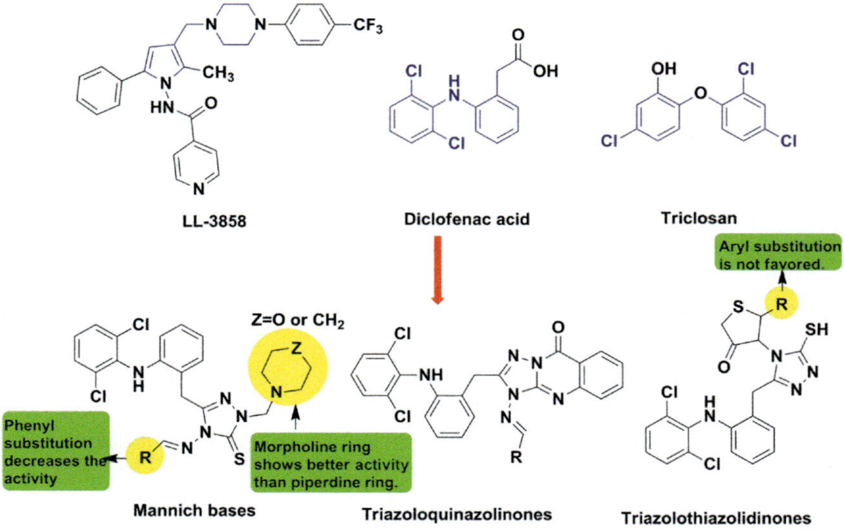

FIGURE 9.7   Design strategy and structure–activity relationship studies of diphenylamines fused with 1,2,4-triazoles.

bearing the chain length of four carbon atoms also show promising antimycobacterial activity. Docking studies of the derivative into the active site CYP121 enzyme that is essential for the viability bacteria also justifies the SAR obtained from biological data. It is observed that nitrogen of the triazole nucleus displays crucial hydrogen bonding with the Gln385 and Ala167 amino acid residues in

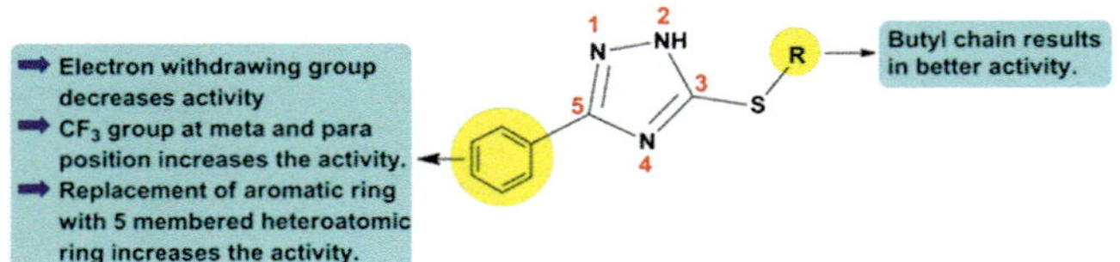

5-(4-aminophenyl)-4-alkyl/aryl-2,4-dihydro-3H-1,2,4-triazole-3-thiones

3-aryl-5-(alkyl-thio)-1H-1,2,4-triazoles derivatives

FIGURE 9.8 Antimycobacterial compounds containing 1,2,4-triazole nucleus.

the active site of enzyme. Also, heteroatomic ring and alkyl substitutions stabilized the ligand in the active site by forming electrostatic interactions with the Phe168 and Arg72 residues (Fig. 9.8) [41].

3.4 Antiinflammatory Activity

Some fused and nonfused 1,2,4-triazoles with (2,4-dichlorophenoxy) moiety prepared from 3-((2,4-dichlorophenoxy)methyl)-4-amino-4H-1,2,4-triazole-5-thiol are screened for their antiinflammatory and molluscicidal activities. The SAR established that the triazolothiadiazole derivatives substituted with 2,4-dichlorophenoxy and 4-methoxyphenyl at C-6 position display better activity in comparison to standard. Moreover, the triazolothiadiazine derivatives having 2,4-dichlorophenoxy at C-3 and phenyl/benzofuryl at C-6 positions also possess significant activity [42] (Fig. 9.9).

Various 4-(substituted benzylideneamino)-5-(substituted phenoxymethyl)-2*H*-1,2,4-triazol-3(4*H*)-thiones and various 2-[4-(substituted benzylideneamino)-5-(substituted phenoxymethyl)-4*H*-1,2,4-triazol-3-yl thio] acetic acid derivatives are reported to be antiinflammatory activity. Among all the active compounds, the one with two chlorine groups at ortho- and para-positions on both the phenoxymethyl ring and imine phenyl ring displays potent activity. This demonstrates that the presence of more than one electron withdrawing group greatly improves the activity (Fig. 9.9) [43].

The series of 3-[1-(4-(2-methylpropyl)phenyl)ethyl]-1,2,4-triazole-5-thione and its condensed derivatives 6-benzylidenethiazolo[3,2-b]-1,2,4-triazole-5(6H)-ones are screened for antiinflammatory activity using carrageenan-induced hind paw edema model. The biological activity data demonstrates that compounds substituted with a chloro group at para-position of the phenyl ring

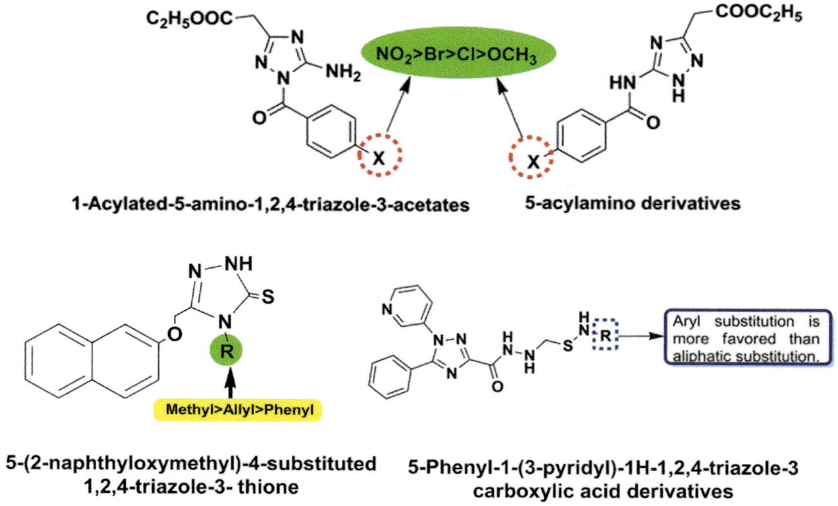

FIGURE 9.9 Various types of 1,2,4-triazole derivatives as antiinflammatory agents.

displays the most prominent activity. Substitution of a methyl group on various positions of the phenyl ring shows that the meta-position is most favored followed by para-position and then ortho-position. Also ortho-substituted methoxy group displays more potent activity than trisubstituted methoxy group on the phenyl ring (Fig. 9.9) [44].

Different acylated 1,2,4-triazole-3-acetates are studied for antiinflammatory activities along with gastric ulcerogenic effects and acute toxicity. Furthermore, molecular modeling studies are performed in order to rationalize the obtained biological results. 1-Acylated-5-amino-1,2,4-triazole-3-acetates displayed potent antiinflammatory activity than the corresponding 5-acylamino derivatives, and among the acyl substituents in the 1-acylated and/or the 5-acylamino derivatives, the order of the antiinflammatory activity was 4-nitrobenzoyl > 4-bromobenzoyl > 4-chlorobenzoyl > 4-methoxybenzoyl > unsubstituted benzoyl (Fig. 9.9) [45].

In another study, 5-(2-naphthyloxymethyl)-4-substituted-1,2,4-triazole-3-thione derivatives are evaluated as orally active antiinflammatory agents with reduced side effects and compared with naproxen, indomethacin, and phenylbutazone by performing carrageenan-induced foot pad edema assay. The compounds displayed excellent antiinflammatory activity and none of the compounds showed significant side effects compared with reference nonsteroidal antiinflammatory drugs. The SAR studies revealed that methyl substitution results in higher activity than allyl or phenyl substitution (Fig. 9.9) [46].

Similarly, 5-phenyl-1-(3-pyridyl)-1H-1,2,4-triazole-3-carboxylic acid derivatives are also synthesized and are found equipotent to indomethacin and celecoxib as reference drugs and displayed no ulcerogenic activity. The substitution on the amide nitrogen by alkyl group resulted in a slight increase of the antiinflammatory activity. On the other hand, replacement of the alkyl group by a phenyl group or *p*-methyl-phenyl group displayed a dramatic increase in activity. The substitution by *p*-bromo-phenyl displayed comparable activity to indomethacin. This means that the aryl substitution on amide nitrogen or N4 of thiosemicarbazide is very important for the activity [47].

3.5 Anticonvulsant Activity

Different 1,2,4-triazole derivatives designed and synthesized by various researchers are also found to possess anticonvulsant activity (Fig. 9.10). A series of 1-(2-naphthyl)-2-(1,2,4-triazol-1-yl)ethanone oximes (**21**) and oxime ethers (**22**) are synthesized as nafimidone analogs. Compounds are screened as anticonvulsant agents using pentylenetetrazole and maximal electroshock seizure (MES) model. Based on the biological data, it is found that size of ortho substitution on the oxime ethers greatly affects the activity. Like compounds with smaller substitutions such as methyl, ethyl, propyl, and allyl displays better activity with respect to the compounds possessing larger substitutions like benzyl and substituted benzyl groups. Introduction of branched or unsaturated alkyl chains on ethers do not affect the activity [48].

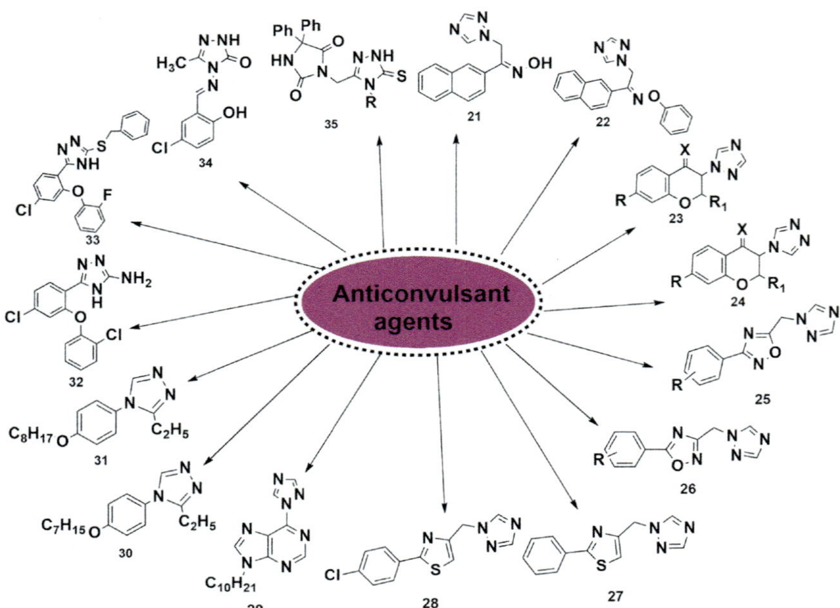

FIGURE 9.10 Various reported anticonvulsant agents containing 1,2,4-triazole moiety.

In another study, triazol-1-ylchromanones (**23**) and triazol-4-ylchromanones (**24**) are also shown to possess anticonvulsant activity. Among all, the compound with X=O and R=R$_1$=H exhibits potent activity in delaying seizures in comparison to valproate [49]. Two sets of aryl/alkyl-azoles (**25**) and (**26**) containing 1,2,4-oxadiazole as a linker are also evaluated for their anticonvulsant activity along with neurotoxicity studies. Based on the biological data, 5-phenyl-3-(1H-1,2,4-triazol-1-ylmethyl)-1,2,4-oxadiazole displays potent anticonvulsant activity and the molecular modeling studies shows that chloro group at the para-position of the phenyl ring is important for the activity [50]. Some novel thiazole incorporated (arylalkyl) azoles consisting of a lipophilic aryl ring linked by an alkylene bridge to nitrogen of an azole ring are also being investigated for their anticonvulsant activity. The biological studies revealed that 1-[(2-phenylthiazol-4-yl)methyl]-1H-1,2,4-triazole (**27**) and its 4-chlorophenyl analog (**28**) showed highest anticonvulsant activity [51].

Several purine-based triazoles also possess anticonvulsant activity. The compound 9-decyl-6-(1H-1,2,4-triazol-1-yl)-9H-purine (**29**) was found to be the most active compound. The SAR studies revealed that the triazole nucleus is important for the activity and its replacement with other heterocycles such as imidazole, methylimidazole, and pyrazole results in compounds with decreased activity and higher neurotoxicity. Linking the purine ring and 1,2,4-triazole via sulfur atom abolishes the activity, indicating that the isolated triazole act as an active centre than the triazole incorporated into the purine [52].

One-pot synthesis of 4-alkoxylphenyl-3-ethyl-4H-1,2,4-triazoles by utilizing three-components (primary amines, acyl hydrazines, and DMF dimethylacetal) are also investigated for their anticonvulsant activities using the MES test. Two compounds, 3-ethyl-4-(4-heptyloxyphenyl)-4H-1,2,4-triazole (**30**) and 3-ethyl-4-(4-octyloxyphenyl)-4H-1,2,4-triazole (**31**), have higher activity in comparison with the standard drug phenytoin. In SAR studies it is observed that chloro substitution is favored over fluoro on the phenyl ring and the activity order of the chloro-substituted derivatives is 2-Cl>2,6-Cl>4-Cl>3-Cl [53]. Some 3-(2-phenoxy) phenyl-4H-1,2,4-triazoles as simple non rigid analogues of estazolam, which is a well-known benzodiazepine agonist, are also reported. In vivo activity results of these compounds in mice revealed that the triazole ring having amino group at C-5 position results in improved anticonvulsant activity and the other substituents resulted in the activity in the order of NH$_2$ > OEt>SMe>SO$_2$Me>SO$_2$Bz>SH>OH. Compounds bearing chloro substitution on C-2 position of the phenoxy group and C-4 position of the phenyl ring displays better activity than the unsubstituted analogues. Compound **32** (2-chlorophenoxy derivative) is found to be the most potent compound in the synthesized series [54].

Various phenoxybenzyl-3-mercapto-1,2,4-triazoles and 5-(4- chloro-2-(2-fluorophenoxy)benzyl)-3-benzylthio-4H-1,2,4-triazoles (**33**) possessed the highest relative binding affinity, thereby increased anticonvulsant activity in comparison to diazepam. The biological studies indicate that ortho substitution on phenol moiety and chloro substitution on the central ring of triazole results in increased binding affinity to the benzodiazepine receptor. Replacement of the chlorine group by fluoro on the phenoxy ring also increases the activity [55]. In another study, 5-substituted 1,2,4-triazol-3-one derivatives are also reported for anticonvulsant activity. Among the tested compounds, the 5-methyl analog (**34**) shows promising anticonvulsant activity [56]. Various triazole-3(4H)-thione phenytoin hybrids (**35**) synthesized via the cyclization reaction of thiosemicarbazide have also been reported, of which four of them with ethyl, phenyl, methoxy-phenyl, and chloro phenyl displayed potent anticonvulsant activity. Compounds having an aromatic ring at the N-4 position of triazole-3(4H)-thione displayed more enhanced activity than the ethyl-substituted analog (R = Et) [57].

3.6 Antioxidant Activity

Various 2,4-dihydro-3H-1,2,4-triazole-3-thiones (**36**), by carrying out dehydrative cyclization of hydrazinecarbothioamide derivatives, are synthesized compounds and screened for their antioxidant and urease inhibition activities. The SAR studies revealed that the compounds substituted with NO_2 and Cl at para-position of ring A and dimethyl group at ortho- and para-positions on aryl ring B contributed to the radical scavenging ability. The mechanism involves the formation of negatively charged adduct, and the presence of an electron withdrawing group stabilizes the negatively charged adduct formed. The presence of a small lipophillic group such as methyl at para-position or a polarizable group like NO_2 at meta-position of the aromatic ring at the C-5 position of triazole results in improved urease inhibitory activity (Fig. 9.11) [58].

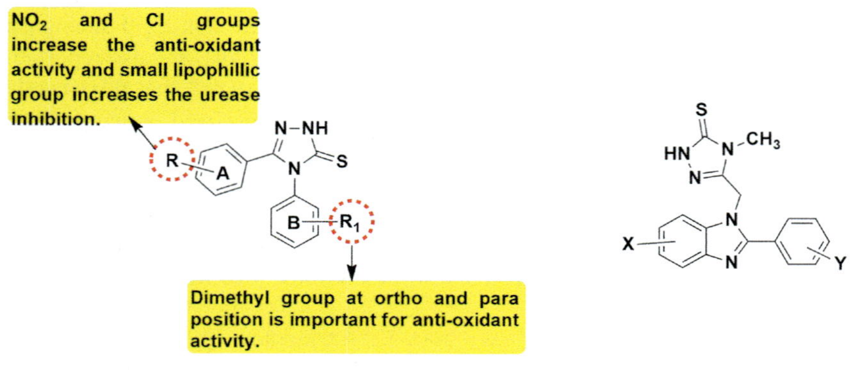

Similar analogues, 4-methyl-2H-1,2,4-triazole-3(4H)-thione (**37**) (Fig. 9.11), derivatives of the benzimidazole class, are also reported for their antioxidant activity. The synthesized compounds displayed moderate antioxidant activity [59].

4. 1,2,4-TRIAZOLES AS MULTITARGETING AGENTS

1,2,4-Triazole possesses a wide array of pharmacological activity as discussed earlier, and plays an important role as a multitargeting agent, which is of therapeutic use. One such report is the designing of a hybrid consisting of diaryl-1,2,4-triazole fused with hydroxamic acid or *N*-hydroxyurea as novel anti-inflammatory agents. The biological studies revealed that compounds displayed dual COX-2/5-LOX inhibitory activities in vitro. The most active compound shown in Fig. 9.12 displayed optimal inhibitory activity at a concentration of 0.15 μM for COX-2 and 0.85 μM for LOX-2. The compound also selectively inhibited COX-2 relative to COX-1 with a selectivity index of 0.012 [60].

Lawrence et al. first reported dual aromatase-sulfatase inhibitors. Aromatase inhibitors are involved in the blockage of the biosynthesis of estrogens. Whereas steroid sulfatase is important for the hydrolysis of estrone 3-sulfate and is an additional source of tumor estrogen, blockade of both enzymes may result in a more effective endocrine therapy. One of the synthesized compounds showed dual inhibitory activity against both enzymes in vivo (Fig. 9.13A) [61].

Jackson et al. also reported a series of triazole-based hybrids as dual inhibitors of aromatase and sulfatase inhibitors (Fig. 9.13B). The in vivo studies

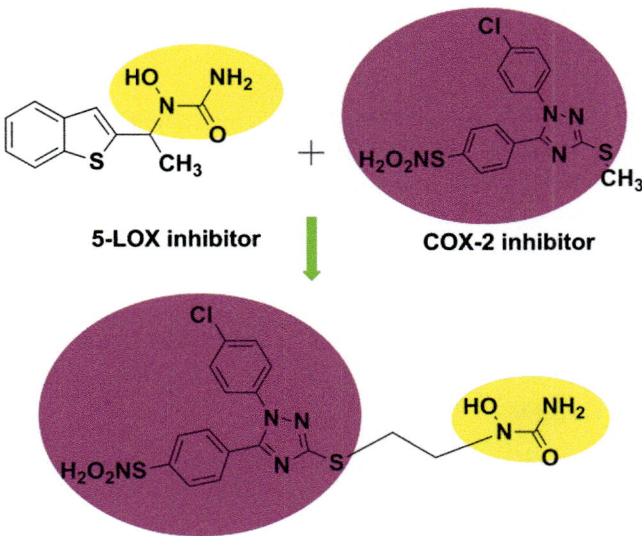

FIGURE 9.12 Dual COX-2 and 5-LOX inhibitor of 1,2,4-triazole as antiinflammatory agents.

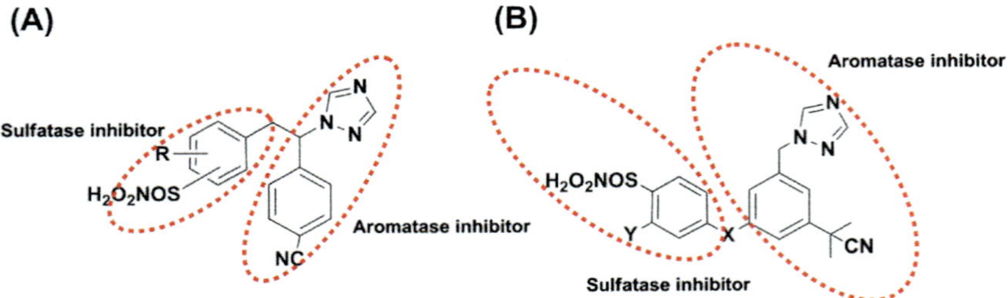

FIGURE 9.13 Different aromatase-sulfatase dual inhibitors reported by Lawrence et al. (A) and Jackson et al. (B) having 1,2,4-triazole moiety.

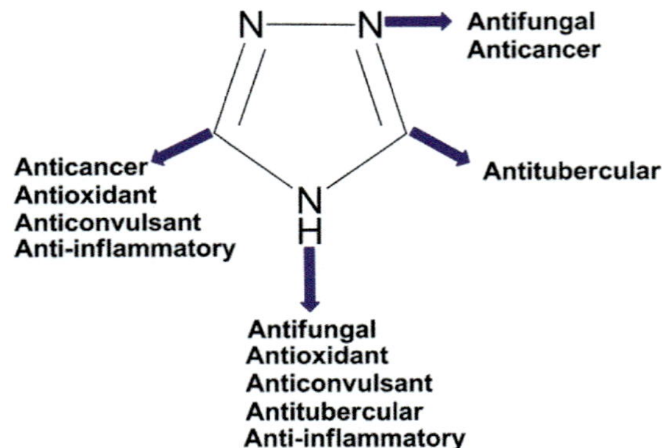

FIGURE 9.14 Different sites of 1,2,4-triazole that are explored for different pharmacological activities.

showed that the compounds displayed dual inhibition and triazole nucleus is important for the enzyme inhibition as it acts as a haem ligating group [62].

1,2,4-Triazole nucleus is an emerging class of heterocyclic compounds that possess a plethora of therapeutic benefits. Extensive research on 1,2,4-triazole nucleus has revealed its various derivatives by exploring various sites that can be used in various pathological conditions (Fig. 9.14).

These derivatives containing 1,2,4-triazole as a core heterocyclic nucleus can be utilized to synthesize new hybrids that can act as multitargeting agents.

5. CONCLUSION

Important heterocyclic compound 1, 2, 4-triazole has been already explored for various diseases, therefore it can be further utilized as a multitargeting agent to provide improved therapeutic activity. By increasing the understanding of triazole inhibitors, various safe and effective multitargeting drugs can be designed that will facilitate rational drug design and could lead to the advancement in the treatment of various critical diseases. Distinct pharmacophores of two targets can be integrated to obtain multitargeting agents containing 1,2,4-triazole as an important substructure.

REFERENCES

[1] L.-P. Guan, et al., Synthesis of some quinoline-2 (1H)-one and 1, 2, 4-triazolo [4, 3-a] quinoline derivatives as potent anticonvulsants, Journal of Pharmacy and Pharmaceutical Sciences 10 (3) (2007) 254–262.

[2] R. Gujjar, et al., Identification of a metabolically stable triazolopyrimidine-based dihydroorotate dehydrogenase inhibitor with antimalarial activity in mice, Journal of Medicinal Chemistry 52 (7) (2009) 1864–1872.

[3] M. Chen, et al., Synthesis and antibacterial activity of some heterocyclic β-enamino ester derivatives with 1, 2, 3-triazole, Heterocyclic Communications 6 (5) (2000) 421–426.

[4] Y.A. Al-Soud, M.N. Al-Dweri, N.A. Al-Masoudi, Synthesis, antitumor and antiviral properties of some 1, 2, 4-triazole derivatives, Il Farmaco 59 (10) (2004) 775–783.

[5] Y. Mohammad, et al., Synthesis and biological evaluation of novel 3-O-tethered triazoles of diosgenin as potent antiproliferative agents, Steroids 118 (2017) 1–8.

[6] M. Huang, et al., Synthesis and biological evaluation of salinomycin triazole analogues as anticancer agents, European Journal of Medicinal Chemistry 127 (2017) 900–908.

[7] K. Karrouchi, et al., Synthesis, antioxidant and analgesic activities of Schiff bases of 4-amino-1, 2, 4-triazole derivatives containing a pyrazole moiety, in: Annales Pharmaceutiques Francaises, Elsevier, 2016.

[8] C. Lass-Flörl, Triazole antifungal agents in invasive fungal infections, Drugs 71 (18) (2011) 2405–2419.

[9] S. Balabadra, et al., Synthesis and evaluation of naphthyl bearing 1, 2, 3-triazole analogs as antiplasmodial agents, cytotoxicity and docking studies, Bioorganic and Medicinal Chemistry 25 (1) (2017) 221–232.

[10] P. Khaligh, et al., Synthesis and in vitro antibacterial evaluation of novel 4-substituted 1-menthyl-1, 2, 3-triazoles, Chemical and Pharmaceutical Bulletin 64 (11) (2016) 1589–1596.

[11] T. Lee, et al., Synthesis and evaluation of 1, 2, 3-triazole containing analogues of the immunostimulant α-GalCer, Journal of Medicinal Chemistry 50 (3) (2007) 585–589.

[12] G. Wang, et al., Synthesis, in vitro evaluation and molecular docking studies of novel triazine-triazole derivatives as potential α-glucosidase inhibitors, European Journal of Medicinal Chemistry 125 (2017) 423–429.

[13] P. Martins, et al., Heterocyclic anticancer compounds: recent advances and the paradigm shift towards the use of nanomedicine's tool box, Molecules 20 (9) (2015) 16852–16891.

[14] O.R. Gautun, P.H. Carlsen, Rearrangement of 4H-triazoles. synthesis and thermolysis of 15N-labelled 4-ethyl-3, 5-diphenyl-4H-1, 2, 4-triazole, Acta Chemica Scandinavica 46 (1992) 469–473.

[15] G.M. Shelke, et al., Microwave-assisted catalyst-free synthesis of substituted 1, 2, 4-triazoles, Synlett 26 (03) (2015) 404–407.

[16] S. Ueda, H. Nagasawa, Facile synthesis of 1, 2, 4-triazoles via a copper-catalyzed tandem addition–oxidative cyclization, Journal of the American Chemical Society 131 (42) (2009) 15080–15081.

[17] H. Huang, et al., Copper-catalyzed oxidative C (sp3)–H functionalization for facile synthesis of 1, 2, 4-triazoles and 1, 3, 5-triazines from amidines, Organic Letters 17 (12) (2015) 2894–2897.

[18] G.M. Castanedo, et al., Rapid synthesis of 1, 3, 5-substituted 1, 2, 4-triazoles from carboxylic acids, amidines, and hydrazines, Journal of Organic Chemistry 76 (4) (2011) 1177–1179.

[19] W.S. Bechara, et al., One-pot synthesis of 3, 4, 5-trisubstituted 1, 2, 4-triazoles via the addition of hydrazides to activated secondary amides, Organic Letters 17 (5) (2015) 1184–1187.

[20] M. Nakka, et al., A simple and efficient synthesis of 3, 4, 5-trisubstituted/N-fused 1, 2, 4-triazoles via ceric ammonium nitrate catalyzed oxidative cyclization of amidrazones with aldehydes using polyethylene glycol as a recyclable reaction medium, Synthesis 47 (04) (2015) 517–525.

[21] A. Gogoi, et al., Synthesis of 1, 2, 4-triazoles via oxidative heterocyclization: selective C–N bond over C–S bond formation, Journal of Organic Chemistry 80 (18) (2015) 9016–9027.

[22] L.-Y. Wang, et al., An effective nitrilimine cycloaddition for the synthesis of 1, 3, 5-trisubstituted 1, 2, 4-triazoles from oximes with hydrazonoyl hydrochlorides, Synlett 2011 (10) (2011) 1467–1471.

[23] B. Wong, et al., A safe synthesis of 1, 5-disubstituted 3-amino-1H-1, 2, 4-triazoles from 1, 3, 4-oxadiazolium hexafluorophosphates, Synthesis 45 (08) (2013) 1083–1093.

[24] P. Yin, et al., Highly efficient cyanoimidation of aldehydes, Organic Letters 11 (23) (2009) 5482–5485.

[25] E. Beutler, Cladribine (2-chlorodeoxyadenosine), The Lancet 340 (8825) (1992) 952–956.

[26] A. Duran, H. Dogan, S. Rollas, Synthesis and preliminary anticancer activity of new 1, 4-dihydro-3-(3-hydroxy-2-naphthyl)-4-substituted-5H-1, 2, 4-triazoline-5-thiones, Il Farmaco 57 (7) (2002) 559–564.

[27] D.-K. Kim, J. Kim, H.-J. Park, Design, synthesis, and biological evaluation of novel 2-pyridinyl-[1, 2, 4] triazoles as inhibitors of transforming growth factor β1 type 1 receptor, Bioorganic and Medicinal Chemistry 12 (9) (2004) 2013–2020.

[28] R. Lin, et al., 1-Acyl-1 H-[1, 2, 4] triazole-3, 5-diamine analogues as novel and potent anticancer cyclin-dependent kinase inhibitors: synthesis and evaluation of biological activities, Journal of Medicinal Chemistry 48 (13) (2005) 4208–4211.

[29] K.S. Bhat, et al., Synthesis and antitumor activity studies of some new fused 1, 2, 4-triazole derivatives carrying 2, 4-dichloro-5-fluorophenyl moiety, European Journal of Medicinal Chemistry 44 (12) (2009) 5066–5070.

[30] M.Z. Milošev, et al., Mannich bases of 1, 2, 4-triazole-3-thione containing adamantane moiety: synthesis, preliminary anticancer evaluation, and molecular modeling studies, Chemical Biology and Drug Design 89 (6) (2017) 943–952.

[31] L. Gu, et al., Copper salt-catalyzed formation of a novel series of triazole–spirodienone conjugates with potent anticancer activity, RSC Advances 7 (16) (2017) 9412–9416.

[32] J. Xu, et al., Design, synthesis and antifungal activities of novel 1, 2, 4-triazole derivatives, European Journal of Medicinal Chemistry 46 (7) (2011) 3142–3148.

[33] S. Eswaran, A.V. Adhikari, N.S. Shetty, Synthesis and antimicrobial activities of novel quinoline derivatives carrying 1, 2, 4-triazole moiety, European Journal of Medicinal Chemistry 44 (11) (2009) 4637–4647.

[34] A. Stana, et al., New thiazolyl-triazole Schiff bases: synthesis and evaluation of the anti-Candida potential, Molecules 21 (11) (2016) 1595.

[35] H. Karaca Gençer, et al., New benzimidazole-1, 2, 4-triazole hybrid compounds: synthesis, anticandidal activity and cytotoxicity evaluation, Molecules 22 (4) (2017) 507.

[36] G.S. Kumar, Y.R. Prasad, S. Chandrashekar, Synthesis and pharmacological evaluation of novel 4-isopropylthiazole-4-phenyl-1, 2, 4-triazole derivatives as potential antimicrobial and antitubercular agents, Medicinal Chemistry Research 22 (2) (2013) 938–948.

[37] N. Seelam, et al., Synthesis and in vitro study of some fused 1, 2, 4-triazole derivatives as antimycobacterial agents, Journal of Saudi Chemical Society 20 (4) (2016) 411–418.

[38] I. Küçükgüzel, et al., Some 3-thioxo/alkylthio-1, 2, 4-triazoles with a substituted thiourea moiety as possible antimycobacterials, Bioorganic and Medicinal Chemistry Letters 11 (13) (2001) 1703–1707.

[39] K.M. Krishna, et al., Design, synthesis and 3D-QSAR studies of new diphenylamine containing 1, 2, 4-triazoles as potential antitubercular agents, European Journal of Medicinal Chemistry 84 (2014) 516–529.

[40] F. Kandemirli, et al., The structure-antituberculosis activity relationships study in a series of 5-(4-aminophenyl)-4-substituted-2, 4-dihydro-3h-1, 2, 4-triazole-3-thione derivatives. A combined electronic-topological and neural networks approach, Medicinal Chemistry 2 (4) (2006) 415–422.

[41] N.D. Rode, et al., Synthesis, biological evaluation and molecular docking studies of novel 3-aryl-5-(alkyl-thio)-1H-1, 2, 4-triazoles derivatives targeting *Mycobacterium tuberculosis*, Chemical Biology and Drug Design 90 (2017) 1206–1214.

[42] M. El Shehry, A. Abu-Hashem, E. El-Telbani, Synthesis of 3-((2, 4-dichlorophenoxy) methyl)-1, 2, 4-triazolo (thiadiazoles and thiadiazines) as anti-inflammatory and molluscicidal agents, European Journal of Medicinal Chemistry 45 (5) (2010) 1906–1911.

[43] R. Hunashal, et al., Synthesis, anti-inflammatory and analgesic activity of 2-[4-(substituted benzylideneamino)-5-(substituted phenoxymethyl)-4H-1, 2, 4-triazol-3-yl thio] acetic acid derivatives, Arabian Journal of Chemistry 7 (6) (2014) 1070–1078.

[44] B. Tozkoparan, et al., 6-benzylidenethiazolo [3, 2-b]-1, 2, 4-triazole-5 (6h)-onessubstituted with ibuprofen: synthesis, characterization and evaluation of anti-inflammatory activity, European Journal of Medicinal Chemistry 35 (7) (2000) 743–750.

[45] A.M. Abdel-Megeed, et al., Design, synthesis and molecular modeling study of acylated 1, 2, 4-triazole-3-acetates with potential anti-inflammatory activity, European Journal of Medicinal Chemistry 44 (1) (2009) 117–123.

[46] E. Palaska, et al., Synthesis and anti-inflammatory activity of 1-acylthiosemicarbazides, 1, 3, 4-oxadiazoles, 1, 3, 4-thiadiazoles and 1, 2, 4-triazole-3-thiones, Il Farmaco 57 (2) (2002) 101–107.

[47] S.M. Rabea, et al., Synthesis of 5-Phenyl-1-(3-pyridyl)-1H-1, 2, 4-triazole-3-carboxylic acid derivatives of potential anti-inflammatory activity, Archiv der Pharmazie 339 (1) (2006) 32–40.

[48] A. Karakurt, et al., Synthesis of some oxime ether derivatives of 1-(2-naphthyl)-2-(1, 2, 4-triazol-1-yl) ethanone and their anticonvulsant and antimicrobial activities, Archiv der Pharmazie 339 (9) (2006) 513–520.

[49] S. Emami, et al., Azolylchromans as a novel scaffold for anticonvulsant activity, Bioorganic and Medicinal Chemistry Letters 16 (7) (2006) 1803–1806.

[50] H.-J. Lankau, et al., New GABA-modulating 1, 2, 4-oxadiazole derivatives and their anticonvulsant activity, European Journal of Medicinal Chemistry 42 (6) (2007) 873–879.

[51] N. Ahangar, et al., 1-[(2-Arylthiazol-4-yl) methyl] azoles as a new class of anticonvulsants: design, synthesis, in vivo screening, and in silico drug-like properties, Chemical Biology and Drug Design 78 (5) (2011) 844–852.

[52] S.-B. Wang, et al., Synthesis and anticonvulsant activity of novel purine derivatives, European Journal of Medicinal Chemistry 84 (2014) 574–583.

[53] J. Chen, et al., Synthesis and anticonvulsant evaluation of 4-(4-alkoxylphenyl)-3-ethyl-4H-1, 2, 4-triazoles as open-chain analogues of 7-alkoxyl-4, 5-dihydro [1, 2, 4] triazolo [4, 3-a] quinolines, Bioorganic and Medicinal Chemistry 15 (21) (2007) 6775–6781.

[54] T. Akbarzadeh, et al., Design and synthesis of 4H-3-(2-phenoxy) phenyl-1, 2, 4-triazole derivatives as benzodiazepine receptor agonists, Bioorganic and Medicinal Chemistry 11 (5) (2003) 769–773.

[55] S. Mashayekh, et al., Synthesis, receptor affinity and effect on pentylenetetrazole-induced seizure threshold of novel benzodiazepine analogues: 3-Substituted 5-(2-phenoxybenzyl)-4H-1, 2, 4-triazoles and 2-amino-5-(phenoxybenzyl)-1, 3, 4-oxadiazoles, Bioorganic and Medicinal Chemistry 22 (6) (2014) 1929–1937.

[56] B. Kahveci, et al., Synthesis of some novel 1, 2, 4-triazol-3-one derivatives bearing the salicyl moiety and their anticonvulsant activities, Archiv der Pharmazie 347 (6) (2014) 449–455.

[57] S. Botros, et al., Synthesis and anticonvulsant activity of new phenytoin derivatives, European Journal of Medicinal Chemistry 60 (2013) 57–63.

[58] I. Khan, et al., Synthesis, antioxidant activities and urease inhibition of some new 1, 2, 4-triazole and 1, 3, 4-thiadiazole derivatives, European Journal of Medicinal Chemistry 45 (11) (2010) 5200–5207.

[59] C. Kuş, et al., Synthesis and antioxidant properties of novel N-methyl-1, 3, 4-thiadiazol-2-amine and 4-methyl-2H-1, 2, 4-triazole-3 (4H)-thione derivatives of benzimidazole class, Bioorganic and Medicinal Chemistry 16 (8) (2008) 4294–4303.

[60] B. Jiang, et al., Discovery of potential anti-inflammatory drugs: diaryl-1, 2, 4-triazoles bearing N-hydroxyurea moiety as dual inhibitors of cyclooxygenase-2 and 5-lipoxygenase, Organic and Biomolecular Chemistry 12 (13) (2014) 2114–2127.

[61] L.L. Woo, et al., First dual aromatase-steroid sulfatase inhibitors, Journal of Medicinal Chemistry 46 (15) (2003) 3193–3196.

[62] T. Jackson, et al., Dual aromatase–sulfatase inhibitors based on the anastrozole template: synthesis, in vitro SAR, molecular modelling and in vivo activity, Organic and Biomolecular Chemistry 5 (18) (2007) 2940–2952.

Chapter 10

Benzoxazolinone: A Scaffold With Diverse Pharmacological Significance

Himanshu Verma, Om Silakari
Punjabi University, Patiala, India

Chapter Outline

1. INTRODUCTION

2-Benzoxazolinone (BOA) is an allelochemical, derived from secondary metabolism of plants, which plays a key role in various biological processes. Its derivatives exhibit diverse properties that are essential for human health via interacting with a number of cellular targets involved in various disease conditions. BOA derivatives have been associated with various types of biological properties. Lespagnol et al. [1] first prepared and reported BOA to possess hypnotic properties. Numerous derivatives of BOA were tested for various activities including anticonvulsant, antipyretic, analgesic, cardiotonic, antiulcer, antineoplastic or antibacterial, antimicrobial, and antifungal effects [2].

In 2009, Hadizadeh et al. [3] synthesized some BOA derivatives that possessed both antioxidant and antidepressant activities, which are analogues of the well-defined antidepressant drug bupropion. Safakish and coworkers [4] designed and synthesized BOA-based derivatives as anti-HIV-1 agents. Despite being synthetic, it is also a naturally occurring phytoconstituent.

Key Heterocycle Cores for Designing Multitargeting Molecules. https://doi.org/10.1016/B978-0-08-102083-8.00010-8

Nature has been generous in inventing a vast number of natural products with medicinal values, like BOA-based compounds. BOA was first discovered in nature by Virtanen et al. as an antifusarium factor of rye seedlings. Later, it was identified in corn, wheat, and all other members of the *Graminae* family. It has also been found in other *Acanthus* species and *Blepharis edulis* of the same family. Usually, BOA is stored in the vacuole of plant cells in the form of D-glucosides. An interesting fact for BOA is that it serves as an important factor of host plant resistance against microbial diseases and insects by acting as allelochemical and endogenous ligands. Thus, with interdisciplinary investigations, it was found that BOA has the potential to act as a natural pesticide, which stimulated many biologists, biochemists, and chemists to derivatize BOA so as to make it useable as a natural pesticide [5].

Structural modification within the core scaffold is affected by the diversified incorporation of a number of related building blocks, which further results in a final compound with different structural features and pharmacological properties [6]. This has always been the main focus of the medicinal chemistry research community, to indulge in developing novel agents against various disease conditions. Throughout this chapter, we aim to shed light upon chemistry and diverse pharmacological activities of BOA, along with some limited discussion about its multitargeting potential.

2. CHEMISTRY OF 2-BENZOXAZOLINONE

BOA is a heterocyclic compound composed of a benzene ring fused with a five-membered ring containing oxygen and nitrogen as the hetero atoms. The numbering of BOA is derived from the parent benzoxazole (I) as given in Fig. 10.1. Initially, as per *Chemical Abstracts*, BOA was assigned the numbering system as shown in II. Under this system, III was referred to as benzoxazolone rather than as BOA. In chemical literature that originated before 1900, o-oxycarbanil and carbonyl-o-aminophenol were very often used to designate III [7]. BOA (III) has structural similarity to urethane (IV); in fact, these are cyclic urethanes [8]. BOA (III) is also a cyclic isostere of coumarin, for which antimicrobial activities have been broadly explored and performed [9].

The corresponding tautomer of BOA is 2-benzoxazolol (2-hydroxybenzoxazole). The observed tautomeric form of BOA is formed as a result of the removal of hydrogen from the nitrogen atom in the ring to O of the –C=O [10].

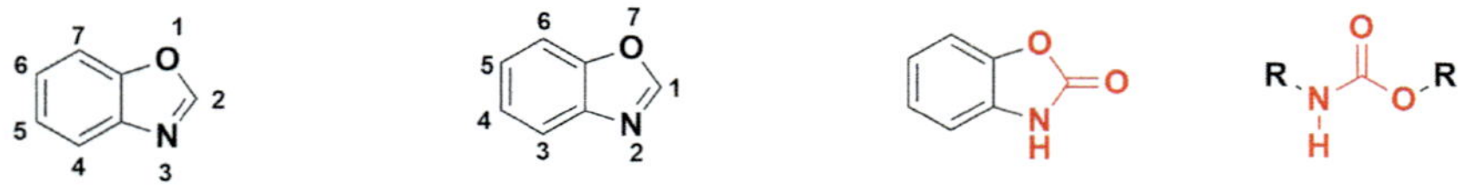

FIGURE 10.1 Nomenclature of 2-benzoxazolinone.

The resulting tautomeric form of BOA is shown in Fig. 10.2. Morais et al. (2005) found that BOA is a planar molecule, and its amide tautomer was found to be more stable than the corresponding enol (2-hydroxybenzoxazole) by about 60 kJ/mol.

Thermodynamic study carried out by Morais et al. (2005) revealed that BOA has more negative enthalpy of formation than its parent heterocyclic compound like benzoxazole by about 255 kJ/mol in the solid and gaseous phases [11]. A compound description is incomplete without its physicochemical property description.

It is important to understand the 3D crystal structure and physicochemical characterization of BOA. Calculated 3D coordinates of a unit cell of BOA (PubChem CID: 6043) are edge lengths of a cell with values 4.3823, 6.5805, and 20.8271 along the x, y, and z axes, respectively, while angles α, β, and γ are 90 degrees each. The melting point of BOA lies in the range of 141–142 degrees with an aqueous solubility of >20.3 μg/mL at pH 7.4 with logP value of 1.16. The molecular weight of BOA is 135.122 g/mol. It has one hydrogen bond donor and two hydrogen bond acceptors with no rotatable bond [12].

Various synonyms of BOA are 2-benzoxazolinone, benzoxalinone, benzoxazolin-2(3H)-one, benzoxazolin-2-one, benzoxazolone, benzoxazolone zinc salt, and benzoxazolone-2 [12]. BOA with modification in local steric and electronic properties could result in more selectivity and affinity toward targets involved in various diseases [4]. BOA can react in several ways, including N-alkylation [4], N-acetylation [13], substitution [14,15], Mannich reaction [16], and titanium mediated aldol reactions of N-acyl-BOA.

Several synthetic methods have been developed to synthesize various derivatives of BOA and modified to obtain products of high yield with desired quality and purity [17].

3. SYNTHESIS

Many drug molecules are discovered serendipitously and have a prolonged tradition in medicine. These drug molecules may be derived from plants, animals, bacteria, and marine organisms, and used for wide spectrum of biological activities. BOA is also a naturally existing cyclic carbamate derived from plant sources. BOA with varied substitutions, were isolated from plant sources, including 6,7-dimethoxy-BOA, 4-acetyl-BOA, and 6-methoxy-BOA. Interestingly, these isolated BOA-based derivatives possess biological activities like leishmaniasis, antioxidant and hepatoprotective effects, antiinflammatory

FIGURE 10.2   Tautomerism in 2-benzoxazolinone.

and antimicrobial activity, and so on. This made researchers explore BOA by synthesizing it in the lab with varied substitutions, as the major bottlenecks that continue to affect natural product drug discovery are supply issues and low yield [1,18–24].

A smart approach to get a compound of our interest is to synthesize it in a cost-efficient manner with fewer steps involved, and making it an easy synthetic route. Various synthetic procedures have been reported to synthesize BOA. Although fewer different synthetic routes possible for synthesizing BOA have been reported. But, being a privileged scaffold, BOA is synthesized by different research community, to explore BOA for the treatment of different disease conditions. Some of the schemes that have been reported are described in Fig. 10.3.

A classic procedure reported for synthesizing BOA is to fuse 2-amino phenol and urea (Scheme 1a). Safakish and associates employed this long-established procedure to synthesize BOA and make some substitutions to obtain a BOA derivative possessing anti-HIV properties [4].

Li and Xia [25] synthesized BOA in a single step using 2-aminophenol as a reaction substrate and Pd/C–NaI as catalyst system in DMF solvent, in 1 h reaction time (Scheme 1b). Pd/C or I_2 alone cannot act as an efficient catalyst for the oxidation of cyclocarbonylation of β-amino alcohols. Thereby, Pd/C and I_2 in combination is used as an active catalyst for this reaction, making it easily synthesizable [25].

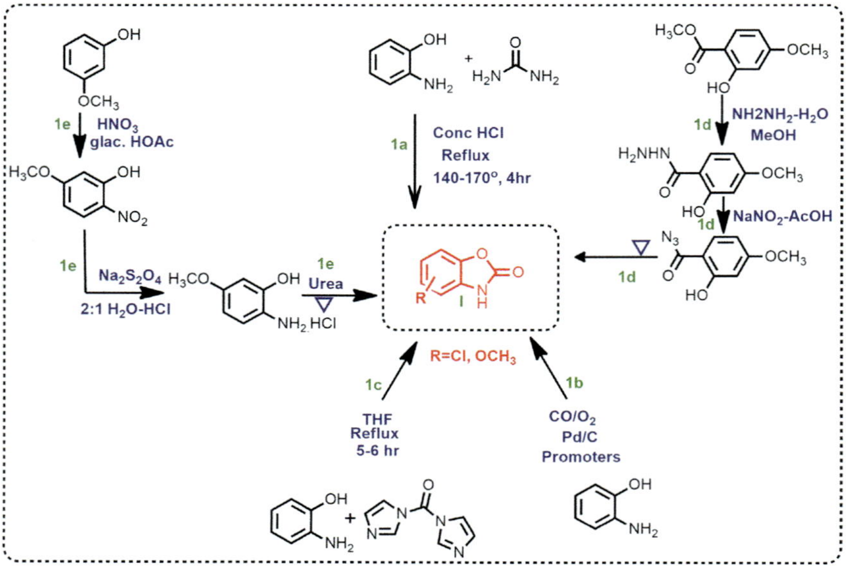

FIGURE 10.3 Synthetic schemes for 2-benzoxazolinone derivatives.

Later, Haider et al. suggested a novel synthetic auxiliary scheme to synthesize BOA by dissolving ortho amino phenol in anhydrous tetrahydrofuran along with 1,1-carbonyldiimidazole and stirring the reaction mixture at 65°C for a period of 5–6h (Scheme 1c) [26].

Kubo and associates utilized commercial 4-methoxy salicylic acid as starting material followed by treatment with diazomethane to get a methyl ester in return, which is converted to the hydrazide with hydrazine hydrate in 60% yield. Diazotization of the resultant intermediate with sodium nitrite gave the crystalline keto azide, which on pyrolysis in xylene afforded the isocyanate intermediate by Curtius rearrangement, which later underwent intramolecular addition of phenolic hydroxyl group to give 6-methoxy-2-benzoxyzolinone (Scheme 1d), resulting in the final product [27].

Richey et al. synthesized substituted BOA by a simplified, improved procedure for the formation and reduction of 5-methoxy-2-nitrophenol and subsequently, fusing the corresponding amine hydrochloride with urea to get 6-methoxy-2-benzoxazolinone as required (Scheme 1e) [28].

4. ACTIVITY PROFILE OF 2-BENZOXAZOLINONE

Among various privileged structures containing a heterocyclic ring that have been explored well for developing novel drug entities, BOA has played a key role in medicinal chemistry. Diversely substituted BOA has a diverse activity profile. A significant amount of research activity has been directed towards the derivatives of this nucleus and they are used as anticancer, antiviral, antinociceptive, antidepressant, antiinflammatory, antioxidant, anti-HIV agents, and for treating neurodegenerative diseases, which are discussed below.

4.1 Leishmanicidal Activity

Leishmaniasis is a disease caused by protozoa of the genus *Leishmania*. Insect bites (generally sandflies) are responsible for causing leishmaniasis. Two sandfly genera have been reported to be directly involved in the transmission of the disease. Researchers are making efforts to search for the novel leishmaniasis chemotherapy. Despite being toxic, the treatment of leishmania is still based on the traditional use of pentavalent antimony. One of the reasons for the hindrance to the discovery of a good, single antileishmanial drug is that parasites are well protected inside the macrophages, hence the compounds that are toxic to the parasite are also toxic to the host. One approach that has been employed to develop novel antileishmanial drugs of improved activity is to find the mode by which a drug can be released into the macrophage cell [29].

BOA isolated from the leaves of *Acanthus illicifolius* possesses a leishmanicidal activity in vitro against *Leishmania donovani* promastigotes with an IC_{50} of 40 µg/mL (296 µM) [29]. In vitro antileishmania activity of BOA was comparable with the standard drug pentamidine [30].

4.2 Antiinflammatory and Analgesic Activity

Inflammation has a noteworthy role in several disease conditions like asthma, atherosclerosis, Alzheimer's disease, rheumatoid arthritis, diabetes mellitus, carcinoma, Crohn's disease, gout, multiple sclerosis, osteoarthritis, psoriasis, bacterial or viral infections, and so on. Prostaglandins, leukotrienes, histamines, serotonin, nitric oxide, leukotrienes, interleukins (IL-1 to IL-16), iNOS production, tumor necrosis factor (TNFα), NF-κB, and chemokines trigger the inflammatory system [17]. These mediators are produced via a diverse signaling pathway involving enzymes like cyclooxygenase (COX), caspases, and kinases. Various analgesic–antiinflammatory agents have been designed to alleviate pain and inflammation by targeting these diverse signaling pathways, especially COX.

Most of the conventional nonsteriodal antiinflammatory agents (NSAIDs) like aspirin and ibuprofen induce gastrointestinal lesions. On the other hand, opioid analgesics results in dependency, thereby reducing their clinical use. To figure out the problems associated with them, some of the BOA-based derivatives containing 1-arylpiperazinyl moiety at 3-position have been reported to possess antiinflammatory and analgesic activity (Fig. 10.4). In 1996, Gökhan and associates [31] disclosed analgesic–antiinflammatory activity possessed by 6-acyl-3-piperazinomethyl-2-BOA-based derivatives. These derivatives were synthesized with substituents at the 6- and 3-positions. Analgesic activity was determined using a modified Koster test, while Peacock dial thickness gauge was used for antiinflammatory activity estimation. Results of Biological activity revealed that compound **1** manifested 60.59% of analgesic activity, which is more than aspirin with 49.56% of analgesic activity. On the other hand, compound 2 had shown better antiinflammatory activity i.e. 46.51%, more in comparison to the reference drug indomethacin with 24.00% of antiinflammatory activity. This clearly indicates that group 6-(4-methoxybenzoyl) as R_3 substituent and 3-((4-(4-nitrophenyl) piperazin-1-yl)methyl) as R_1 substituent are necessary for analgesic activity, whereas 6-(4-bromobenzoyl) group favored antiinflammatory activity [31].

Another attempt was made by Köksal et al. [32] to explore BOA substituted with 1-arylpiperazinyl moiety at 3-position and nitro group substituted at 5-position. The designed Mannich bases of 5-nitro-3-substituted piperazino-methyl-2-BOA were examined for their antiinflammatory and analgesic activity by using two different bioassays, carrageenan-induced hind paw edema and p-benzoquinone-induced abdominal constriction tests in mice, respectively. In addition, ulcerogenic activity was also performed to identify any gastric side effects they possess. Most reassuring activity results were observed for the compounds bearing electron-withdrawing substituents (–F, –Cl) at the ortho-para-position of the phenyl nucleus on the piperazine ring at 3-position of BOA scaffold. The analgesic activities of all compounds were higher than their anti-inflammatory activities [32]. This certainly indicates compounds **3–6** deserve attention and may further be evaluated.

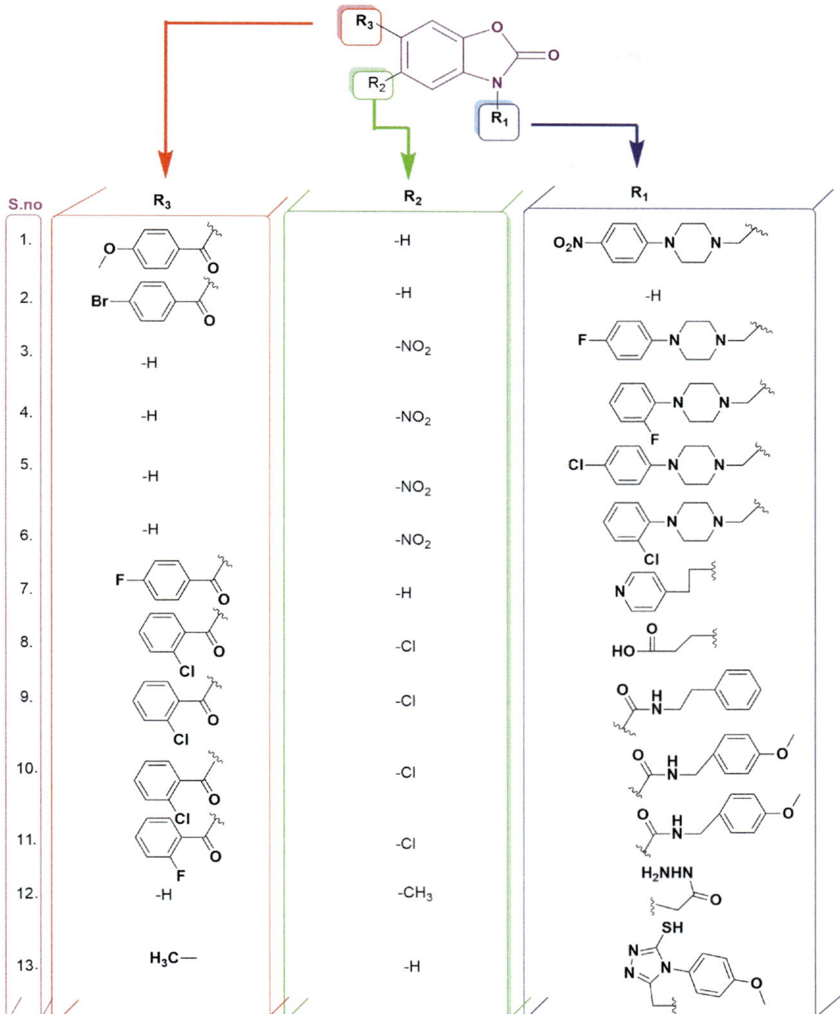

FIGURE 10.4   List of 2-benzoxazolinone-based compounds possessing analgesic–antiinflammatory activity.

The structural variations imposed on the BOA molecule by researchers have usually been in 3-, 5-, and 6-positions, and the biologic effects resulting from these changes have been studied. Palaska and coworkers synthesized a series of 3-[2-(2- and/or 4-pyridyl)ethyl] benzoxazolinone derivatives, for which biological activity revealed that these compounds are antiinflammatory drugs with analgesic properties. Interestingly, none of the compounds manifested gastrointestinal bleeding at 100 mg/kg dose level. In addition to this, activity studies also revealed that the synthesized compounds are more active than indomethacin. Among the series of BOA-based compounds, Compound **7** has a superior analgesic profile with low gastric ulceration incidence, making it favorable for development as a clinically useful analgesic. The structure–activity relationship (SAR) of pyridyl-ethylated BOA revealed that analgesic activity depends on the pyridyl group at 3-position of the benzoxazolinone ring. In general, the 4-pyridyl group is more favorable than the 2-pyridyl group for analgesic activity. Moreover, compounds containing 4-fluorobenzoyl and 4-pyridyl groups are the most potent analgesic series [33].

Uenlue and associates synthesized 6-acyl-2-BOA derivatives to study their analgesic–antiinflammatory activity. SAR disclosed an overall increase in the analgesic–antiinflammatory activity of the compounds containing propanoic acid residue, while acetic acid derivatives do not exhibit comparable satisfactory features. In vivo analgesic–antiinflammatory activities revealed that compound **8** is the most potent BOA-based derivative that possessed both analgesic and antiinflammatory activity to the highest level [34].

Following Uenlue and associates' published report over analgesic and antiinflammatory activity of 6-acyl -2-BOA-based derivatives, Banoglu et al. [35] explored amide derivatives of [5-chloro-6-(2-chloro/fluorobenzoyl)-2-benzoxazolinone-3-yl]acetic acids to investigate their potential as analgesic and antiinflammatory compounds. Their study was aimed at maintaining the highest analgesic and antiinflammatory activity and preventing any gastric side effects common with traditional NSAIDs. Among the series of these derivatives, four were found to be very potent analgesic and antiinflammatory agents without gastric lesions in tested animals. Moreover, COX selectivity of the active compounds was also scrutinized by using in vitro human whole blood assay. From these four active molecules, three indicated selectivity for COX-2 to some extent, although did not show potent inhibitory activity against COX-2. Compounds **9–11** favored the activity results. This clearly demonstrated that derivatives with 4-bromo and 4-methoxyphenyl substituent exhibited more superior activity than that of aspirin. Moreover, there was decrease in activity observed with incorporation of an ethylene spacer between the phenyl and the amide nitrogen [35].

Another effort by Göksen et al. [9] to explore acetic acid hydrazide containing 5-methyl-2-BOA for analgesic–antiinflammatory activity was also considerable. Since some of the molecules among these series exhibited potent activity, compounds containing thiadiazole moeity and benzylidene hydrazine substituted molecules possessed strong analgesic activity. But 2-(5-methyl-2-benzoxazolinone-3-yl)acetylhydrazine **(12)** was found to be the most potent antiinflammatory and analgesic drug [9].

The 2-BOA nucleus has been the keynote due to its diverse potent pharmacological properties. In an attempt to discover novel derivatized BOA, Haider and coworkers [26] synthesized a focused library of 20 compounds that is composed of BOA and 1,2,4-triazole moieties linked through a methylene as a linker. These molecules are antiinflammatory agents, as revealed by activity studies. Compound **13** was highly selective for COX-2 with $IC_{50} = 2.6 \mu M$, while showing $IC_{50} = 110\,\mu M$ for COX-1. Moreover, compound **13** was found to be a more selective COX-2 inhibitor as compared to indomethacin (COX-1 $IC_{50} = 3.80\,\mu M$; COX-2 $IC_{50} = 7.20\,\mu M$). This undoubtedly indicates that 6-CH_3 substitution on the BOA ring manifested better activity in comparison to the unsubstituted 2-BOA ring. Additionally, para-substituted $-OCH_3$ on the phenyl ring attached to the triazolyl ring also favored activity [26].

4.3 Anti-HIV Reverse Transcriptase Activity

HIV (human immunodeficiency syndrome) is a major health-linked problem prevailing in countries with low or middle income. The cells of the immune system are the bull's-eye for retroviruses like HIV, which will impair and destroy the proper functioning of the immune cells. Person infected with HIV virus for a period of 10–15 years may reach the advance stage called acquired immunodeficiency syndrome (AIDS). Patient with AIDS remain infected with this virus throughout life. Scientists hope to develop the most effective treatment for AIDS by expanding alternative drugs in the pipeline by stimulating research efforts globally. Several papers were published with BOA-based compounds as novel anti-HIV agents [36].

In 2002, Wang and Ng [37] reported an anti-HIV potential of BOA by carrying out anti-HIV reverse transcriptase activity using a nonradioactive ELISA kit (Boehringer Mannheim). 6-methoxy-2-benzoxazolinone (6-MBOA) **(14)** was found to be the most potent compound compared to pineal indoles [37].

In 2010, He and associates [38] had designed and synthesized novel pseudopeptide incorporated with benzheterocycles like BOA for their activity as HIV-1 protease inhibitors. It was clear that pseudopeptide incorporated with BOA **(15)** was the most potent final compound, as it possesses an ability to inhibit the enzyme activity with an IC_{50} value of 5 nM. SAR study revealed that benzheterocycles linked to the carbonyl group via a prolonged chain resulted in reduction of binding affinity. Second, addition of a small alkyl substituent with befitting size to the $-CH_2-$ of P_1-P_2 linkage as a side chain resulted in enhanced inhibitory potency, and in this study, isopropyl was found to be a suitable side chain. Decrease in the inhibitory potency was observed with replacement of the isobutyl substituent at P_1 group with a phenyl substituent [38].

Safakish et al. introduced a 1,3,4-thiadiazole ring to the BOA scaffold that resulted in greater anti-HIV-1 activity and less cytotoxicity. Later, these compounds were tested for their potential to inhibit p24 expression in HeLa cell cultures, and it was observed that the presence of 2-amino-1,3,4-thiadiazole ring as a spacer significantly improved the antiviral activity in comparison to other linkers like benzohydrazide or carbothioamide. Aminophenyl at C-2 position of 1,3,4-thiadiazole ring was most favorable and showed the highest inhibitory effect with 84% inhibition rate among all the synthesized compounds. It was reasoned that the heteroatoms of the thiadiazole ring would have been involved in chelation of Mg^{2+} ions present in the integrase (IN) active site. A molecular docking study revealed that 2-BOA exhibits π–π stacking interaction with an

IN active site. Substituted phenyl ring in these designed structures inserted well into the hydrophobic pocket of the IN active site. Moreover, phenyl ring substitution with either electron-withdrawing or electron-donating group in compounds possessing thiadiazole or benzohydrazide linker decreased the potency, whereas cell viability remained unaffected by this replacement [4]. The most active compound was found to be compound **16**.

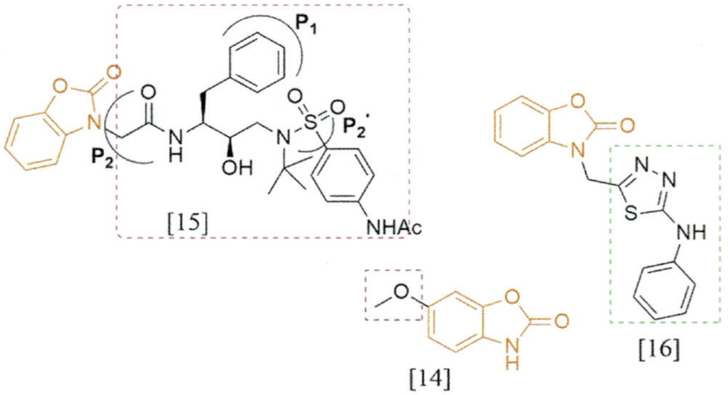

4.4 Anticancer Agents

Curing cancer and saving the patients suffering from it has always been a challenge for most of the researchers around the world involved in identifying the most potent anticancer drug. Though researchers consider it a big challenge, but they never miss any opportunity to face this challenge. Considering the current scenario, extensive efforts are made to improve oncologic therapies so that such advancements can make considerable improvements in prognosis and survival [39].

BOA-based derivatives were reported to possess anticancer activity. Several recent reports have evidenced the role of BOA-based derivatives in cancer therapies.

Erstwhile, benzimidazole ribosides have been reported to inhibit both RNA and DNA, but are not active as antiviral and anticancer agents. Therefore, to obtain a series of potent ribosides, Advani and associates studied the anticancer activity of BOA ribosides. The basis of designing BOA ribosides was its isosteric relationship to benzimidazoles and purines. Advani et al. reported varied substituted 3-[1′(2′3′4′-tri-O-benzoyl-β-D-ribopyranosyl)]-2-benzoxazolinones as potential anticancer agents [40]. Different substitutions were made on BOA riboside to obtain various derivatives (Fig. 10.5) [39].

Murty and coworkers [41] synthesized and evaluated 3-[(3-substituted) propyl]-1,3-benzoxazol-2(3H)-one derivatives against the cancer cell lines HeLa, MCF-7,A549, and SW-480. Activity result revealed that compound **17** was the most potent among the series of BOA-based derivatives. This clearly manifested that 3-(4-(3-chlorophenyl) piperazin-1-yl) propyl substitution at 3-position and chloro substitution at 5-position of BOA is necessary to display

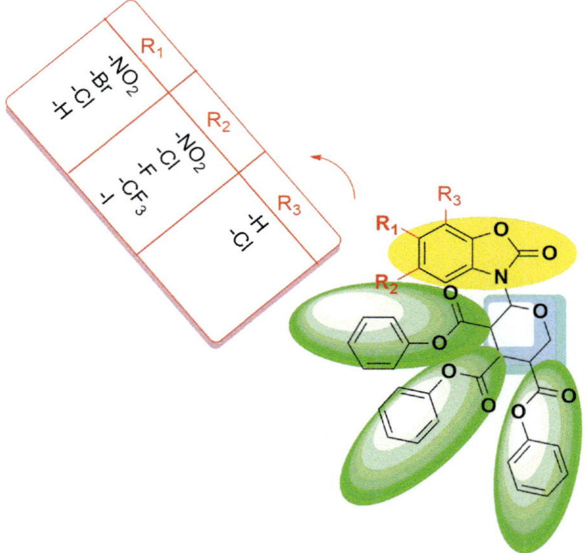

FIGURE 10.5   2-Benzoxazolinone ribosides with varied possible substitutions.

antiproliferative activity. Thereby, BOA-based compounds have the potential to act as potential anticancer agents [41].

Schadt et al. [42] filled a patent that disclosed novel 2-oxo-3-benzylbenzox-azol-2-one derivatives as MET kinase inhibitors for the treatment of tumors.

[17]

Benzoxazolone carboxamides were reported as potent acid ceramidase (AC) inhibitors. In vivo activity revealed that compound **18** was the most potent AC inhibitor (Fig. 10.6). In spite of potent activity, it was found to be highly stable benzoxazolone carboxamide with optimum solubility profile. SAR revealed the importance of the electron withdrawing group at positions 5–7 for activity enhancement. Benzoyl group at 6-position was involved in enhancement of stability. On the other hand, *p*-methoxyphenyl at 6-position was the least efficient for the activity. The most favorable group for potent activity was obtained with *p*-fluoro phenyl group at 6-position [43].

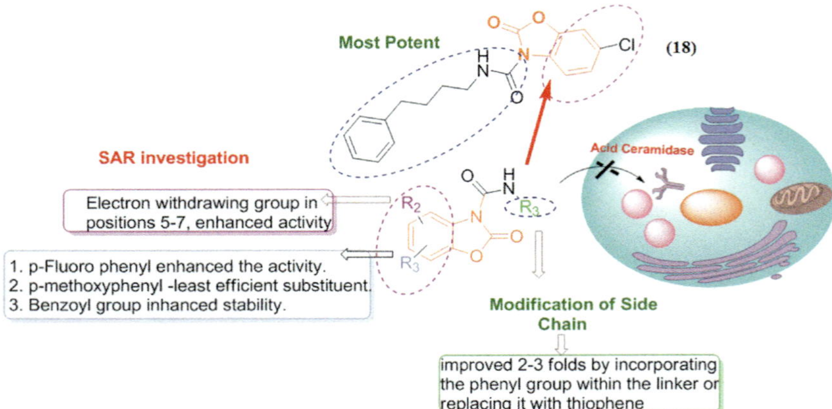

FIGURE 10.6 Structure–activity relationship studies of benzoxazolone carboxamide.

Gerova and coworkers [44] designed and synthesized a novel class of heterocyclic analogues of natural cis-stilbene combretastatin A-4 with BOA scaffold. The synthesized molecules were further evaluated for activity studies against cancer cell lines (HepG2, EA·hy 926 and K562 cells). Among all the series of molecules, compound **19** was found to be the most potent anticancer agent, and the potent growth inhibition activity observed with compound **19** was found to be in the range of 50–750 nM. It was revealed from SAR studies that Z-stilbene was found to be highly potent isomers, while E-isomer has shown little or no activity. In addition to this, styryl fragment on the BOA ring plays a key role in this series of compounds. Also, a number of methoxy groups in ring A as well as 4-, 5-, 6-, or 7-positions of styryl fragments on BOA heterocycles plays a key role in imparting potent antiproliferative and proapoptotic effects in liver cancer cells, being similar or better compared to CA-4 (compound-**20**). Moreover, compound **19** was also tested against other cell lines like MCF-7, MDA-MD-231, MCF-10A, HaCaT, NHEK, HT-29, Colon-26, and A-549. The activity results revealed that the bioisosteric replacement of 3-hydroxy-4-methoxyphenyl moiety of CA-4 (ring B) with BOA nucleus would be a useful approach to search for novel anticancer agents [44].

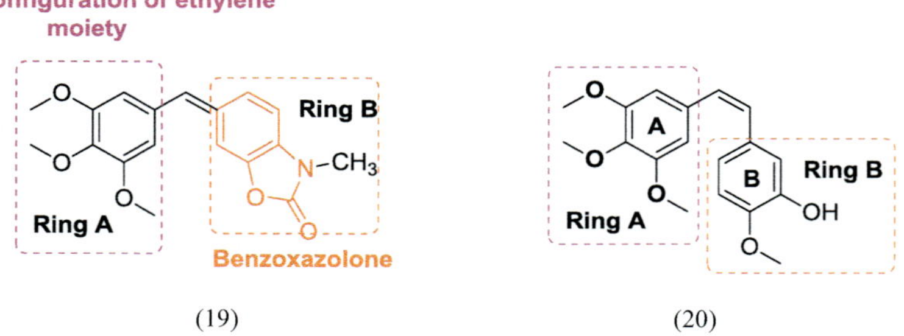

Novel benzo[d]oxazol-2(3H)-one derivatives bearing 7-substituted-4-enthoxyquinoline moieties as c-Met kinase inhibitors have been reported. The designed BOA-based inhibitors were evaluated against EBC-1 cell lines. The cell line study disclosed compound **21** to be the most potent c-Met kinase inhibitor with an IC_{50} value of 5 nM against proliferation of the EBC-1 cell line, while manifested IC_{50} value of 1 nM for c-Met kinase. A molecular docking study manifested that both N-methylpyrazole groups fit well into hydrophilic regions of c-Met kinase, where they enjoy supporting hydrophilic interactions. Moreover, it also revealed weakening of π-π stacking interactions with Tyr1230 of c-Met, with an introduction of a substituent at position C-5. Among these series of molecules, compounds possessing both sterically and electron donating groups like methyl and amino groups are found to be the most potent anticancer drugs. On the other hand, analogues with electron withdrawing groups like fluoro and methyl formate groups are not well tolerated for anticancer activity. Thus, SAR is well explored for 2-BOA and quinolone ring substituted analogues, indicating their potential as chemotherapeutic agents [45].

(21)

4.5 Antidepressant Activity

The area of pharmacotherapy of depression was established in the 1950s, flooded with marked publications and discoveries that still govern the manner in which we tackle depression. Depression has been noticeable as the common cold psychopathology [3]. Affected people feel powerlessness and are the most vulnerable to suicidal thoughts [46]. The pathophysiology behind depression is the network dysfunctioning with imbalance in the levels of glutamate (Glu) and gammaaminobutyric acid, which has been identified in both human and animal studies of depression [47]. As per Owens and coworkers, any alteration in serotonergic neuronal function is a cause of major depression. Several reports include well-evident causes of depression and have mentioned that patients suffering from depression have reduced levels of 5-hydroxyindoleacetic acid, the major metabolite of serotonin (5-HI) in cerebrospinal fluid of drug-free depressed patients. Decreased plasma concentration of tryptophan in depressed patients has also been reasoned for depression [48].

Thus it was necessary to come up with appropriate treatment for depression by prompting research efforts. Evidence has been found that indicates BOA derivatives possess antidepressant properties.

Hadizadeh et al. [3] designed and synthesized BOA-based bupropion (well-known antidepressant drug) analogues and evaluated for their antidepressant activity using a forced swimming test in mice. Results obtained from the biological activity study revealed that all analogues possessed activity comparable to that of standard drug bupropion at a dose range of 2.5–20 mg/kg. Compounds **22** and **23** were found to be the most active [3].

(22) (23)

Manikandan et al. [49] proposed BOA derivatives as therapeutic lead compounds for use in the case of disorders associated with depression and pain. The proposed compounds were found to be potential dual inhibitors that could favorably target both human serotonin and norepinephrine transporters. The study was justified via computational approach [49].

4.6 Antinociceptive Agent

Nociceptive pain is the result of threatened damage to nonneural tissue followed by activation of nociceptors [50]. Thus, there are demands for the need of antinociceptive agents that could suppress activity in the neurons, which respond specifically to nociceptive pain. Several reports have been published that claim BOAs to be antinociceptive agents.

Novel (5-Chloro-2(3H)- Benzoxazolon-3-yl) propanamide derivatives were synthesized, specifically evaluated for antinociceptive activity via tail clip, tail flick, hot plate, and writhing methods. In comparison to the standard drugs like dipyrone and aspirin, compound **24** (1-[3-(5-Chloro-2(3H)-benzoxazolon-3-yl) propanoyl]-4-(4-chlorophenyl)piperazine was a highly active antinociceptive agent, which clearly indicates that chloro at 5-position is favorable for the activity. Some of the compounds were found to be active in all the tests when compared with control group and standard compounds. Thus it is purposeful to say that these three compounds can possibly act as central acting analgesic agents and peripheral acting analgesic agents [51].

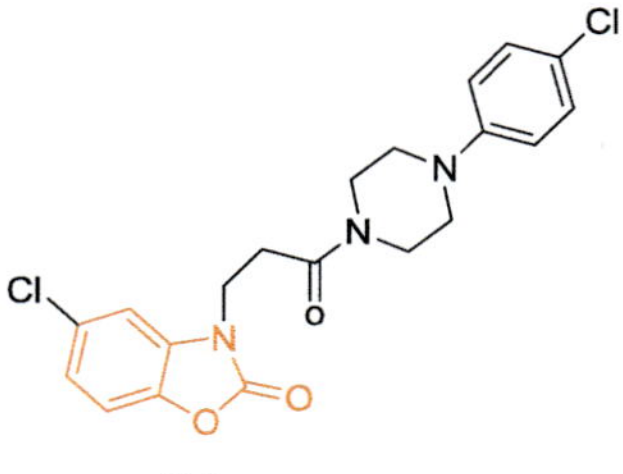

(24)

Novel 1,2,3-triazole-based BOA derivatives have been reported to possess antinociceptive activity. The antinociceptive activity was evaluated on the basis of writhing test and tail immersion methods. As a resultant of the writhing test, compound **25** displayed the most potent antinociceptive activity with 41.83% in comparison to the reference drug indomethacin (44.69%) inhibition. The antinociceptive activity of the active compound was evaluated via both chemical and peripheral mechanisms of analgesia. As revealed by SAR studies, weak electron-withdrawing halogen atoms on the aromatic ring attached to the triazollyl ring manifested greater activity in comparison to compounds having NO_2 group at para- and meta-position, whereas electron-donating groups like OCH_3, OC_2H_5 as well as compounds constituting pyridyl substitutions were not tolerated [52].

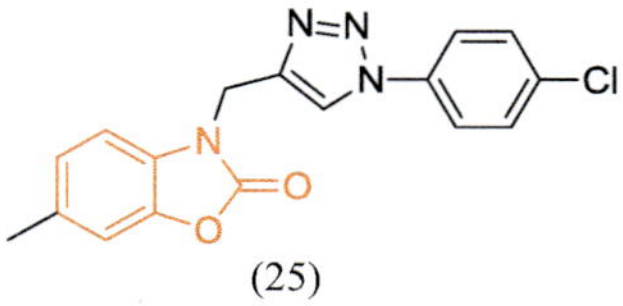

(25)

Moreover, 2-benzoxazolinone-based 1,2,4-triazoles also have been explored as potential proinflammatory cytokine inhibitors that were synthesized and screened for their in vivo antiinflammatory and antinociceptive activities. Among all the designed compounds, compound **26** was found to be most potent antinociceptive activity with 56.70% inhibition of analgesia as a result of a tail immersion activity, indicating the necessity of a nitro group at 5-position and a p-methoxy group on an aromatic ring attached to 1,2,4-triazoles moiety for the activity in comparison to the unsubstituted phenyl ring. Additionally, among –Br, –Cl, –F substitutions on the phenyl ring attached to the triazolyl ring, bromo was more favored, followed by the chloro group [28].

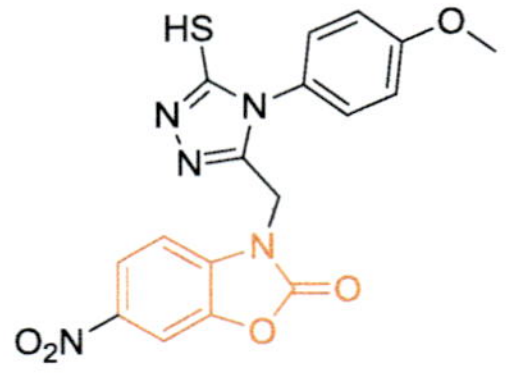

(26)

Pero et al. published BOA-based aryl sulfonamides as potent, selective $Na_v1.5$ inhibitors. Pero and associates also carried out mouse formalin paw assay to study in vivo antinociceptive activity. As a result of the assay, compound **27** was found to possess favorable antinociceptive activity [53].

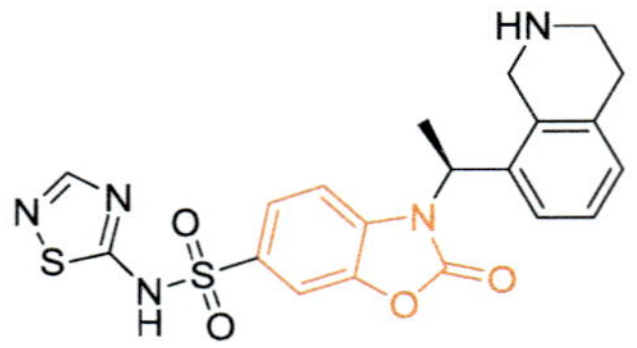

(27)

4.7 Antioxidant Activity

Antioxidants and free-radical scavengers have gained importance, being involved in inhibiting low density lipoprotein oxidation and atherosclerotic development. With time, the role of antioxidants has been reviewed. Several diseases like cancer, cardiovascular diseases, arteriosclerosis, neural disorders, skin irritations, and inflammations occur due to accumulation of high levels of free radicals in various tissues. These high levels of free radicals are able to oxidize biomolecules, making lipid membrane susceptible to oxidative damage. Generation of carbon radicals and peroxy radical production (the cause of lipid peroxidation) are the marked outcome of its increased levels. The single radical initiates lipid peroxidation chain reactions that cause damage to many molecules [17]. The body has a defense mechanism to undo the effect of increased levels of free radicals by making these free radicals available to superoxide dismutase, copper, and iron transport proteins, as well as lipid-soluble and water-soluble antioxidants. It is a well-known fact that any deviation in the normal physiological conditions of a body leads to several diseases; similarly, any imbalance between the level of free radicals and antioxidant mechanisms leads to many human diseases. Most of the antioxidants trigger their antioxidant action either by preventing the initiation of oxidation or by acting as chain-breaking antioxidants.

In the new paradise, free radicals become more ruinous, which plays a role in the pathogenesis of many diseases. Thus, it is worthwhile to design novel antioxidants that will prevent radical induced damage. It is evident from several reports that antioxidants possess anticancer, antiinflammatory, anticardiovascular, and many other activities.

The pharmacological effect of BOA-based derivatives has been extensively studied by virtue of their antineoplastic and antiinflammatory properties. Furthermore, BOA is endowed with antioxidant properties mediated by their free-radical scavenging ability. On the other hand, 1,6-di-t-butylphenol has been reported to possess most potent antioxidant properties by a free-radical scavenging ability, in order to design a molecule possessing both analgesic and antiinflammatory and radical scavenger properties by linking BOA heterocycle to the 1,6-di-t-butylphenol moiety via a chalcone linker. Thus, Aichaoui et al. [54] synthesized antioxidant chalcone derivatives of BOA and evaluated them for antioxidant properties by carrying out in vitro studies. Biological results revealed that compound **28** showed the most potent antioxidant activity, which is 10 times more than the standard antioxidant probucol. The scheme for designing BOA-based chalcones is given in Scheme 10.1 [54].

The antidepressant bupropion is a well-known norepinephrine and dopamine reuptake inhibitor. Several reports have been published that claim 2-BOA analogues of bupropion possess both antidepressant and antioxidant properties. The forced swimming test was employed for carrying out antidepressant activity, while employing DPPH radical scavenging activity for investigating antioxidant activity. Activity results displayed that all analogues are effective molecules in comparison to control at doses 2.5–20 mg/kg. It is evident from SAR studies that BOA-based bupropion analogue with piperidino moiety is the most potent among the series, followed by diethylamino, tert-butylamino, and morpholino, respectively, when arranged in descending order of biological activity [3]. Compounds **29** and **30** were most potent among the series.

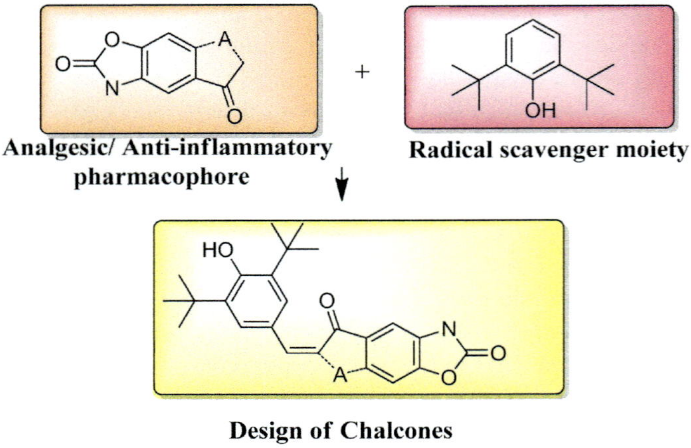

Analgesic/ Anti-inflammatory pharmacophore

+

Radical scavenger moiety

Design of Chalcones

SCHEME 10.1 Designing 2-benzoxazolinone-based chalcones as antioxidizing agents.

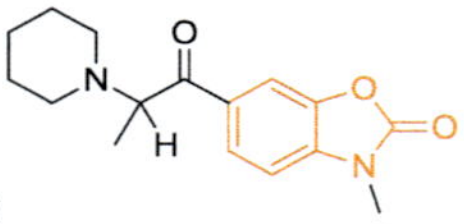

(29)

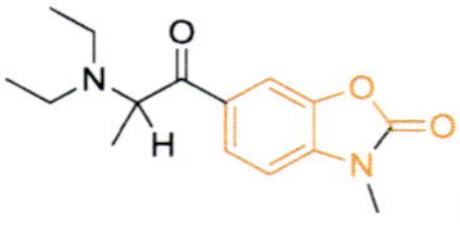

(30)

2-Benzoxazolinone derivatives containing thiosemicarbazide, triazole, thiadia-zole, and hydrazone units have also been reported to exhibit antioxidant activity. The antioxidant activity study results disclosed that compound **31** possessed remarkable dose-dependent antioxidant properties among the synthesized compounds. The SAR disclosed that hydrazone carrying hydroxyl substituents at the aromatic ring is favorable for antioxidant activity [55]. Hydrazones served as the potent metal chelators by exhibiting relatively high reducing power with free radical scavenging capacity, thus possessing an ability to lower lipid peroxidation. The most active compound among the series of these synthesized compounds was compound **31**.

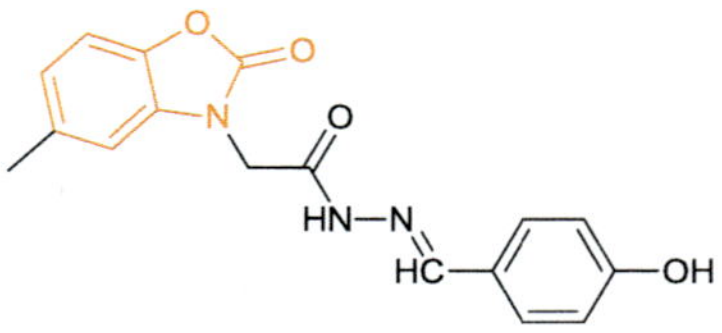

(31)

5. 2-BENZOXAZOLINONE ANALOGS AS MULTITARGETING THERAPY

Several published reports confirmed the superiority of multiple targeting thera-pies over the single targeting drug regimen whenever a question of drug effi-cacy, side-effects and adverse compensatory mechanisms arises, especially in complex diseases like diabetes, inflammation, rheumatoid arthritis, atherosclerosis, neurodegeneration, allergy, infection, and cancer.

In the past decades researchers involved in drug development focused on a limited number of key targets crucial for disease treatment. Tremendous efforts have been made to obtain highly potent drugs with less side effects. To treat asymptomatic and chronic diseases, earlier combination therapy was employed that reduced patient compliance. Therefore it was worthwhile to use the concept of multitargeted drug therapy, where the biological profile has been rationally designed to enhance efficacy and safety in the therapy of a particular disease [56].

Following the trend of multitargeted drug design, BOA analogues were efficiently optimized with multiple biological activities that resulted in discovery of polyfunctional agents for the treatment of multifactorial diseases or coexisting disease, or disease with multiple pathological indications or symptoms [17].

It is difficult to treat inflammatory conditions associated with infection, especially in patients suffering from drug–drug interaction. This provoked researchers to find an agent with dual effect; that is, analgesic–antiinflammatory and antimicrobial effect. The first attempt to use BOA for polyfunctionality was made by Gökhan et al. [57] by designing and synthesizing total 16 6-Acyl-3-piperazinomethyl-based BOA derivatives possessing both analgesic and antimicrobial activities. Among the series of synthesized compounds, compound **32** with 4-(2-pyridyl) piperazine moiety at 3-position of BOA manifested the highest antifungal activities toward *Candida krusei* and analgesic activity as compared to the standard drugs fluconazole and aspirin, respectively. The synthesized molecules were not only evaluated for antifungal activity but also were evaluated for antibacterial activity using the microdilution method. Among all synthesized derivatives, only one compound **33** possessed significant antibacterial action at 16 μg/mL concentration against *Staphylococcus aureus*. Compound **33** possessed both analgesic and moderate antibacterial activities against *S. aureus* [57]. Thus, compounds **32** and **33** were the most potent molecules among the designed series. This indicates that the difluorobenzoyl group is important for analgesic activity. Thus BOA-related derivatives can be therapeutic lead compounds for the treatment of infection associated inflammatory conditions.

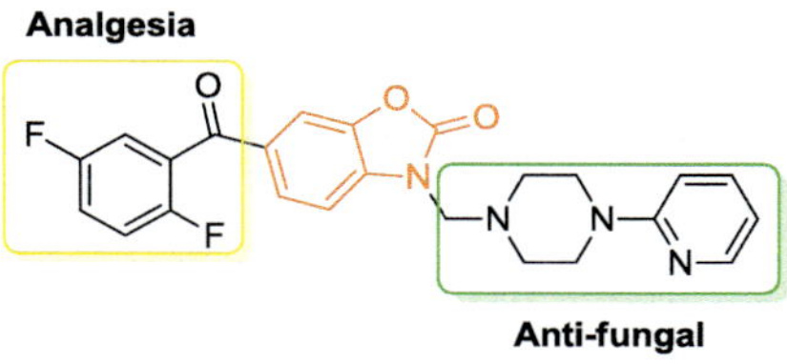

(32)

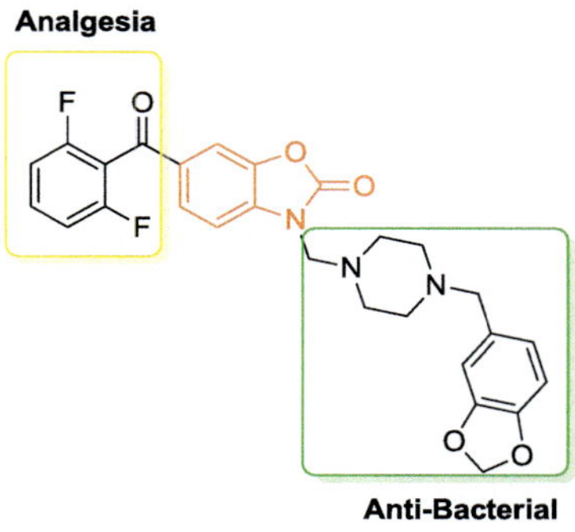

(33)

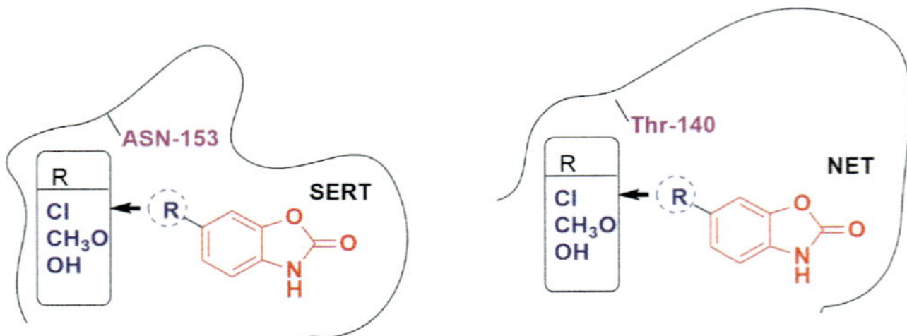

FIGURE 10.7 2-Benzoxazolinone-based derivatives within the active pocket of selective serotonin transporter and norepinephrine transporter.

Serotonin (5-Hydroxytryptamine-5-HT) regulates psychological and behavior processes. Among various 5-HT receptor subtypes, 5-HT$_{1A}$R and 5-HT$_7$R are attractive targets in drug discovery. In 2014 Salerno et al. [58] made an effort to design arylpiperazinylalkyl-based BOAs for their affinity toward both 5-HT7 and 5-HT1A receptors (well-known serotonin receptors). They performed molecular docking simulations to carry out the structural interaction fingerprint analysis, and the result manifested that these dual inhibitors retained interactions with key amino acid residues within the binding pockets of both serotonin receptors. Among the series of BOA-based inhibitors, compound **34** was found to be most potent dual inhibitor of 5-HT7 and 5-HT1A receptors with K$_i$ 5-HT$_{1A}$/K$_i$ 5-HT$_7$=0.02 [58].

(34)

Serotonin and norepinephrine reuptake inhibitors (SNRIs) are two therapeutic targets for treating a wide range of biological functions including pain, anxiety, and depression. It has been reported that dual inhibitors of human serotonin and norepinephrine transporters (NETs) and human selective serotonin transporters (SERTs) are effective analgesics and derivatives of 2-BOA possess antidepressant activity. Thus, it was hypothesized to design BOA-based derivatives having both antidepressant and analgesic activity by exploring 6-position with varied functional groups as shown in Fig. 10.7. With computational studies, Manikandan et al. proposed these derivatives as potential candidates to treat disorders associated with depression and pain. The molecular docking study revealed that oxygen in the

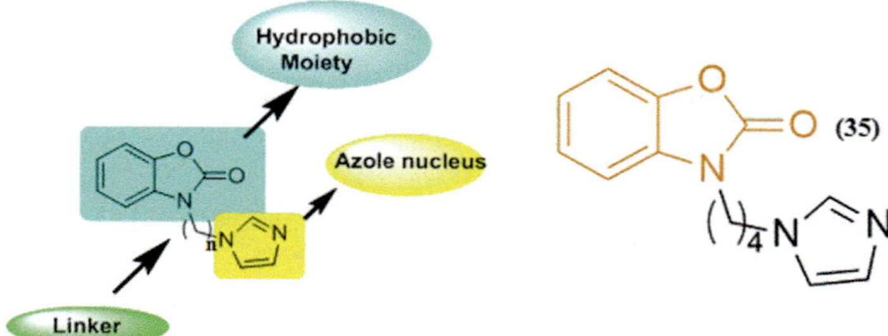

FIGURE 10.8 Designed BOA-based HO-1 and HO-2 dual inhibitor with three crucial features necessary for activity.

ring manifested hydrogen bonding with NH of key amino acid Asn153 in case of SERT. On the other hand, NH in the ring interacted with Thr140 of NET via hydrogen bonding, while benzene ring in both cases fit well in the hydrophobic pocket. Thus, BOA-based drugs can contribute to the development of novel analgesics with fewer adverse side effects compared to conventional selective SNRIs [59].

In some pathological conditions, including certain cancers, overexpression of a cytoprotective enzyme heme oxygenase (HO) occurs. Reports indicate that downregulation of HO-1 and HO-2 systems results in tumor regression. For antiproliferative activity, novel BOA-based derivatives were designed and synthesized as inhibitors of both HO-1 and HO-2. The designed structure has three key chemical moieties for the inhibition of HO are an azolyl nucleus, a hydrophobic portion, and a central alkyl linker (central alkyl length may vary) (Fig. 10.8). Among these three chemical moieties imidazole rings play a crucial role connected to a hydrophobic group represented by 2-BOA by means of alkyls of different lengths [60]. Imidazole serves as an anchor by coordinating with heme iron of the enzymatic complex. The hydrophobic moiety stabilizes the enzyme-inhibitor complex by interacting with important residues in the hydrophobic pocket [Phe33, Met34, Phe37, Val50, Leu54, Leu147, Phe167, and Phe214]. Among the series, Compound **35** was found to be HO-1 and HO-2 inhibitor, with an HO-1 of IC_{50} $17.7 \pm 1.2\,\mu M$, and HO-2 IC_{50} of $35.5 \pm 1.3\,\mu M$.

6. CONCLUSION

It was evident from several reports that BOA is extensively explored at 3- and 6-position by making varied substitutions. BOA-based derivatives exhibit a wide range of biological activities. This clearly indicates that BOA is a privileged substructure. To some extent it is used as a multitargeting drug regimen. This discussion/multitargeting report envisaged the research community to further explore this privileged structure for multifactorial diseases.

REFERENCES

[1] A. Lespagnol, M. Durbet, G. Mongy, Sur le pouvoir hypnotique de la benzoxazolone, Comp Rend Soc Biol 135 (1941) 1255–1257.

[2] N. Gökhan, H. Erdoğan, N.T. Durlu, R. Demirdamar, Analgesic activity of acylated 2-benzoxazolinone derivatives, Il Farmaco 54 (1) (1999) 112–115.

[3] F. Hadizadeh, M.A. Ebrahimzadeh, H. Hosseinzadeh, V. Motamed-Shariaty, S. Salami, A.R. Bekhradnia, Antidepressant and antioxidant activities of some 2-benzoxazolinone derivatives as bupropion analogues, Pharmacologyonline 1 (2009) 331–335.

[4] M. Safakish, Z. Hajimahdi, R. Zabihollahi, M.R. Aghasadeghi, R. Vahabpour, A. Zarghi, Design, synthesis, and docking studies of new 2-benzoxazolinone derivatives as anti-HIV-1 agents, Medicinal Chemistry Research (2017) 1–9.

[5] D. Sicker, M. Frey, M. Schulz, A. Gierl, Role of natural benzoxazinones in the survival strategy of plants, International Review of Cytology 198 (2000) 319–346.

[6] J.W. Park, S.J. Nam, Y.J. Yoon, Enabling techniques in the search for new antibiotics: combinatorial biosynthesis of sugar-containing antibiotics, Biochemical Pharmacology 134 (2017) 56–73.

[7] J. Sam, J.L. Valentine, Preparation and properties of 2-benzoxazolinones, Journal of Pharmaceutical Sciences 58 (9) (1969) 1043–1054.

[8] S. Ilieva, B. Galabov, D.G. Musaev, K. Morokuma, Computational study of the aminolysis of 2-benzoxazolinone, Journal of Organic Chemistry 68 (9) (2003) 3406–3412.

[9] U. Salgın-Gökşen, N. Kelekçi, K.Y. Göktaş Ö, E. Kılıç, Ş. Işık, G. Aktay, M. Özalp, 1-Acylthiosemicarbazides, 1, 2, 4-triazole-5 (4H)-thiones, 1, 3, 4-thiadiazoles and hydrazones containing 5-methyl-2-benzoxazolinones: synthesis, analgesic-anti-inflammatory and antimicrobial activities, Bioorganic and Medicinal Chemistry 15 (17) (2007) 5738–5751.

[10] B. Jakob, 5, 6-dichlorobenzoxazolinone-2 and 5, 6-dichlorobenzothiazolinone-2 (J.R. Geigy Ag, Invent. assignee), United States patent US 2,922,794; January 26, 1960.

[11] V.M. Morais, M.S. Miranda, M.A. Matos, J.F. Liebman, Experimental and computational thermochemistry of three nitrogen-containing heterocycles: 2-benzimidazolinone, 2-benzoxazolinone and 3-indazolinone, Molecular Physics 104 (3) (2006) 325–334.

[12] PubChem. https://pubchem.ncbi.nlm.nih.gov/search/-collection=compounds.

[13] L. Liu, H. Fan, P. Qi, Y. Mei, L. Zhou, L. Cai, X. Lin, J. Lin, Synthesis and hepatoprotective properties of *Acanthus ilicifolius* alkaloid A and its derivatives, Experimental and Therapeutic Medicine 6 (3) (2013) 796–802.

[14] K.Z. Łączkowski, K. Misiura, A. Biernasiuk, A. Malm, I. Grela, Synthesis and antimicrobial activities of novel 6-(1, 3-thiazol-4-yl)-1, 3-benzoxazol-2 (3H)-one derivatives, Heterocyclic Communications 20 (1) (2014) 41–46.

[15] H. Zhang, K. Liu, R. Liu, Q. Li, Y. Li, Q. Wang, S. Liu, Design, synthesis and herbicidal activities of tetrahydroisoindoline-1, 3-dione derivatives containing alkoxycarbonyl substituted 2-benzoxazolinone, Chinese Journal of Chemistry 33 (7) (2015) 749–755.

[16] Y. Mulazim, C. Berber, H. Erdogan, M.H. Ozkan, B. Kesanli, Synthesis and analgesic activities of some new 5-chloro-2 (3H)-benzoxazolone derivatives, The EuroBiotech Journal 1 (3) (2017) 235–240.

[17] M. Singh, M. Kaur, O. Silakari, Flavones: an important scaffold for medicinal chemistry, European Journal of Medicinal Chemistry 84 (2014) 206–239.

[18] R.K. Goel, D. Singh, A. Lagunin, V. Poroikov, PASS-assisted exploration of new therapeutic potential of natural products, Medicinal Chemistry Research 20 (9) (December 1, 2011) 1509–1514.

[19] J.A. Klun, C.L. Tipton, J.F. Robinson, D.L. Ostrem, M. Beroza, Isolation and identification of 6, 7-dimethoxy-2-benzoxazolinone from dried tissues of Zea mays and evidence of its cyclic hydroxamic acid precursor, Journal of Agricultural and Food Chemistry 18 (4) (1970) 663–665.

[20] D.A. Fielder, F.W. Collins, B.A. Blackwell, C. Bensimon, J.W. ApSimon, Isolation and characterization of 4-acetyl-benzoxazolin-2-one (4-ABOA), a new benzoxazolinone from Zea mays, Tetrahedron Letters 35 (4) (1994) 521–524.

[21] C.S. Tang, S.H. Chang, D. Hoo, K.H. Yanagihara, Gas chromatographic determination of 2(3)-benzoxazolinones from cereal plants, Phytochemistry 14 (9) (1975) 2077–2079.

[22] B.H. Babu, B.S. Shylesh, J. Padikkala, Antioxidant and hepatoprotective effect of *Acanthus ilicifolius*, Fitoterapia 72 (3) (2001) 272–277.

[23] K.M.S. Kumar, B. Gorain, D.K. Roy, S.K. Samanta, M. Pal, P. Biswas, A. Roy, D. Adhikari, S. Karmakar, T. Sen, Anti-inflammatory activity of *Acanthus ilicifolius*, Journal of Ethnopharmacology 120 (1) (2008) 7–12.

[24] S. Bose, A. Bose, Antimicrobial activity of *Acanthus ilicifolius* (L.), Indian Journal of Pharmaceutical Sciences 70 (6) (2008) 821.

[25] F. Li, C. Xia, Synthesis of 2-oxazolidinone catalyzed by palladium on charcoal: a novel and highly effective heterogeneous catalytic system for oxidative cyclocarbonylation of β-aminoalcohols and 2-aminophenol, Journal of Catalysis 227 (2) (2004) 542–546.

[26] S. Haider, M.S. Alam, H. Hamid, A. Dhulap, S. Umar, M.S. Yar, S. Bano, S. Nazreen, Y. Ali, C. Kharbanda, Design, synthesis and docking studies of 2-benzoxazolinone-based 1, 2, 4-triazoles as proinflammatory cytokine inhibitors, Medicinal Chemistry Research 23 (9) (2014) 4250–4268.

[27] I. Kubo, T. Kamikawa, Identification and efficient synthesis of 6-methoxy-2-benzoxazolinone (MBOA), an insect antifeedant, Cellular and Molecular Life Sciences 39 (4) (1983) 355.

[28] D.R. John, J.S. Allen, L.C. Albert, J.N. BeMiller, On the synthesis of 6-Methoxy-2-benzoxazolinone, Agricultural and Biological Chemistry 39 (3) (1975) 683–685.

[29] P.B. de Carvalho, E.I. Ferreira, Leishmaniasis phytotherapy. Nature's leadership against an ancient disease, Fitoterapia 72 (6) (2001) 599–618.

[30] D. Singh, V. Aeri, Phytochemical and pharmacological potential of *Acanthus ilicifolius*, Journal of Pharmacy and Bioallied Sciences 5 (1) (2013) 17.

[31] N. Gökhan, H. Erdogan, B. Tel, R. Demirdamar, Analgesic and antiinflammatory activity screening of 6-acyl-3-piperazinomethyl-2-benzoxazolinone derivatives, European Journal of Medicinal Chemistry 31 (7–8) (1996) 625–628.

[32] M. Köksal, N. Gökhan, E. Küpeli, E. Yesilada, H. Erdogan, Analgesic and antiinflammatory activities of some new mannich bases of 5-nitro-2-benzoxazolinones, Archives of Pharmacal Research 30 (4) (2007) 419–424.

[33] E. Palaska, N. Gökhan, H. Erdoğan, B.C. Tel, R. Demirdamar, Synthesis of some new pyridyl-ethylated benzoxazolinones with analgesic and anti-inflammatory activities, Arzneimittelforschung 49 (09) (1999) 754–758.

[34] S. Uenlue, T. Önkol, Y. Duendar, B. Oekcelik, E. Kuepeli, E. Yeşilada, N. Noyanalpan, M.F. Şahin, Synthesis and analgesic and anti-inflammatory activity of some new (6-acyl-2-benzoxazolinone and 6-acyl-2-benzothiazolinone derivatives with acetic acid and propanoic acid residues, Archiv der Pharmazie 336 (8) (2003) 353–361.

[35] E. Banoglu, B. Okcelik, E. Kupeli, S. Uenlue, E. Yeşilada, M. Amat, J.F. Caturla, M.F. Sahin, Amide derivatives of [5-Chloro-6-(2-chloro/fluorobenzoyl)-2-benzoxazolinone-3-yl] acetic acids as potential analgesic and anti-inflammatory compounds, Archiv der Pharmazie 336 (4–5) (2003) 251–257.

[36] R. Verma, P. Khanna, S. Chawla, M. Dhankar, HIV vaccine: can it be developed in the 21st century? Human Vaccines and Immunotherapeutics 12 (1) (2016) 222–224.

[37] H.X. Wang, T.B. Ng, Demonstration of antifungal and anti-human immunodeficiency virus reverse transcriptase activities of 6-methoxy-2-benzoxazolinone and antibacterial activity of the pineal indole 5-methoxyindole-3-acetic acid, Comparative Biochemistry and Physiology - Part C: Toxicology and Pharmacology 132 (2) (2002) 261–268.

[38] M. He, H. Zhang, X. Yao, M. Eckart, E. Zuo, M. Yang, Design, biologic evaluation, and SAR of novel pseudo-peptide incorporating benzheterocycles as HIV-1 protease inhibitors, Chemical Biology and Drug Design 76 (2) (2010) 174–180.

[39] D. Cardinale, G. Biasillo, C.M. Cipolla, Curing cancer, saving the heart: a challenge that cardioncology should not miss, Current Cardiology Reports 18 (6) (2016) 51.

[40] S.B. Advani, J. Sam, Potential anticancer and antiviral agents. Substituted 3-[1′(2′, 3′, 4′-tri-O-benzoyl-β-d-ribopyranosyl)]-2-benzoxazolinones, Journal of Heterocyclic Chemistry 5 (1) (1968) 119–122.

[41] M.S. Murty, K.R. Ram, R.V. Rao, J.S. Yadav, J.V. Rao, V.T. Cheriyan, R.J. Anto, Synthesis and preliminary evaluation of 2-substituted-1, 3-benzoxazole and 3-[(3-substituted) propyl]-1, 3-benzoxazol-2 (3H)-one derivatives as potent anticancer agents, Medicinal Chemistry Research 20 (5) (2011) 576–586.

[42] O. Schadt, D. Dorsch, F. Stieber, A. Blaukat, 2-oxo-3-benzylbenzoxazol-2-one derivatives and related compounds as Met kinase inhibitors for the treatment of tumours (Merck Patent Gesellschaft Mit Beschrankter Haftung, Invent. assignee), United States patent US 8,431,572; 2013.

[43] A. Bach, D. Pizzirani, N. Realini, V. Vozella, D. Russo, I. Penna, L. Melzig, R. Scarpelli, D. Piomelli, Benzoxazolone carboxamides as potent acid ceramidase inhibitors: synthesis and structure–activity relationship (SAR) studies, Journal of Medicinal Chemistry 58 (23) (2015) 9258–9272.

[44] M.S. Gerova, S.R. Stateva, E.M. Radonova, R.B. Kalenderska, R.I. Rusew, R.P. Nikolova, C.D. Chanev, B.L. Shivachev, M.D. Apostolova, O.I. Petrov, Combretastatin A-4 analogues with benzoxazolone scaffold: synthesis, structure and biological activity, European Journal of Medicinal Chemistry 120 (2016) 121–133.

[45] D. Lu, A. Shen, Y. Liu, X. Peng, W. Xing, J. Ai, M. Geng, Y. Hu, Design and synthesis of novel benzo [d] oxazol-2 (3H)-one derivatives bearing 7-substituted-4-enthoxyquinoline moieties as c-Met kinase inhibitors, European Journal of Medicinal Chemistry 115 (2016) 191–200.

[46] P. Gilbert, Depression: The Evolution of Powerlessness, Routledge, 2016.

[47] M.S. Lener, M.J. Niciu, E.D. Ballard, M. Park, L.T. Park, A.C. Nugent, C.A. Zarate, Glutamate and gamma-aminobutyric acid systems in the pathophysiology of major depression and antidepressant response to ketamine, Biological Psychiatry 81 (10) (2017) 886–897.

[48] M.J. Owens, C.B. Nemeroff, Role of serotonin in the pathophysiology of depression: focus on the serotonin transporter, Clinical Chemistry 40 (2) (1994) 288–295.

[49] S. Manikandan, A.R. Saranya, T. Ramanathan, K.S. Kesavanarayanan, L.K. Teh, M.Z. Salleh, Identification of benzoxazolinone derivatives based inhibitors for depression and pain related disorders using human serotonin and norepinephrine transporter as dual therapeutic target: a computational approach, International Journal of Drug Delivery 6 (4) (2014) 389–395.

[50] J. Zaki, T.D. Wager, T. Singer, C. Keysers, V. Gazzola, The anatomy of suffering: understanding the relationship between nociceptive and empathic pain, Trends in Cognitive Sciences 20 (4) (2016) 249–259.

[51] T. Önkol, M.F. Sahin, E. Yildirim, K. Erol, S. Ito, Synthesis and antinociceptive activity of (5-chloro-2 (3H)-benzoxazolon-3-yl) propanamide derivatives, Archives of Pharmacal Research 27 (11) (2004) 1086.

[52] S. Haider, M.S. Alam, H. Hamid, S. Shafi, A. Nargotra, P. Mahajan, S. Nazreen, A.M. Kalle, C. Kharbanda, Y. Ali, A. Alam, Synthesis of novel 1, 2, 3-triazole based benzoxazolinones: their TNF-α based molecular docking with in-vivo anti-inflammatory, antinociceptive activities and ulcerogenic risk evaluation, European journal of medicinal chemistry 70 (2013) 579–588.

[53] J.E. Pero, M.A. Rossi, H.D. Lehman, M.J. Kelly, J.J. Mulhearn, S.E. Wolkenberg, M.J. Cato, M.K. Clements, C.J. Daley, T. Filzen, E.N. Finger, Benzoxazolinone aryl sulfonamides as potent, selective Na v. 1.7 inhibitors with in vivo efficacy in a preclinical pain model, Bioorganic and Medicinal Chemistry Letters 27 (12) (2017) 2683–2688.

[54] H. Aichaoui, F. Guenadil, C.N. Kapanda, D.M. Lambert, C.R. McCurdy, J.H. Poupaert, Synthesis and pharmacological evaluation of antioxidant chalcone derivatives of 2 (3H)-benzoxazolones, Medicinal Chemistry Research 18 (6) (2009) 467–476.

[55] S.Y. Çiftci, N.G. Kelekçi, U.S. Gökşen, G. Uçar, Free-radical scavenging activities of 2-benzoxazolinone derivatives containing thiosemicarbazide, triazole, thiadiazole and hydrazone units, Hacettepe University Journal of the Faculty of Pharmacy 31 (1) (2011) 27–50.

[56] A. Koeberle, O. Werz, Multi-target approach for natural products in inflammation, Drug Discovery Today 19 (12) (2014) 1871–1882.

[57] N. Gökhan, H. Erdoğan, N.T. Durlu, R. Demirdamar, M. Oezalp, Synthesis and evaluation of analgesic, anti-inflammatory and antimicrobial activities of 6-acyl-3-piperazinomethyl-2-benzoxazolinones, Arzneimittelforschung 53 (02) (2003) 114–120.

[58] L. Salerno, V. Pittalà, M.N. Modica, M.A. Siracusa, S. Intagliata, A. Cagnotto, M. Salmona, R. Kurczab, A.J. Bojarski, G. Romeo, Structure–activity relationships and molecular modeling studies of novel arylpiperazinylalkyl 2-benzoxazolones and 2-benzothiazolones as 5-HT 7 and 5-HT 1A receptor ligands, European Journal of Medicinal Chemistry 85 (2014) 716–726.

[59] T. Ramanathan, S. Manikandan, A.R. Saranya, K.S. Kesavanarayanan, L.K. Teh, M.Z. Salleh, Identification of benzoxazolinone derivatives based inhibitors for depression and pain related disorders using human serotonin and norepinephrine transporter as dual therapeutic target: a computational approach, International Journal of Drug Delivery 6 (4) (2015) 389–395.

[60] L. Salerno, V. Pittala, G. Romeo, M.N. Modica, A. Marrazzo, M.A. Siracusa, V. Sorrenti, C. Di Giacomo, L. Vanella, N.N. Parayath, K. Greish, Novel imidazole derivatives as heme oxygenase-1 (HO-1) and heme oxygenase-2 (HO-2) inhibitors and their cytotoxic activity in human-derived cancer cell lines, European Journal of Medicinal Chemistry 96 (2015) 162–172.

Glossary

Acute Present or experienced to a severe or intense degree.

Acute toxicity Adverse effects of a substance that result either from a single exposure or from multiple exposures in a short period of time (usually less than 24 h). To be described as acute toxicity, the adverse effects should occur within 14 days of the administration of the substance.

Acquired immune deficiency syndrome (AIDS) An infectious and usually fatal human disease caused by a retrovirus, **HIV**, which attacks T cells. The virus multiplies within and kills individual T cells, until no T cells remain, leaving the affected individual helpless in the face of microbial infections because his or her immune system is now incapable of defending against them.

Acetylcholinesterase (AChE) An enzyme that catalyzes the breakdown of acetylcholine and other choline esters that function as neurotransmitters.

Activation energy Extra energy that must be possessed by atoms or molecules in addition to their ground-state energy in order to undergo a particular chemical reaction. Activation energy can be applied externally as heat but this is inappropriate for living organisms. Instead, they rely on biological catalysts (enzymes), which decrease the activation energy needed for the reaction to take place.

Active site The region of an enzyme molecule that contains the substrate binding site and the catalytic site for converting the substrate(s) into product(s).

Activin-like kinase 5 (ALK5) A type I receptor of transforming growth factor (TGF)-beta that mediates abnormal proliferation of vascular smooth muscle cells.

Adenocarcinoma A malignant tumor formed from glandular structures in epithelial tissue.

Adenovirus Any of a group of DNA viruses first discovered in adenoid tissue, most of which cause respiratory diseases.

Adipose tissue A fatty connective tissue, the matrix of which contains large, closely packed, fat-filled cells; occurs either around the liver and kidneys where it stores energy, or in the dermis of the skin where it stimulates the body from heat loss.

ADME Absorption, distribution, metabolism, and excretion.

Adrenaline A hormone secreted by the adrenal glands that increases rates of blood circulation, breathing, and carbohydrate metabolism and prepares muscles for exertion.

Adrenochrome A chemical compound with the molecular formula $C_9H_9NO_3$ produced by the oxidation of adrenaline (epinephrine). The derivative carbazochrome is a hemostatic medication. Despite a similarity in chemical names, it is unrelated to chrome or chromium.

Advanced glycation end product (AGE) A protein or lipid that becomes glycated as a result of exposure to sugars. AGEs can be a factor in aging and in the development or worsening of many degenerative diseases such as diabetes, atherosclerosis, chronic kidney disease, and Alzheimer's disease.

Adverse effect; adverse event; adverse drug event (In medicinal chemistry) Undesirable reaction in response to the administration of a drug or test compound.

Affinity The tendency of a molecule to associate with another. The affinity of a drug is its ability to bind to its biological target (receptor, enzyme, transport system, etc.). For pharmacological receptors it can be thought of as the frequency with which the drug, when brought into the proximity of a receptor by diffusion, will reside at a position of minimum free energy within the force field of that receptor.

African sleeping sickness Sleeping sickness, also known as African trypanosomiasis, is a disease caused by infection with the parasite *Trypanosoma brucei*.

Agar plate diffusion method A test of the antibiotic sensitivity of bacteria. It uses antibiotic discs to test the extent to which bacteria are affected by those antibiotics.

Agonist An agonist is an endogenous substance or a drug that can interact with a receptor and initiate a physiological or a pharmacological response characteristic of that receptor (contraction, relaxation, secretion, enzyme activation, etc.).

Albino mice A small mammal of the order Rodentia, which is bred and used for scientific research.

Aldol reactions When the enolate of an aldehyde or a ketone reacts at the α-carbon with the carbonyl of another molecule under basic or acidic conditions to obtain β-hydroxy aldehyde or ketone. Aldol is an abbreviation of aldehyde and alcohol.

Aldose reductase (ALR2) A cytosolic NADPH-dependent oxidoreductase that catalyzes the reduction of a variety of aldehydes and carbonyls, including monosaccharides.

Alkaloids A class of nitrogenous organic compounds of plant origin with pronounced physiological actions on humans. They include many drugs (morphine, quinine) and poisons (atropine, strychnine).

Allelochemical A chemical produced by a living organism that exerts a detrimental physiological effect on individuals of another species when released into the environment.

Allodynia Refers to central pain sensitization (increased response of neurons) following normally nonpainful, often repetitive, stimulation. Allodynia can lead to the triggering of a pain response from stimuli that do not normally provoke pain.

Allosteric modulator A drug that binds to a receptor at a site distinct from the active site. Induces a conformational change in the receptor, which alters the affinity of the receptor for the endogenous ligand. Positive allosteric modulators increase the affinity, whereas negative allosteric modulators decrease the affinity.

Allosteric site A site that allows molecules to either activate or inhibit (turn off) enzyme activity. It's different than the active site on an enzyme, where substrates bind. When allosteric activators bind to the allosteric site, the enzyme binds the substrate better, and the reaction becomes faster.

Alzheimer's disease A progressive disease that destroys memory and other important mental functions.

Amastigote A protist cell that does not have visible external flagella or cilia. The term is used mainly to describe a certain phase in the life cycle of trypanosomes.

Amino acids Organic compounds containing amine ($-NH_2$) and carboxyl ($-COOH$) functional groups, along with a side chain (R group) specific to each amino acid.

Amyloid beta (Aβ) Denotes peptides of 36–43 amino acids that are crucially involved in Alzheimer's disease as the main component of the amyloid plaques found in the brains of Alzheimer patients.

Anabolic agents; Anabolic steroids The familiar name for synthetic substances related to the male sex hormones (e.g., testosterone). They promote the growth of skeletal muscle (anabolic effects) and the development of male sexual characteristics (androgenic effects) in both males and females.

Analgesics An analgesic or painkiller is any member of the group of drugs used to achieve analgesia, relief from pain.

Analogue; analog Chemical compound having structural similarity to a reference compound.

Androgen receptor The androgen receptor, also known as NR3C4, is a type of nuclear receptor that is activated by binding either of the androgenic hormones, testosterone, or dihydrotestosterone in the cytoplasm and then translocating into the nucleus.

Angina pectoris A type of chest pain caused by reduced blood flow to the heart (ischemic chest pain).

Angiogenesis The formation of new blood vessels.

Angiotensin I A peptide hormone that causes vasoconstriction and an increase in blood pressure. It is part of the renin-angiotensin system, which is a major target for drugs that raise blood pressure.

Angiotensin converting enzyme (ACE) A central component of the renin-angiotensin system (RAS), which controls blood pressure by regulating the volume of fluids in the body. It converts the hormone angiotensin I to the active vasoconstrictor angiotensin II. Therefore, ACE indirectly increases blood pressure by causing blood vessels to constrict.

Antagonist An antagonist is a drug or a compound that opposes the physiological effects of another. At the receptor level, it is a chemical entity that opposes the receptor-associated responses normally induced by another bioactive agent.

Anthelmintic; antihelminthic A group of antiparasitic drugs that expel parasitic worms and other internal parasites from the body by either stunning or killing them and without causing significant damage to the host.

Antibacterial Active against bacteria.

Antibiotic Also called antibacterial, a type of antimicrobial drug used in the treatment and prevention of bacterial infections. They may either kill or inhibit the growth of bacteria.

Antibody An antibody, also known as an immunoglobulin, is a large, Y-shaped protein produced mainly by plasma cells that is used by the immune system to neutralize pathogens such as pathogenic bacteria and viruses.

Anticholinergic A substance that blocks the neurotransmitter acetylcholine in the central and the peripheral nervous system.

Anticoagulant Chemical substances that prevent or reduce coagulation of blood, prolonging the clotting time. Commonly referred to as blood thinners.

Antidepressant Drug used for the treatment of major depressive disorder and other conditions, including dysthymia, anxiety disorders, obsessive–compulsive disorder, eating disorders, chronic pain, neuropathic pain, and in some cases, dysmenorrhea, snoring, migraine, attention-deficit hyperactivity disorder (ADHD), addiction, dependence, and sleep disorders.

Antidiabetic Drug used to treat diabetes mellitus by lowering glucose levels in the blood. With the exceptions of insulin, exenatide, liraglutide, and pramlintide, all are administered orally and are thus also called oral hypoglycemic agents or oral antihyperglycemic agents.

Antiemetic A drug that is effective against vomiting and nausea. Antiemetics are typically used to treat motion sickness.

Antifungal An antifungal medication, also known as an antimycotic medication, is a pharmaceutical fungicide or fungistatic used to treat and prevent mycoses such as athlete's foot, ringworm, candidiasis (thrush), serious systemic infections such as cryptococcal meningitis, and others.

Antifusarium Drugs active against fungal plant pathogens that cause plant diseases, especially wilting like Panama disease of banana (Musa spp.), also known as fusarium wilt of banana.

Antigen A foreign molecule (antigen) triggering the production of an antibody (also called immunoglobulin) by the immune system of an organism.

Antihypertensive A class of drugs used to treat hypertension. Antihypertensive therapy seeks to prevent the complications of high blood pressure, such as stroke and myocardial infarction.

Antiinflammatory The property of a substance or treatment that reduces inflammation or swelling.

Antineoplastic Inhibiting or preventing the growth and spread of tumors or malignant cells.

Antimigraine A medication intended to reduce the effects or intensity of migraine headache. Examples include triptans such as zolmitriptan and ergot alkaloids such as methysergide.

Antioxidant A substance such as vitamin C or vitamin E that removes potentially damaging oxidizing agents in a living organism.

Antiproliferative Of or relating to a substance used to prevent or retard the spread of cells, especially malignant cells, into surrounding tissues.

Antiseptic An antimicrobial substance that is applied to living tissue/skin to reduce the possibility of infection, sepsis, or putrefaction.

Antitumor Inhibiting the growth of a tumor or tumors.

Antiviral A class of medication used specifically for treating viral rather than bacterial infections.

Anxiety A mental health disorder characterized by feelings of worry or fear that are strong enough to interfere with one's daily activities.

Anxiolytic A drug used to relieve anxiety.

Apoptosis A process of programmed cell death that occurs in multicellular organisms. Biochemical events lead to characteristic cell changes and death.

Arenavirus A virus that is a member of the family Arenaviridae. These viruses infect rodents and occasionally humans; arenaviruses that infect snakes have also been discovered. At least eight arenaviruses are known to cause human disease.

Arrhythmia Improper beating of the heart, whether irregular, too fast, or too slow.

Aromaticity Chemical property of conjugated rings that results in unusual stability.

Assay An investigative procedure in laboratory medicine, pharmacology, environmental biology, and molecular biology for qualitatively assessing or quantitatively measuring the presence, amount, or functional activity of a target entity.

Asthma A condition in which a person's airways become inflamed and narrow, and swell and produce extra mucus, which makes it difficult to breathe.

Asymptomatic disease When a patient is a carrier for a disease or infection but experiences no symptoms. A condition might be asymptomatic if it fails to show the noticeable symptoms with which it is usually associated. Asymptomatic infections are also called subclinical infections.

Atherogenesis A disorder of the artery wall that involves adhesion of monocytes and lymphocytes to the endothelial cell surface; migration of monocytes into the subendothelial space and differentiation into macrophages; ingestion of low density lipoproteins and modified or oxidized low density lipoproteins or the process of forming atheromas, plaques in the inner lining (the intima) of arteries.

Atherosclerosis The build-up of fats, cholesterol, and other substances in and on the artery walls.

Attenuated An attenuated virus is a weakened, less vigorous virus. A vaccine against a viral disease can be made from an attenuated, less virulent strain of the virus, a virus capable of stimulating an immune response and creating immunity but not causing illness.

ATP (adenosine triphosphate) A nucleotide that contains a large amount of chemical energy stored in its high-energy phosphate bonds. It releases energy when it is broken down (hydrolyzed) into adenosine diphosphate (ADP). The energy is used for many metabolic processes.

ATP binding pocket A 250-residue sequence within an ATP binding protein's primary structure. It is the environment in which ATP catalytically actives the enzyme and, as a result, is hydrolyzed to ADP.

Aurora kinase A (AurA) An enzyme that in humans is encoded by the AURKA gene. Aurora A is a member of a family of mitotic serine/threonine kinases. It is implicated with important processes during mitosis and meiosis whose proper function is integral for healthy cell. Also known as serine/threonine-protein kinase 6.

Autoimmune disorder A disease in which the body's immune system attacks healthy cells.

Auxin A class of plant hormones with some morphogenlike characteristics. Auxins have a cardinal role in coordination of many growth and behavioral processes in the plant's life cycle and are essential for plant body development.

BACE-1 (beta-secretase 1) An enzyme that in humans is encoded by the *BACE1*. BACE1 is an aspartic-acid protease important in the formation of myelin sheaths in peripheral nerve cells. Also known as beta-site amyloid precursor protein cleaving enzyme 1, beta-site APP cleaving enzyme 1, membrane-associated aspartic protease 2, memapsin-2, aspartyl protease 2, and ASP2.

Baker-Venkatraman rearrangement The chemical reaction of 2-acetoxyacetophenones with base to form 1,3-diketones. This rearrangement reaction proceeds via enolate formation followed by acyl transfer. It is named after the scientists Wilson Baker and Krishnasami Venkataraman and is often used to synthesize chromones and flavones.

Baker's yeast The common name for the strains of yeast commonly used as a leavening agent in baking bread and bakery products, where it converts the fermentable sugars present in the dough into carbon dioxide and ethanol.

Bartoli indole synthesis The chemical reaction of ortho-substituted nitroarenes with vinyl Grignard reagents to form substituted indoles.

Base peak The most intense (tallest) peak in a mass spectrum, due to the ion with the greatest relative abundance (relative intensity; height of peak along the spectrum's y-axis).

Basicity The number of hydrogen atoms replaceable by a base in a particular acid.

Bacteria Bacteria constitute a large domain of prokaryotic microorganisms. Typically a few micrometers in length, bacteria have a number of shapes, ranging from spheres to rods and spirals.

Basophill A basophilic white blood cell.

Bending Can be applied to certain molecules to describe their molecular geometry.

Bile acid A steroid acid found predominantly in the bile of mammals and other vertebrates. Different molecular forms of bile acids can be synthesized in the liver by different species.

Binding affinity Binding affinity is the strength of the binding interaction between a single biomolecule (e.g., protein or DNA) to its ligand/binding partner (e.g., drug or inhibitor).

Binding conformation The orientation of the ligand relative to the receptor as well as the conformation of the ligand and receptor when bound to each other.

Binding constant The binding constant, or association constant, is a special case of the equilibrium constant K, and is the inverse of the dissociation constant. It is associated with the binding and unbinding reaction of receptor (R) and ligand (L) molecules, which is formalized as $R + L \rightleftharpoons RL$.

Binding site; binding pocket A depression, tunnel, or cleft where ligands bind on a protein/receptor surface.

Bischler indole synthesis A chemical reaction that forms a 2-aryl-indole from an α-bromo-acetophenone and excess aniline.

Bioactive agent A compound that has an effect on a living organism, tissue, or cell. In the field of nutrition bioactive compounds are distinguished from essential nutrients.

Bioavailability The proportion of a drug or other substance that enters the circulation when introduced into the body and so is able to have an active effect.

Biodistribution study Tracks where compounds of interest travel in an experimental animal or human subject. For example, in the development of new compounds for positron emission tomography (PET) scanning, a radioactive isotope is chemically joined with a peptide (subunit of a protein).

Bioisostere A compound resulting from the exchange of an atom or group of atoms with another, broadly similar atom or group of atoms. The objective of a bioisosteric replacement is to create a new compound with similar biological properties to the parent compound. The bioisosteric replacement may be physicochemically or topologically based.

Biological membrane; biomembrane An enclosing or separating membrane that acts as a selectively permeable barrier within living things.

Bivalent ligand Two druglike molecules (pharmacophores or ligands) connected by an inert linker.

Blood–brain barrier (BBB) A semipermeable membrane separating the blood from the cerebrospinal fluid, and constituting a barrier to the passage of cells, particles, and large molecules.

Bone mineral density (BMD) A test uses X-rays to measure the amount of minerals—namely calcium—in your bones. This test is important for people who are at risk for osteoporosis, especially women and older adults. The test is also referred to as dual energy X-ray absorptiometry (DXA).

Boron analysis A test method to determine if the boron concentration is within acceptable limits.

Boron neutron capture therapy (BNCT) A noninvasive therapeutic modality for treating locally invasive malignant tumors such as primary brain tumors and recurrent head and neck cancer, with the nonradioactive isotope boron-10.

Bovine viral diarrhea virus (BVDV) Significant economic disease of cattle that is endemic in the majority of countries throughout the world. The causative agent is a member of the *Pestivirus* genus of the family Flaviviridae.

Bradycardia Abnormally slow heart action.

Bronchitis Inflammation of the lining of bronchial tubes, which carry air to and from the lungs.

Buffer Either a weak acid and its salt or else a weak base and its salt that form an aqueous solution that resists pH changes.

Butyrylcholinesterase (BChE) A nonspecific cholinesterase enzyme that hydrolyzes many different choline-based esters. Also known as BuChE, pseudocholinesterase, or plasma esterase.

Cachexia Weakness and wasting of the body due to severe chronic illness.

Calpain A protein belonging to the family of calcium-dependent, nonlysosomalcysteine proteases (proteolytic enzymes) expressed ubiquitously in mammals and many other organisms. Calpains constitute the C2 family of protease clan CA in the MEROPS database. The calpain proteolytic system includes the calpain proteases, the small regulatory subunit CAPNS1, also known as CAPN4, and the endogenous calpain-specific inhibitor, calpastatin.

Cannabinoid receptor Located throughout the body, it is part of the endocannabinoid system, which is involved in a variety of physiological processes including appetite, pain-sensation, mood, and memory. Cannabinoid receptors are of a class of cell membrane receptors in the G protein-coupled receptor superfamily.

Cancer A disease in which abnormal cells divide uncontrollably and destroy body tissue.

Candidiasis A fungal infection typically on the skin or mucous membranes caused by candida.

Capsid The protein shell of a virus. It consists of several oligomeric structural subunits made of protein called protomers. The observable three-dimensional morphological subunits, which may or may not correspond to individual proteins, are called capsomeres. The capsid encloses the genetic material of the virus.

Carbonyl reductase A NADPH-dependent, mostly monomeric, cytosolic enzyme with broad substrate specificity for many endogenous and xenobiotic carbonyl compounds. They catalyze the reduction of endogenous prostaglandins, steroids, and other aliphatic aldehydes and ketones. Carbonyl reductase (secondary-alcohol: NADP(+) oxidoreductase) belongs to the family of short chain dehydrogenases/reductases (SDR).

Carcinoma A type of cancer that starts in cells that make up the skin or the tissue lining organs, such as the liver or kidneys. Like other types of cancer, carcinomas are abnormal cells that divide without control. They are able to spread to other parts of the body, but don't always.

Carrageenan-induced paw edema assay Used to detect antiinflammatory activity that suppresses prostaglandin production, but other inflammatory mediators are important in the pathogenesis of the lesion.

Catalyst A substance that changes the rate of a chemical reaction (usually accelerating it) without itself undergoing a change. Enzymes are biological catalysts and are proteinaceous in nature.

Catalytic anionic site (CAS) Located at the active site of AChE.

Catalytic domain The part of the protein chain that contains the region where the catalyzed chemical reaction takes place.

Cataract Clouding of the normally clear lens of the eye.

Catecholamine Any of a class of aromatic amines, which includes a number of neurotransmitters such as adrenaline and dopamine.

Cathepsin B A lysosomal cysteine protease of the papain family. It functions in intracellular protein catabolism and in certain situations may also be involved in other physiological processes such as processing of antigens in the immune response, hormone activation, and bone turnover.

CDK (cyclin-dependent kinase) A family of sugar kinases first discovered for their role in regulating the cell cycle. They are also involved in regulating transcription, mRNA processing, and the differentiation of nerve cells.

Cell differentiation The process by which a cell becomes specialized in order to perform a specific function, as in the case of a liver cell, a blood cell, or a neuron.

Cell line Cells that grow and replicate continuously outside the living organism and are maintained in vitro for medical and/or research purposes.

Cell proliferation The process that results in an increase of the number of cells, and is defined by the balance between cell divisions and cell loss through cell death or differentiation. Cell proliferation is increased in tumors.

Cell signaling Part of any communication process that governs basic activities of cells and coordinates all cell actions. The ability of cells to perceive and correctly respond to their microenvironment is the basis of development, tissue repair, and immunity, as well as normal tissue homeostasis.

Cerebrospinal fluid A clear, colorless body fluid found in the brain and spinal cord. It is produced in the choroid plexuses of the ventricles of the brain, and absorbed in the arachnoid granulations. It provides buoyancy and protection to the brain.

Chagas' disease An infectious disease caused by a parasite found in the feces of the triatomine bug.

Chemical database Specific electronic repository for storage and retrieval of chemical information.

Chemical library A collection of distinct, defined and characterized molecules or mixtures thereof.

Chemical probe A small molecule that is used to study and manipulate a biological system such as a cell or an organism by reversibly binding to and altering the function of a biological target (most commonly a protein) within that system.

Chemiluminescence; chemoluminescence The emission of light (luminescence) resulting from a chemical reaction. There may also be limited emission of heat.

Chemopreventive agent A drug or compound that interferes with a disease process; for example, cancer chemopreventive agents used to inhibit, delay, or reverse carcinogenesis.

Chemosensitizing agent A drug that makes tumor cells more sensitive to the effects of chemotherapy.

Chemotactic Agents that cause chemotaxis.

Chemotherapy A category of cancer treatment that uses one or more anticancer drugs as part of a standardized chemotherapy regimen. Chemotherapy may be given with a curative intent, or it may aim to prolong life or to reduce symptoms.

Chemokines Chemokines are a family of small cytokines, or signaling proteins secreted by cells. Their name is derived from their ability to induce directed chemotaxis in nearby responsive cells; they are chemotactic cytokines.

Chicken pox A highly contagious viral infection that causes an itchy, blisterlike rash on the skin.

Chemical shift The resonant frequency of a nucleus relative to a standard in a magnetic field.

Chemotaxis Movement of a motile cell or organism, or part of one, in a direction corresponding to a gradient of increasing or decreasing concentration of a particular substance.

Cholesterol A waxy, fatlike substance that's found in all the cells in the body. The body needs some cholesterol to make hormones, vitamin D, and substances that help in food digestion.

Chromosome A threadlike structure of nucleic acids and protein found in the nucleus of most living cells, carrying genetic information in the form of genes.

Chromophore An atom or group whose presence is responsible for the color of a compound.

Chronic disease Disease persisting for a long time or constantly recurring like chronic bronchitis.

Cirrhosis Chronic liver damage from a variety of causes leading to scarring and liver failure.

CK (creatine kinase) An enzyme expressed by various tissues and cell types. CK catalyzes the conversion of creatine and utilizes adenosine triphosphate (ATP) to create phosphocreatine (PCr) and adenosine diphosphate (ADP). Also known as creatine phosphokinase (CPK) or phosphocreatine kinase.

Claisen–Schmidt condensation The reaction between an aldehyde/ketone and an aromatic carbonyl compound lacking an alpha-hydrogen (cross aldol condensation).

Clinical trial An experiment or observation done in clinical research. Such prospective biomedical or behavioral research studies on human participants are designed to answer specific questions.

c-Met kinase A protein in humans that is encoded by the MET gene. The protein possesses tyrosine kinase activity. Also called tyrosine-protein kinase Met or hepatocyte growth factor receptor.

Colorectal cancer A cancer of the colon or rectum, located at the digestive tract's lower end.

Compound A defined and discreet molecule. A compound can either be a small molecule, a protein or antibody, or an oligonucleotide.

Computer-assisted drug design (CADD) Use of computational tools and algorithms for the visualization, design, and optimization of leads and drug molecules.

Conformation The three-dimensional or spatial orientation of bonds and atoms in a molecule. A certain molecule usually exists in a number of possible conformations.

Congener Substance structurally related to another and linked by origin or function. Note: Congeners may be analogues or vice versa but not necessarily. The term "congener," while most often a synonym for "homologue," has become somewhat more diffuse in meaning so that the terms "congener" and "analogue" are frequently used interchangeably in the literature. See also analogue, follow-on drug.

Complex disease A disorder in which the cause is considered to be a combination of genetic effects and environmental influences.

Congestive heart failure A chronic condition in which the heart doesn't pump blood as well as it should.

Conjugate A compound formed by joining two or more chemical compounds.

Conjugation In organic chemistry terms, it is used to describe the situation that occurs when π systems (e.g., double bonds) are linked together. An isolated π (pi) system exists only between a single pair of adjacent atoms (e.g., C=C). Derived from a Latin word that means "to link together."

Conjugative stabilization The stabilization of dienes by conjugation.

Convulsion A sudden, violent, irregular movement of the body caused by involuntary contraction of muscles and associated especially with brain disorders such as epilepsy, the presence of certain toxins or other agents in the blood, or fever in children.

Coronary heart disease Damage or disease in the heart's major blood vessels.

Counterpart Molecule that corresponds to or has the same function as another in different conditions.

Coxsackie B3 virus A group of six serotypes of Coxsackievirus, a pathogenic enterovirus, that trigger illnesses ranging from gastrointestinal distress to full-fledged pericarditis and myocarditis (Coxsackievirus-induced cardiomyopathy). The virus has approximately 7400 base pairs.

Crohn's disease A chronic inflammatory bowel disease that affects the lining of the digestive tract.

Crystal structure In crystallography, crystal structure is a description of the ordered arrangement of atoms, ions, or molecules in a crystalline material.

Cyclin-dependent kinase (CDK) A protein complex (CDKC, cyclin-CDK) formed by the association of an inactive catalytic subunit of a protein kinase, cyclin-dependent kinase (CDK), with a regulatory subunit, cyclin. Once cyclin-dependent kinases bind to cyclin, the formed complex is in an activated state.

Cyclooxygenase (COX) An enzyme responsible for formation of prostanoids, including thromboxane and prostaglandins such as prostacyclin. Officially known as prostaglandin-endoperoxide synthase.

Cytochrome P450 A protein of the superfamily containing heme as a cofactor and, therefore, a hemoprotein. CYPs use a variety of small and large molecules as substrates in enzymatic reactions.

Cytoprotective Chemical compound that provides protection to cells against harmful agents. For example, a gastric cytoprotectant is any medication that combats ulcers not by reducing gastric acid but by increasing mucosal protection.

Cytostatic A cellular component or medicine that inhibits cell growth.

Cytotoxicity Toxic to cells.

Degranulation A cellular process that releases antimicrobial cytotoxic or other molecules from secretory vesicles called granules found inside some cells. It is used by several different cells involved in the immune system, including granulocytes (neutrophils, basophils, and eosinophils) and mast cells.

Delayed gastric emptying (gastroparesis) A condition that affects the stomach muscles and prevents proper stomach emptying.

Dengue A mosquito-borne viral disease occurring in tropical and subtropical areas.

Dengue virus (DENV) The cause of dengue fever. It is a mosquito-borne single positive-stranded RNA virus of the family Flaviviridae; genus *Flavivirus*. Five serotypes of the virus have been found, all of which can cause the full spectrum of disease.

Depression A mental health disorder characterized by persistently depressed mood or loss of interest in activities, causing significant impairment in daily life.

Deshielding effect (downfield) The opposite of shielding; when we say that an atom is deshielded, we mean a nucleus whose chemical shift has been increased due to removal of electron density, magnetic induction, or other effects.

Designed multiple ligands Compounds conceived and synthesized to act on two or more molecular targets.

Deoxyribonucleic acid (DNA) The chemical building block of the genetic information in the cell, specifying the characteristics of most living organisms. DNA is usually in the form of two complementary strands.

Diabetes mellitus (DM) A group of metabolic disorders in which there are high blood sugar levels over a prolonged period. Symptoms of high blood sugar include frequent urination, increased thirst, and increased hunger. Commonly referred to as diabetes.

Diabetic complication Microvascular complications include damage to eyes (retinopathy), leading to blindness; to kidneys (nephropathy), leading to renal failure; and to nerves (neuropathy), leading to impotence and diabetic foot disorders (which include severe infections leading to amputation); and include macrovascular complications, including heart diseases.

Diazotization The nitrosation of primary aromatic amines with nitrous acid (generated in situ from sodium nitrite and a strong acid, such as hydrochloric acid, sulfuric acid, or HBF_4) leads to diazonium salts, which can be isolated if the counterion is nonnucleophilic.

Dielectric An electrical insulator that can be polarized by an applied electric field. When a dielectric is placed in an electric field, electric charges do not flow through the material as they do in an electrical conductor but only slightly shift from their average equilibrium positions causing dielectric polarization.

Diels–Alder reaction An organic chemical reaction between a conjugated diene and a substituted alkene, commonly termed the dienophile, to form a substituted cyclohexene derivative.

Dihydroorotate dehydrogenase (DHODH) An enzyme that in humans is encoded by the DHODH gene on chromosome 16. The protein encoded by this gene catalyzes the fourth enzymatic step, the ubiquinone-mediated oxidation of dihydroorotate to orotate, in de novo pyrimidine biosynthesis.

Dipole moment The polarity of a polar covalent bond.

Diversity-oriented synthesis (DOS) A strategy for quick access to molecule libraries with an emphasis on skeletal diversity.

DNA binding affinity Strength of the binding interaction between DNA and its binding site.

DNA binding site A type of binding site found in DNA where other molecules may bind. DNA binding sites are distinct from other binding sites in that (1) they are part of a DNA sequence (e.g., a genome) and (2) they are bound by DNA-binding proteins. DNA binding sites are often associated with specialized proteins known as transcription factors, and are thus linked to transcriptional regulation.

DNA minor groove binder Molecule that binds to the minor groove of DNA where backbones of DNA are close together; not a sequence-specific binding.

Dopamine (DA) An organic chemical of the catecholamine and phenethylamine families that plays several important roles in the brain and body.

Downfield See **deshielding**.

DPPH A common abbreviation for the organic chemical compound 2,2-diphenyl-1-picrylhydrazyl. It is a dark-colored crystalline powder composed of stable free-radical molecules.

Dreadful disease Disease causing or involving great suffering.

Drug cocktail or drug combination (In drug therapy) Administration of two or more distinct pharmacological agents to achieve a combination of their individual effects.

Drug delivery Process by which a drug is administered to its intended recipient. Note: Examples include administration orally, intravenously, or by inhalation. See also **drug distribution, targeted drug delivery**.

Drug discovery The process by which new candidate medications are discovered.

Drug–drug interaction A situation in which a substance (usually another drug) affects the activity of a drug when both are administered together.

Drug regimen A systematic plan (as of diet, therapy, or medication) especially when designed to improve and maintain the health of a patient.

Drug repositioning The application of known drugs and compounds to treat new indications (i.e., new diseases). Also known as drug repurposing, reprofiling, retasking, or therapeutic switching.

Drug resistance Reduction in effectiveness of a medication.

Drug safety Assessment of the nontolerable biological effects of a drug.

Drug target Biological molecule or molecules that are inhibited, activated, or modulated by a drug molecule. Drug targets can be any form or combinations of (glyco) proteins, DNA, or RNA.

Dual activity Having two activities at the same time.

Dual binding site Presence of two distinct ligand binding sites on the same molecular target.

Duplex sequencing A tag-based error correction method to improve sequencing accuracy. In this method, adapters (with primer sequences and random 12 bp indices) are ligated onto the template and amplified using PCR.

Dynamic equilibrium A state of balance between continuing processes.

Ebola virus A virus that causes severe bleeding, organ failure, and can lead to death.

EC$_{50}$ Represents the plasma concentration/AUC required for obtaining 50% of the maximum effect in vivo.

ED$_{50}$ The dose of a drug that is pharmacologically effective for 50% of the population exposed to the drug or a 50% response in a biological system that is exposed to the drug.

Efficacy Power or capacity to produce a desired, or in the context of disease, beneficiary effect. It can be measured as the maximum response a compound is capable of producing.

Efflux pump A proteinaceous transporter localized in the cytoplasmic membrane of all kinds of cells. Efflux pumps are active transporters, meaning that they require a source of chemical energy to perform their function.

Efflux pump inhibitor (EPI) Efflux occurs due to the activity of membrane transporter proteins widely known as multidrug efflux systems (MES). One plausible alternative is the combination of conventional antimicrobial agents/antibiotics with small molecules that block MES known as multidrug efflux pump inhibitors (EPIs).

Eicosanoid A signaling molecule made by the enzymatic or nonenzymatic oxidation of arachidonic acid or other polyunsaturated fatty acids that is similar to arachidonic acid, 20 carbon units in length.

Electron releasing group In organic chemistry, an electron donating group (EDG) or electron releasing group (ERG) (+I effect) is an atom or functional group that donates some of its electron density into a conjugated π system via resonance or inductive effects, thus making the π system more nucleophilic.

Electron sink An atom on a molecule or ion that can accept a new bond or lone pair of electrons.

Electron withdrawing group (EWG) Draws electrons away from a reaction center. When this center is an electron-rich carbanion or an alkoxide anion, the presence of the electron-withdrawing substituent has a stabilizing effect. Examples include halogens (F, Cl), nitriles, carbonyls, nitro group.

Electrophile In organic chemistry, a reagent attracted to electrons. An electrophile is a positively charged or neutral species having vacant orbitals that are attracted to an electron-rich center. It participates in a chemical reaction by accepting an electron pair in order to bond to a nucleophile.

Electrophilic substitution Chemical reactions in which an electrophile displaces a functional group in a compound, which is typically, but not always, a hydrogen atom.

Electrophoresis A technique used in laboratories in order to separate macromolecules based on size. The technique applies a negative charge so proteins move toward a positive charge. This is used for both DNA and RNA analysis.

Electronegativity A chemical property that describes the tendency of an atom to attract a shared pair of electrons (or electron density) toward itself.

Enantioselective Relating to or being a chemical reaction in which one enantiomer of a chiral product is preferentially produced.

Enzyme A macromolecule, usually a protein that functions as a (bio) catalyst by increasing the reaction rate.

Empirical design strategy An empirical approach is one that is based on the results of experiments or experience. Generally, it requires a number of observations to be made in order to ascertain the relationships between input variables and outcomes. It is not necessary to firmly establish the scientific basis for the relationships between variables and outcomes as long as the limitations with such an approach are recognized.

Enantiomers Each of a pair of molecules that are mirror images of each other.

Encapsidated DNA DNA enclosed in a protein shell.

Endothelin A Endothelin-1 (ET-1) is a potent endogenous vasoconstrictor, mainly secreted by endothelial cells. It acts through two types of receptors ETA and ETB. Apart from a vasoconstrictive action, ET-1 causes fibrosis of the vascular cells and stimulates production of reactive oxygen species.

Endogenous ligand An endogenous ligand is produced in the body, not introduced into the body, such as certain drugs.

Enteritis Inflammation of the intestine, especially the small intestine, usually accompanied by diarrhea.

Enzyme A protein catalyzing a chemical reaction. Biological processes are to a high degree driven by enzymatic reactions.

Epilepsy A disorder in which nerve cell activity in the brain is disturbed, causing seizures.

Epidermal growth factor receptor (EGFR) A transmembrane receptor protein in humans.

Epstein–Barr virus early antigen (EBV-EA) activation The presence of antibody to the early antigen (EA) of Epstein–Barr virus (EBV) indicates that EBV is actively replicating. Generally, this antibody can be detected only during active EBV infection, such as in patients with infectious mononucleosis.

Esterification Conversion of carboxylic acids to esters using acid and alcohols (Fischer esterification). When a carboxylic acid is treated with an alcohol and an acid catalyst, an ester is formed (along with water).

Extracellular signal-regulated kinase (ERK) In molecular biology, ERKs or classical MAP kinases are widely expressed protein kinase intracellular signaling molecules that are involved in functions including the regulation of meiosis, mitosis, and postmitotic functions in differentiated cells.

Etiological agents Relating to the etiology of a disease.

Familial amyloid polyneuropathy (FAP) A rare group of autosomal dominant diseases wherein the autonomic nervous system and/or other nerves are compromised by protein aggregation and/or amyloid fibril formation. Also familial amyloidotic neuropathies, neuropathic heredofamilial amyloidosis, familial amyloid polyneuropathy.

Farnesoid X receptor (FXR) A nuclear receptor that is encoded by the *NR1H4* gene in humans and highly expressed in liver and intestine. Also known as bile acid receptor (BAR) or NR1H4 (nuclear receptor subfamily 1, group H, member 4).

FDA Food and Drug Administration. US regulatory agency (http://www.fda.gov) responsible for the evaluation and approval of drugs and medical devices.

Fibrinogen A glycoprotein that in vertebrates circulates in the blood. During tissue and vascular injury it is converted enzymatically by thrombin to fibrin and subsequently to a fibrin-based blood clot.

Fibroblast A cell in connective tissue that produces collagen and other fibers.

Fibroblast growth factor receptor (FGFR) As the name implies, a receptor that binds to members of the fibroblast growth factor family of proteins. Some of these receptors are involved in pathological conditions. For example, a point mutation in FGFR3 can lead to achondroplasia.

Fibromyalgia Widespread muscle pain and tenderness.

Fibrosarcoma A sarcoma in which the predominant cell type is a malignant fibroblast.

First pass metabolism Also presystemic metabolism, a phenomenon of drug metabolism whereby the concentration of a drug is greatly reduced before it reaches the systemic circulation.

Fischer indole synthesis A chemical reaction that produces the aromatic heterocycle indole from a (substituted) phenylhydrazine and an aldehyde or ketone under acidic conditions.

Flavivirus A genus of viruses in the family Flaviviridae. This genus includes the West Nile virus, dengue virus, tick-borne encephalitis virus, yellow fever virus, Zika virus, and several other viruses.

Flavoenzyme (Organic chemistry) Any oxidoreductase that requires flavin adenine dinucleotide (FAD) as a prosthetic group that functions in electron transfers.

Fluorescence The emission of light by a substance that has absorbed light or other electromagnetic radiation. It is a form of luminescence.

Fluorescence imaging (confocal microscopy) The visualization of fluorescent dyes or proteins as labels for molecular processes or structures. It enables a wide range of experimental observations including the location and dynamics of gene expression, protein expression, and molecular interactions in cells and tissues.

Fluorescence quenching Any process that decreases the fluorescence intensity of a sample. A variety of molecular interactions can result in quenching. These include excited-state reactions, molecular rearrangements, energy transfer, ground-state complex formation, and collisional quenching.

Fluorescent probe A molecule that absorbs light of a specific wavelength and emits light of a different, typically longer, wavelength (a process known as fluorescence), used to study biological samples.

Focal adhesion kinase (FAK) A protein that, in humans, is encoded by the PTK2 gene. PTK2 is a focal adhesion-associated protein kinase involved in cellular adhesion (how cells stick to each other and their surroundings) and spreading processes (how cells move around). It has been shown that when FAK was blocked, breast cancer cells became less metastatic due to decreased mobility. Also known as PTK2 protein tyrosine kinase 2 (PTK2).

Follicular lymphoma A type of non-Hodgkin lymphoma. It develops when the body makes abnormal B-lymphocytes.

Forced swimming test A rodent behavioral test used for evaluation of antidepressant drugs, antidepressant efficacy of new compounds, and experimental manipulations that are aimed at rendering or preventing depressivelike states.

Formalin paw assay A nociception assay to evaluate the ability of an animal, usually a rodent, to detect a noxious stimulus such as the feeling of pain, caused by stimulation of nociceptors. These assays measure the existence of pain through behaviors such as withdrawal, licking, immobility, and vocalization.

Free-radical An uncharged molecule (typically highly reactive and short-lived) having an unpaired valency electron.

Free radical scavenger Compound that neutralizes free radicals to help prevent damage to the body's cells. Free radicals have been linked to aging, weakened immune systems, and most chronic disease processes, including heart disease, diabetes, stroke, Alzheimer's disease, arthritis, and DNA damage leading to mutations and some types of cancer.

FRET (fluorescence resonance energy transfer) A distance-dependant radiationless transfer of energy from an excited donor fluorophore to a suitable acceptor fluorophore. One of few tools available for measuring nanometer scale distances and the changes in distances, both in vitro and in vivo.

Friedel–Crafts reaction A set of reactions developed by Charles Friedel and James Crafts in 1877 to attach substituents to an aromatic ring. There are two main types of Friedel–Crafts reactions: alkylation reactions (alkylation of an aromatic ring with an alkyl halide using a strong Lewis acid catalyst) and acylation reactions (acylation of aromatic rings with an acyl chloride using a strong Lewis acid catalyst). Both proceed by electrophilic aromatic substitution.

Fries rearrangement Named for the German chemist Karl Theophil Fries, a rearrangement reaction of a phenolic ester to a hydroxy aryl ketone by catalysis of Lewis acids. It involves migration of an acyl group of phenol ester to the aryl ring. The reaction is ortho- and para-selective.

Fukuyama indole synthesis A versatile tin-mediated chemical reaction that results in the formation of 2,3-disubstituted indoles. A practical one-pot reaction that can be useful for the creation of disubstituted indoles. Most commonly tributyltin hydride is utilized as the reducing agent.

Fungus (fungi, pl.) Any member of the group of eukaryotic organisms that includes microorganisms such as yeasts and molds, as well as the more familiar mushrooms.

Fused pharmacophore An approach that involves overlapping of two selective pharmacophores, to develop a single molecule effective against both targets.

GABA receptor A class of receptors that respond to the neurotransmitter gamma-aminobutyric acid (GABA), the chief inhibitory compound in the mature vertebrate central nervous system. There are two classes of GABA receptors GABAA and GABAB.

Gassman indole synthesis A series of chemical reactions used to synthesize substituted indoles by addition of aniline and a ketone bearing a thioether substituent.

Gastroenteritis (stomach flu) An intestinal infection marked by diarrhea, cramps, nausea, vomiting, and fever.

Gastroprotective Drugs that counteract gastric mucosal damage through mechanisms unrelated to inhibition of acid secretion.

Gate keeper As protein, monitors transfer of a protein from the endoplasmic reticulum to the Golgi apparatus and prevents transfer of newly synthesized proteins with inappropriate conformations or with unpaired thiol groups. As residue, partially or fully blocks a hydrophobic region deep in the ATP binding pocket. The gatekeeper residue contributes to the selectivity of kinases for small molecule inhibitors.

Gene transcription factor In molecular biology, a protein that controls the rate of transcription of genetic information from DNA to messenger RNA by binding to a specific DNA sequence. Also sequence-specific DNA-binding factor.

Genome The haploid set of chromosomes in a gamete or microorganism, or in each cell of a multicellular organism.

Genotype The genetic constitution of an individual organism.

Glaucoma A group of eye conditions that can cause blindness.

Glucagon receptor (GCGR) A 62 kDa protein that is activated by glucagon and is a member of the class BG-protein coupled family of receptors, coupled to G alpha i, G_s, and to a lesser extent, G alpha q.

Glutathione reductase (GR) An enzyme that in humans is encoded by the GSR gene. Glutathione reductase (EC 1.8.1.7) catalyzes the reduction of glutathione disulfide (GSSG) to the sulfhydryl form glutathione (GSH), which is a critical molecule in resisting oxidative stress and maintaining the reducing environment of the cell. Also known as glutathione-disulfide reductase (GSR).

Glycoprotein Any of a class of proteins with carbohydrate groups attached to the polypeptide chain.

Glycoside In chemistry, a molecule in which a sugar is bound to another functional group via a glycosidic bond. Glycosides play numerous important roles in living organisms. Many plants store chemicals in the form of inactive glycosides.

Gonadotropin Glycoprotein polypeptide hormone secreted by gonadotrope cells of the anterior pituitary of vertebrates.

Gout A form of arthritis characterized by severe pain, redness, and tenderness in joints.

G-quadruplex In molecular biology, structures formed in nucleic acids by sequences that are rich in guanine.

Gram-negative bacteria A group of bacteria that do not retain the crystal violet stain used in the gram-staining method of bacterial differentiation.

Gram-positive bacteria Bacteria that give a positive result in the gram stain test, which is traditionally used to quickly classify bacteria into two broad categories according to their cell wall.

Hemorrhagic fever virus (HFV) A diverse group of animal and human illnesses in which fever and hemorrhage are caused by a viral infection.

Half-life (1) For a chemical reaction, the time at which half of the substance has been consumed and turned into product. (2) In biochemistry, the time required for the disappearance or decay of one-half of a given component in a system. Half-lives vary from isotope to isotope, some being less than a millionth of a second and some more than a million years; symbol, *T1/2*; also called *half-time*.

Helicase An enzyme that breaks hydrogen bonds between complementary base pairs of DNA, thereby causing separation of two strands in a DNA molecule before replication.

Helminth Large multicellular organism that can generally be seen with the naked eye when mature. Also commonly known as a parasitic worm.

Hemagglutinin A glycoprotein found on the surface of influenza viruses. It is responsible for binding the virus to cells with sialic acid on the membranes, such as cells in the upper respiratory tract or erythrocytes.

Hematin polymerization A parasite-specific process that enables the detoxification of heme following its release in the lysosomal digestive vacuole during hemoglobin degradation, and represents both an essential and a unique pharmacological drug target.

Hematologic malignancy A form of cancer that begins in the cells of blood-forming tissue, such as the bone marrow, or in the cells of the immune system. Examples of hematologic cancer are acute and chronic leukemias, lymphomas, multiple myeloma, and myelodysplastic syndromes.

Hemetsberger indole synthesis A chemical reaction that thermally decomposes a 3-aryl-2-azido-propenoic ester into an indole-2-carboxylic ester. Yields are typically above 70%. Also called the Hemetsberger–Knittel synthesis.

Hemolytic index A tool to measure hemolysis in vitro.

Hemorrhagic cystitis The sudden onset of hematuria combined with bladder pain and irritative bladder symptoms. Hematuria is blood in the urine.

Hepatitis C virus (HCV) A small (55–65 nm in size), enveloped, positive-sense single-stranded RNA virus of the family Flaviviridae. Hepatitis C virus is the cause of hepatitis C and some cancers such as liver cancer (hepatocellular carcinoma, or HCC) and lymphomas in humans.

Hepatocarcinogenesis The production of cancer of the liver.

Hepatocellular carcinoma The most common form of liver cancer.

Herpes simplex virus (HSV) A virus causing contagious sores, most often around the mouth or on the genitals.

Heterocyclic compound Cyclic organic compounds in which at least one carbon atom is substituted by another element, the heteroatom (frequently N, O, and S and less frequently B, Sn, As, and Se).

High density lipoprotein (HDL) One of the five major groups of lipoproteins, which are complex particles composed of multiple proteins that transport all fat molecules (lipids) around the body within the water outside cells.

Highly active antiretroviral therapy (HAART) The standard treatment consists of a combination of at least three drugs that suppress HIV replication.

High-throughput screening (HTS) Method for the rapid assessment of the activity of samples from large compound collections.

Hinsberg reaction Test for the detection of primary, secondary, and tertiary amines. In this test, the amine is shaken well with the Hinsberg reagent in the presence of aqueous alkali (either KOH or NaOH). A reagent containing an aqueous sodium hydroxide solution and benzenesulfonyl chloride is added to a substrate.

Histamine A small molecule derived from the amino acid histidine, released from mast cells and basophils in allergic reactions; causes irritation, dilation of blood vessels, and contraction of smooth muscle.

Histamine receptor 1 (H_1) Belongs to the family of rhodopsin-like G-protein-coupled receptors. This receptor is activated by the biogenic amine histamine. It is expressed in smooth muscles, on vascular endothelial cells, in the heart, and in the central nervous system.

Histone deacetylases (HDAC) A class of enzymes that remove acetyl groups ($O{=}C{-}CH_3$) from an ε-N-acetyl lysine amino acid on a histone, allowing the histones to wrap the DNA more tightly.

Hodgkin's lymphoma Cancer of part of the immune system called the lymphatic system.

Homogeneous time resolved fluorescence assay (HTRF) Most frequently used generic assay technology to measure analytes in a homogenous format, which is the ideal platform used for drug target studies in high-throughput screening (HTS). This technology combines fluorescence resonance energy transfer technology (FRET) with time-resolved measurement (TR).

HMG-CoA reductase The rate-controlling enzyme of the mevalonate pathway, the metabolic pathway that produces cholesterol and other isoprenoids.

5-HT 5-hydroxytryptamine receptors or serotonin receptors, are a group of G protein-coupled receptor and ligand-gated ion channels found in the central and peripheral nervous systems.

Huckel's rule Rule of aromaticity that states that if a cyclic, planar molecule has $4n+2$ π electrons, it is considered aromatic. Named for German chemist and physicist Erich Hückel.

Human cytomegalovirus (HCMV) Human cytomegalovirus is a species of the *Cytomegalovirus* genus of viruses, which in turn is a member of the viral family known as Herpesviridae or herpesviruses. Typically abbreviated as HCMV or, commonly but more ambiguously, CMV. Also known as human herpesvirus-5 (HHV-5).

Human immunodeficiency virus (HIV) A lentivirus that causes HIV infection and over time, acquired immunodeficiency syndrome (AIDS).

Human leukotriene A4 hydrolase (LTA4H-h) A bifunctional enzyme that converts leukotriene A4 to leukotriene B4 and acts as an aminopeptidase.

Human serum albumin A protein found in human blood. It is the most abundant protein in human blood plasma; it constitutes about half of serum protein. Produced in the liver, it is soluble and monomeric.

Human rhinovirus The most common viral infectious agent in humans and the predominant cause of the common cold. Rhinovirus infection proliferates in temperatures between 33 and 35°C, the temperatures found in the nose.

Hybridization In chemistry, the concept of mixing atomic orbitals into new hybrid orbitals (with different energies, shapes, etc., than the component atomic orbitals) suitable for the pairing of electrons to form chemical bonds in valence bond theory.

Hybrid molecule Defined as chemical entities with two or more structural domains having different biological functions and dual activity, indicating that a hybrid molecule acts as two distinct pharmacophores.

Hydrogen bond An electrostatic attraction between two polar groups that occurs when hydrogen (H) atom, covalently bound to a highly electronegative atom such as nitrogen (N), oxygen (O), or fluorine (F), experiences the electrostatic field of another highly electronegative atom nearby.

Hydrogen bond acceptor The atom to which the hydrogen atom participating in the hydrogen bond is covalently bonded, and is usually a strongly electronegative atom such as N, O, or F.

Hydrogen bond donor The neighboring electronegative ion or molecule; must possess a lone electron pair in order to form a hydrogen bond.

Hydrogen-bond potential Often implicitly parameterized as a combination of Lennard-Jones (L-J) and electrostatic terms. In force fields that use an explicit hydrogen bonding term, the hydrogen bond potential is typically a distance-dependent function without any directional component.

Hydrolysis The chemical breakdown of a compound due to reaction with water.

Hydrophilicity The tendency of a molecule to be solvated by water.

Hydrophobicity The association of nonpolar groups or molecules in an aqueous environment that arises from the tendency of water to exclude nonpolar molecules.

Hydrophobic (or hydrophobicity) pocket Simply regions of proteins where clusters of amino acids with hydrophobic side-chains coexist.

Hyperemesis gravidarus A severe type of nausea and vomiting during pregnancy.

Hyperglycemia Abnormally high blood glucose (blood sugar) level; the hallmark sign of both type 1 and type 2 diabetes. The main symptoms of hyperglycemia are increased thirst and a frequent need to urinate.

Hyperproliferation In biology, an abnormally high rate of proliferation of cells by rapid division; substantial overproliferation.

Hypogonadism A failure of the gonads, testes in men and ovaries in women, to function properly.

Hypothetical binding domain In biochemistry, a hypothetical protein whose existence has been predicted, but for which there is a lack of experimental evidence that it is expressed in vivo.

Hypoglycemic agent Drug used to treat diabetes mellitus by lowering glucose levels in the blood. With the exceptions of insulin, exenatide, liraglutide, and pramlintide, all are administered orally and are thus also called oral hypoglycemic agents.

IC50 (inhibitory concentration 50) The concentration of an enzyme inhibitor or receptor antagonist that reduces the enzyme activity or agonist response by 50%.

Immunomodulator A chemical agent (as methotrexate or azathioprine) that modifies the immune response or the functioning of the immune system (as by the stimulation of antibody formation or the inhibition of white blood cell activity).

Impaired cognitive functions (ICF) Slight but noticeable and measurable decline in cognitive abilities, including memory and thinking skills. A person with ICF is at an increased risk of developing Alzheimer's or another dementia.

Inflammation A localized physical condition in which part of the body becomes reddened, swollen, hot, and often painful, especially as a reaction to injury or infection.

Inflammatory bowel disease (IBD) Ongoing inflammation of all or part of the digestive tract.

Influenza A common viral infection that can be deadly, especially in high-risk groups.

Inhibitor A substance that binds to an enzyme and decreases the enzyme's activity.

Inodilators Agents with inotropic effects that also cause vasodilation leading to decreased systemic and/or pulmonary vascular resistance (SVR, PVR); for example, milrinone, levosimendan.

Insomnia Persistent problems falling and staying asleep.

Insulin sensitizers Agents increasing muscle, fat, and liver sensitivity to insulin. TZDs are referred to as insulin sensitizers and also are blood sugar normalizing or euglycemics (drugs that help return the blood sugar to normal range without the risk of low blood sugars).

Integrase (IN) Also called retroviral integrase, an enzyme produced by a retrovirus that enables its genetic material to be integrated into the DNA of the infected cell.

Intercalation In biochemistry, the insertion of molecules between the planar bases of DNA. This process is used as a method for analyzing DNA and it is also the basis of certain kinds of poisoning.

Interferon A group of signaling proteins made and released by host cells in response to the presence of several pathogens such as viruses, bacteria, parasites, and tumor cells.

Interstitial nephritis; tubulo-interstitial nephritis A form of nephritis affecting the interstitium of the kidneys surrounding the tubules; that is, is inflammation of the spaces between renal tubules.

In silico Computationally as opposed to in vitro or in vivo.

In silico screening Evaluation of compounds using computational methods.

Insulin A peptide hormone produced by beta cells of the pancreatic islets, considered to be the main anabolic hormone of the body.

Ionic interaction Electrostatic attraction between two groups of opposite charge.

Isomer Each of two or more compounds with the same formula but a different arrangement of atoms in the molecule and different properties.

Isostere Atoms, molecules, or ions of similar size containing the same number of atoms and valence electrons.

Isozyme An enzyme that differs in amino acid sequence but catalyzes the same chemical reaction. These enzymes usually display different kinetic parameters, or different regulatory properties.

JAK3 Tyrosine-protein kinase enzyme that in humans is encoded by the *JAK3* gene.

Julia indole synthesis Involves a [3,3]-sigmatropic rearrangement. The [3,3]-sigmatropic rearrangement of sulfinamide 2 to indole 5 also gives sulfenic acid, which decomposes presumably to sulfur, hydrogen sulfide, sulfur dioxide, and sulfuric acid.

Juvenile arthritis Arthritis or inflammation of the joints, in children. The most common symptoms of juvenile arthritis are joint swelling, pain, and stiffness that don't go away. Juvenile arthritis is usually an autoimmune disorder.

Kala-azar A disease caused by infection with leishmania parasites.

Keratinocytes An epidermal cell that produces keratin.

Keratoconjuctivitis Inflammation ("-itis") of the cornea and conjunctiva.

Kinase phosphotransferase Enzyme that transfers a phosphate group from high-energy donor molecules, such as ATP, to specific target molecules.

Kinetoplast A mass of mitochondrial DNA lying close to the nucleus in some flagellate protozoa.

Knoevenagel condensation (Modification of aldol condensation.) A nucleophilic addition of an active hydrogen compound to a carbonyl group followed by a dehydration reaction in which a molecule of water is eliminated (hence *condensation*). The product is often an αβ-unsaturated ketone (a conjugatedenone).

Larock indole synthesis Heteroannulation reaction that uses palladium as a catalyst to synthesize indoles from an ortho-iodoaniline and a disubstituted alkyne.

Larvicide (larvacide) An insecticide that is specifically targeted against the larval life stage of an insect.

Latency The state of existing but not yet being developed.

Lead compound In drug discovery, a chemical compound that has pharmacological or biological activity likely to be therapeutically useful, but may still have suboptimal structure that requires modification to fit better to the target.

Lead discovery The process of identifying active new chemical entities, which by subsequent modification may be transformed into a clinically useful drug.

Lead optimization The synthetic modification of a biologically active compound, to fulfill all stereoelectronic, physicochemical, pharmacokinetic, and toxicologic requirements for clinical usefulness.

Lehmstedt–Tanasescu reaction A method in organic chemistry for the organic synthesis of acridone derivatives from a 2-nitrobenzaldehyde and an arene compound.

Leimgruber–Batcho indole synthesis A series of organic reactions that produce indoles from o-nitrotoluenes. The first step is the formation of an enamine using N,N-dimethylformamide dimethyl acetal and pyrrolidine. The desired indole is then formed in a second step by reductive cyclization.

Leishmania A genus of trypanosomes that are responsible for the disease leishmaniasis. They are spread by sandflies.

Leukemia A cancer of blood-forming tissues, hindering the body's ability to fight infection.

Leukotriene Any of a group of biologically active compounds, originally isolated from leucocytes. They are metabolites of arachidonic acid, containing three conjugated double bonds.

Leukotriene A4 hydrolase (LTA4H) A human gene; the protein encoded by this gene is a bifunctional enzyme that converts leukotriene A4 to leukotriene B4 and acts as an aminopeptidase.

Ligand Ion or molecule that binds to a molecular target to elicit, block, or attenuate a biological response.

Ligand-based drug design An approach used in the absence of the receptor 3D information; it relies on knowledge of molecules that bind to the biological target of interest.

Ligand efficiency (LE) Measure of the free energy of binding per heavy atom count (i.e., nonhydrogen) of a molecule.

Linker; spacer Used to join two distinct moieties in prodrugs or hybrid designing.

Lipids A loosely defined group of small biomolecules that are insoluble in water but dissolve readily in nonpolar organic solvents, and contain fatty acids, sterols, or isoprenoid compounds. Oils, such as olive and coconut, as well as waxes, such as beeswax and earwax, are all lipids. One class, the phospholipids, forms the structural basis of biological membranes.

Lipid kinase Phosphorylates lipids in the cell, both on the plasma membrane as well as on the membranes of the organelles (e.g., phosphatidylinositol-3-OH kinase (PI(3)K) and sphingosine kinase). The addition of phosphate groups can change the reactivity and localization of the lipid and can be used in signal transmission.

Lipid peroxidation The oxidative degradation of lipids. It is the process in which free radicals "steal" electrons from the lipids in cell membranes, resulting in cell damage. This process proceeds by a free radical chain reaction mechanism.

Lipophilicity Represents the affinity of a molecule or a moiety for a lipophilic environment. It is commonly measured by its distribution behavior in a biphasic system, either liquid–liquid (e.g., partition coefficient in octan- 1 -ol/water) or solid–liquid (retention on reversed-phase high performance liquid chromatography (RP-HPLC) or thin-layer chromatography (TLC) system).

Liver X receptor (LXR) A member of the nuclear receptor family of transcription factors, closely related to nuclear receptors such as the PPARs, FXR, and RXR. LXRs are important regulators of cholesterol, fatty acid, and glucose homeostasis.

logP Measure of the lipophilicity of a compound by its partition coefficient between an a polar solvent (e.g., 1-octanol) and an aqueous buffer. Thus, P is the quotient of the concentration of nonionized drug in the solvent divided by the respective concentration in buffer.

Low density lipoproteins (LDL) One of the five major groups of lipoprotein that transport all fat molecules around the body in the extracellular water.

Luteinizing hormone A hormone produced by gonadotropic cells in the anterior pituitary gland. In females, an acute rise of LH triggers ovulation and development of the corpus luteum.

Lymphatic flariasis A tropical, parasitic disease that affects the lymph nodes and lymph vessels.

Lymphocyte A form of small leucocyte (white blood cell) with a single round nucleus, occurring especially in the lymphatic system.

Lypoxygenase (LOX) A family of (nonheme), iron-containing enzymes, most of which catalyze the dioxygenation of polyunsaturated fatty acids in lipids containing a cis,cis-1,4-pentadiene into cell signaling agents that serve diverse roles as autocrine signals that regulate the function of their parent cells, paracrine signals that regulate the function of nearby cells, and endocrine signals that regulate the function of distant cells.

Macrophage A large phagocytic cell found in stationary form in the tissues or as a mobile white blood cell, especially at sites of infection.

Macro-vascular pathology Disease of any large (macro) blood vessels in the body, including the coronary arteries, the aorta, and the sizable arteries in the brain and limbs. This sometimes occurs when a person has had diabetes for an extended period of time.

Madelung indole synthesis Chemical reaction that produces (substituted or unsubstituted) indoles by the intramolecular cyclization of N-phenylamides using strong base at high temperature. The Madelung synthesis was reported in 1912 by Walter Madelung.

Malaria A disease caused by a plasmodium parasite, transmitted by the bite of infected mosquitoes.

Mammal Vertebrate within the class Mammalia, a clade of endothermic amniotes distinguished from reptiles by the possession of a neocortex, hair, three middle ear bones, and mammary glands.

Mannich reaction An organic reaction that consists of an amino alkylation of an acidic proton placed next to a carbonyl functional group by formaldehyde and a primary or secondary amine or ammonia. The final product is a β-amino-carbonyl compound, also known as a Mannich base.

MAP kinase (mitogen-activated protein kinase) A type of protein kinase that is specific to the amino acids serine and threonine.

Mast cell A cell filled with basophil granules, found in numbers in connective tissue and releasing histamine and other substances during inflammatory and allergic reactions.

Mastocytosis A condition that occurs when mast cells accumulate in skin and/or internal organs such as the liver, spleen, bone marrow, and small intestines. The signs and symptoms vary based on which part(s) of the body are affected. There are two main forms of mastocytosis: cutaneous and systemic.

Matrix metalloproteinase-9 (MMP-9) A class of enzymes that belong to the zinc-metalloproteinases family involved in the degradation of the extracellular matrix. Also known as 92 kDa type IV collagenase, 92 kDa gelatinase, or gelatinase B (GELB).

Maximal electroshock seizure (MES) Model that can effectively test the potential efficacy of a compound for inhibiting tonic seizure.

Medicinal chemistry A chemistry-based discipline, also involving aspects of biological, medical, and pharmaceutical sciences. It is concerned with the invention, discovery, design, identification, and preparation of biologically active compounds, the study of their metabolism, the interpretation of their mode of action at the molecular level, and the construction of structure-activity relationships.

Mediterranean diet A diet of a type traditional in Mediterranean countries, characterized especially by a high consumption of vegetables and olive oil and moderate consumption of protein, and thought to confer health benefits.

Meerwein–Eschenmoser–Claisen rearrangement An organic reaction where an allylic alcohol is heated with N,N-dimethylacetamide dimethyl acetal to produce a γ,δ-unsaturated amide.

Meiosis Specialized type of cell division that reduces the chromosome number by half, creating four haploid cells, each genetically distinct from the parent cell that gave rise to them.

Melanin A broad term for a group of natural pigments found in most organisms. Melanin is produced by the oxidation of the amino acid tyrosine, followed by polymerization.

Melanin-concentrating hormone (MCH) A cyclic 19-amino acid orexigenic hypothalamic peptide originally isolated from the pituitary gland of teleost fish where it controls skin pigmentation. In mammals it is involved in the regulation of feeding behavior, mood, sleep-wake cycle, and energy balance.

Melanoma The most serious type of skin cancer.

Melatonin A hormone made by the pineal gland, a small gland in the brain. Melatonin helps control your sleep and wake cycles.

Metal chelators (chelating agents) Chemical compounds whose structures permit the attachment of their two or more donor atoms (or sites) to the same metal ion simultaneously and produce one or more rings. For example, EDTA is a versatile chelating agent.

Metastasis (metastases, pl.; metastatic, adj.) Transfer of disease from one organ or part of the body to another not directly connected with it, due either to transfer of pathogenic organisms or to transfer of cells; all malignant tumors are capable of metastasizing.

Microorganism A microscopic organism, especially a bacterium, virus, or fungus.

Microvascular pathology Disease of small blood vessels such as neuropathy, nephropathy, and retinopathy.

Minimum inhibitory concentration (MIC) The lowest concentration of an antimicrobial (like an antifungal, antibiotic, or bacteriostatic) drug that will inhibit the visible growth of a microorganism after overnight incubation.

Mitochondria Organelles, or parts of a eukaryote cell. They are in the cytoplasm, not the nucleus. They make most of the cell's supply of adenosine triphosphate (ATP), a molecule that cells use as a source of energy. Their main job is to convert energy.

Miosis Excessive constriction of the pupil of the eye.

Mitosis In cell biology, part of the cell cycle when replicated chromosomes are separated into two new nuclei.

Mitotic catastrophe Mechanism of delayed mitosis-linked cell death, a sequence of events resulting from premature or inappropriate entry of cells into mitosis that can be caused by chemical or physical stresses.

Molecular ion peak In the mass spectrum, the peak corresponding to heaviest ion (the one with the greatest m/z value) is likely to be the molecular ion peak.

Molecular probe A molecular probe is a group of atoms or molecules used in molecular biology or chemistry to study the properties of other molecules or structures.

Monoamine oxidase An enzyme (present in most tissues) that catalyzes the oxidation and inactivation of monoamine neurotransmitters.

Monocyte A type of leukocyte, or white blood cell.

Monotherapy The treatment of a disease with a single drug.

Morbidity The rate of disease in a population.

Morita–Baylis–Hillman (MBH) reaction Carbon–carbon bond forming reaction between the α-position of an activated alkene and an aldehyde, or generally a carbon electrophile. Employing a nucleophilic catalyst, such as tertiary amine and phosphine, this reaction provides a densely functionalized product.

Morning sickness Nausea in pregnancy, typically occurring in the first few months. Despite its name, the nausea can affect pregnant women at any time of day.

Mortality rate The number of deaths in a given area or period, or from a particular cause.

mTOR (mechanistic target of rapamycin) A kinase that in humans is encoded by the *MTOR* gene. Also known as the mammalian target of rapamycin and FK506-binding protein 12-rapamycin-associated protein 1 (FRAP1).

MTT assay A colorimetric assay for assessing cell metabolic activity. NAD (P) H-dependent cellular oxidoreductase enzymes may, under defined conditions, reflect the number of viable cells present.

Multicomponent reaction (MCR) Reaction in which three or more starting materials react to form a product, where basically all or most of the atoms contribute to the newly formed product.

Multicopy simultaneous search method (MCSS) A method that determines energetically favorable positions of different functional groups in a binding site of interest. It provides functionality maps of binding site.

Multidrug resistance (MDR) Characteristic of cells that confers resistance to the effects of several different classes of drugs. Note: There are several forms of drug resistance. Each is determined by genes that govern how cells will respond to chemical agents. One type of multidrug resistance involves the ability to eject several drugs out of cells (e.g., efflux pumps such as P-glycoprotein).

Multifactorial disease In medicine, referring to multiple factors in heredity or disease. For example, traits and conditions that are caused by more than one gene occurring together are multifactorial, and diseases that are caused by more than one factor interacting (e.g., heredity and diet in diabetes) are multifactorial.

Multiple myeloma A cancer of plasma cells.

Multiple sclerosis A disease in which the immune system eats away at the protective covering of nerves.

Multiplet An NMR signal that is split, but is too complex to interpret easily. This might arise from non–first-order splitting, or two or more overlapping signals.

Multipotent drug A drug having a strong physiological or chemical effect.

Multitarget drug Ligand acting on more than one distinct molecular target. Targets may be of the same or different mechanistic classes.

Multitarget drug discovery (MTDD) Deliberate design of compounds that act on more than one molecular target.

Multitargeting The ability of ligand to act on more than one target.

Mur ligases A set of four Mur ubiquitin ligase enzymes—MurC, MurD, MurE, MurF—that catalyze the addition of a short polypeptide to UDP-D-acetylmuramic acid in the process of bacterial cell wall buildup from peptidoglycans.

Muscarinic acetylcholine receptor (mAChR) Acetylcholine receptor that forms G protein-receptor complexes in the cell membranes of certain neurons and other cells.

Myasthenia gravis A weakness and rapid fatigue of muscles under voluntary control.

Myocardial infarction (heart attack) A blockage of blood flow to the heart muscle.

Myocardial ischemia (atherosclerotic heart disease) Damage or disease in the heart's major blood vessels.

Myocardial postischemic reperfusion Tissue damage caused when blood supply returns to tissue (reperfusion) after a period of ischemia or lack of oxygen (anoxia or hypoxia).

Mycobacterium A genus of *Actinobacteria*, given its own family, the Mycobacteriaceae. This genus includes pathogens known to cause serious diseases in mammals, including tuberculosis (*Mycobacterium tuberculosis*) and leprosy (*Mycobacterium leprae*) in humans.

Myelodysplastic syndrome A group of disorders caused when something disrupts the production of blood cells.

Myelofibrosis A serious bone marrow disorder that disrupts the body's normal production of blood cells. The result is extensive scarring in your bone marrow, leading to severe anemia, weakness, fatigue, and often an enlarged spleen.

Myelogenous leukemia A slowly progressing and uncommon type of blood-cell cancer that begins in the bone marrow.

Myeloma (Kahler's disease) A cancer of plasma cells.

N-acetylation Refers to the process of introducing an acetyl group (resulting in an acetoxy group) at nitrogen, namely the substitution of an acetyl group for an active hydrogen atom.

N-alkylation A type of organic reaction between an alkyl halide and ammonia or an amine. The reaction is called nucleophilic aliphatic substitution (of the halide), and the reaction product is a higher substituted amine.

Nagana cattle disease A form of the disease trypanosomiasis, occurring chiefly in cattle and horses and caused by several species of the protozoan *Trypanosoma*.

Natural killer cell A type of lymphocyte (a white blood cell) and a component of innate immune system. NK cells play a major role in the host-rejection of both tumors and virally infected cells. Also known as NK cells, K cells, and killer cells.

Negative enthalpies of formation Indicates the reactants have greater enthalpy or that it is an exothermic reaction (heat is produced).

Nenitzescu indole synthesis A chemical reaction that forms 5-hydroxyindole derivatives from benzoquinone and β-aminocrotonic esters.

Neoplasm A new and abnormal growth of tissue in a part of the body, especially as a characteristic of cancer.

Nephroblastoma A type of cancer that starts in the kidneys. It is the most common type of kidney cancer in children.

Nephropathy Nephropathy means kidney disease or damage. Diabetic nephropathy is damage to your kidneys caused by diabetes.

Network analysis It is the mathematical description of physical contacts between proteins in the cell.

Neuroblastoma A cancer that is commonly found in the adrenal glands.

Neurodegenerative disorders Progressive loss of structure or function of neurons, including death of neurons. Many neurodegenerative diseases—including amyotrophic lateral sclerosis, Parkinson's, Alzheimer's, and Huntington's—occur as a result of neurodegenerative processes.

Neurokinin-1 The tachykinin receptor 1 (TACR1) also known as neurokinin 1 receptor (NK1R) or substance P receptor (SPR) is a G protein coupled receptor found in the central nervous system and peripheral nervous system. The endogenous ligand for this receptor is Substance P, although it has some affinity for other tachykinins. The protein is the product of the *TACR1* gene.

Neuropathy Damage to or disease affecting nerves.

Neutrophils The most abundant type of granulocytes and the most abundant (40%–70%) type of white blood cells in most mammals. They form an essential part of the innate immune system. Also known as neutrocytes.

Neurotransmitters Endogenous chemicals that enable neurotransmission. Also known as chemical messengers.

Nicotinic receptor; nicotinic acetylcholine receptor; nAChR A receptor protein that responds to the neurotransmitter acetylcholine. Nicotinic receptors also respond to drugs, including the nicotinic receptor agonist nicotine.

Nitric oxide synthase A family of enzymes catalyzing the production of nitric oxide from L-arginine. NO is an important cellular signaling molecule.

NMDA (*N*-methyl-D-aspartate) receptor (NMDAR) A glutamate receptor and ion channel protein found in nerve cells.

Nomenclature A set of rules to generate systematic names for chemical compounds. The nomenclature used most frequently worldwide is the one created and developed by the International Union of Pure and Applied Chemistry (IUPAC).

Non-Hodgkin's lymphoma Cancer that starts in the lymphatic system.

Non–insulin-dependent diabetes mellitus (NIDDM) Type 2 diabetes; a chronic condition that affects the way the body processes blood sugar (glucose).

Nonnucleoside reverse transcriptase inhibitor (NNRTI) A class of antiretroviral drugs used to treat HIV infection or AIDS and in some cases hepatitis B.

Nonradioactive ELISA kit Used for the detection of retroviral reverse transcriptase (RT) activity associated with retroviral SIV and HIV.

Non–small cell lung cancer (NSCLC) One of two major types of lung cancer that can affect smokers and nonsmokers.

Nonsteroidal antiinflammatory drug (NSAID) Drug class that reduces pain, decreases fever, prevents blood clots and, in higher doses, decreases inflammation.

Nuclear hormone receptor Proteins form a class of ligand activated proteins that, when bound to specific sequences of DNA, serve as on-off switches for transcription within the cell nucleus. These switches control the development and differentiation of skin, bone, and behavioral centers in the brain.

Nucleophile A chemical species that donates an electron pair to an electrophile to form a chemical bond in relation to a reaction.

Nucleophilic substitution Nucleophilic substitution is a fundamental class of reactions in which an electron-rich nucleophile selectively bonds with or attacks the positive or partially positive charge of an atom or a group of atoms to replace a leaving group.

Oculocutaneous albinism (OCA) A group of rare inherited disorders characterized by a reduction or complete lack of melanin pigment in the skin, hair, and eyes. These conditions are caused by mutations in specific genes that are necessary for the production of melanin pigment in specialized cells called melanocytes.

Oncology A branch of medicine that deals with the prevention, diagnosis, and treatment of cancer.

Orexigenic A drug, hormone, or compound that increases appetite; appetite stimulant.

Organometallic reagent Compounds that contain a metal-carbon bond, R-M, are known as organometallic reagents. Organometallic compounds of Li, Mg (Grignard reagents) are among some of the most important organic reagents.

Oriental sore; tropical sore The most common form of leishmaniasis affecting humans. It is a skin infection caused by a single-celled parasite that is transmitted by the bite of a phlebotomine sandfly.

Orthostatic hypotension Decrease in systolic blood pressure of 20 mm Hg or a decrease in diastolic blood pressure of 10 mm Hg within 3 min of standing when compared with blood pressure from the sitting or supine position.

Osteoarthritis A type of arthritis that occurs when flexible tissue at the ends of bones wears down.

Osteoblast A cell that secretes the substance of bone.

Osteoporosis A condition in which bones become weak and brittle.

Ovarian cancer A cancer that begins in the female organs that produce eggs (ovaries).

Overexpression In biology, to make too many copies of a protein or other substance. Overexpression of certain proteins or other substances may play a role in cancer development.

Oxidation The process or result of oxidizing or being oxidized.

Oxidative coupling A coupling reaction of two molecular entities through an oxidative process, usually catalyzed by a transition metal compound and involving dioxygen as the oxidant. A relevant aliphatic coupling reaction is the oxidative coupling of methane.

Oxidative stress Reflects an imbalance between the systemic manifestation of reactive oxygen species and a biological system's ability to readily detoxify the reactive intermediates or to repair the resulting damage.

Paget's disease A disease that disrupts the replacement of old bone tissue with new bone tissue.

Pancreatic cancer Cancer that begins in the organ lying behind the lower part of the stomach (pancreas).

Pan inhibitor A general description of a nonspecific inhibitor.

Paper disc diffusion technique Used to test the effectiveness of antibiotics on a specific microorganism. An agar plate is first spread with bacteria, and then paper disks of antibiotics are added. The bacteria is allowed to grow on the agar media, and then observed.

Parasite A symbiotic relationship in which one organism benefits and the other is harmed.

Partial atomic charge Used to quantify the degree of ionic versus covalent bonding of any compound across the periodic table.

Parkinson's disease A disorder of the central nervous system that affects movement, often including tremors.

Patent A set of exclusive rights granted by a sovereign state or intergovernmental organization to an inventor or assignee for a limited period of time in exchange for detailed public disclosure of an invention.

Pathogen An organism or other agent that causes disease, such as viruses or bacteria.

Pathogenesis The manner of development of a disease.

Pathophysiology The disordered physiological processes associated with disease or injury.

Peptide A short chain of amino acid monomers linked by peptide bonds.

Peripheral anionic site (PAS) This lies at the entrance of the binding pocket of AChE and has a putative regulatory function.

Peripheral blood mononuclear cell (PBMC) A peripheral blood cell having a round nucleus. These cells consist of lymphocytes (T cells, B cells, NK cells) and monocytes, whereas erythrocytes and platelets have no nuclei, and granulocytes (neutrophils, basophils, and eosinophils) have multilobed nuclei.

Peroxisome proliferator activated receptor (PPARγ) In the field of molecular biology, the peroxisome proliferator-activated receptors are a group of nuclear receptor proteins that function as transcription factors regulating the expression of genes.

Pesticide A substance meant to control pests. The term pesticide includes all of the following: herbicide, insecticides nematicide, molluscicide, piscicide, avicide, rodenticide, bactericide.

P-glycoprotein An important protein of the cell membrane that pumps many foreign substances out of cells. Also known as multidrug resistance protein 1 or ATP-binding cassette subfamily B member 1 or cluster of differentiation 243.

pH Negative of the base 10 logarithm of the activity of the hydrogen ion.

Plaque A sticky deposit on teeth in which bacteria proliferate.

Plaque forming unit (PFU) In virology, a measure of the number of particles capable of forming plaques per unit volume, such as virus particles.

Platelet Small enucleated cells that arise from cells called megakaryocytes in the bone marrow and are found in large numbers in the bloodstream; they help initiate blood clotting when blood vessels are injured; also known as thrombocytes.

Platelet-derived growth factor receptor (PDGFR) These are cell surface tyrosine kinase receptors for members of the platelet-derived growth factor (PDGF) family. There are two forms of the PDGF-R, alpha and beta each encoded by a different gene. Depending on which growth factor is bound, PDGF-R homo- or heterodimerizes.

Pocketomics The discipline concerned with the shape, size, and composition of binding sites. Some authors have turned to pocketomics to identify meaningful target combination.

Polarizability Relative tendency of a charge distribution.

Poly (ADP-ribose) polymerase (PARP) A family of proteins involved in a number of cellular processes such as DNA repair, genomic stability, and programmed cell death.

Polymerase chain reaction A technique used in molecular biology to amplify a single copy or a few copies of a segment of DNA across several orders of magnitude, generating thousands to millions of copies of a particular DNA sequence.

Polymorphism The ability of a solid material to exist in more than one form or crystal structure.

Polyol pathway A two-step process that converts glucose to fructose and is involved in pathology of diabetic complications. Also called the sorbitol-aldose reductase pathway.

Polypharmacology The design or use of pharmaceutical agents that act on multiple targets or disease pathways. In recent years even with remarkable scientific advancements and significant increase of global R&D spending, drugs are frequently withdrawn from markets.

Polypharmacy (combination therapy) The concurrent use of multiple medications by a patient.

Positive ionotropic effect Increase in the strength of muscular contraction.

Phagocytosis The ingestion of bacteria or other material by phagocytes and amoeboid protozoans.

Pharmacokinetic The study of absorption, distribution, metabolism, and excretion (ADME) of bioactive compounds in a higher organism.

Pharmacodynamic The study of how a drug affects an organism.

Pharmacophore The ensemble of steric and electronic features that is necessary to ensure the optimal supramolecular interactions with a specific biological target structure and to trigger (or to block) its biological response.

Pharmacophore mapping Very important part of drug design for prediction of small molecule binding to a target macromolecule.

Pharmacotherapy Medical treatment by means of drugs.

Phenotype The set of observable characteristics of an individual resulting from the interaction of its genotype with the environment.

Phosphoinositide 3-kinase (PI3Kγ) A family of enzymes involved in cellular functions such as cell growth, proliferation, differentiation, motility, survival, and intracellular trafficking, which in turn are involved in cancer.

Phospholipase A2 (PLA2) An enzyme that releases fatty acids from the second carbon group of glycerol. This particular phospholipase specifically recognizes the sn-2 acyl bond of phospholipids and catalytically hydrolyzes the bond releasing arachidonic acid and lysophosphatidic acid.

Phosphatase An enzyme that uses water to cleave a phosphoric acid monoester into a phosphate ion and an alcohol. Because a phosphatase enzyme catalyzes the hydrolysis of its substrate, it is a subcategory of hydrolases.

Phosphorylation A biochemical process that involves the addition of phosphate to an organic compound. Examples include the addition of phosphate to glucose to produce glucose monophosphate and the addition of phosphate to adenosine diphosphate (ADP) to form adenosine triphosphate (ATP).

Physicochemical Relating to physics and chemistry or to physical chemistry.

Phytoconstituents Chemical compounds that occur naturally in plants (*phyto* means "plant" in Greek).

Photodegradation Degradation of a photodegradable molecule caused by the absorption of photons, particularly those wavelengths found in sunlight, such as infrared radiation, visible light, and ultraviolet light.

Pim kinase Proto-oncogene serine/threonine-protein kinase Pim-1 is an enzyme that in humans is encoded by the PIM1 gene. Pim-1 is a proto-oncogene which encodes for the serine/threonine kinase of the same name.

pKa It is a measure of acid strength. It depends on the identity and chemical properties of the acid. pH is a measure of [H+] in a solution. For acids, the smaller the pKa, the more acidic the substance is (the more easily a proton is lost, thus the lower the pH).

Platelet activating factor Platelet-activating factor, also known as PAF, PAF-acether or AGEPC (acetyl-glyceryl-ether-phosphorylcholine) is a potent phospholipid activator and mediator of many leukocyte functions, platelet aggregation and degranulation, inflammation, and anaphylaxis. It is also involved in changes to vascular permeability, the oxidative burst, chemotaxis of leukocytes, as well as augmentation of arachidonic acid metabolism in phagocytes.

Polarizability The ability to form instantaneous dipoles. It is a property of matter. Polarizabilities determine the dynamical response of a bound system to external fields, and provide insight into a molecule's internal structure.

PPAR-γ Peroxisome proliferator-activated receptor gamma (PPAR-γ or PPARG); a type II nuclear receptor that in humans is encoded by the *PPARG* gene. Also known as the glitazone receptor, or NR1C3 (nuclear receptor subfamily 1, group C, member 3).

PPAR-δ Peroxisome proliferator-activated receptor beta or delta (PPAR-β or PPAR-δ); a nuclear receptor that in humans is encoded by the *PPARD* gene. Also known as NR1C2 (nuclear receptor subfamily 1, group C, member 2).

Privileged scaffold A molecular framework or chemical moiety that is statistically recurrent among known drugs or among a specific array of biologically active compounds. These privileged elements can be used as a basis for designing new active biological compounds or compound libraries.

Privileged structure Substructural feature that confers desirable (often druglike) properties on compounds containing that feature. They often consist of a semirigid scaffold that presents multiple hydrophobic residues without undergoing hydrophobic collapse.

Proapoptotic effect Promoting or causing apoptosis.

Prodrug A medication or compound that, after administration, is metabolized into a pharmacologically active drug.

Prognosis Medical term for predicting the likelihood of a person's survival.

Proinflammatory Capable of promoting inflammation.

Proliferation Rapid reproduction of a cell, part, or organism.

Promastigote The motile, elongated, extracellular form in the life cycle of some protozoans (family Trypanosomatidae, and especially genus *Leishmania*) that is characterized by a single anterior flagellum and no undulating membrane.

Prophylactic/Prophylaxis A medicine or course of action used to prevent disease.

Prostaglandins A class of lipid-soluble, hormonelike, regulatory molecules derived from arachidonate and other polyunsaturated fatty acids by virtually all cells; stimulate contraction or expansion of smooth muscles and contraction of blood vessels, have also been used in the induction of labor and abortion.

Prostate cancer A cancer in a man's prostate, a small walnut-sized gland that produces seminal fluid.

Protease An enzyme that breaks down proteins and peptides.

Protein A large biomolecule composed of one or more chains of amino acids in a specific order. Proteins are required for the structure, function, and regulation of cells, tissues, and organs.

Protein binding pocket See **binding pocket**.

Protein kinase C (PKC) A family of protein kinase enzymes that are involved in controlling the function of other proteins through the phosphorylation of hydroxyl groups of serine and threonine amino acid residues on these proteins, or a member of this family.

Psoriasis A condition in which skin cells build up and form scales and itchy, dry patches.

PTP1B Tyrosine-protein phosphatase nonreceptor type 1; an enzyme that is the founding member of the protein tyrosine phosphatase (PTP) family. Also known as protein-tyrosine phosphatase 1B.

Pulmonary fibrosis Interstitial lung disease that can result in lung scarring. As the lung tissue becomes scarred, it interferes with a person's ability to breathe. In some cases, the cause of pulmonary fibrosis can be found. But most cases of pulmonary fibrosis have no known cause. These cases are called idiopathic pulmonary fibrosis (IPF).

Psychopathology The scientific study of mental disorders.

QSAR or 3D-QSAR CoMFA CoMFA (comparative molecular field analysis) is a 3D QSAR technique based on data from known active molecules. CoMFA can be applied, as it often is, when the 3D structure of the receptor is unknown.

Quantum yield The quantum yield (Φ) of a radiation-induced process is the number of times a specific event occurs per photon absorbed by the system. The event is typically a kind of chemical reaction.

Radiotherapy Radiation therapy using ionizing radiation, generally as part of cancer treatment to control or kill malignant cells and normally delivered by a linear accelerator. Often abbreviated RT, RTx, or XRT.

Raynaud's disease A condition in which some areas of the body feel numb and cool in certain circumstances.

Reactive oxygen species (ROS) Chemically reactive chemical species containing oxygen. Examples include peroxides, superoxide, hydroxyl radical, and singlet oxygen.

Rearranged during transfection (RET) Proto-oncogene encodes a receptor tyrosine kinase for members of the glial cell line-derived neurotrophic factor (GDNF) family of extracellular signaling molecules. *RET* loss of function mutations are associated with the development of Hirschsprung's disease, while gain of function mutations are associated with the development of various types of human cancer, including medullary thyroid carcinoma, multiple endocrine neoplasias type 2A and 2B, pheochromocytoma, and parathyroid hyperplasia.

Receptor A protein molecule that receives chemical signals from outside a cell.

Recurrent glioblastoma multiforme Repeatedly occurring malignant tumor affecting the brain or spine.

Reduction Chemical reaction that involves the gaining of electrons by one of the atoms involved in the reaction.

Regiochemical problem When several factors influence the regioselectivity of a reaction.

Regioisomer Molecules that have the same molecular formula but have different connectivities or order in which they are put together.

Reference drug A standardized substance used as a measurement base for similar substances. Where the exact active substances of a new drug are not known, a reference drug provides a calibrated level of biological effects against which new preparations of the drug can be compared.

Reflux Lab technique that involves the boiling and condensing of a solution.

Refractive index In optics, a dimensionless number that describes how light propagates through that medium.

Reissert indole synthesis A series of chemical reactions designed to synthesize indole or substituted-indoles from ortho-nitrotoluene and diethyl oxalate.

Renal cell carcinoma A disease in which malignant (cancer) cells are found in the lining of tubules (very small tubes) in the kidney.

Renin–Angiotensin system (RAS) A hormone system that regulates blood pressure and fluid balance.

Replication In molecular biology, the biological process of producing two identical replicas of DNA from one original DNA molecule. This process occurs in all living organisms and is the basis for biological inheritance.

Resistance The degree of unresponsiveness of a disease-causing microorganism or cell to drugs (for example, penicillin-resistant bacteria).

Resistance index A measure of pulsatile blood flow that reflects the resistance to blood flow caused by microvascular bed distal to the site of measurement.

Retinoblastoma An eye cancer that begins in the back of the eye (retina), most commonly in children.

Retinopathy A complication of diabetes that affects the eyes.

Retrosynthetic Technique for solving problems in the planning of organic syntheses. This is achieved by transforming a target molecule into simpler precursor structures without assumptions regarding starting materials.

Reverse transcriptase An enzyme used to generate complementary DNA from an RNA template, a process termed reverse transcription.

Rhabdomyosarcoma A type of sarcoma, cancer of soft tissue (such as muscle), connective tissue (such as tendon or cartilage), or bone.

Rheumatoid arthritis A chronic inflammatory disorder affecting many joints, including those in the hands and feet.

Rho-associated protein kinase A kinase belonging to the AGC family of serine-threonine kinases. It is involved mainly in regulating the shape and movement of cells by acting on the cytoskeleton.

RNA (ribonucleic acid) A polymeric molecule essential in various biological roles in coding, decoding, regulation, and expression of genes.

Rotatable bond Any single bond, not in a ring, bound to a nonterminal heavy atom. The number of rotatable bonds (RBN) is the number of bonds that allow free rotation around themselves.

SAR (structure–activity relationship) The relationship between the chemical or 3D structure of a molecule and its biological activity.

Sarcoma A tumor that affects connective tissue. Soft tissue sarcomas, as the name suggests, affect the soft tissues; these include fat, muscle, blood vessels, deep skin tissues, cartilage, tendons, and ligaments. Sarcomas are a relatively rare type of cancer.

Schiff's bases A compound with the general structure $R_2C=NR'$. They can be considered a subclass of imines, being either secondary ketimines or secondary aldimines depending on their structure.

Schizontocide An agent selectively destructive of the schizont of a sporozoan parasite.

Schotten–Baumann reaction The reaction between an acid chloride and an alcohol to form an ester. The reaction was first described in 1883 by German chemists Carl Schotten and Eugen Baumann.

Secondary plant metabolism A term for pathways and small molecule products of metabolism that are not absolutely required for the survival of the organism. Examples include antibiotics and pigments.

Sedative (tranquillizer) A substance that induces sedation by reducing irritability or excitement. They are central nervous depressants and interact with brain activity causing its deceleration.

Seizure A disorder in which nerve cell activity in the brain is disturbed, causing seizures.

Selective index The relative effectiveness of the investigational product in inhibiting viral replication compared to inducing cell death is defined as the therapeutic or selectivity index (i.e., CC50 value/EC50 value).

Senile dementia Dementia occurring in old age as a result of progressive brain degeneration.

Senile systemic amyloidosis (SSA) Characterized by infiltration of amyloid transthyretin fibrils in the myocardium, SSA occurs mainly (but not always) in elderly men.

Serotonin (5-hydroxytryptamine) A monoamine neurotransmitter. Biochemically derived from tryptophan, serotonin is primarily found in the gastrointestinal tract, blood platelets, and the central nervous system of animals, including humans.

Serotonin and norepinephrine reuptake inhibitors (SNRIs) A class of antidepressant drugs used in the treatment of major depressive disorder (MDD). They are sometimes also used to treat anxiety disorders, obsessive-compulsive disorder (OCD), attention-deficit hyperactivity disorder (ADHD), chronic neuropathic pain, and fibromyalgia syndrome (FMS), and for the relief of menopausal symptoms.

Serotonin transporter (SERT or 5-HTT) A protein that in humans is encoded by the SLC6A4 gene. SERT is a type of monoamine transporter protein that transports serotonin from the synaptic cleft to the presynaptic neuron. Also known as the sodium-dependent serotonin transporter and solute carrier family 6 member 4.

Sigmatropic (Claisen) rearrangement In organic chemistry, a pericyclic reaction wherein the net result is one σ-bond changed to another σ-bond in an uncatalyzed intramolecular process.

Signaling pathway A set of chemical reactions in a cell that occurs when a molecule, such as a hormone, attaches to a receptor on the cell membrane. The pathway is actually a cascade of biochemical reactions inside the cell that eventually reach the target molecule or reaction.

Simian immunodeficiency virus A retrovirus that causes persistent infections in at least 45 species of African nonhuman primates.

Similarity ensemble approach (SEA) Relates proteins based on the setwise chemical similarity among their ligand. It can be used to rapidly search large compound databases and to build cross-target similarity maps.

Single targeting Action of ligand on one molecular target.

Spectrum (spectra, pl.) The characteristic wavelength of electromagnetic radiation (or a portion thereof) that is emitted or absorbed by an object or substance, atom, or molecule.

Spectrometry An analytical technique that measures the mass-to-charge ratio of charged particles. Rutherford backscattering spectrometry is an analytical technique used to determine the structure and composition of materials by measuring the back-scattering of a beam of high energy ions impinging on a sample.

Spectroscopy The analysis of the interaction between matter and any portion of the electromagnetic spectrum. Traditionally, spectroscopy involved the visible spectrum of light, but X-ray, gamma, and UV spectroscopy also are valuable analytical techniques.

Spondylitis An inflammatory arthritis affecting the spine and large joints.

Stacking π–π interaction In chemistry, pi stacking refers to attractive, noncovalent interactions between aromatic rings, since they contain pi bonds. These interactions are important in nucleobase stacking within DNA and RNA molecules, protein folding, template-directed synthesis, materials science, and molecular recognition. Also called π–π stacking.

Stereocenter (stereogenic center) Any point in a molecule, though not necessarily an atom, bearing groups, such that an interchanging of any two groups leads to a stereoisomer. The term stereocenter was introduced in 1984 by Kurt Mislow and Jay Siegel.

Steric effect In chemistry, an influence on a reaction's course or rate determined by the fact that all atoms within a molecule occupy space, thus certain collision paths are either disfavored or favored.

STD (sexually transmitted disease) An infection transmitted through sexual contact, caused by bacteria, viruses, or parasites.

Stromal cell Connective tissue cell of any organ, for example in the uterine mucosa (endometrium), prostate, bone marrow, lymph node, and ovary. They are cells that support the function of the parenchymal cells of that organ. Fibroblasts and pericytes are among the most common types of stromal cells.

Structural interaction fingerprints analysis A novel method for analyzing 3D protein-ligand binding interactions.

Structure-based design A drug design strategy based on the 3D structure of the target obtained by X-ray or NMR.

Substituents An atom or group of atoms taking the place of another atom or group or occupying a specified position in a molecule.

Substrate (1) The specific compound acted upon by an enzyme molecule. (2) The medium on which an organism (especially a microorganism) can grow.

Sundberg indole synthesis This reaction can be traced back to the inceptive work of Smith in 1951, for the synthesis of carbazoles by the pyrolysis or photolysis of o-azido biphenyl1 and the thermolysis of o-(α-thienyl)-phenyl azide to 4-thieno[3,2-b]indole and α- and β-(o-azidophenyl)-pyridine to α- and γ-carbolines, respectively.

Superoxide dismutase Superoxide dismutase is an enzyme that alternately catalyzes the dismutation of the superoxide radical into either ordinary molecular oxygen or hydrogen peroxide.

Support vector machines (SVM) A reliable virtual screening tool for prioritizing molecules with the required biological activity and minimum toxicity.

Susceptibility The degree to which an organism is sensitive to a therapeutic or a disease.

Synergistic inhibition An effect arising between two or more agents, entities, factors, or substances that produce an effect greater than the sum of their individual effects. It is opposite of antagonism.

Tachycardia Rapid beating of the heart.

Target Protein (e.g., receptor, enzyme, or ion channel), RNA, or DNA that is implicated in a clinical disorder or the propagation of any untoward event.

Targeted drug delivery Approach to target a drug to a specific tissue or molecular target using a prodrug or antibody recognition systems.

Targeted radionuclide therapy Targeted therapy that uses peptides radiolabeled with either ^{90}Y or ^{177}Lu to deliver radiation to cancer cells that express somatostatin receptors.

Tau protein (τ protein) A protein that stabilizes microtubules. They are abundant in neurons of the central nervous system and are less common elsewhere, but are also expressed at very low levels in CNS astrocytes and oligodendrocytes.

Tautomer Structural isomer that can readily convert to another form that differs only by the attachment position of a hydrogen atom and the location of double bond(s).

Tautomerism Reversible interconversion of two different tautomers.

Template Core portion of a molecule common to all members of a chemical library or compound series.

Therapeutic index (TI) A comparison of the amount of a therapeutic agent that causes the therapeutic effect to the amount that causes toxicity. Also referred to as therapeutic ratio.

Thermal hyperalgesia Injury or inflammation releases a range of inflammatory mediators that increase the sensitivity of sensory neurons to noxious thermal or mechanical stimuli.

Thermodynamics A branch of physics concerned with heat and temperature and their relation to other forms of energy and work.

Thrombin An enzyme formed in the blood of vertebrates that acts upon fibrinogen to form fibrin; it is hence, essential to the process of blood clotting; formed from a blood protein called prothrombin.

Thrombosis The formation of a blood clot inside a blood vessel, obstructing the flow of blood through the circulatory system. When a blood vessel (a vein or an artery) is injured, the body uses platelets (thrombocytes) and fibrin to form a blood clot to prevent blood loss.

Thromboxane A hormone of the prostacyclin type released from blood platelets, which induces platelet aggregation and arterial constriction.

Thymocytes A lymphocyte within the thymus gland.

Topical Relating or applied directly to a part of the body.

Topoisomerase or telomerase An enzyme that participates in the overwinding or underwinding of DNA. The winding problem of DNA arises due to the intertwined nature of its double-helical structure.

Toxicity The potential for a molecule to produce harmful effects. Toxicity can be measured as TD50, the median toxic dose, or LD50, the median lethal dose.

Transthyretin amyloid fibrils Insoluble protein fibrils, resulting from the self-assembly of a conformational intermediate are implicated to be the causative agent in several human amyloid diseases including familial amyloid polyneuropathy (FAP) and senile systemic amyloidosis (SSA).

Transcutaneous Existing, applied, or measured across the depth of the skin.

Transition state The state corresponding to the highest energy along the reaction coordinates. It has more free energy in comparison to the substrate or product; thus, it is the least stable state. The specific form of the transition state depends on the mechanisms of the particular reaction.

Transmembrane binding pocket Membrane receptors, transmembrane receptors, or binding pockets are receptors that are embedded in the membranes of cells. They act in cell signaling by receiving (binding to) extracellular molecules. They are specialized integral membrane proteins that allow communication between the cell and the extracellular space.

Transplant rejection Occurs when transplanted tissue is rejected by the recipient's immune system, which destroys the transplanted tissue. Transplant rejection can be lessened by determining the molecular similitude between donor and recipient and by use of immunosuppressant drugs after transplant.

Trolox Trolox is a water-soluble analog of vitamin E sold by Hoffman-LaRoche. It is an antioxidant like vitamin E and it is used in biological or biochemical applications to reduce oxidative stress or damage.

Tropical Related to the region of the earth surrounding the equator.

Tropomyosin-receptor-kinase B (TrkB) A protein that in humans is encoded by the NTRK2 gene. TrkB is a receptor for brain-derived neurotrophic factor (BDNF), which are small protein growth factors that induce the survival and differentiation of distinct cell populations. Also known as tyrosine receptor kinase B, or BDNF/NT-3 growth factors receptor or neurotrophic tyrosine kinase, receptor, type 2.

Trypanothione reductase (TR) In enzymology, an enzyme that catalyzes the chemical reaction trypanothione + NADP$^+$ trypanothione disulfide + NADPH + H$^+$.

Tuberculosis A disease caused by *Mycobacterium tuberculosis*.

Tubule A minute tube, especially as an anatomical structure (i.e., kidney tubule).

Tubulin In molecular biology, can refer to either the tubulin protein superfamily of globular proteins or one of the member proteins of that superfamily.

Tumor A mass of cells, growing in an uncontrolled manner.

Tumor necrosis factor (TNF-α) A cell-signaling protein involved in systemic inflammation; one of the cytokines that make up the acute phase reaction.

Tumor regression The partial or complete disappearance of a malignant tumor in the absence of all treatment, or in the presence of therapy, that is considered inadequate to exert significant influence on neoplastic disease.

Tyrosinase An oxidase that is the rate-limiting enzyme for controlling the production of melanin.

Tyrosine kinase An enzyme that can transfer a phosphate group from ATP to a protein in a cell. It functions as an on-or-off switch in many cellular functions. Tyrosine kinases like Src, VEGFR, and Hck are a subclass of protein kinase. The phosphate group is attached to the amino acid tyrosine on the protein.

Ulcer A sore that develops on the lining of the esophagus, stomach, or small intestine.

Ullmann condensation (Ullmann ether synthesis) A variation of the Ullmann reaction, in which a phenol is coupled to an aryl halide to create a diaryl ether in the presence of a copper compound. Named after Fritz Ullmann.

Ultrastructural The architecture of cells that is visible at higher magnifications than found on a standard optical light microscope.

UDP-N-acetyl muramoylalanine D-glutamate Ligase (MurD ligase) Protein involved in cell wall formation; catalyzes the addition of glutamate to the nucleotide precursor UDP-N-acetylmuramoyl-L-alanine (UMA).

Vacuole A membrane-bound organelle present in all plant and fungal cells and some protist, animal, and bacterial cells.

Vagus nerve Historically cited as the pneumogastric nerve, it is the 10th cranial nerve or CN X, and interfaces with parasympathetic control of the heart, lungs, and digestive tract.

Vascular endothelial growth factor receptor (VEGFR) There are three main subtypes of VEGFR, numbered 1, 2, and 3. Also, they may be membrane-bound (mbVEGFR) or soluble (sVEGFR), depending on alternative splicing.

Vasoconstriction A narrowing of the blood vessels, often in response to cold, through a contraction of involuntary muscles in the walls of the vessels brought about by a stimulus from the sympathetic nervous system.

Vascular sclerosis Arteriolosclerosis and arteriosclerosis, a thickening and hardening of arterioles and arteries, respectively, of systemic organs, such as kidney, heart, brain, and eyes; account for and sustain arterial hypertension.

Vasodilation The expansion of blood vessels by relocation of muscles, mainly controlled by the sympathetic nervous system.

Vasopressin A posterior pituitary hormone that regulates the kidney's retention of water. Also referred to as antidiuretic hormone (ADH).

Vasorelaxant Any agent that reduces tension in the blood vessel walls.

Vasospasmic disease Peripheral vascular disorder caused by vasospasm, which means a reversible localized or diffuse vasoconstriction of arteries or smaller blood vessels. Vasospastic syndromes include Raynaud's disease, acrocyanosis, and livedo reticularis.

Vector In medicine, a carrier of disease or of medication. For example, in malaria a mosquito is the vector that carries and transfers the infectious agent. In molecular biology, a vector may be a virus or a plasmid that carries a piece of foreign DNA to a host cell.

Vilsmeier–Haack reaction The chemical reaction of a substituted amide with phosphorus oxychloride and an electron-rich arene to produce an aryl aldehyde or ketone.

Viral adsorption The first step in the viral life cycle followed by penetration, uncoating, synthesis (transcription if needed, and translation), and release.

Viral internalization The entering of cells by viruses following virus attachment. This is achieved by endocytosis, by direct membrane fusion of the viral membrane with the cell membrane, or by translocation of the whole virus across the cell membrane.

Virtual chemical library Collection of chemical structures constructed solely in electronic form or on paper.

Virtual screening Evaluation of compounds using computational methods.

Virus A small infectious agent that replicates only inside the living cells of other organisms.

Vitamin B (vitamin B$_{12}$) Any of a group of substances (the vitamin B complex) essential for the working of certain enzymes in the body and, although not chemically related, are generally found together in the same foods. They include thiamine (vitamin B$_1$), riboflavin (vitamin B$_2$), pyridoxine (vitamin B$_6$), and cyanocobalamin (vitamin B$_{12}$).

V$_{max}$ The maximum velocity of an enzymatic reaction when the binding site is saturated with substrate.

WHO The World Health Organization is a specialized agency of the United Nations that is concerned with international public health. It was established on April 7, 1948, headquartered in Geneva, Switzerland.

Wittig reaction (olefination) A chemical reaction of an aldehyde or ketone with a triphenyl phosphonium ylide to give an alkene and triphenylphosphine oxide.

Xanthine oxidase (XO) A form of xanthine oxidoreductase, a type of enzyme that generates reactive oxygen species. These enzymes catalyze the oxidation of hypoxanthine to xanthine and can further catalyze the oxidation of xanthine to uric acid.

X-ray crystallography A technique for determining the 3D arrangement of atoms in a molecule, based on the diffraction pattern of X-rays passing through a crystal of the molecule.

Yeast A common term for many families of unicellular fungi; includes species used for brewing beer and making bread, as well as pathogenic species; contains enzymes to convert grape sugar to wine and other products of fermentation.

Yellow fever A viral infection spread by a particular species of mosquito.

Zone of inhibition An area around the wafer where bacteria have not grown enough to be visible because an antibiotic stops the bacteria from growing or kills the bacteria.

Zwitterion (German, for "ions of both kind") A dipolar ion, with spatially separated positive and negative charges.

B

Flavone *(Continued)*
    antitumor agents, 138–141
    antiulcer agents, 146–148
    cardiovascular diseases, 153–155
    cosmetic agents, 159–160
    GABA antagonists, 162–163
    immunomodulators, 160–161
    lipid-lowering agents, 156
    neuroprotective agents, 143–145
    photoprotectants, 161–162
    spasmolytic agents, 156–157
    vasorelaxants, 157–158
    XO inhibitors, 155–156
    role as multitargeting agents in
        multifactorial diseases, 163–166
    synthetic schemes for flavone
        synthesis, 136f
Flavonoid(s), 133–135, 134f, 143
    glycosides, 134–135
Flavonolignans, 134–135
Flavonols, 134–135
Floxacrine, 101–103
Fluconazole, 327
Fluorescent energy transfer system
    (FRET system), 111
Fluorescent probes of acridones, 111–112
7-Fluoro-3,4-dihydro-2H-1,4-benzothiazine
    derivatives, 262
3(4-Fluorophenylthio) propyl group, 38
Fluvastatin, 285–287, 304–305
FMS-related tyrosine kinase-3 (FLT3), 309
Focal adhesion kinase (FAK), 290–292
Fosdevirine. *See* GSK2248761 inhibitor
Fragment-based approach. *See* Pharmacophore
    approach
Free heme, 115
Free-radical scavengers, 358
Free –NH moiety of TZD, 178
Frova. *See* Frovatriptan
Frovatriptan, 299–300
Fungal infections, 257–258
Furanoflavonoids, 142
Fused molecules, 9
Fused pharmacophores, 10–11
fVIIa/TF complex. *See* Factor VIIa/Tissue
    Factor complex (fVIIa/TF complex)
fXa. *See* Factor Xa (fXa)

G

G-quadruplexes, 81
Gabapentin, 311
GAMG, 222

Gammaaminobutyric acid, 355
Gassman synthesis, 213
Gastric ulcers, 212
Gastrointestinal toxicity (GI toxicity), 82–83
Gemcitabine, 325
Gene expression patterns, 140
Gilurytmal. *See* Ajmaline
Glioblastomas (GBMs), 309
Glitazones
    as anticancer agents, 195–196
    as antihypertensive agents, 196
    drug repurposing for, 195–196
    as neuroprotective agents, 195
Glu. *See* Glutamate (Glu)
Glucagon receptor (GCGR), 45–46
7-(O-β-Glucosyloxy)oxindole-3-acetic acid
    (GOA), 228
GLUT-4 modulator, 198–199
Glutamate (Glu), 355
Glutamic-acid-based selective MurD
    inhibitors, 197
Glutathione reductase (GR), 104
Glycofolinine, 95–96
Glycogen synthase kinase-3 (GSK3), 112–113
Glycosyltransferase, 62–76
Glyfoline congeners, 62
Go6976 inhibitor, 292–293
GOA. *See* 7-(O-β-Glucosyloxy)oxindole-3-
    acetic acid (GOA)
Golotimod, 295, 297
Gonadotropin-releasing hormone (GnRH), 46
GR. *See* Glutathione reductase (GR)
GSK2248761 inhibitor, 295, 297
GSK2606414 inhibitor, 292–293
GSK2656157 inhibitor, 292–293
GSK3. *See* Glycogen synthase kinase-3 (GSK3)

H

H3 receptors in central nervous system, 43
4-Halo-6-[2-(4-arylpiperazin-1-yl)ethyl]-1H-
    benzimidazoles, 44
Haloalkoxyacridones, 101–103
Halogen-substituted 7-chloro-1,3-
    dihydroxyacridone, 96–98
Halogen-substituted acridone derivative, 108
α-Halogenated acid chloride, 213
Halogenation reaction, 34
HCV NS3/4A protease, 20
Hemagglutinin, 294
Hemagglutinin type 1 (H1), 274–275
Heme oxygenase (HO), 363
Heme-interacting acridone derivatives, 115

Printed in the United States
By Bookmasters